C0-AKZ-589

ALLE ZEIT WACH
1842

Neurobiology of Arachnids

Edited by

Friedrich G. Barth

Contributors

K. S. Babu · F. G. Barth · A. D. Blest · B. Claas
Gerta Fleissner · Günther Fleissner · R. F. Foelix
L. Forster · P. Görner · M. F. Land · R. Legendre
H. Mittelstaedt · A. Reißland · T. M. Root
E.-A. Seyfarth · R. G. Sherman · P. Weygoldt
S. Yamashita

With 162 Figures

Springer-Verlag
Berlin Heidelberg New York Tokyo

Professor Dr. FRIEDRICH G. BARTH
Zoologisches Institut
Fachbereich Biologie
J.-W.-Goethe-Universität
Siesmayerstraße 70
D-6000 Frankfurt am Main 1

Cover illustrations:

Poecilotheria regalis an Indian bird-spider
Scorpio maurus an Algerian scorpion, natural length 6 cm

both redrawn from

PIERRE-P. GRASSÉ (ed.): Traité de Zoologie. Tome VI, Masson et Cie. Paris 1949

ISBN 3-540-15303-9 Springer-Verlag Berlin Heidelberg New York Tokyo
ISBN 0-387-15303-9 Springer-Verlag New York Heidelberg Berlin Tokyo

Library of Congress Cataloging in Publication Data. Main entry under title: Neurology of arachnids. Bibliography: p. Includes index. 1. Nervous system–Arachnida. I. Barth, Friedrich G., 1940– . II. Babu, K. S. QL459.2.N48 1985 595.4′04182 85-4753

Typesetting, printing and bookbinding: Konrad Triltsch, Graphischer Betrieb, D-8700 Würzburg.

2131/3130-543210

Dedicated to all those who
– for no good reason –
dislike animals with eight legs

Preface

Arachnids rarely come to mind when one discusses arthropod neurobiology. In fact much more is now known and written about the nervous systems of insects and crustaceans. Several arguments have led us to conclude, however, that the time has come to document important aspects of the neurobiology of spiders, scorpions, and their kin, as well.

Studies of arachnid neurobiology have made considerable progress since the last comprehensive treatment by Bullock and Horridge in their monumental monograph on invertebrate nervous systems published in 1965. This is especially true for research performed in the last decade. Several problems related to the structure and function of arachnid nervous and sensory systems have now been studied in considerable depth but have so far not been given adequate space under one cover.

A particular incentive to produce this book has been the importance attributed to comparative approaches in neurobiology. Neglecting a large taxonomic group such as the arachnids – which comprises some 60,000 species living a wide range of different lives – would mean ignoring an enormous potential source of knowledge. In writing the chapters of this book we have striven to present some of the unique features of the arachnids. But the result of our efforts is not just meant to contribute to an understanding of the particularities of the arachnids. Knowledge of the adaptive radiation among at least all the large arthropod groups can also sharpen our insight into the basic traits found again and again, and likely to be shaped by similar demands for carrying out a particular task.

Having committed myself to work with spiders – which are not just "honorary insects" – and having been intrigued by their sensory and behavioral capacities over the years, I am hopeful that our book will lure some of our colleagues into work with arachnids. As the papers in this volume amply testify, there are excellent experimental animals among them, and there is a wealth of unanswered questions, many of which are addressed here in the various chapters.

The scope of the book ranges from the neuroanatomy of the central nervous system, sensory physiology and neuroethology to cybernetics and the circadian clock. We feel that none of the specializations of modern neurobiology – valuable and necessary as they may be – should be considered in isolation. Also, neurobiology has become a truly interdisciplinary science. It will be obvious to the reader that we have not striven for an encyclopedia listing all known details. Rather we intended – wherever possible – to focus on principles and mechanisms and to add the necessary framework and background.

I am very grateful to the authors for their generous cooperation, to Dr. D. Czeschlik and the editorial staff of Springer-Verlag for their advice and patience, to Mrs. U. Ginsberg for competent secretarial work, to Mrs. H. Hahn for help with the figures, and to Dr. E.-A. Seyfarth for valuable comments and criticism.

Special thanks are due to my wife and children. They charmingly tolerated my many manuscripts even under the Mediterranean sun during our vacations.

Frankfurt am Main, Spring 1985 FRIEDRICH G. BARTH

Contents

C Senses and Behavior

D The Motor System

E Neurobiology of a Biological Clock

Contributors

You will find the addresses at the beginning of the respective contribution

A The Central Nervous System: Structure and Development

I Patterns of Arrangement and Connectivity in the Central Nervous System of Arachnids

K. SASIRA BABU

CONTENTS

1 Introduction

In arthropods the organization of the central nervous system is related to the body segmentation and to the degree of development of segmental appendages and sense organs. In arachnids the body is divided into a prosoma and opisthosoma. Arachnids do not have antennae or appendages on the opisthosoma apart from spinnerets and the appendages they have are concentrated on the prosoma. In addition, arachnids have developed special sensory structures like pectines, malleoli, flagella, modified sensory legs, slits and other sense organs. These and other characteristic features are reflected by the structure of the arachnid central nervous system (CNS). This chapter gives a comparative account of the external morphology and internal anatomy of the CNS of five well-known arachnid orders (scorpions, whip scorpions, tailless whip scorpions, wind scorpions and spiders). What we know of the major features of the arachnid CNS is mostly due to the work of a few authors (Saint-Remy 1890, Borner 1904; Gottlieb 1926; Hanström 1928; Kaestner 1932, 1933, 1940; Millot 1949; Babu 1965, Babu and Barth 1984). A brief review of its less-known functions is included.

Department of Zoology, S. V. University P. G. Centre, Kavali – 524202, Nellore Dt. (A. P.) India

2 External Morphology

The arachnid central nervous system consists of a dorsal, anterior brain or supraesophageal ganglion with circumesophageal connectives joining it to the subesophageal mass, which in turn may be connected to a ventral cord of segmental ganglia.

Although the segmentation of the arachnid brain is still a topic of discussion (see Weygoldt Chap. II, this Vol.), it is generally accepted that the supraesophageal ganglion (brain) consists of a protocerebrum and tritocerebrum. The absence of antennae implies the absence of a deutocerebrum as found in crustaceans and insects. The tritocerebrum is the anterior-most part of the ventral chain of ganglia embryologically, but during development it migrates forward, fuses with the preoral brain and constitutes the cheliceral ganglion. Thus the brain of an adult arachnid is made up of a protocerebral and tritocerebral ganglion.

Located underneath the brain is the subesophageal nerve mass, the foremost part of the ventral nerve cord. Fusion of subesophageal ganglia is characteristic of arachnids. The degree of this fusion varies in different orders. This nerve mass is formed of 9 ganglia in Scorpionidea (*Heterometrus*), 10 in Solifugida (*Galeodes*), 12 in Uropygida (*Thelyphonus*), 16 in Araneae (*Poecilotheria*) and 17 in Amblypygida (*Phrynichus*).

An arthropod-type long double ventral nerve cord with seven ganglia is present in scorpions. In whip and wind scorpions a small single abdominal ganglion is located in the opisthosoma. Abdominal ganglia are absent in most other arachnids.

A feature of arachnids not so pronounced in other arthropods is the presence of coverings around the central nervous system. The cephalothoracic nerve mass is surrounded externally by a tissue, called the neural lamella, which is especially thick on the ventral side between the sternum and the subesophageal ganglia. This tissue is also present in thin layers at other parts of the CNS. The subesophageal mass is also covered on its dorsal side by an extensively developed endosternite.

The neural lamella is composed of four to six compactly arranged cell layers. A cellular perineurium beneath the neural lamella like that found in insects is absent. Neuroglial cells occupy spaces left by the nerve cells, and in the central fibrous core. The entire dorsal surface of the brain, the subesophageal ganglia, and main nerve trunks are traversed by blood vessels and sinuses.

2.1 Nerves Arising from the Central Nervous System

The only nerves arising from the protocerebrum are the optic nerves (Fig. 1). Most arachnids have four pairs of eyes. The median optic nerves form one (whip scorpions) to two pairs (scorpions, wind scorpions, spiders) arising from the dorsal and anterior-most part of the protocerebrum. Ventral to these are the lateral optic nerves: one pair in scorpions (*Heterometrus*) and wind scorpions (*Galeodes*), and three to four pairs in spiders (*Cupiennius*), whip scorpions (*Thelyphonus*) and tailless whip scorpions (*Phrynichus*).

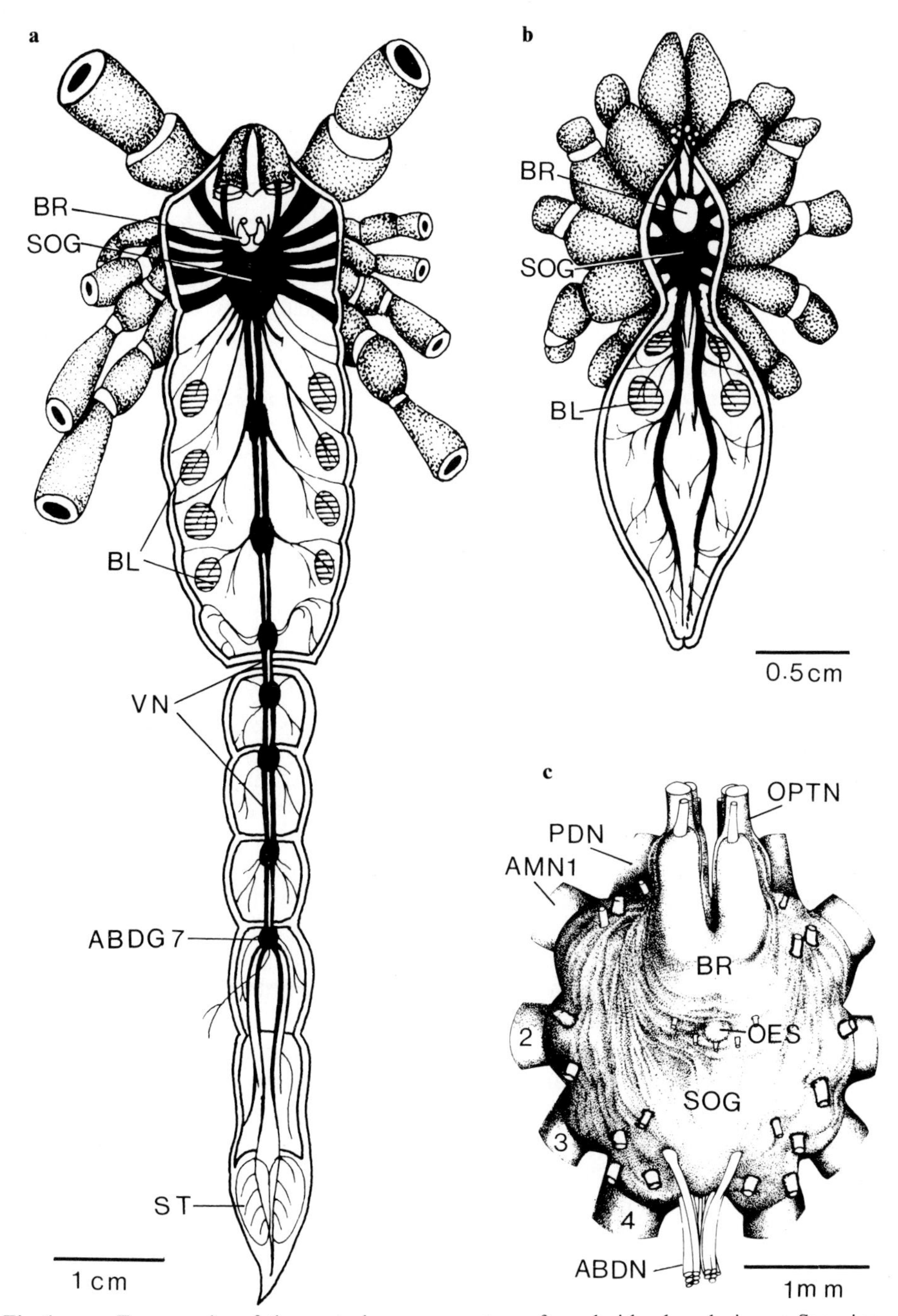

Fig. 1 a – c. Topography of the central nervous system of arachnids, dorsal view. **a** Scorpion (*Heterometrus*); note anterior fused nerve mass, and ventral nerve cord with seven free abdominal ganglia. **b** Spider (*Poecilotheria*), showing the single anterior fused nerve mass and lack of ganglia in the opisthosoma. **c** Spider (*Cupiennius*), cephalothoracic nerve mass: dorsal brain or supraesophageal ganglion and ventral composite subesophageal ganglia. *ABDG 7* seventh abdominal ganglion; *ABDN* abdominal nerves; *AMN1–4* first to fourth leg nerves; *BL* book lungs; *BR* brain or supraesophageal ganglion; *OES* esophagus; *OPTN* optic nerves; *PDN* pedipalpal nerve; *SOG* subesophageal ganglia; *ST* sting; *VN* ventral nerve cord. (**a, b** after Babu 1965; **c** Babu and Barth 1984)

A pair of thick nerves innervate the chelicerae from the tritocerebrum. This part of the brain also gives off one (scorpions) to three pairs (whip scorpions) of small rostral or sympathetic nerves that constitute part of the sympathetic nervous system of arachnids (see Legendre Chap. III, this Vol.).

The large size of the subesophageal mass is primarily related to the origin of one pair of pedipalpal and four pairs of leg nerves (Fig. 1c). Several pairs of abdominal nerves also arise from the subesophageal mass, and their number varies in different groups of arachnids. The anterior four mesosomatic segmental nerves in *Heterometrus*, five in *Galeodes*, seven in *Thelyphonus*, and all the opisthosomatic nerves of *Phrynichus* and *Cupiennius* arise from the subesophageal ganglia.

The special types or arrangements of sensory receptors concentrated either on the first pair of walking legs (*Thelyphonus* and *Phrynichus*), or the malleoli (racket organs) on the fourth pair of walking legs (*Galeodes*) or the pectines (comb plates) in the second mesosomatic segment (scorpions) are innervated by especially large nerves.

2.2 Ventral Nerve Cord

In amblypygids, araneids (only Mygalomorphae have one free terminal ganglion; Hanström 1928), and phalangids free abdominal ganglia are absent. They have migrated forward and fused with the subesophageal ganglia.

Among arachnids, only scorpions have a long double ventral nerve cord, connected to the subesophageal ganglia (Fig. 1a). There are seven free ganglia, three in the mesosoma and four in the metasoma, joined by double longitudinal connectives. Two ganglia fuse to form the last (seventh) abdominal ganglion that innervates the fourth and fifth metasomatic segments and also the terminal sting apparatus. Typically, each ganglion gives off two pairs of segmental nerves of equal size, of which the anterior one innervates the dorsal part of the body and the posterior one its ventral region.

In the solifugid *Galeodes*, a single abdominal ganglion is present in the first abdominal segment. It is connected to the posterior part of the subesophageal mass by a pair of thin connectives. This small ganglion gives off five pairs of nerves that innervate the posterior six abdominal segments.

The uropygids (*Thelyphonus*) also have single small abdominal ganglion, located between the eigth and ninth abdominal segments, and connected with the subesophageal mass by a pair of connectives. The free abdominal ganglion gives off seven pairs of nerves that innervate the eigth to twelfth abdominal segments as well as the mobile multisegmented sensory flagellum.

3 Internal Morphology

3.1 Types and Distribution of Nerve Cells

As in other arthropods, in all ganglia the cells are arranged in the periphery, whereas the central parts of the ganglia are fibrous (Fig. 7c). In the brain, the

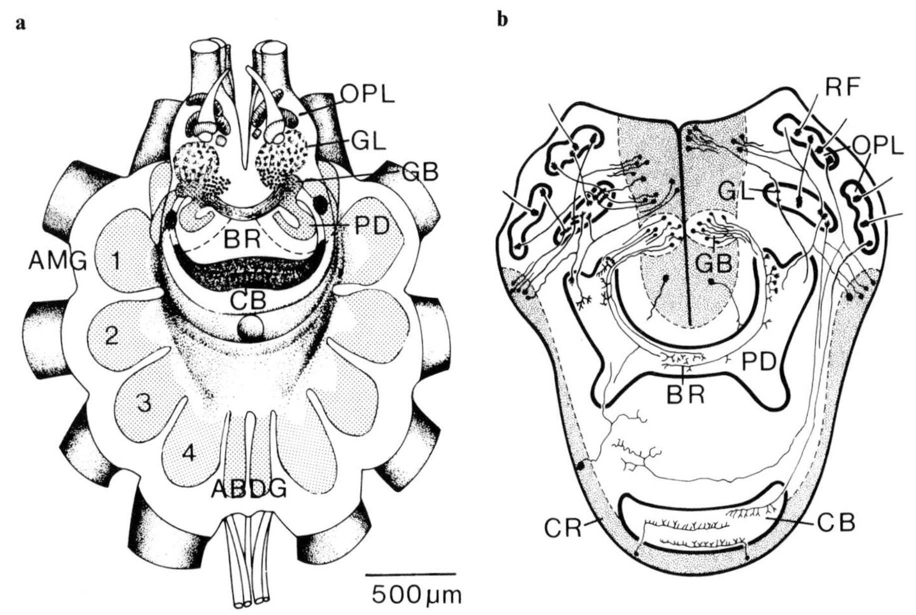

Fig. 2a, b. Diagramatic view of the various important centers of the brain of *Cupiennius* (Babu and Barth 1984). **b** Brain of a wandering spider, dorsal view (Golgi method, after Hanström 1921). *ABDG* abdominal ganglion; *AMG1–4* first to fourth leg ganglion; *BR* bridge; *CB* central body; *CR* cellular layer; *GB* globuli cells; *GL* glomeruli; *OPL* optic lamella; *PD* peduncle; *RF* primary receptor fibers

cells are packed in the frontal, dorsal, and lateral areas, but are absent from the posterior and ventral regions. In the subesophageal and abdominal ganglia the cells are restricted to the ventral and ventrolateral regions. Synaptic contacts are known to occur in the central fibrous core and in special neuropile masses. They are never found in the cellular cortex.

The nerve cells are typical monopolar neurons: from their cell body arises one neurite, which gives off several short highly branched processes and a single axon (Fig. 5a). Single-cell characterizations by electrical recordings and morphological identification with CoS preparations are still fragmentary. Four types of cells are known only on the basis of histological studies applying classical reduced silver stains (Babu 1965, 1969).

Globuli Cells (Type A Cells). A common feature among arachnids is the presence of a compactly arranged mass of globuli cells in the protocerebrum (Figs. 2, 3, and 6c). These cells give off fine parallel bundles of axons (diameter 0.4 µm) arranged in characteristic shapes, called the mushroom bodies. Each globuli cell has a large, chromatin-rich nucleus (6 to 8 µm) and poorly staining cytoplasm.

Type B Cells. This second category of small cells is numerous in the brain, subesophageal and abdominal ganglia. Their cytoplasm is clear and the nucleus has granular chromatin with or without nucleoli. The cell diameter ranges from 10 to 15 μm, with a nucleus of 8 to 12 μm.

Neurosecretory Cells (Type C Cells). The third category of cells was histologically demonstrated as neurosecretory. More of these cells are found in the protocerebrum than in other ganglia. Their average size ranges from 20 to 30 μm, with a nucleus of 12 to 20 μm.

Motor/Interneurons (Type D Cells). These are the largest cells in the central nervous system of arachnids (Fig. 6). They function as motor or interneurons. They are prominent in all ganglia except in the protocerebrum. In some spiders type D cells are as large as 140 μm. Their average size, however, varies from 30 to 80 μm.

3.2 Fibrous Mass

The central fibrous mass of both the brain and the subesophageal mass is totally devoid of nerve cell bodies (Fig. 7c). It consists only of processes of neurons and the terminal arborizations of peripheral sensory neurons. In the brain and other ganglia, fibrous mass is highly organized into longitudinal and transverse tracts which lack synaptic contacts. Instead, the areas of synaptic contacts are special structured neuropile masses of the protocerebrum like the optic ganglia, the mushroom bodies, and the central body, as well as the dense fine fibrous neuropile regions of other ganglia.

3.2.1 Protocerebrum

In arachnids the important and complex neuropile masses are located in the protocerebrum.

Globuli and Mushroom Bodies. A pair of compact masses of globuli cells is located in the anterior mid-central part of the protocerebrum (Figs. 2 and 6). The degree of their development varies widely in different groups of arachnids. The globuli and associated mushroom bodies (corpora pedunculata) are completely absent in sedentary spiders (Agelenidae, Drassidae and Dysderidae; Hanström 1921), whereas in amblypygids they are very well developed and occupy the entire mid-dorsal and even the lateral parts of the protocerebrum. They occupy 3.4% of the brain in wind scorpions (*Galeodes*), 3.7% in scorpions (*Heterometrus*), 12.6% in whip scorpions (*Thelyphonus*) and 48% in tailles whip scorpions (*Phrynichus*). In *Phrynichus* there are two independent bodies which are highly convoluted (Babu 1965). Similarly well-developed structures were found in *Neophrynus* (49%) and *Limulus* (79%) (Hanström 1928).

The shape of the mushroom bodies is characteristic of each group. A single large haft formed by densely packed, fine and short axons and dendrites arising mostly from globuli cells is present in *Galeodes* and *Cupiennius.* In scorpions, whip scorpions, and tailless whip scorpions, these form smaller secondary hafts.

The functions of these mushroom bodies are still largely unknown, which is also the case even with regard to insects and crustaceans. The shape and arrangement and their connections with the brain and other ganglia suggest that they are associated with complex behavioral activities. In insects they are known to have an inhibitory effect on locomotor, respiratory, and reflex activities (Huber 1967). Among different castes and species of bees and termites their high development is related to complex social behavior (Wheeler 1910). In *Phrynichus*, *Limulus*, ants, and in some decapods, polychaetes, and onychophores, the mushroom bodies are particularly well developed but have no known functional significance that we can as yet detect (Bullock and Horridge 1965).

Optic Mass. Compared to insects and crustaceans, the optic masses of arachnids are poorly developed. Often these are small masses of neuropile, without any obvious specialization. Also in arachnids fewer types of neurons are present in association with optic neuropile, as compared to insects and crustaceans.

The optic lobes in the anterior part of the protocerebrum are composed of several distinct neuropile masses, each surrounded by small type B cells. Details of neuron connections are known from Golgi preparations (Hanström 1921) and electron microscope studies (Trujillo-Cenóz 1965). The axons from the eye end in the first optic neuropile, where each of them makes multiple synaptic contacts with visual ganglion cells. These ganglion cells in turn relay information to the deeper optic neuropile, which itself is connected to the central body and the corpora pedunculata (Fig. 2b).

Like the degree of development of the mushroom bodies, that of the eyes and the optic centers varies widely among the arachnids and even within the spiders. Some hunting spiders have exceptionally well-developed eyes and associated ganglionic masses (Fig. 6e) (see also Chaps. IV to VI, this Vol.). Thus the optic neuropiles of *jumping spiders* (Salticidae) and *wolf spiders* (Lycosidae) are well developed, whereas they appear much less refined in sedentary spiders (Araneae) (Hanström 1921). This is also reflected by the total mass of the optic neuropiles which in the Araneida occupy only 3 to 4% of the total brain volume, compared to 25% in the Salticidae. Similarly, in web-building and sedentary spiders, axons of the lateral eyes do not end as a palisade in the first optic neuropile and there are no glomeruli in their second optic neuropile. In Salticidae and Lycosidae, however, the degree of development of the optic neuropiles associated with the lateral eyes is highest among arachnids. The median eyes have two simple masses of optic neuropiles on each side of the brain and are posteriorly interconnected with the central body. Both the optic neuropiles of the lateral eyes are better developed. Axons from the lateral eyes terminate as a palisade in the large horseshoe-shaped first neuropile mass. The two optic masses are joined through a well-formed chiasma. The second optic mass, which also contains glomeruli, receives fibers from the globuli cells, apart from fibers coming from the surrounding type B cells. The second optic mass of the lateral eyes also is linked with the central body through a well-formed large optic tract (Fig. 2a).

In *scorpions* (*Vejovis*, Hanström 1923), the median eyes are associated with one and the lateral eyes with two optic neuropiles on each side; both meet at a common third optic neuropile. The optic masses of the median eyes are one on each side in *Galeodes* and *Thelyphonus* and two in *Phrynichus*. The optic masses of the lateral eyes are one in *Thelyphonus* and *Poecilotheria*, two in *Phrynichus*, and absent in *Galeodes*.

Central Body. In all arachnids the central body appears to be the final integrating center for visual input from both median and lateral eyes. Apart from this, axons from other parts of the protocerebrum and subesophageal ganglia also terminate here, suggesting that it is a motor and association center. Neither electron microscope studies nor ablation, or lesion or stimulation experiments have been done to understand the functional significance of the central body in arachnids.

Although present in all arthropods, the central bodies in different groups may not be homologous structures or even have similar functions. Their position and structure is characteristic of each group. In arachnids the central body is a flat, crescent-shaped mass of neuropile which lies across the postero-dorsal part of the protocerebrum (Fig. 6e). It is covered by layers of small type B cells on its dorsal and posterior sides. As in other arthropods the central body is divided into lamellae (Fig. 7a, b, and c) and consists of several lobes (two in *Galeodes*, three in *Phrynichus* and *Poecilotheria*, and four in *Heterometrus* and *Thelyphonus*). In different arachnids its relative volume varies: 4.3% of the total volume of the brain in the spider, 3.5% in the scorpion, 3.1% in whip scorpion, 2% in wind scorpion, 0.8% in tailless whip scorpion.

The absence or poorly developed nature of one or more of the important centers of the protocerebrum appear to be compensated by another well-developed nerve center. Thus, for example, the highest development of the central body in the spider (Theraphosidae) is accompanied by poorly represented globuli cells and the absence of mushroom bodies. Similarly, the extremely well-developed globuli and associated structures in the amblypygid go along with a much reduced central body. Such inverse relationships are also known in some other spiders (Hanström 1928), insects and crustaceans (Bullock and Horridge 1965).

3.2.2 *Subesophageal Mass*

Although the subesophageal mass is composite and consists of several individual ganglia, it presents a unified structure integrated by connections between these ganglia, which are seen as well-formed longitudinal and transverse tracts (Fig. 3).

The subesophageal mass is traversed in the antero-posterior direction by six to seven pairs of longitudinal tracts, as in insects. In each ganglion the tracts pass through the fibrous mass at different levels, maintaining characteristic distances between each other. The dorsal tracts are larger and contain fibers of larger diameter (6 to 10 μm) arising from motor or interneurons of each ganglion. The ventral tracts, on the other hand, contain smaller fibers primarily contributed from incoming sensory axons. Anteriorly, these tracts ascend into the

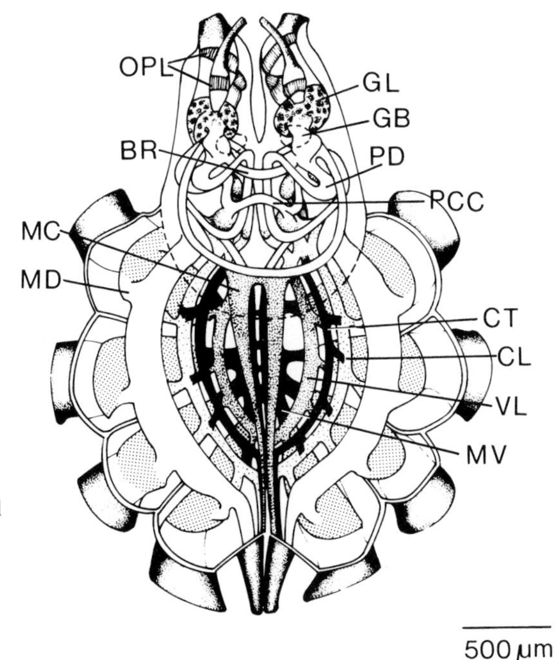

Fig. 3. Dorsal view of the cephalothoracic nerve mass showing important centres and fiber tracts of the brain and subesophageal nerve mass of *Cupiennius salei* (Babu and Barth 1984). Central body not represented for reasons of clarity. *BR* bridge; *CL* centro-lateral tract; *CT* central tract; *GB* globuli cells; *GL* glomeruli; *OPL* optic lamella; *PCC* protocerebral commissure; *PD* peduncle; *MC* mid-central tract; *MD* mid-dorsal tract; *MV* mid-ventral tract; *VL* ventro-lateral tract

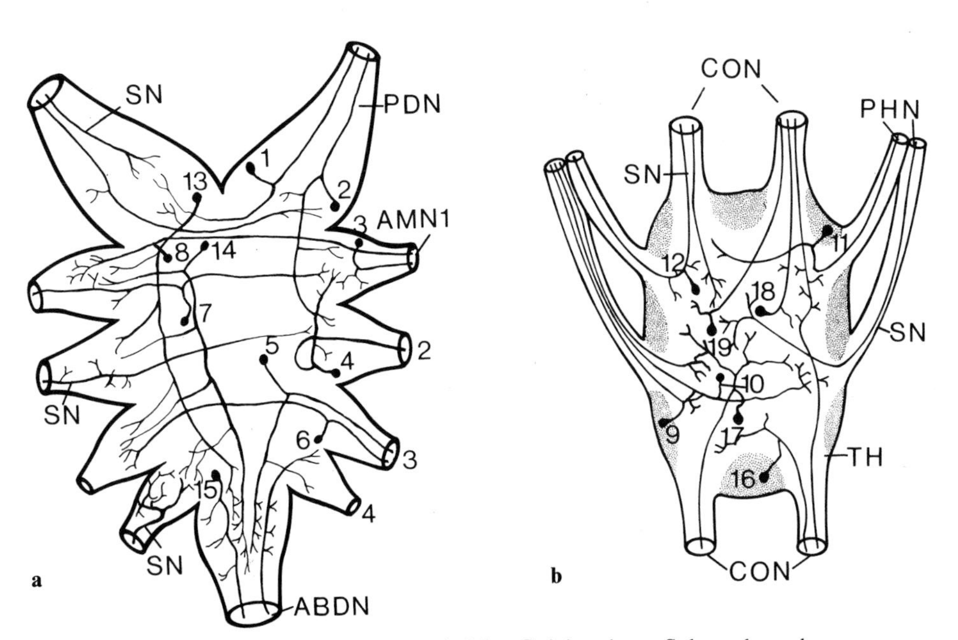

Fig. 4a, b. Examples of *neurons* as revealed by Golgi stain. **a** Subesophageal nerve mass. **b** First free abdominal ganglion of scorpion (*Vejovis*, aster Hanström 1923). *ABDN* abdominal nerve; *AMN 1–4* first to fourth leg nerves; *CON* connectives; *PHN* peripheral nerves; *1–12* motor neurons; *13–19* interneurons; *SN* sensory neurons; *TH* through fibers

brain through esophageal connectives and terminate either in the mushroom bodies or the central body.

Around the pedipalpal and four pairs of leg ganglia, the axons of type B and D cells are arranged into bundles and form conspicuous commissures interconnecting the opposite ganglia. Axons from these bundles enter the longitudinal tracts or peripheral nerves or disappear in the central mass of the neuropile. The motor pathways (with large fibers) in general are dorsally located in a ganglion. The sensory pathways (many fine fibers contributed mainly by the peripheral nerves) occupy the ventral regions of the ganglion.

The distribution pattern of some motor and interneurons from Golgi preparations is shown in Fig. 4a (Hanström 1923). An important feature of neuronal organization is the wide *ramification* of some interneurons, which spread into several ganglia of one side (scorpion) and some into many ganglia of both sides (spiders) of the subesophageal mass. There are also intraganglionic neurons. Motor neurons were identified on the basis of a large cell body and peripheral axon. According to the distribution of their dendrites, they were classified into different types of ipsilateral and contralateral motor neurons. The majority seem to have the cell body, dendrites, and the axon on the same side of a ganglion (Hanström 1923). With minor changes, these observations were confirmed by intracellular dye injections (cobalt stain) and electrophysiological recordings in a scorpion (Bowerman and Burrows 1980). In this latter study eight classes of motor neurons have been characterized physiologically and the different motor neurons innervating a particular muscle were found to have their cell bodies widely separated in the ganglia. Sensory axons, as they enter, terminate on the same side of the ganglion and in some cases on the opposite side (Hanström 1923, Babu 1969).

Anatomically specialized neuropile masses have developed in the subesophageal mass in relation to some of the highly specialized sense organs of various arachnids (Babu 1965). Thus the first pair of modified sensory legs in *Phrynichus* and *Thelyphonus*, which are used as a sort of antennae, are well represented in the subesophageal mass by particularly well-developed ventral sensory association centers. In *Galeodes*, the malleoli, which are racket-shaped sense organs of undetermined function, also form a large and conspicuous sensory mass in the ventral region of the subesophageal mass. Similarly, the pectines of scorpions, known to function as mechano- and chemoreceptors, are connected to special pectinal sensory mass.

According to the analysis of the post-embryonic development of the nervous system of a spider (*Argiope*), the number of neurons remains the same at all stages of growth. The total cephalothoracic nerve mass increases 24-fold in volume from the first instar to the adult stage. This growth is due to an increase in cell volume and the number of glial cells. Most of the neural growth is due to the growth of fibrous mass, that is an increase in fiber size and ramification and number of incoming sensory fibers (Babu 1975). A recent cell count in *Cupiennius* showed that the two ganglia of the brain contain ca. 50,900 cells, whereas all 16 ganglia of the subesophageal mass together have ca. 49,000 cells (Babu and Barth 1984). In *Argiope*, which exhibits great sexual dimorphism, the number of cells is higher in females than in males by 11% in the subesophageal mass and by 58% in the brain. During postnatal growth of the nervous system the volume of type B cells (somata) increases 50-fold and of type D cells 600-fold in female spiders.

3.2.3 "Free" Abdominal Ganglia

The basic structure of abdominal ganglia as found in scorpions and wind scorpions is the same as that of the subesophageal ganglia (Babu 1965). Again, the cellular cortex is located on their ventral and ventro-lateral sides, while glial tissue is present on the dorsal and dorso-lateral regions. Type B and D cells are present and the former are numerous. The central region of the ganglion is occupied by the fibrous mass, which again presents an organized arrangement of longitudinal and transverse tracts. In the scorpion, traversing along the entire length there are seven pairs of major longitudinal tracts starting from the subesophageal ganglia and passing through all the abdominal ganglia. In each ganglion they maintain the same spatial distance between each other.

In the scorpion, the transverse tracts are well represented as dorsal commissures in each ganglion. Similar tracts are also present on the ventral side, and these are dominated by sensory fibers coming from the paired segmental

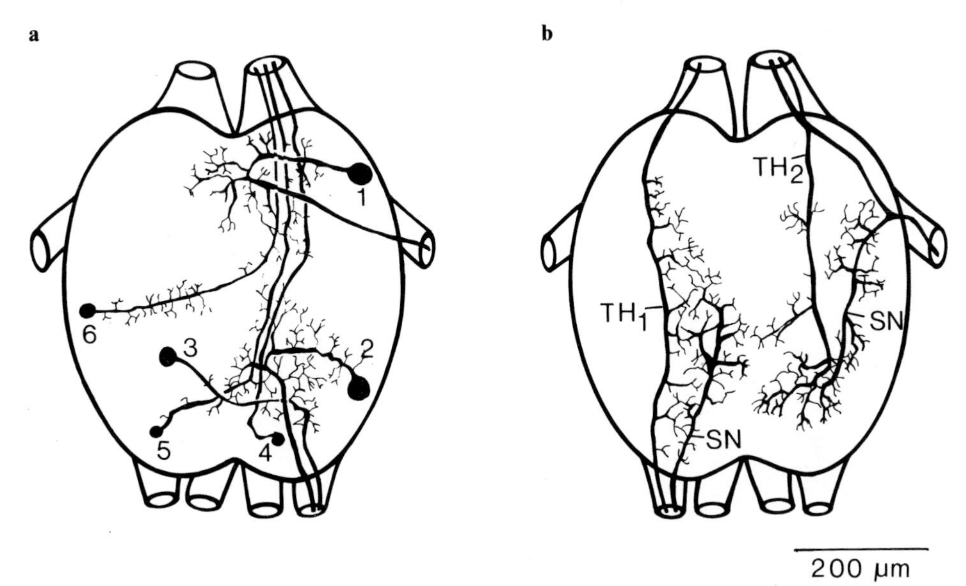

Fig. 5a, b. Seventh *abdominal ganglion* of scorpion (*Heterometrus*). (After Yellamma et al. 1979). **a** *Motor neurons* of fourth and fifth metasomatic segmental nerves and ascending *interneurons. 1* and *2* ipsilateral motor neurons; *5* and *6* contralateral ascending interneurons. **b** *Through fiber* and *sensory* endings. *SN* sensory fibers; TH_1 ascending sensory fiber, giving off collaterals in the seventh ganglion; TH_2 descending neuron with terminations on the dendritic neuropile of motor neurons. CoS method

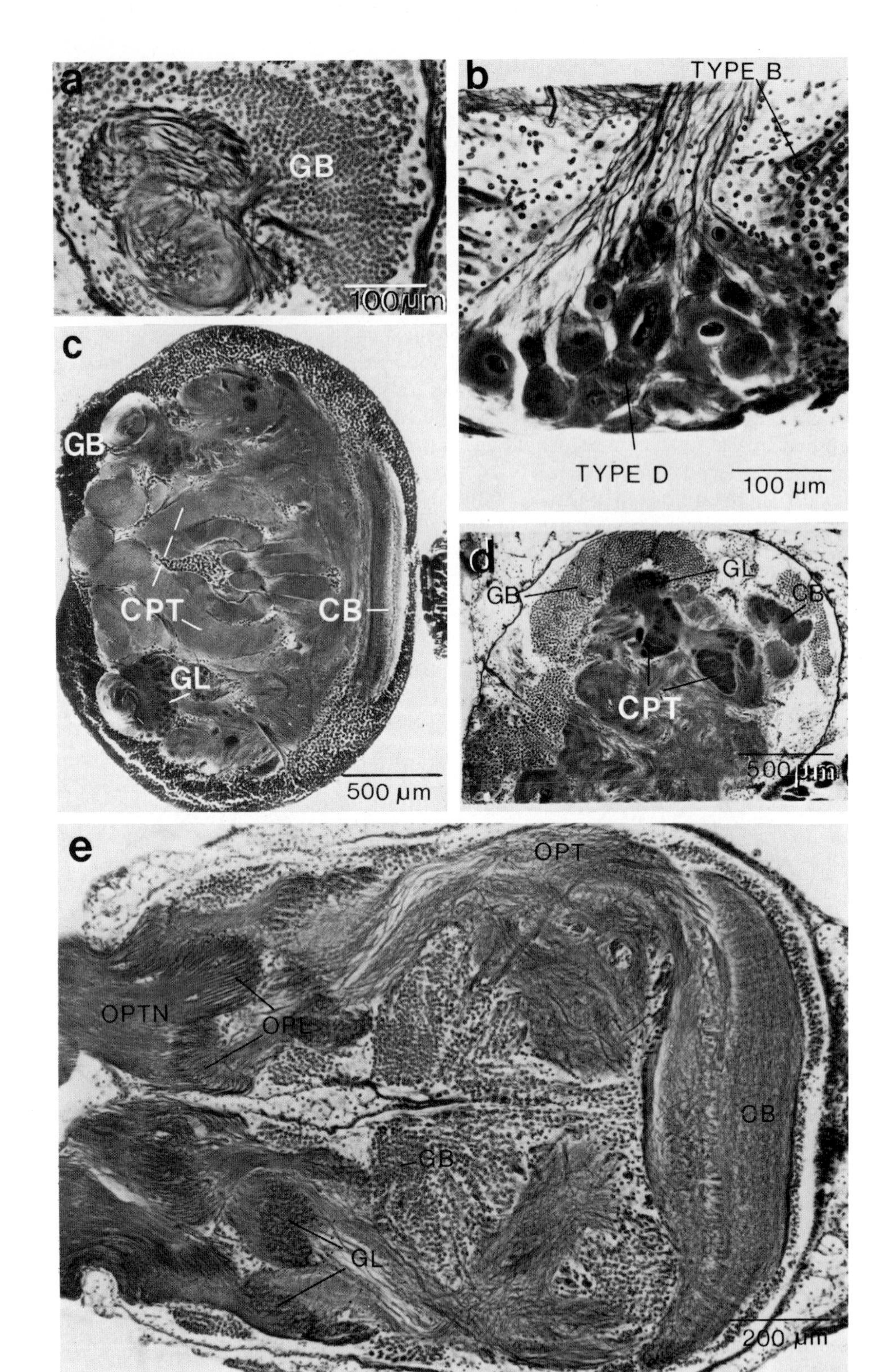
a
GB
100 µm
b
TYPE B
TYPE D
100 µm
c
GB
CPT
CB
GL
500 µm
d
GL
GB
CB
CPT
500 µm
e
OPT
OPTN
OPL
CB
GB
GL
200 µm

nerves. Groups of cells on each side containing both type B and D cells send their processes into the ipsilateral segmental nerves. A few of them also enter the contralateral segmental nerves.

In the abdominal ganglion, there are ipsi- and contralateral motor neurons and ascending and descending interneurons. Sensory axons either terminate on the same side or send arborizations to the opposite side (Fig. 4b).

The large musculature of the fifth metasomatic segment in scorpions, which operates the sting, is innervated by two ipsilateral and two contralateral groups of motor neurons located in the posterior region of the ganglion (Yellamma et al. 1979). One ipsilateral group of neurons gives rise to giant axons (diameter 40 μm) with conduction velocities of 6 to 7 ms^{-1} (Fig. 5). The diameter of other axons ranges between 5 to 20 μm. Sensory fibers from the fourth and fifth metasomatic segmental nerves, connectives, and telsonic nerve terminate ipsilaterally on the dendritic neuropile of the motor neurons. Electrical recordings show monosynaptic connections between the sensory fibers and these giant motor neurons. Some of the sensory fibers, while passing through the seventh ganglion, send collaterals that activate the contralateral interneurons (Sanjeeva-Reddy and Rao 1970).

4 Physiology

Knowledge of the physiological properties of the central nervous system of arachnids is very scarce, and much behind of that in insects and crustaceans.

Neurochemistry. According to histochemical studies, there is cholinergic synaptic transmission in the spider CNS. In the neuropile mass of araneid and agelenid spiders, acetylcholinesterase was shown to be particularly rich and less so in the optic ganglia and the central body (Meyer and Pospiech 1977). In addition, the presence of neurohumorally active amines was demonstrated (Meyer and Jehner 1980).

The modulation of the electrical activity in the central nervous system is thought to be due to neurohormones that regulate the circadian locomotor rhythm in scorpions (see also Fleissner, Chap. XX, this Vol.). From the few biochemical and electrophysiological studies on scorpions we know that the isolated ventral nerve cord exhibits maximal spontaneous electrical activity between 16.00 h and 24.00 h. Extracts of the cephalothoracic nerve mass and blood isolated at 17.00 h and 20.00 h enhanced the electrical activity of the ventral nerve cord. Similar extracts taken at 23.00 h and at 02.00 h, however, dimin-

Fig. 6a–e. *Globuli* cell mass (**a**), *small* cells (type B) and *large* (type D) motor neurons (**b**) in a hunting spider (*Cupiennius*) (Babu and Barth 1984). *Corpora pedunculata* and *central body* of a tailless whip scorpion (*Phrynichus*) in horizontal (**c**) and of a scorpion (*Heterometrus*) in sagittal (**d**) sections (from Babu 1965). Note the large lobulated stalks of the mushroom bodies of the tailless whip scorpion. **e** Horizontal section of the protocerebrum of the spider, *Cupiennius*. (Babu and Barth 1984). *CB* central body; *CPT* corpora pedunculata; *GB* globuli cells; *GL* glomeruli; *OPL* optic lamella; *OPTN* optic nerves

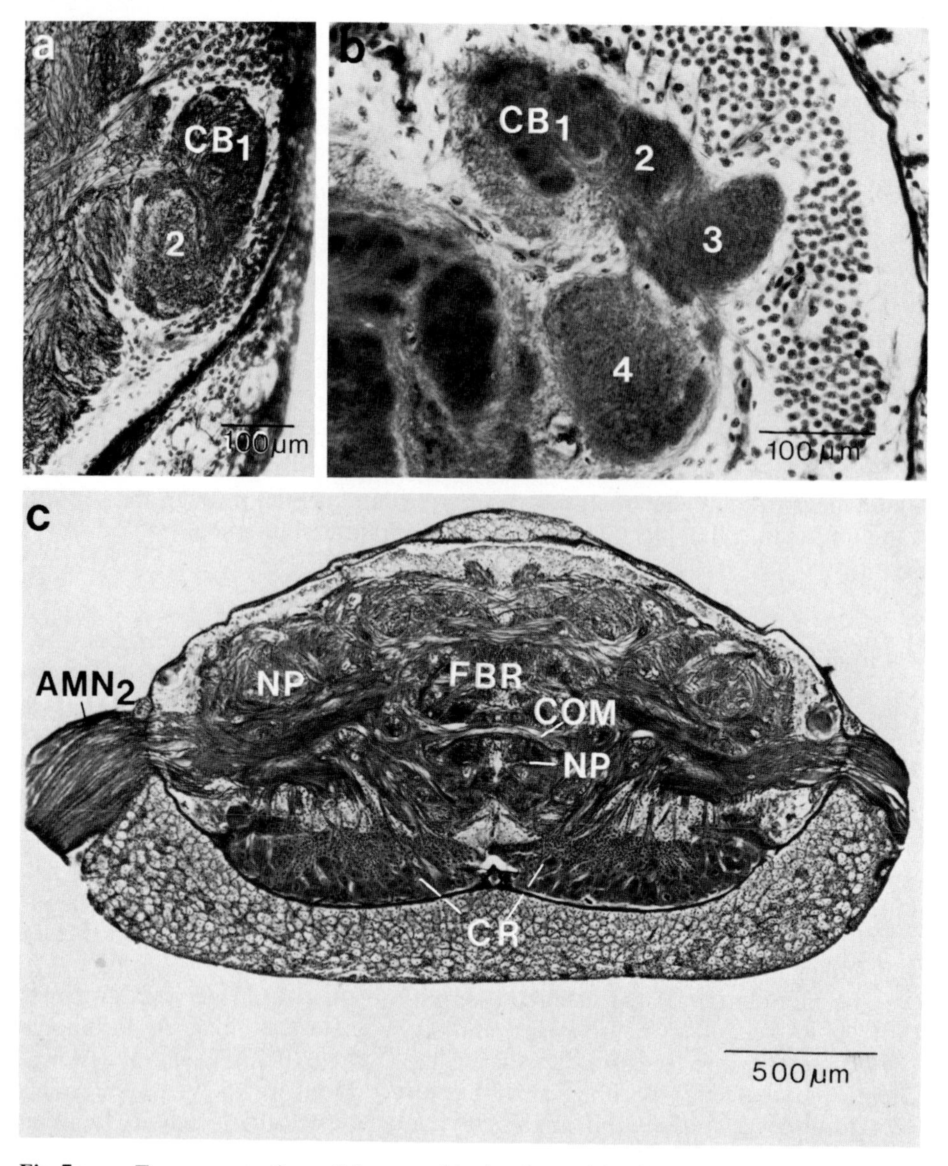

Fig. 7 a – c. Transverse sections of the central body of **a** a spider (*Cupiennius*) and **b** a scorpion (*Heterometrus*). **c** Transverse section of subesophageal nerve mass to show the distribution of cell layers with respect to the central fibrous core in spider (*Cupiennius*). (Palmgren 1948 silver technique) (**a** and **c** Babu and Barth 1984). AMN_2 second leg nerve; *CB 1–4* lobes of central body; *COM* commissure; *CR* cellular layer; *FBR* fibrous core; *NP* neuropile

ished the electrical activity (Rao and Gopalakrishna Reddy 1967). The chemical factors involved are presumed to be neurohormonal and produced by the neurosecretory cells of the subesophageal ganglia. A diel rhythm of the ACh content and the AChE activity also was reported in the scorpions (Venkatachari and Muralikrishna Dass 1968, Vasantha et al. 1975).

Electrophysiology. The basic physiological properties are comparable to those found in insects and crustaceans. Intracellular recordings from the somata of motor neurons in the subesophageal mass of scorpions suggested that the spike-initiating site is distant from the soma. Two motor neurons innervating synergistic muscles share many of their synaptic inputs (Bowerman and Burrows 1980).

Motor fibers arising from the abdominal ganglion of the scorpion showed a resting discharge varying from regular to intermittent bursts. Electric shocks or mechanical stimuli elicited multiple responses, postburst depression, inhibition, excitation and activation. Interneurons in the ventral nerve cord showed spatial summation, synaptic lability, activation of silent units, after discharge and inhibition (Venkatachari and Babu 1970, Babu and Venkatachari 1972).

Behavioral Analysis. In the scorpion, *Heterometrus*, the defensive strike response was analyzed by behavioral and lesion studies of the subesophageal mass (Palka and Babu 1967). It is quite different from the evasive responses of cockroach and earthworm. The strike pathway is characterized by strong resistance to habituation, graded response and longitudinal lability, as indicated by independent movement of the tail and pedipalps. It was found to have a response time of 20 to 150 ms, and appears to be initiated by small receptors scattered over the surface of the legs and other parts of the body. The strike response can be unilaterally abolished by lesions in the subesophageal nerve mass while sparing locomotor and leg withdrawal movements.

In the spider (*Cupiennius*), predatory or escape behavior can be elicited by both substrate and air-borne vibrations. The metatarsal (Hergenröder and Barth 1983a and b) and pretarsal (Speck and Barth 1982) slit sense organs and trichobothria (Görner and Andrews 1969) together determine these responses (see also Barth, Chap. VIII and Reißland and Görner, Chap. XI, this Vol.). In scorpions, the direction and distance of prey is sensed by substrate vibrations through mechanoreceptors on its tarsal leg segments (Brownell and Farley 1979a and b). Details of conclusions drawn on the circuitary in the CNS from these and other behavioral studies are given in Chapter VIII (Barth) and Chapter XV (Mittelstaedt), this Volume.

5 Conclusions

Obviously the central nervous system of arachnids is a rich unexplored field, ready for harvesting. Our knowledge of the finer anatomy and physiology of the CNS of arachnids is much behind what is known in insects and crustaceans. The urgent need is to adopt modern staining procedures like fluorescent dyes, cobalt chloride, and HRP, as well as electrophysiological recording techniques. Features of particular attraction for such studies are those which set the arachnids apart from other arthropods. Examples are the lack of a deutocerebrum found in insects and crustaceans and the large fused subesophageal ganglionic mass which receives most of the sensory input from the legs and pedipalps instead of antennae.

In particular from the comparative point of view, special neuropile masses like mushroom bodies, optic masses, central body, and sensory neuropiles deserve much more attention than they have so far received. These important association centers are all located in the protocerebrum and show great variation among different groups of arachnids. Thus the *central body*, a large conspicuous mass of neuropile in the postero-dorsal region of the protocerebrum consists of two (spider) to four (scorpion) lobes. The *mushroom bodies* are highly developed in some arachnids, such as tailless whip scorpions, but altogether absent in others, such as sedentary spiders. The *optic masses* are in general poorly developed in arachnids as compared to insects and crustaceans, the jumping and wolf spiders, however, being notable exceptions (see also Chap. IV to VI, this Vol.) on vision.

Another feature worth more detailed study are neuropile masses in the subesophageal ganglia developed in relation to sensory structures only found in some arachnids such as the comb plates of scorpions, the racket organs of wind scorpions, and the modified antenna-like first legs of whip scorpions.

Acknowledgments. I thank H. Hahn and G. Kreuder for valuable help with the preparation of the illustrations and the Deutsche Forschungsgemeinschaft for support during my stay at the Zoology Institute of the University of Frankfurt (grant SFB 45/A4 to F. G. Barth).

References

Babu KS (1965) Anatomy of the central nervous system of arachnids. Zool Jahrb Anat 82:1–154

Babu KS (1969) Certain histological and anatomical features of the central nervous system of a large Indian spider, *Poecilotheria*. Am Zool 9:113–119

Babu KS (1975) Post-embryonic development of the central nervous system of the spider, *Argiope aurantia* (Lucas). J Morphol 146:325–342

Babu KS, Barth FG (1984) Neuroanatomy of the central nervous system of the wandering spider, *Cupiennius salei* Keys. Zoomorphology 104:344–359

Babu KS, Venkatachari SAT (1972) Activity patterns of interneurons in the ventral nerve cord of the scorpion, *H. fulvipes*. Indian J Exp Biol 10:49–58

Borner C (1904) Beiträge zur Morphologie der Arthropoden. 1. Ein Beitrag zur Kenntnis der Pedipalpen. Zoologica 17:1–174

Bowermann RF, Burrows M (1980) The morphology and physiology of some walking leg motor neurons in a scorpion. J Comp Physiol 140:31–42

Brownell P, Farley RD (1979a) Detection of vibrations in sand by tarsal sense organs of the nocturnal scorpion, *Paruroctonus mesaensis*. J Comp Physiol 131:23–30

Brownell P, Farley RD (1979b) Orientation to vibrations in sand by the nocturnal scorpion *Paururoctonus mesaensis:* mechanism of target localization. J Comp Physiol 131:31–38

Bullock TH, Horridge A (1965) Structure and function in the nervous systems of invertebrates. Freeman, San Francisco

Görner P, Andrews P (1969) Trichobothrien, ein Ferntastsinnesorgan bei Webespinnen. Z Vergl Physiol 64:301–317

Gottlieb K (1926) Über das Gehirn des Skorpions. Z Wiss Zool 127:185–243
Hanström B (1921) Über die Histologie und vergleichende Anatomie der Sehganglien und Globuli der Araneen. K Sven Vet Akad Handl 61:1–39
Hanström B (1923) Further notes on the central nervous system of arachnids: scorpions, phalangids and trap-door spiders. J Comp Neurol 35:249–272
Hanström B (1928) Vergleichende Anatomie des Nervensystems der wirbellosen Tiere. Springer, Berlin Heidelberg New York
Hergenröder R, Barth FG (1983a) Vibratory signals and spider behavior: How do the sensory inputs from the eight legs interact in orientation? J Comp Physiol 152:361–371
Hergenröder R, Barth FG (1983b) The release of attack and escape behavior by vibratory stimuli in a wandering spider (*Cupiennius salei* Keys). J Comp Physiol 152:347–358
Huber F (1967) Central control of movements and behavior of invertebrates. In: Wiersma CAG (ed) Invertebrate nervous systems. Univ Chicago Press, Chicago
Kaestner A (1932) Pedipalpi. In: Kükenthal W, Krumbach T (eds) Handbuch der Zoologie, vol III. De Gruyter, Berlin, pp 17–22
Kaestner A (1933) Solifugae. In: Kükenthal W, Krumbach T (eds) Handbuch der Zoologie, vol III. De Gruyter, Berlin, pp 225–234
Kaestner A (1940) Scorpiones. In: Kükenthal W, Krumbach T (eds) Handbuch der Zoologie, vol III. De Gruyter, Berlin, pp 131–140
Meyer W, Jehnen R (1980) The distribution of monoamine oxidase and biogenic monoamines in the central nervous system of spiders (Arachnida, Araneida). J Morphol 164:69–81
Meyer W, Pospiech B (1977) The distribution of acetylcholinesterase in the central nervous system of web-building spiders (Arachnida, Araneae). Histochemistry 51:201–208
Millot J (1949) Chelicerates. In: Grassé PP (ed) Traîté de Zoologie, vol VI. Masson, Paris, pp 263–743
Palka J, Babu KS (1967) Towards the physiological analysis of defensive responses of scorpions. Vergl Physiol 55:286–298
Palmgren A (1948) A rapid method for selective silver staining of nerve fibres and nerve endings in mounted paraffin sections. Acta Zool 29:377–392
Rao KP, Gopalakrishna Reddy T (1967) Blood-borne factors in circadian rhythms of activity. Nature (London) 213:1047–1048
Saint-Remy G (1890) Contribution à l'étude du cerveau chez les arthropodes trachéates. Arch Zool Exp 5:1–274
Sanjeeva-Reddy P, Rao KP (1970) The central course of the hair afferents and the pattern of contralateral activation in the central nervous system of the scorpion. *Heterometrus fulvipes*. J Exp Biol 53:165–169
Speck J, Barth FG (1982) Vibration sensitivity of pretarsal slit sensilla in the spider leg. J Comp Physiol 148:187–194
Trujillo-Cenóz O (1965) Some aspects of the structural organization of the arthropod eye. Cold Spring Harbor Symp Quant Biol 30:371–382
Vasantha N, Venkatachari SAT, Murali Mohan P, Babu KS (1975) On the acetylcholine content in the scorpion, *Heterometrus fulvipes* C. Koch. Experientia 31:451–452
Venkatachari SAT, Babu KS (1970) Activity of motor fibers in the scorpion, *H. fulvipes*. Indian J Exp Biol 8:102–111
Venkatachari SAT, Muralikrishna Dass P (1968) Cholinesterase activity rhythm in the ventral nerve cord of scorpion. Life Sci 7:617–621
Wheeler WM (1910) Ants, their structures, development and behavior. Columbia Univ Press, New York
Yellamma K, Murali Mohan P, Babu KS (1979) Morphology and physiology of giant fibres in the seventh abdominal ganglion of the scorpion, *Heterometrus fulvipes*. Proc Indian Acad Sci (Ani Sci) 89:29–38

II Ontogeny of the Arachnid Central Nervous System

PETER WEYGOLDT

CONTENTS

1 Introduction

Our knowledge of the development of nervous systems has greatly increased during the last few years. Specific cell markers like horseradish peroxidase and monoclonal antibodies have made possible studies of the cell lineages leading to the formation of neuromeres in insects and leeches. In both cases, the number of neuroblasts that form a particular ganglion is known; in leeches it was even possible to trace the history of these neuroblasts back to individual blastomeres. Similar studies are under way for vertebrates (Barald 1982; Goodman 1982; Jacobsen 1982; Stent et al. 1982). In the grasshopper, Goodman and coworkers have shown the events by which – by a fixed pattern of cell divisions and differentiation – neurons are formed by particular neuroblasts. These authors have been able to demonstrate that the neurons derived from a particular neuroblast share certain features, such as transmitters, and vary with respect to others, such as electrical properties, which, however, are shared by the progeny of a given birth position (Goodman et al. 1980; Goodman and Spitzer 1981 a, b). In addition, the problem of pathfinding by growing axons and of the formation of synaptic connections has been studied with success. In insects, pioneer neurons and muscle pioneers have been identified that are formed early in development and later guide axons to their targets (Edwards 1982; Flaster et al. 1982; Goodman et al. 1981, 1982; Ho et al. 1983).

Biologisches Institut I (Zoologie), Albert-Ludwigs-Universität, Albertstr. 21 a, D-7800 Freiburg i. Br., Federal Republic of Germany

Such exciting studies do not exist for arachnids. The development of the nervous system of this group of arthropods has been studied mainly by authors interested in morphological problems.

In particular, data on brain development have been used to learn about the metameric composition of the arthropod forehead, segment formation, and the homology of various parts and appendages of the anterior body regions. Orders studied in this respect include the scorpions (Brauer 1895; Mathew 1956) amblypygids (Weygoldt 1975), uropygids (Kaestner 1951; Yoshikura 1961), spiders (Legendre 1958, 1959, 1979; Pross 1966; Rempel 1957; Yoshikura 1955, 1958), and, to a lesser extent, opilionids (Moritz 1957), solpugids (Junqua 1966), and mites (Aeschlimann 1958). Orders in which development has not been studied include the palpigradids and ricinuleids. Studies in which changes in behavior are related to developmental processes in the nervous system, or studies demonstrating the development of integration of various parts of the nervous system are, as far as I know, nearly nonexistent. Only the papers by Meier (1967) and Babu (1975) address such questions to some extent.

The larger part of this chapter will therefore deal with the embryology of the chelicerate nervous system and its bearing on morphological questions.

2 Development of Ganglia in Arachnids

The developmental events leading to the formation of an arachnid segmental ganglion differ from the corresponding events in crustaceans and insects. In the grasshopper (Goodman 1982) and in the crustacean *Diastylis* (Dohle 1976), a number of neuroblasts differentiate from the embryonic ectoderm. By a fixed pattern of cell divisions perpendicular to the ventral surface of the embryo, the neuroblasts form rows of ganglion mother cells, the uppermost of which divide to form ganglion cells. Thus, in the grasshopper or crustacean, a developing ganglion consists of cell rows arranged in a regular pattern and starting with a neuroblast at the outer surface.

Developing ganglia of arachnids look different. Multiplications of cells result in invaginations instead of cell rows (Fig. 6). The underlying cellular events have not been studied in detail. Neuroblasts have been mentioned only for the scorpion *Heterometrus* (Mathew 1956) and for the liphistiid spider *Heptathela* (Yoshikura 1955). Most studies on arachnid embryology do not mention neuroblasts (Anderson 1973). It seems that small neuroblasts or ganglion mother cells divide to form cell clusters which finally invaginate. The innermost cells of the invaginations are the first ganglion cells and soon start to grow fibers.

Invaginations during ganglion formation have also been observed in pycnogonids (Winter 1980), myriapods (Tiegs 1940, 1947; Dohle 1964) and in onychophorans (Pflugfelder 1948). In these arthropods, each ganglion is formed from a single invagination called ventral organ.

In pycnogonids, the ventral organs are composed of invaginated neuroblasts which then divide to form ganglion cells or ganglion mother cells (Winter 1980).

3 Embryology of the Chelicerate Nervous System and its Bearing on Head Morphology

3.1 The Problem

Numerous chelicerates have been studied in order to obtain information on the segmental composition of the anterior part of the prosoma and on its homology with the forehead of other arthropods. The following discussion is only meaningful if there is reason to believe that the chelicerates are related to other arthropods. In fact, all studies applying the methods developed by Remane (1956) and Hennig (1950) have shown that the Euarthropoda form a natural taxon including the Arachnata with trilobites and chelicerates and the Mandibulata with crustaceans and tracheates (Boudreaux 1979a, b; Lauterbach 1973, 1980a, b; Paulus 1979; Weygoldt 1979). The studies of Manton (1973, 1977, 1978) seem to indicate separate origins for the different arthropod subphyla. They are masterpieces on functional morphology but have little bearing on taxonomic questions.

The arthropod body is composed of a number of segments or metameres, each originally having its own pair of appendages, a pair of ganglia (more adequately called neuromeres), and a pair of coelomic cavities. Anterior to the first segment there is the acron, a presegmental region, homologous to the annelid prostomium. It lacks coelomic cavities and appendages, but contains a ganglion, the archicerebrum.

The primitive arthropod head bears five pairs of appendages, one pair of antennae and four pairs of legs in trilobites, and two pairs of antennae and three pairs of mouth parts in crustaceans. Chelicerates lack antennae; their first pair of appendages are the chelicerae. The cheliceral ganglion has been homologized to the tritocerebrum of the Mandibulata – and the chelicerae to the second antennae – since it retains a postesophageal commissure in some (not all) arachnids, and since it is connected to the stomatogastric system (Holmgren 1916; Hanström 1928; Snodgrass 1952, 1960).

There is general agreement that the region bearing the second antennae or, in chelicerates, the chelicerae, is a true segment or metamere, the tritocephalon. There are, however, different interpretations of the region anterior to the tritocephalon. According to Holmgren (1916), Hanström (1928), and Snodgrass (1952, 1960), everything in front of the tritocephalon belongs to the presegmental acron. Many embryologists, however, have observed that during development the region of the first antennae does not differ from that of the second antennae. It bears a pair of appendages, a pair of ganglia (the deutocerebrum), and a pair of mesodermal somites, sometimes even with coelomic cavities. It is therefore considered a true segment, the deutocephalon, serially homologous to the tritocephalon. Embryologists have even found an additional pair of coelomic cavities, the preantennulary coelom, indicating a further vestigial segment, the prosocephalon (Heymons 1901; Manton 1928, 1934, 1949, 1960; Tiegs 1940, 1947; Weber 1952, 1954; Siewing 1963, 1969; Weygoldt 1959, 1961; Scholl 1963, 1969, 1977; Larink 1969, 1970; Malzacher 1968; Knoll 1974). According to these authors, the arthropod forehead is composed of a presegmental acron, the preantennulary segment or prosocephalon, the antennulary segment or deutocephalon and the antennary segment or tritocephalon (Table 1).

If the chelicerates which lack antennae are the sister group of the trilobites and if the Arachnata are the sister group of the Mandibulata, then it seems reasonable to search for vestigial antennular structures, particularly for a vestigial deutocerebrum, in the chelicerate brain.

Various authors have expressed different opinions concerning the composition of the arachnid forehead (Legendre 1979). Johannson (1933) tried to find traces of antennular ganglia in *Limulus*, and Pross (1966, 1977) even claimed to have discovered the homologs of all parts of the mandibulate brain in the spider *Pardosa*. Legendre (1959, 1979) found even more segments and neuromeres in the spider brain. Other authors have failed to find structures which can, with certainty, be considered ganglia of reduced segments. Part of this controversy is due to the fact that the termini "segment" and "ganglion" have been used differently, sometimes meaning any region of the body or any accumulation of ganglion cells.

In the following I will concentrate on the description of brain development in two groups of arachnids in which the development of the nervous system shows different degrees of complexity, that is in pseudoscorpions on the one hand, and amplypygids and spiders on the other.

3.2 Pseudoscorpions

Pseudoscorpions are small arachnids with a specialized development. Their embryos develop in a brood pouch and are fed by a fluid secreted by the ovary. Many pseudoscorpions inhabit leaf litter, tree bark, or similar dark places, and have reduced eyes. Other species have two or four lateral eyes. Median eyes and their ganglia are always lacking, as are also corpora pedunculata. The development of pseudoscorpions is characterized by a precocious development of a large embryonic pharynx and hollow appendage rudiments filled by diverticula of the embryonic gut (Fig. 1 A) (Weygoldt 1964, 1965, 1968).

3.2.1 Development of Neuromeres and Brain

The segmental ganglia, in pseudoscorpions, arise as paired invaginations (Figs. 1B, C). In the opisthosoma, each invagination represents a developing ganglion. During further development the invaginations separate from the epidermis and the thickness of their interior walls increases due to numerous cell divisions (Fig. 1D). Thereby, the ganglionic cavities decrease in size and finally disappear (Figs. 2, 3). At the same time, before the complete disappearance of the cavities, the uppermost cells differentiate into neurons and grow dendritic fibers, and the developing neuropil forms commissures and connectives.

In the prosoma, as well as in the opisthosoma, there is a stage in which each developing neuromere is easily recognized by the presence of the slowly nar-

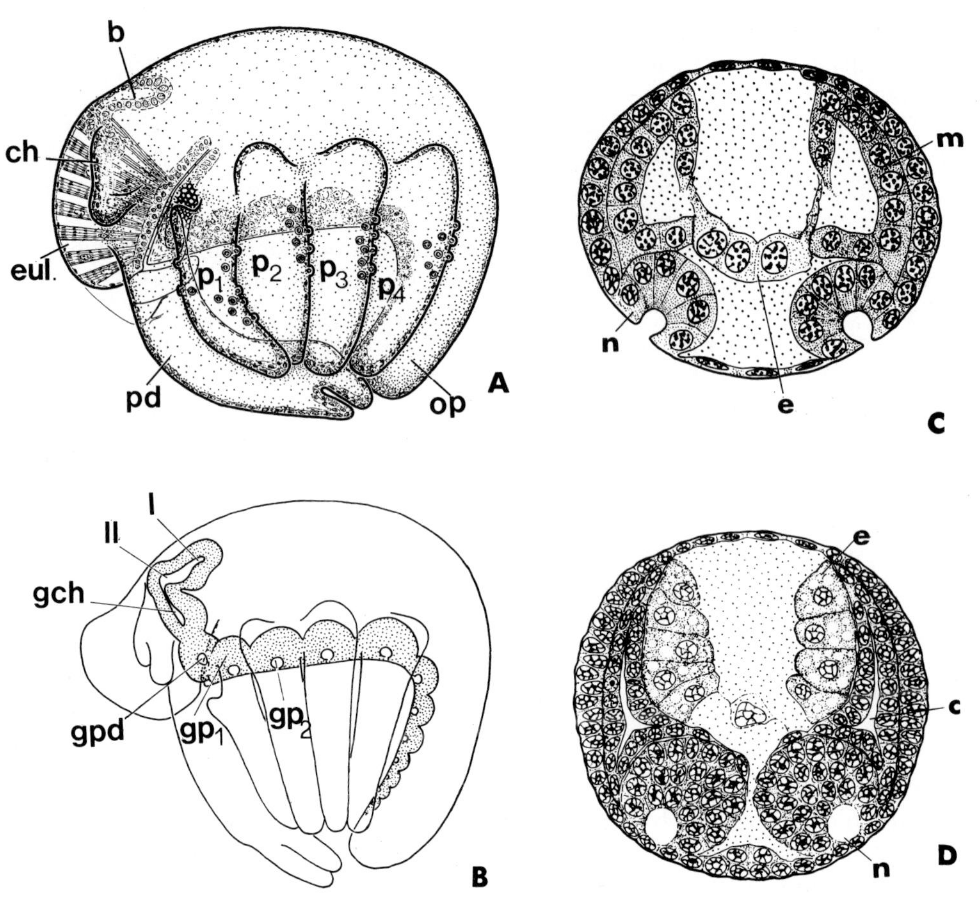

Fig. 1. A – D The development of neuromeres in the pseudoscorpion, *Neobisium muscorum.* **A** Side view of a transparent embryo. Except for the pharynx with its huge embryonic upper lip, the only internal structures visible are the invaginating rudiments of the nervous system. **B** The developing ganglionic anlagen drawn into the outline of a slightly older embryonic stage. **C** and **D** Cross-sections through the developing opisthosoma showing the invaginations of neuromeres. *b* early brain rudiment; *c* coelomic cavity; *ch* chelicera; *e* entoderm; *eul* upper lip (labrum) of the embryonic pharynx; *gch* ganglion of the cheliceral segment; gp_1, gp_2 ganglia of the first and second leg segments; *gpd* ganglion of the pedipalpal segment; *m* mesoderm; *n* developing neuromere; *op* opisthosoma; $p_1 - p_4$ anlagen of the first to fourth legs; *pd* pedipalpus; *I* and *II* brain rudiment during the process of division into two parts with two cavities. (Weygoldt 1964, 1965)

rowing ganglionic cavity and the developing commissure. There are six such neuromeres in the prosoma, representing the neuromeres of the appendage-bearing segments. Anterior to the chelicerae, however, there are two such ganglionic cavities on either side (I and II on Fig. 1 B and 3), which have developed from a single pair of invaginations.

Further development of the nervous system involves drastic changes. The opisthosomal ganglia, once they have separated from the epidermis, start mov-

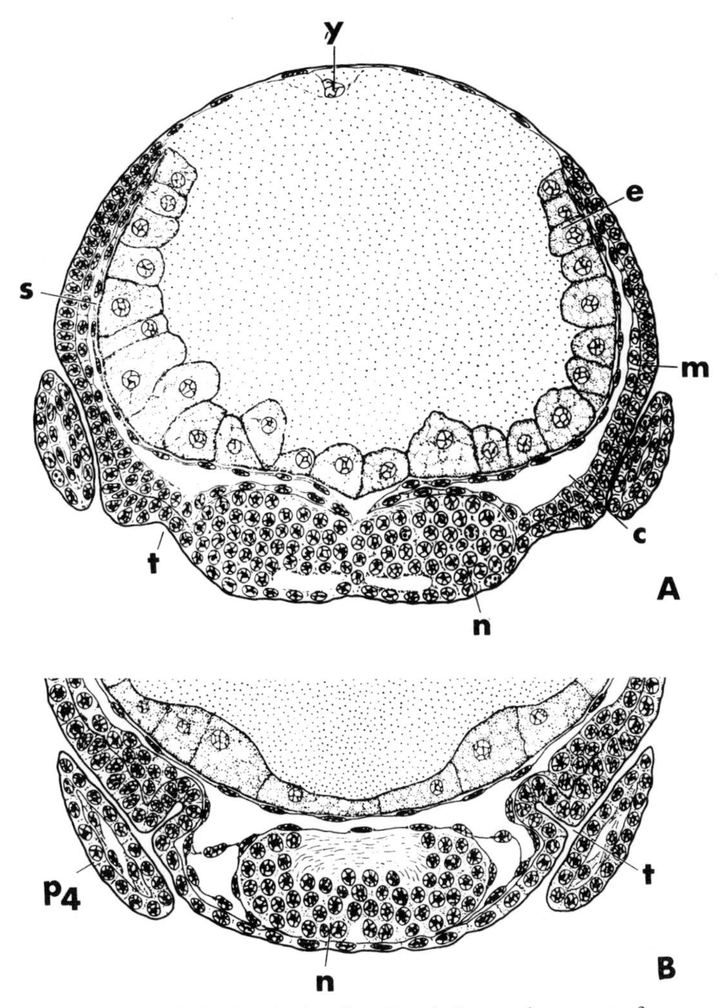

Fig. 2. A and **B** Cross-sections through the developing fourth opisthosomal segment of an embryo of the pseudoscorpion, *Chernes cimicoides,* showing later stages of the developing neuromeres with the disappearing ganglionic cavities. In **A** these cavities are still present, in **B** the neuromere has fully separated from the epidermis and the cavities are obliterated. *c* coelomic cavity; *e* entoderm; *m* mesoderm; *n* neuromere; p_4 fourth leg; *s* splanchnic part of mesoderm; *t* invaginating trachea; *y* yolk entoderm. (Weygoldt 1964)

ing forward, and all postcheliceral ganglia merge into one large subesophageal mass. The brain develops from the cheliceral neuromere and from the anlagen in front of these. As stated above, the anlagen early divide into two parts, each of which contains a ganglionic cavity. During the elevation of the anterior body region, the anterior-most part of the brain anlage, together with the anterior-most pair of ganglionic cavities, is shifted upward and finally backward (Fig. 3B). Later it occupies the posterior-most part of the supraesophageal ganglionic mass and gives rise to the developing central body.

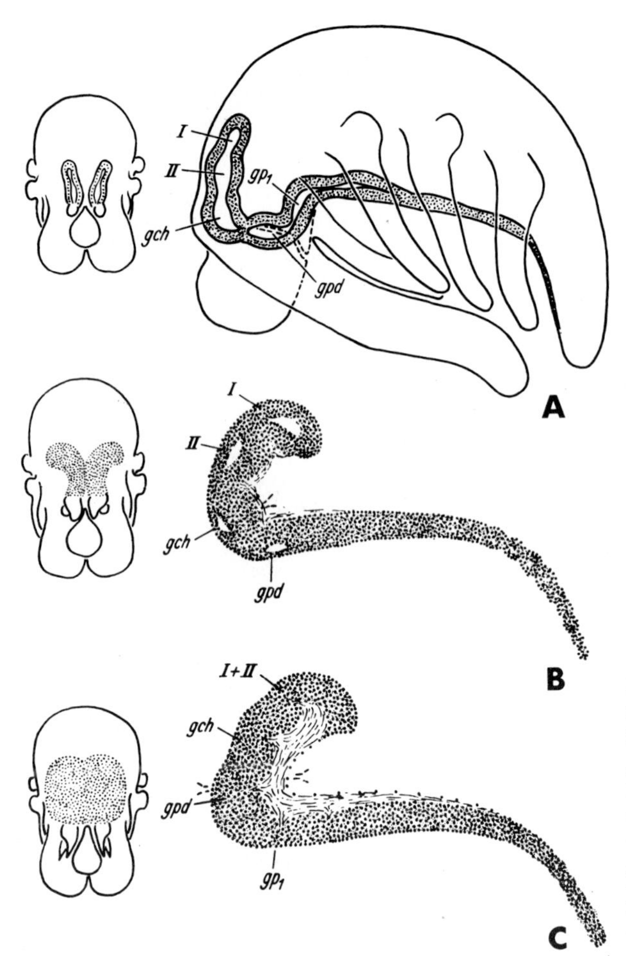

Fig. 3 A – C. The development of the nervous system in the pseudoscorpion, *Pselaphochernes scorpioides*, as seen in diagrammatic parasagittal sections (*right*) and front views (*left*). *gch* cheliceral ganglion; gp_1 first leg ganglion; *gpd* pedipalpal ganglion; *I* and *II* the two cavities of the developing brain. (Weygoldt 1964)

3.2.2 Discussion

Now the question of the segmental composition of the pseudoscorpion forehead and its brain is open for speculation. There can be no doubt that the cheliceral ganglionic rudiment represents a neuromere. The anlagen in front of it may be interpreted in three ways. They could be the archicerebrum, that is the ganglion (not neuromere) of the presegmental acron (homologous to the annelid prostomium). The early appearance of two pairs of ganglionic cavities may be an early subdivision of the archicerebrum for functional reasons. They could also be the anlagen of two neuromeres, represented by the rudiments with their ganglionic cavities; an archicerebrum would then not be present. Finally, they may be interpreted as an archicerebrum and a neuromere of a precheliceral segment, the first rudiment, with its cavity representing the archicerebrum, and the second representing the neuromere of a precheliceral metamere.

3.3 Amblypygids and Araneids

3.3.1 Segmental Ganglia

In these arachnids, the development of the CNS is more complex than in pseudoscorpions. Each neuromere is formed by a number of invaginations (Fig. 4A), not by a single pair of invaginations as in pseudoscorpions. Cells rapidly divide and form small clusters which invaginate. They continue to proliferate when the innermost cells are differentiating into neurons and growing axons. Thus, sections through developing ganglia show proliferating cells at the periphery of the neuromeres and growing axons which form commissures and connectives at the interior side.

3.3.2 Brain

Brain development is even more complex (Lambert 1909, Pross 1966; Weygoldt 1975; Yoshikura 1955). At first, a pair of large cephalic lobes are formed in front of the cheliceral segment (Fig. 5A–D). The lobes quickly thicken by divisions of many cells which form numerous invaginations of different sizes (Fig. 4B). At the same time two crescent-shaped grooves appear on the anterior

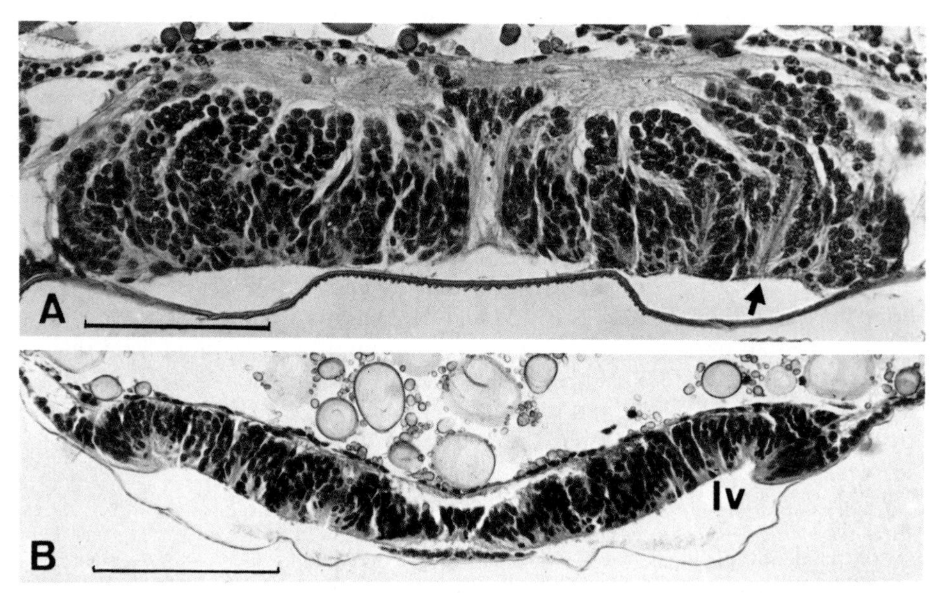

Fig. 4. A Cross-section through a developing prosomal ganglion of the whip spider, *Tarantula marginemaculata*. *Arrow* points to one of the numerous invaginations which develop along with an increase of the number of neurons. The interior-most neurons grow dendritic processes which form the neuropile. *Bar*=0.2 mm. **B** Cross-section through the brain anlagen of an embryo of *Tarantula marginemaculata*. The brain is in the developmental stage shown in Fig. 5B. *lv* points to the initial stage of the formation of the lateral vesicles. *Bar*=0.4 mm. (Weygoldt 1975)

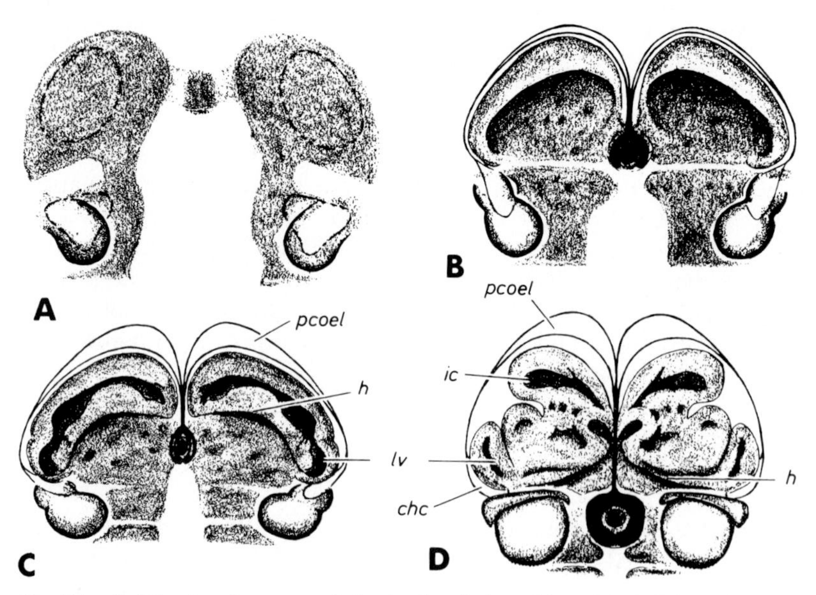

Fig. 5A–D. The development of the brain of the whip spider, *Tarantula marginemaculata,* drawn from whole mounts of the brain anlagen of four different stages. in **A** the brain anlagen is only a pair of epithelial thickenings in front of the anlagen of the chelicerae. *Broken lines* indicate the borders of the precheliceral and cheliceral coeloms. In **B** a hood is starting to overgrow the thickenings. In **C** the hood is growing farther backward and the lateral vesicles are separating from the interior brain cavities. **D** shows a later stage. The tissue surrounding the interior brain cavities represents the anlagen of the central body, and the lateral vesicles will form the optic masses of the lateral eyes. *chc* cheliceral coelom; *h* hood; *ic* interior brain cavity; *lv* lateral vesicle; *pcoel* precheliceral coelom. (Weygoldt 1975)

part of the lobes, and each groove is overgrown by a hood (Fig. 5B, C; 6B, C). Soon afterward each groove divides into a large anterior and a smaller lateral part. The lateral parts form the lateral vesicles, each of which is later split into an anterior and a posterior part, i.e., the anlagen of the anterior and posterior optic masses of the lateral eyes (Fig. 5C, D).

The median parts of the crescent-shaped grooves are separated from the exterior when their hoods fuse with their bases. By this process the interior brain cavities are formed. These and their surrounding cells soon move backward and fuse medially thereby forming the rudiment of the central body (Fig. 6B).

Fig. 6 A–C. Sections through advanced stages of the developing brain of the whip spider, *Tarantula marginemaculata.* **A** The cross-section shows that even in this advanced stage there are still many invaginations of different sizes (most of them seen in cross-section). **B** and **C** Sagittal sections through the same stage. **B** is a more lateral section which clearly shows the hood covering a large part of the brain rudiment and forming the exterior brain cavity which is still in open connection with the exterior. The interior brain cavity is nearly obliterated and forms the anlagen of the central body. **C** shows a more medial section with the developing corpora pedunculata. *amc* anterior optic mass of the median eye; *bs* blood sinus; *cb* central body; *cp* corpora pedunculata; *ec* exterior brain cavity; *h* hood; *lv* lateral vesicle; *pc* precheliceral coelom; *pmc* posterior optic mass of the median eye; *vs* ventral sinus. Bar=0.1 mm. (Weygoldt 1975)

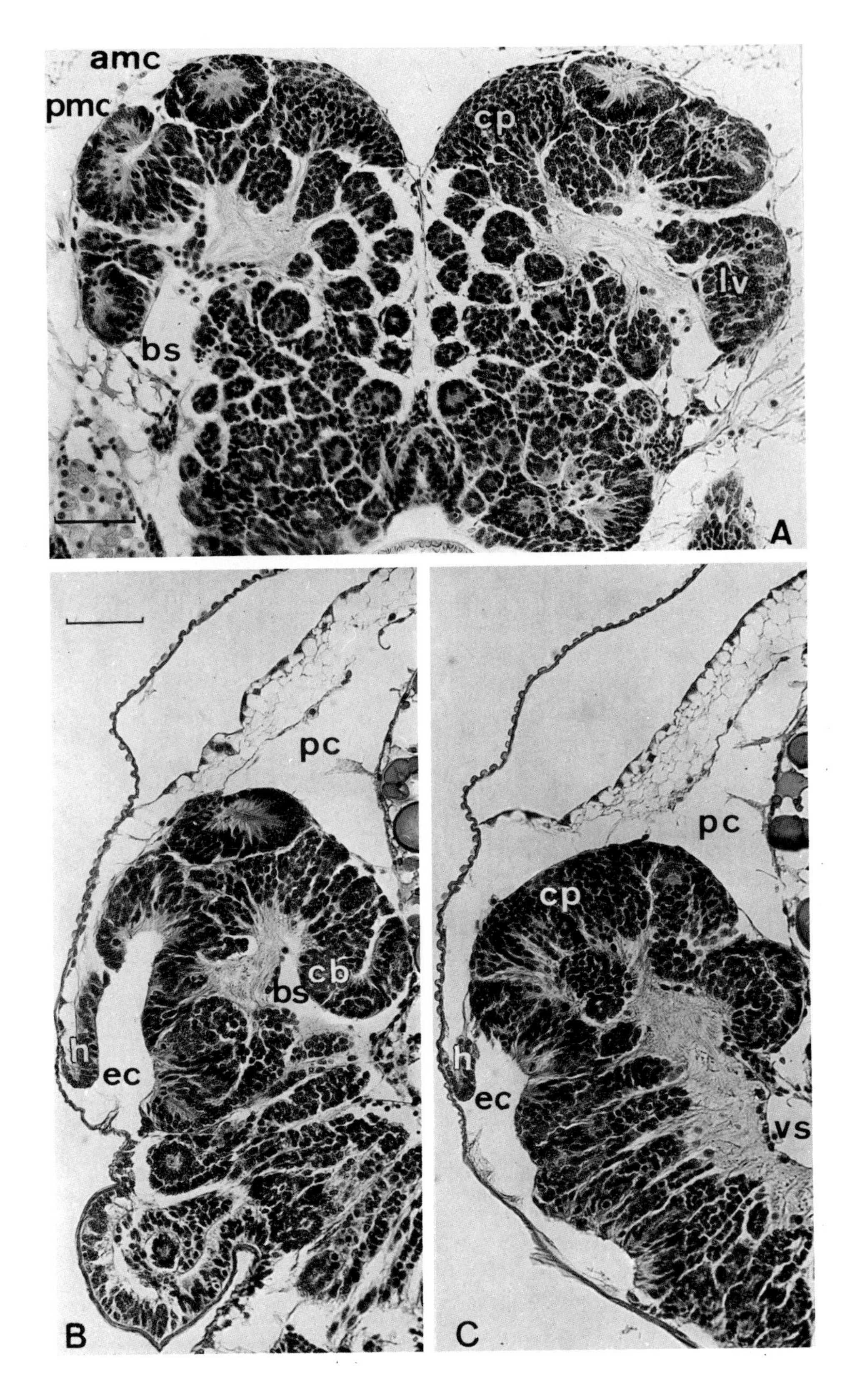
amc
pmc
cp
lv
bs
A
pc
cb
bs
h
ec
B
pc
cp
h
ec
vs
C

Table 1. Composition of the mandibulate and chelicerate head as seen by many embryologists

		Mandibulata			Chelicerata		
Region		Appendage	Part of Brain		Appendage	Part of brain	
	Acron	–	Archicerebrum	Protocerebrum	–	Archicerebrum	Protocerebrum
1st Segm.	Prosocephalon (preantennulary segment)	Labrum?	Prosocerebrum		Labrum?	Prosocerebrum	
2nd Segm.	Deutocephalon	Antennules	Deutocerebrum		Lost	Deutocerebrum lost or fused to above	
3rd Segm.	Tritocephalon	Antennae	Tritocerebrum		Chelicera	Cheliceral neuromere	
4th Segm.	Segm. of mandibles	Mandibles	Mandibular neuromere		Pedipalpus	Pedipalpal neuromere	
5th Segm.	Segm. of maxillules	Maxillules	Maxillulary neuromere		1st leg	1st leg neuromere	
6th Segm.	Segm. of maxillae	Maxillae	Maxillary neuromere		2nd leg	2nd leg neuromere	

Thus, the central body which later occupies the posterior-most part of the brain again develops from the anterior-most part of the cephalic lobes.

The hood continues to overgrow the brain anlagen and, when the rudiments of the central body have separated from the remaining parts of the cephalic lobes, it forms an anterior or exterior pair of brain cavities (Fig. 6B, C). From their bases many small invaginations continue to thicken the brain anlage and to increase the number of neurons. The further development of some of these invaginations can easily be followed, e.g., of those giving rise to the optic masses of the median eyes (Fig. 6A). The hood later forms the median eyes.

3.3.3 *Discussion*

Although there are differences between orders, development of the central nervous system of other arachnids is similar to that described here for whip spiders and spiders. The two pairs of cavities in the pseudoscorpion brain rudiment are probably homologous to the interior and exterior brain cavities of other arachnids. As is evident from many descriptions, brain development is rather complicated in arachnids and there are no easily recognizable intersegmental boundaries or delineations between possible neuromeres. Therefore, different parts of the brain anlagen have been claimed by various authors to represent neuromeres indicative of reduced segments. For example, Legendre (1979) assumed that the lateral vesicles are vestigial appendages of a preantennulary segment, and Pross (1966) believes the same structures to represent the deutocerebrum. As stated above, the lateral vesicles are the anlagen of the optic masses of the lateral eyes. According to Siewing (1969), Pross (1966, 1977), and Winter (1980), the brain of arachnids contains all the neuromeres supposed to constitute the mandibulate brain (Table 1). They suppose the preantennulary ganglion or prosocerebrum to be represented by the central body (which, in arachnids, develops from the anterior-most part of the embryonic brain anlagen, and which may or may not be homologous to the central body of crustaceans and insects), the protocerebral bridge, and the accessory lobes. Their argument is that in crustaceans and insects these structures develop in the region hypothesized to represent the prosocerebrum (g_1 in Fig. 7B). This is arguing in circles.

The question of the *segmental composition* of the arthropod forehead cannot be answered by studies of the nervous system alone. There are preantennulary coelomic cavities in many arthropods, clearly indicating reduced segments or a segment tightly fused to the acron. Therefore, the search for a preantennulary ganglion seems justified. None of the structures observed during brain development in spiders and whip spiders and other arachnids, however, can be homologized with segmental neuromeres with certainty. In particular the central body is very unlikely to indicate the existence of a neuromere following the archicerebrum, since it develops from the anterior-most part of the cephalic lobes. This does not imply that the arachnid brain has evolved from the archicerebrum alone. In fact, none of the various hypotheses on head development and segmental composition of the arthropod and arachnid forehead

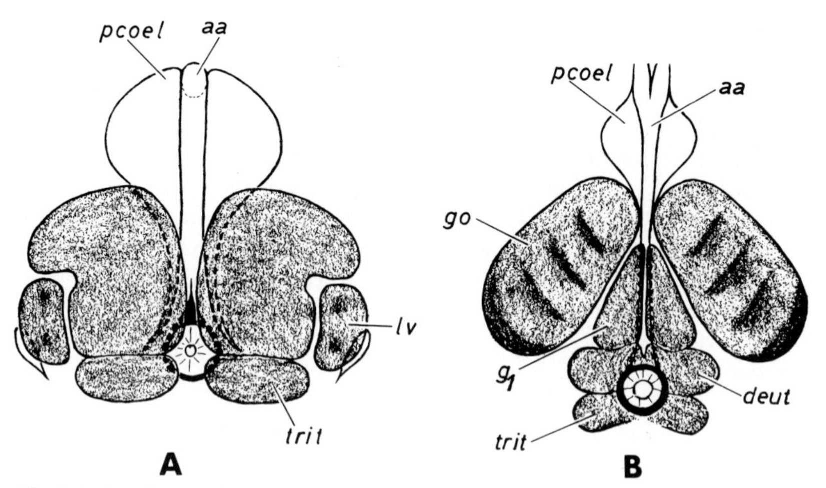

Fig. 7 A, B. Comparison of the brain anlagen of an arachnid as exemplified by the whip spider, *Tarantula marginemaculata* (**A**), and a crustacean as exemplified by the shrimp, *Palaemonetes varians* (**B**). The brain rudiment of the arachnid does not show a clear division into lobes; there is one large protocerebral mass which, in most arachnids, develops from two pairs of brain cavities. The lateral vesicles are not homologous to neuromeres but may be homologous to the optic ganglia of the crustacean brain; they form the optic masses of the lateral eyes. The brain anlagen of the crustacean is early divided into a number of lobes. The largest of these are the optic lobes (*go*). The protocerebrum is formed by these and by the lobes g_1. *aa* anterior aorta; *deut* deutocerebrum; g_1 protocerebrum; *go* optic ganglia; *lv* lateral vesicles; *pcoel* in the arachnid: precheliceral coelom, in the crustacean: preantennulary coelom; *trit* tritocerebrum. (Weygoldt 1975)

can be falsified with the available data. The idea that the arthropod brain is composed of an archicerebrum, a prosocerebrum, a deutocerebrum, and the tritocerebrum, and that all of these except the first one are homologous to segmental neuromeres is probably correct. The attempts to homologize certain parts of the arachnid brain to any of these neuromeres, however, are not convincing. I prefer to consider the arachnid brain as composed of a protocerebrum of unknown segmental composition and of the tritocerebrum. The deutocerebrum has either been lost or is completely fused to the protocerebrum.

The brain development in mandibulates and chelicerates shows typical differences (Fig. 7A, B). The mandibulate, exemplified by a decaped crustacean (Fig. 6B), develops large optic lobes and a small protocerebrum anlage (Fig. 7, g_1) which may or may not contain a prosocerebrum and is followed by the deutocerebrum. In the arachnid, exemplified by a whip spider, vision appears relatively poorly developed and the homologa of the optic lobes are the small lateral vesicles. The bulk of the brain rudiment may be homologous to the small protocerebrum anlage of the mandibulate.

Johannson (1933), when studying the development of *Limulus*, found structures which he hypothesized to represent the *antennal glomeruli* of a reduced deutocerebrum. It seems that what he saw were small ganglionic lobes of unknown significance, but not glomeruli. The antennal glomeruli are association centers processing the sensory inputs coming from the antennules. They are quite similar in insects and crustaceans. In some isopods which tend to reduce the antennules, antennal glomeruli develop in the tritocerebrum, processing sensory inputs from the antennae, and in whip spiders, similar glomeruli are found in the ganglion of the first pair of legs which, in these animals, are highly specialized sensory structures resembling antennae.

4 Postembryonic Development of the Spider CNS

At hatching, all ganglia and all neurons have been formed and the process of concentration which leads to the fusion of all opisthosomatic ganglia with the subesophageal ganglion is complete or nearly so. In spite of this, the CNS of a freshly hatched spiderling or whip spider is quite different from that of an adult spider (Babu 1975). In *Argiope*, at this early stage the CNS constitutes more than 40% of the prosomal volume. The cell bodies of the neurons are of nearly equal size and form a nearly uniform cortex surrounding the neuropile.

During postembryonic development the CNS continues to grow, although this growth is negatively allmetric if compared to the volume of the prosoma. In *Argiope*, the adult brain constitutes only 4% of the prosomal volume in the female and 10% in the male. *Postembryonic growth* of the CNS is due to the following processes (Babu 1975): (1) Increase of cell size in different types of neurons. Some motor neurons increase by up to 600%. (2) Increase of the neuropile mass by continuous growth of dendritic and axonic fibers. (a) Throughout postembryonic development the complexity of the CNS increases by addition of more and new connections among different parts of the neuropile, thus permitting more complex behavior. (b) Motor fibers not only grow longer but also thicken, in some motor neurons from a diameter of 1 μm in the first instar to a maximum of 16 μm in adults. (c) Glial cells form neural lamellae surrounding part of the CNS and also individual envelopes of nerve fibers. The number and size of glial cells increase, as well as the thickness of the neural lamellae. (d) The neuropile further increases by the addition of new afferent fibers from sense organs differentiating during postembryonic development. Trichobothria, slit sensilla, chemotactile hairs and other sensilla increase in number during postembryonic growth. In *Limulus* the number of ommatidia of the compound eyes also increases during development. In arachnids, most of the sensory input comes from the pedipalps and legs; the subesophageal ganglionic mass therefore is its main recipient. The brain only receives information coming from the eyes and from the chelicerae. The growing importance of the sensory inputs from the pedipalps and legs is reflected by the relative growth of the brain and subesophageal mass. In *Argiope*, the brain, during postembryonic development, increases about 10-fold, the subesophageal mass 36-fold.

The brain contains the most important *association centers*, such as the corpora pedunculata and the central body. In particular, the corpora pedunculata receive strong fiber bundles coming from the subesophageal sensory centers, such as the glomeruli of the antenniform legs in whip spiders (Weygoldt 1975). Both the corpora pedunculata and the central body receive fibers from all sensory centers including those of the eyes, and they are connected with the motor centers of the subesophageal mass (Babu 1965). Thus, although the subesophageal ganglion is the major recipient of sensory input as well as the main place of motor centers, the brain remains the most important association center processing incoming information and controlling motor activities (see also Babu, Chap. I, this Vol.).

There are, however, great differences in the morphology of the brain of *various arachnids* (Meier 1967). Arachnids without eyes lack optic ganglia. Some arachnids in which vision is of minor importance also lack corpora pedunculata, e.g., pseudoscorpions. In still others, like whip spiders, whip scorpions, and opilionids, which also do not seem to largely depend on vision, corpora pedunculata are well developed. In spiders, corpora pedunculata are well developed in fast-moving species, e.g., lycosids and thomisids, but missing in dysderids and web-building spiders like agelenids and argiopids. Postembryonic brain development also varies in different spiders. In dysderids, for example, motor neurons and motor centers are well-differentiated at hatching, whereas the development of the association centers is delayed. As early as after the first postembryonic molt (that is, the first molt after shedding of the chorion) the spiderlings of these species start to move in the female tube web. In agelenids, on the other hand, the spiderlings remain in the egg cocoon and the development of motor centers is delayed. In *Argiope*, web-building starts in the late second and early third instar. This behavior seems to be correlated with the final differentiation of the central body (Babu 1975) which, in this species, is the only but well-developed association center.

5 Conclusions

As is evident from the short outlines given, the study of the ontogeny of the arachnid central nervous system is still in its infancy. This is partly due to the fact that, because of their predatory habits, it is laborious to breed arachnids in large numbers. Furthermore, the highly developed parental care of many arachnids makes a continuous supply of eggs for experimental studies difficult.

There are unique characters in the development of the nervous system of arachnids, not shared by insects and crustaceans. The formation of the neuromeres, as well as the development of the gross morphology of the brain, are different. During the development of the arachnid neuromeres, invaginations are the most conspicuous processes. The same is true for the developing brain, and no clearly demarcated lobes are apparent. The developing crustacean and insect brain is early divided into a constant number of lobes which invite speculations about their nature as neuromeres of fused segments. The phylogenetic significance of these differences is still obscure.

References

Aeschlimann A (1958) Développement embryonnaire d'*Ornithodorus moubata* (Murray) et transmission ovarienne de *Borrelia duttoni*. Acta Trop 15:15–64

Anderson DT (1973) Embryology and phylogeny in annelids and arthropods. Pergamon Press, Oxford New York

Babu KS (1965) Anatomy of the central nervous system of arachnids. Zool Jahrb Anat 82:1–154

Babu KS (1975) Postembryonic development of the central nervous system of the spider *Argiope aurantia* (Lucas). J Morphol 146:325–342

Barald KF (1982) Monoclonal antibodies to embryonic neurons: Cell-specific markers for chick ganglion. In: Spitzer NC (ed) Neuronal development. Plenum Press, New York, 101–119

Boudreaux HB (1979a) Significance of intersegmental tendon system in arthropod phylogeny and a monophyletic classification of arthropods. In: Gupta AP (ed) Arthropod phylogeny. Van Nostrand Reinhold, New York, pp 551–586

Boudreaux HB (1979b) Arthropod phylogeny with special reference to insects. Wiley, New York

Brauer A (1895) Beiträge zur Kenntnis der Entwicklungsgeschichte des Skorpions II. Z Wiss Zool 59:351–435

Dohle W (1964) Die Embryonalentwicklung von *Glomeris marginata* (Villers) im Vergleich zur Entwicklung anderer Diplopoden. Zool Jahrb Anat 81:241–310

Dohle W (1976) Die Bildung und Differenzierung des postnauplialen Keimstreifs von *Diastylia rathkei* (Crustacea, Cumacea). II. Die Differenzierung und Musterbildung des Ektoderms. Zoomorphologie 84:235–277

Edwards JS (1982) Pioneer fibers. The case for guidance in the embryonic nervous system of the cricket. In: Spitzer NC (ed) Neuronal development. Plenum Press, New York, pp 255–266

Flaster MS, Macagno ER, Schehr RS (1982) Mechanisms for the formation of synaptic connections in the isogenic nervous system of *Daphnia magna*. In: Spitzer NC (ed) Neuronal development. Plenum Press, New York, pp 267–296

Goodman CS (1982) Embryonic development of identified neurons in the grasshopper. In: Spitzer NC (ed) Neuronal development. Plenum Press, New York, pp 171–212

Goodman CS, Bate CM, Spitzer NC (1981) Embryonic development of identified neurons: Origin and transformation of the H cell. J Neurosci 1:94–102

Goodman CS, Pearson KG, Spitzer NC (1980) Electrical excitability: A spectrum of properties in the progeny of a single embryonic neuroblast. Proc Natl Acad Sci USA 77:1676–1680

Goodman CS, Raper JS, Ho RK, Chang S (1982) Pathfinding by neuronal growth cones in grasshopper embryos. In: Cytochemical methods in neuroanatomy. Liss, New York, 461–491

Goodman CS, Spitzer NC (1981a) The mature electrical properties of identified neurons in grasshopper embryos. J Physiol (London) 313:369–384

Goodman CS, Spitzer NC (1981b) The development of electrical properties of identified neurons in grasshopper embryos. J Physiol (London) 313:385–403

Hanström B (1928) Das Nervensystem der wirbellosen Tiere. Springer, Berlin Heidelberg New York

Hennig W (1950) Grundzüge einer Theorie der phylogenetischen Systematik. Deutscher Zentralverlag, Berlin

Heymons R (1901) Die Entwicklungsgeschichte der Scolopender. Zoologica (Stuttgart) 13:1–244

Ho RK, Ball EE, Goodman CS (1983) Muscle pioneers: Large mesodermal cells that erect a scaffold for developing muscles and motor neurons in grasshopper embryos. Nature (London) 301:66–69

Holmgren N (1916) Zur vergleichenden Anatomie des Gehirns von Polychaeten, Onychophoren, Xiphosuren, Arachniden, Crustaceen, Myriapoden und Insekten. Vet Akad Handl Stockholm 56:1–303

Jacobsen M (1982) Origins of the nervous system in amphibians. In: Spitzer NC (ed) Neuronal development. Plenum Press, New York, 45–99

Johansson G (1933) Beiträge zur Kenntnis der Morphologie und Entwicklung des Gehirns von *Limulus polyphemus*, Acta Zool (Stockholm) 14:1–100

Junqua C (1966) Recherches biologiques et histophysiologiques sur un solifuge saharien *Othoes saharae* Panouse. Thèse Fac Sci Univ Paris Ser A, 4689, 124 pp

Kaestner A (1951) Zur Entwicklungsgeschichte von *Thelyphonus caudatus* L. (Pedipalpi). 3. Teil. Die Entwicklung des Zentralnervensystems. Zool Jahrb Anat 71:1–55

Knoll HJ (1974) Untersuchungen zur Entwicklungsgeschichte von *Scutigera coleoptrata* L. (Chilopoda). Zool Jahrb Anat 92:47–132

Lambert AE (1909) History of the procephalic lobes of *Epeira cinerea.* A study on arachnid embryology. J Morphol 20:413–459

Larink O (1969) Zur Entwicklungsgeschichte von *Petrobius brevistylis* (Thysanura, Insecta). Helgol Wiss Meeresunters 19:111–155

Larink O (1970) Die Kopfentwicklung von *Lepisma saccharina* L. (Insecat, Thysanura). Z Morphol Tiere 67:1–15

Lauterbach K-E (1973) Schlüsselereignisse in der Evolution der Stammgruppe der Euarthropoda. Zool Beitr NS 19:251–299

Lauterbach K-E (1980a) Schlüsselereignisse in der Evolution des Grundplans der Mandibulata (Arthropoda). Abh Naturwiss Ver Hamburg NS 23:105–161

Lauterbach K-E (1980b) Schlüsselereignisse in der Evolution des Grundplans der Arachnata (Arthropoda). Abh Naturwiss Ver Hamburg NF 23:163–327

Legendre R (1958) Contribution à l'étude du système nerveux des aranéides. Ann Biol 34:193–223

Legendre R (1959) Contribution à l'étude du système nerveux des aranéides. Ann Sci Nat Zool 1, 12:339–473

Legendre R (1979) La ségmentation de la région antérieure des Arachnides: Histoire et perspectives actuelles. Bull Soc Zool Fr 104:277–287

Malzacher P (1968) Die Embryogenese des Gehirns paurometaboler Insekten. Untersuchungen an *Carausius morosus* und *Periplaneta americana.* Z Morphol Tiere 62:103–161

Manton SM (1928) On the embryology of a mysid crustacean, *Hemimysis lamornae.* Philos Trans R Soc London Ser B 216:363–463

Manton SM (1934) On the embryology of the crustacean *Nebalia pipes.* Philos Trans R Soc London Ser B 223:163–238

Manton SM (1949) Studies on the Onychophora VII. The early embryonic stages of *Peripatopsis* and some general considerations concerning the morphology and phylogeny of the arthropoda. Philos Trans R Soc London Ser B 233:483–580

Manton SM (1960) Concerning head development in the arthropods. Biol Rev 35:265–282

Manton SM (1973) Arthropod phylogeny – a modern synthesis. J Zool (London) 171:111–130

Manton SM (1977) The arthropods. Habits, functional morphology, and evolution. Clarendon Press, Oxford

Manton SM (1978) Habits, functional morphology and the evolution of pycnogonids. Zool J Linn Soc 63:1–21

Mathew AP (1956) Embryology of *Heterometrus scaber* (Thorell), Arachnida: Scorpionidae. Zool Mem Univ Travancore 1:1–96

Meier F (1967) Beiträge zur Kenntnis der postembryonalen Entwicklung der Spinnen, Araneida, Labidognatha. Unter besonderer Berücksichtigung der Histogenese des Zentralnervensystems. Rev Suisse Zool 74:1–127

Moritz M (1957) Zur Embryonalentwicklung der Phalangiiden (Opiliones, Palpatores) unter besonderer Berücksichtigung der äußeren Morphologie, der Bildung des Mitteldarms und der Genitalanlage. Zool Jahrb Anat 76:331–370

Paulus HF (1979) Eye structure and the monophyly of the Arthropoda. In: Gupta AP (ed) Arthropod phylogeny. Van Nostrand Reinhold, New York, pp 299–383

Pflugfelder O (1948) Entwicklung von *Paraperipatus amboinensis* n. sp. Zool Jahrb Anat 69:443–492

Pross A (1966) Untersuchungen zur Entwicklungsgeschichte der Araneae [*Pardosa hortensis* (Thorell)] unter besonderer Berücksichtigung des vorderen Prosomaabschnittes. Z Morphol Oekol Tiere 58:38–108

Pross A (1977) Diskussionsbeitrag zur Segmentierung des Cheliceraten-Kopfes. Zoomorphologie 86:183–196

Remane A (1956) Die Grundlagen des natürlichen Systems, der vergleichenden Anatomie und der Phylogenetik: Theoretische Morphologie und Systematik 1. Akad Verlagsges Geest & Protig, Leipzig

Rempel JG (1957) The embryology of the black widow spider, *Latrodectus mactans* (Fabr.). Can J Zool 36:35–74

Scholl G (1963) Embryologische Untersuchungen an Tanaidaceen (*Heterotanais oerstedi* Kröyer). Zool Jahrb Anat 80:500–554

Scholl G (1969) Die Embryonalentwicklung des Kopfes und Prothorax von *Carausius morosus* Br. (Insecta, Phasmidae). Z Morphol Tiere 65:1–142

Scholl G (1977) Beiträge zur Embryonalentwicklung von *Limulus polyphemus* L. (Chelicerata, Xiphosura). Zoomorphologie 86:99–154

Siewing R (1963) Zum Problem der Arthropodenkopfsegmentierung. Zool Anz 170:429–468

Siewing R (1969) Lehrbuch der vergleichenden Entwicklungsgeschichte der Tiere. Parey, Hamburg Berlin

Snodgrass RE (1952) A textbook of arthropod anatomy. Comstock, New York

Snodgrass RE (1960) Facts and theories concerning the insect head. Smithson Misc Collect 142:1–61

Stent GS, Weiblat DA, Blair SS, Zackson SL (1982) Cell lineage in the development of the leech nervous system. In: Spitzer NC (ed) Neuronal development. Plenum Press, New York, pp 1–44

Tiegs, OW (1940) The embryology and affinities of the Symphyla, based on a study of *Hanseniella agilis*. Q J Microsc Sci 82:1–115

Tiegs, OW (1947) The development and affinities of Pauropoda, based on a study of *Pauropus silvaticus*. Q J Microsc Sci 88:165–267

Weber H (1952) Morphologie, Histologie und Entwicklungsgeschichte der Articulaten. Fortschr Zool 9:1–231

Weber H (1954) Grundriß der Insektenkunde, 3rd edn. Fischer, Stuttgart

Weygoldt P (1959) Die Embryonalentwicklung des Amphipoden *Gammarus pulex pulex* (L.). Zool Jahrb Anat 77:51–110

Weygoldt P (1961) Beitrag zur Kenntnis der Ontogenie der Dekapoden: Embryologische Untersuchungen an *Palaemonetes varians* (Leach). Zool Jahrb Anat 79:223–270

Weygoldt P (1964) Vergleichend-embryologische Untersuchungen an Pseudoscorpionen (Chelonethi). Z Morphol Oekol Tiere 54:1–106

Weygoldt P (1965) Vergleichend-embryologische Untersuchungen an Pseudoscorpionen. III. Die Entwicklung von *Neobisium muscorum* Leach (Neobisiinea, Neobisiidae). Z Morphol Oekol Tiere 55:321–382

Weygoldt P (1968) Vergleichend-embryologische Untersuchungen an Pseudoscorpionen VI. Die Entwicklung von *Chthonius tetrachelatus* Preyssl., *Chthonius ischnocheles* Hermann (Chthoniinea, Chthoniidae) und *Verrucaditha spinosa* (Chthoniinea, Tridenchthoniidae). Z Morphol Tiere 63:111–154

Weygoldt P (1975) Untersuchungen zur Embryologie und Morphologie der Geißelspinne *Tarantula marginemaculata* C. L. Koch (Arachnida, Amblypygi, Tarantulidae). Zoomorphologie 82:137–199

Weygoldt P (1979) Significance of later embryonic stages and head development in arthropod phylogeny. In: Gupta AP (ed) Arthropod phylogeny. Van Nostrand Reinhold, New York, pp 107–135

Winter G (1980) Beiträge zur Morphologie und Embryologie des vorderen Körperabschnittes (Cephalosoma) der Pantopoda Gerstaecker, 1963. I. Entstehung und Struktur des Zentralnervensystems. Z Zool Syst Evolutionsforsch 18:27–61

Yoshikura M (1955) Embryological studies on the liphistiid spider, *Heptathela kimurai*, part II. Kumamoto J Sci B 2:1–86

Yoshikura M (1958) On the development of a purse-web spider, *Atypus karschi* Dönitz. Kumamoto J Sci B 3:73–86

Yoshikura M (1961) The development of a whip scorpion, *Typopeltis stimpsonii* Wood. Acta Arachnol (Osaka) 17:19–24

III The Stomatogastric Nervous System and Neurosecretion

Roland Legendre

CONTENTS

1 Introduction

The stomatogastric nervous system seems to be built according to the same pattern in all the arachnids. The general scheme apparently even holds for all arthropods. In all these cases the anterior sympathetic ganglia, in the mouth region, lose part of their nervous function to become endocrine organs similar to the neurosecretory cells of the central nervous system, with which they are closely related functionally.

In the following, the structure of the stomatogastric system is first compared among the recent arthropods and then described in detail for the spiders, which are the best-studied arachnid case. Finally, its anatomy in the other arachnid orders will be surveyed.

2 Comparison Among Arthropods

The nervous system of all annelids and arthropods (Annulata) is basically composed of five primitive nerve trunks: two double nerve trunks and a single trunk, the origin of which can be traced to the nervous system of archaic Turbellaria (Chaudonneret 1978).

In the Arthropoda we have theoretically the following nerve trunks:

a) The main double nerve cord with paired segmental ganglia which forms the central nervous system. It folds upon itself during ontogeny, and many

Laboratoire de Zoologie, Université des Sciences et Techniques du Languedoc (Montpellier II), Place E. Bataillon, F-34060 Montpellier, France

neuromeres fuse, especially in the anterior part. The degree of flexion and fusion differs between the different classes and corresponds to evolutionary steps which are not completely analyzed as yet.
b) A lateral double nerve cord running along the central nervous system. This nerve cord may be very small or split into several fine branches; it is particularly visible in the head region. The most anterior of its ganglia may fuse with the brain or remain independent. In general they lose their nervous function and become neuroendocrine organs; their evolution differs from taxon to taxon.
c) A dorsal single nerve trunk running along the anterior part of the digestive tract. This unpaired structure is very small and may be absorbed by the other parts of the nervous system. It can disappear in some taxa. Its function is purely nervous: innervation of the stomodeal part of the digestive tract.

In *arachnids* the lateral paired cord is present in many orders like spiders, pseudoscorpions and whipspiders (Amblypygi), and the anterior ganglia lose their nervous function by different degrees to become neuroendocrine organs with secondary connections to the hindpart of the brain. The unpaired nerve cord is present in all classes: it consists of the stomodeal bridge, rostral nerve, and recurrent nerve, which innervate the muscles of the pharynx, rostrum, and esophagus, respectively. An autonomous sympathetic heart system as known in many other arthropods (Bullock and Horridge 1965) can also be seen in some arachnids like scorpions and spiders (see Sherman Chap. XVI, this Vol.). The particularities of arachnids become clearer by a comparison with other arthropods.

In the *insects* the anterior ganglia of the paired lateral cord fuse to form the corpora cardiaca. In the Apterygota each part of the corpus cardiacum has a nervous connection with the corresponding neuromere of the central nervous system. The underlying embryogenetic processes are very complex, due to morphogenetic acceleration and deceleration. The lateral ganglia of the mandibulary and maxillary metamere give rise to the corpora allata. In the more advanced insects, the corpora allata are no longer connected to the corresponding neuromere, but there is a secondary connection to the brain through the corpora cardiaca. It is possible that the following lateral ganglia (labial and mesothoracic) form the molting gland, whereas the others may contribute to the formation of the lateral perisympathetic organs.

In *myriapods* the anterior ganglia of the lateral paired cord form two neuroendocrine organs: the paraesophageal gland and the cerebral gland (Gabe's organ). The initial nervous connection with the corresponding neuromeres disappears and Gabe's organ makes a new secondary connection with the protocerebrum. It is not possible at this time to say what happens with the rest of the lateral cord.

In *crustaceans* the fundamental organization persists but can be much modified in each taxon. In the Decapoda, the sinus gland, the "deutocerebral" organ, the lateral organ, and the Y-organ may belong to the lateral paired cord.

Neurosecretory cells are present in practically all Invertebrata (Gabe 1966), including arachnids, of which the spiders are the best-studied group (Gabe 1954a; Legendre 1954a). The neurosecretory material moves along the axon to a storage organ, near to or far from the site of its production (paraganglionic plates, neurohemal organs). The neurosecretory material can, during its migration, transit through a retrocerebral endocrine organ where new secretions will be added before the combined material is kept in a storage organ until released into the hemolymph (Juberthie 1983). The activity of the neurosecretory cells is a cyclic one and known to be involved in different physiological functions. Some authors had thought that the activity of the neurosecretory cells of arachnids was linked to the elaboration of the new cuticle during the molt cycle (review by Gabe 1966), but new observations establish that the elaboration of the products of neurosecretion coincides with the activity of the gonads in both sexes (in spiders, Araneae: Gabe 1955, Kühne 1959, Legendre 1959, 1971, Streble 1966, Sadana 1975; in harvestmen, Opiliones: Juberthie 1964; in ixodid ticks, Acari: Eichenberger 1970). During the hibernation of adult spiders of the genus *Tegenaria*, the cerebral groups of neurosecretory cells show no activity (Kühne 1959). During a long fasting period in adult spiders of the genus *Coelotes*, the activity of the cerebral neurosecretory cells is not disturbed (Streble 1966). It is now clear that the neurosecretion of the central nervous system of arachnids has no direct implication in the molt cycle, but an indirect one (Bonaric 1980).

3 The Recent Arachnid Orders

3.1 Spiders

Anatomy. The anatomy of the stomatogastric nervous system is well known in spiders. Schneider (1892) described a pair of sympathetic ganglia behind the brain. Their endocrine functions were demonstrated by Legendre (1953, 1954b). He also named them Schneider's organs I, together with a pair of ganglia situated behind the first organs, near the sucking stomach, called Schneider's organs II. This neuroendocrine retrocerebral system was found in many species of spiders. Kühne (1959), discovered the "Tropfenkomplex" behind and connected with the first organ of Schneider. This is a third organ filled with secretory products. Other authors (Streble 1966; Babu 1973) showed that the Tropfenkomplex is a storage organ and connected to the first organ of Schneider.

This neurohemal organ has no own secretion but it stores (1) the cerebral neurosecretion (after transit through the first organ of Schneider), (2) the secretion of Schneider's organ I itself and (3) the products of neurosecretion of the subesophageal nerve mass, which are collected in the subesophageal paraganglionic plates. The latter partly follow the segmentary dorsal nerves through the endosternite to reach the neurohemal organ independently (Babu 1973; Bonaric 1980; Bonaric et al. 1980). Some species of spiders (*Meta menardi, Araneus quadratus, A. cornutus, Argiope bruennichi*) have no neurohemal organ (Tropfenkomplex) and the neurosecretions of the brain and the secretion of the first organ of Schneider itself are stored in the lateral part of Schneider's organ I (Streble 1966). A recent study (Bonaric and Juberthie 1983) describes the ultrastructure of this neuroendocrine complex.

The single stomatogastric nerve trunk is formed by the rostral nerve, in front of the brain, a recurrent nerve running on the dorsal part of the esophagus, and sometimes a supra-esophageal ganglion. The stomodeal bridge also belongs to this system, but is secondarily incorporated in the cheliceral ganglion

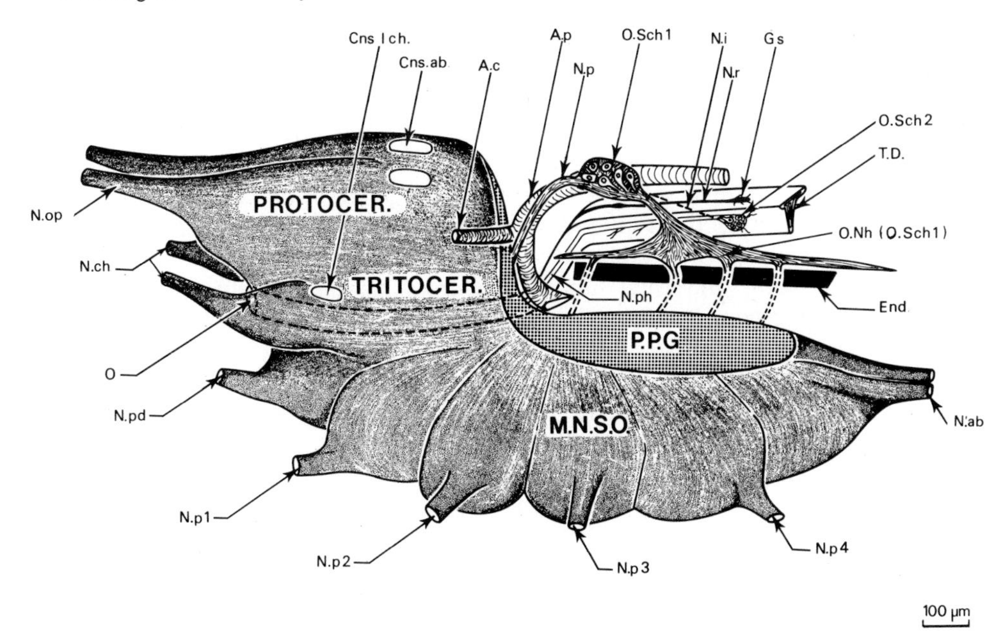

Fig. 1. Retrocerebral neuroendocrine complex and neurohemal formations in the spider *Pisaura mirabilis* Cl. (After Bonaric et al. 1980) *A.c.* Aorta cephalica; *A.p.* Aorta perigastrica; *Cns.ab* protocerebral neurosecretory cells; *Cns. I ch* tritocerebral neurosecretory cells; *End* endosternite; *Gs* supraesophageal ganglion; *M.N.S.O.* subesophageal nerve mass; *N.ab* abdominal nerve; *N.ch* cheliceral nerve; *N.i* interganglionic nerve; *N.p1, p2, p3, p4* nerves of the walking legs; *N.p* principal nerve of the first organ of Schneider; *N.pd* nerve of the pedipalp; *N.ph* pharyngeal nerve; *N.op* optical nerve; *N.r* recurrent nerve; *O* esophagus; *O.Nh* neurohemal organ (Tropfenkomplex); *O.Sch 1, O.Sch 2* first and second organ of Schneider; *P.P.G.* paraganglionic plate; *PROTOCER.* protocerebrum; *T.D.* digestive tract; *TRITOCER.* tritocerebrum (axillary ganglia and lateral nerve not represented)

(Legendre 1956). A prolongation of the paired sympathetic nervous system was described in some spiders (Legendre 1959): the axillary ganglia with the lateral nerves, which are localized between the neuromere of the appendage and the segmental nerve. The axillary ganglia are connected through the lateral nerve, which enters the brain in the posterior part of the cheliceral ganglion. The function of this complex (observed in *Dolomedes fimbriatus*) is unknown. The paired sympathetic nervous system is interpreted as being formed by the axillary ganglia and Schneider's organ I and perhaps partially by Schneider's organ II as well.

Embryologically, the first organ of Schneider develops from an epiblastic zone between the metamere of the rostrum and the metamere of the chelicera, whereas the second organ of Schneider appears as an evagination of the esophagus (Legendre 1959). The connection with the brain is realized through the *principal nerve.* The roots of this nerve are tritocerebral. A second nerve, the *accessory nerve,* is protocerebral, but its existence is still partly hypothetical. An *interganglionic nerve* runs from the first to the second organ; it is a prolongation of the principal nerve. The second organ is directly innervated through the *pharyngeal nerve* which runs along the lateral side of the sucking stomach. The roots of this nerve are in the tritocerebrum and the stomodeal bridge. Nerve ramifications leave the second organ and innervate the mid-gut.

The first organ of Schneider lies behind the brain near the endosternite, but its position varies in detail in different families of spiders. Typically, this organ contains both glandular cells and neurons. In Mygalomorphae all transitions between pure secretory function and pure nervous function are known (Legendre 1958; Streble 1966; Yoshikura and Takano 1972).

In each genus and each family, the gross anatomy of this sympathetic neuroendocrine complex is subject to variations. In *Polybetes* (Clubionidae), the first organ of Schneider includes two groups of secretory cells separated by a constriction (Serna de Esteban 1973); in *Latrodectus mirabilis* (Theridiidae) a second neurohemal organ behind the second organ of Schneider has been described (Serna de Esteban 1981).

Ultrastructural study (Fig. 2) shows that the neurosecretory cells in Schneider's organs have the same structure as that found in the neurosecretory cells in the central nervous system (Bonaric and Juberthie 1983). The neurosecretory cells of the brain lie in the protocerebrum (two groups), in the tritocerebrum (one group) and in each ganglion (pedipalps and four pairs of walking legs and cauda equina) of the subesophageal nerve mass (for a recent survey, see Juberthie 1983).

Recently, Bonaric (1980) discovered another storage organ in spiders, the *paraganglionic plates:* they lie in dorsal and dorsolateral parts of the subesophageal nerve mass and in the posterior part of the brain. They specifically collect the neurosecretory material from the subesophageal ganglia. Four pairs of metameric nerves run from the subesophageal paraganglionic plates through the neurilemma and the endosternite to the prosomatic neurohemal organ. The products of neurosecretion (neurosecretion of the brain and neurosecretion of the first organ of Schneider and neurosecretion of the subesophageal nerve mass) are stored in this prosomatic neurohemal organ and released from it into the blood sinus through exocytosis (Bonaric et al. 1980).

Physiology. The physiological aspects of the stomatogastric nervous system were comprehensively treated by Bonaric (1980, 1981), who demonstrated experimentally the presence and the role of ecdysteroids in the molt cycle and the diapause. Several organs are involved in the molt cycle and winter diapause: the neurosecretory cells of the central nervous system, the neurosecretory cells of the first organ of Schneider and the molting gland. This molting gland was originally described as "endocrine tissue" by Millot (1930). Legendre (1959) suspected an endocrine role of this tissue during molting, but the experimental

Fig. 2a, b. Ultrastructure of the neuroendocrine complex in *Pisaura mirabilis* cl. (from Bonaric 1980). **a** A secretory cell of the first organ of Schneider. *G* Golgi apparatus; *Gs* granular secretion; *m* mitochondria; *N* nucleus; *Pc (CgI)* portion of glial cell; *Reg* reticulum (scale 1 μm). **b** Penetration of the principal nerve (*Npp*) into the prosomatic neurohemal organ (*ONh*) which is filled with neurosecretory granules. *Al* axon terminal; *Ln* neural lamella; *Sh* hemolymphatic sinus (scale 1 μm)

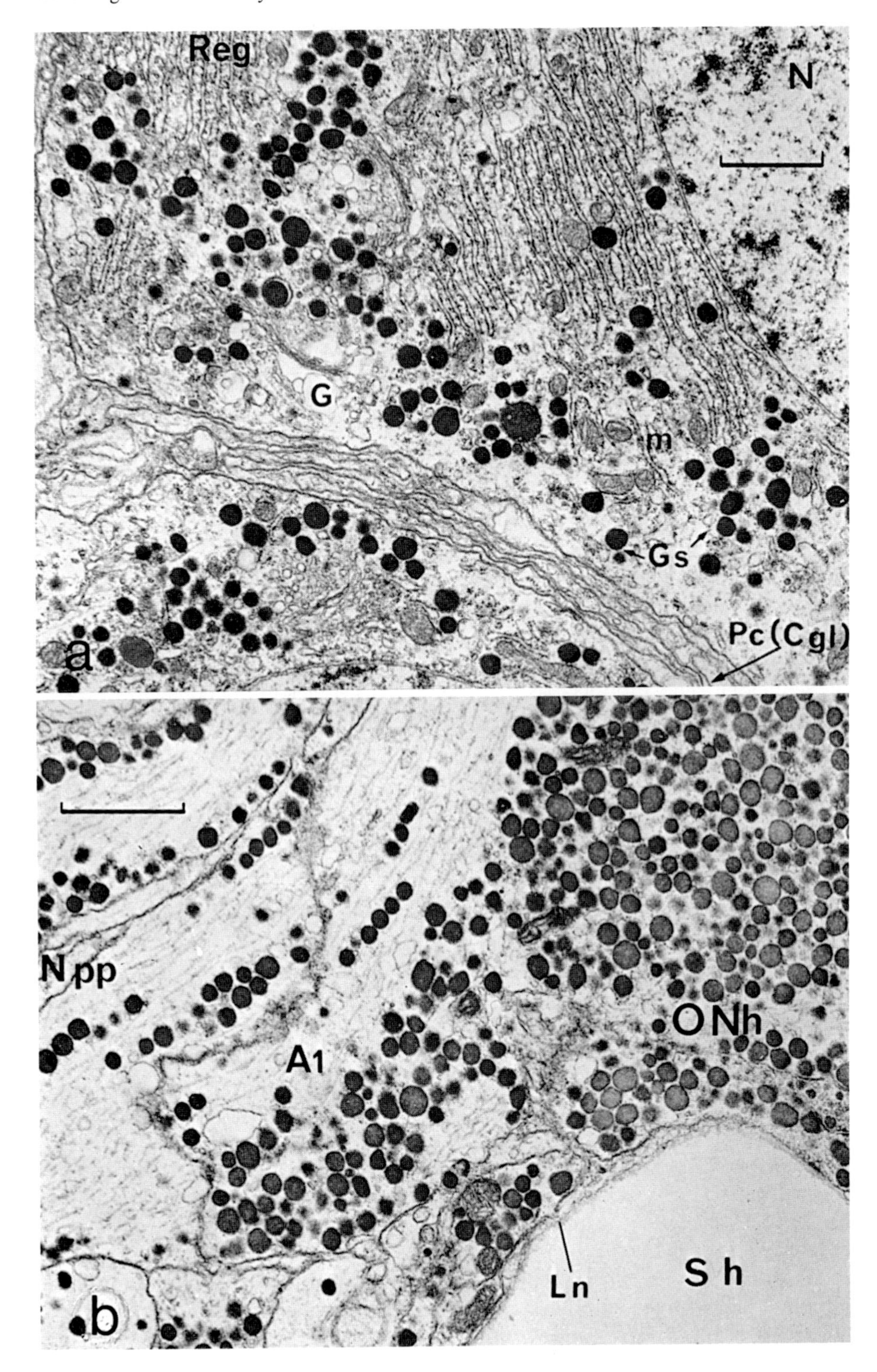
Reg
N
G
m
Gs
Pc (Cgl)
a
Npp
A1
ONh
Ln
Sh
b

attempts of Streble (1966) have not confirmed this conjecture. This organ is generally formed by cells localized in the hemolymph sinus in front of the brain and of the pedipalp ganglion, near the rostrum. It shows no innervation. In *Pisaura mirabilis* (Pisauridae), it is made up of lateral symmetric islets in a metameric arrangement at both sides of the ventral nervous mass and by a hind "ring" of endocrine cells which surrounds the cauda equina. The ultrastructure of the glandular cells shows the characteristics of a steroidogenous tissue and the periodical secretion of this gland can be related with the molt cycle (Bonaric and Juberthie 1980). The precise dosing of endogenous ecdysteroids shows a peak at the last third of the molt cycle, which is accompanied by an activation of the epidermis cells. During ecdysis, the rate of hormone secretion is low. After ecdysis, a brief increase of the rate goes along with the formation of the endocuticle. Introduction of exogenous ecdysteroids induces perturbations of the molt cycle.

In the nymphs of *Pisaura mirabilis*, the neurosecretory cells of the protocerebrum are active during the first half of the molt cycle. At the same time, many mitoses occur in the cells of the molting gland. During the second half of the molt cycle, the neurosecretory cells of the first organ of Schneider are active, and the endocrine cells of the molting gland show great secretory activity. A correspondingly high amount of ecdysteroids is found in the hemolymph.

During the winter diapause, the spider nymph does not molt, its weight remains stable, its respiratory metabolism is low and the secretory activity of the neurosecretory cells is low both in the central nervous system and in the organs of Schneider. The products of neurosecretion are accumulated in the prosomatic neurohemal organ and in the paraganglionic plates.

3.2 Other Orders

Ricinulei. No data have been published except for a short mention of the existence of a paraganglionic plate in a neotropical species (Juberthie 1983).

Solifugae, windscorpions. Anatomical studies in *Othoes saharae* (Junqua 1966) show that the paired stomatogastric nervous system is absent. In *Eremobates*, the unpaired stomatogastric nervous system is formed by a rostral nerve and a recurrent nerve (Henry 1954). Three groups of neurosecretory cells are present in the brain periphery. The axons of these cells carry products of neurosecretion to the paraganglionic plate which is localized in the typical arachnid position. They are active during the intermolt cycle in *Othoes saharae*. Neurosecretory cells of the subesophageal nerve mass are active during the nonfeeding period.

Uropygi, whipscorpions. One median nerve with a paired root runs forward (rostral nerve) and another backward (recurrent nerve) from the base of the brain, in *Mastigoproctus* sp. (Henry 1954). The presence of a paired sympathetic system is probable.

Palpigrada, microwhipscorpions. Little is known about the stomatogastric nervous system of the Palpigradi. Three groups of neurosecretory cells are present in the brain of *Eukoenenia austriaca*, and a metameric group exists in each ganglion of the ventral nerve mass. Products of neurosecretion have been seen in the neuropil of the brain and of the ventral ganglia, but nothing is known about the destination of these products (Juberthie and Juberthie-Jupeau 1963).

Scorpionidae, scorpions. They have been relatively well investigated and their stomatogastric system was first mentioned by Patten in 1890. Police (1903) described one pair of sympathetic ganglia situated behind the syncerebrum in *Euscorpius italicus.* These ganglia, also called "ganglia of Police", have a direct nervous connection with the hind part of the brain. Gabe (1955) reported the presence of neurosecretory cells in the brain of *Euscorpius italicus, E. carpathicus* and *Buthus occitanus,* where he also observed the pathways of the neurosecretory product to the ganglion of Police.

Following these descriptions, other authors have studied the scorpions (Habibulla 1961, 1971; Gabe 1966; Streble 1966; Kwartinikow 1980). Neurosecretory cells are found in groups in the protocerebrum and the tritocerebrum. Still other groups are metamerically arranged in the subesophageal nerve mass. A rostral nerve, in front of the brain, and a recurrent nerve behind it (in *Uroctonus*) are the rest of the dorsal single nerve trunk (Henry 1949). The lateral double nerve cord is well represented: a first pair of nerves (lateral nerves of Police) run from the hind part of the brain to the ganglia of Police (equivalent to the first organ of Schneider of the spiders); a second pair of nerves (intestinal nerves of Police) run backward along the lateral part of the digestive tract and show one or two ganglia corresponding to the second organs of Schneider in spiders.

All these ganglia contain secretory products, which have the same tinctorial affinities as the neurosecretory cells of the brain. They can be considered as neurohemal storage organs. Electron microscopy shows that the ganglia of Police have a neurohemal part with its own secretory products and its own secretory cells (Stockmann, personal communication).

According to Police (1903) a pair of cardiac nerves leaves the hind part of the brain, and, after fusion, forms a single epicardial nerve on the dorsal wall of the heart.

Amblypygi, whipspiders. They have neurosecretory cells in the ventral cortex of the subesophageal ganglion. In *Tarantula marginemaculata* the product of these cells is stored in the dorsal part of the neuropil (neurohemal organ) (Weygoldt 1975). In *Phrynus whitei,* the neurosecretion of the brain is localized in a retrocerebral paraganglionic plate, located at the junction of the syncerebrum and the subesophageal ganglion in the hind part of the brain. The single sympathetic tract is formed by the stomodeal bridge, with subesophageal commissures (Babu 1965). In front of the stomodeal bridge lies a small frontal (rostral) ganglion. The rostral nerve has a paired root.

The paired sympathetic system is well developed. A pair of pharyngeal nerves runs backward along the esophagus and into two small intestinal ganglia, which are secretory (this formation seems to correspond to the second organ of Schneider of spiders). Behind the brain lies one pair of *retrocerebral glands* connected with the brain through a tritocerebral nerve. These glands make contact with a typical neurohemal organ, which stores their secretion. On the other part, they are connected to other small sympathetic nerves which innervate the musculature of the foregut. At the origin of each appendicular nerve from the ventral ganglia, there are one, two, or three small ganglia. These ganglia are not strictly metameric, and not interconnected through a "lateral nerve" (Heurtault 1978). Their fibers innervate the muscles of the appendage.

Pseudoscorpiones, pseudoscorpions. Initially described by Gabe (1955) in *Garypus beauvoisi* and *Chelifer cancroides*, neurosecretory cells in the central nervous system have been found by other authors in other species (Heurtault 1969, 1973; Boissin and Cazal 1969). These cells are localized in the cellular cortex of the brain near the central body, and are active during the egg-laying period (Heurtault 1971). The neurosecretory pathway runs to the paraganglionic plates at the posterior side of the brain and above the anterior part of the subesophageal nerve mass (Juberthie and Heurtault 1975). Neurosecretory cells are also present in the subesophageal nerve mass.

The sympathetic stomatogastric nervous system has been well studied in *Neobisium caporiaccoi* (Heurtault 1973). The single nerve cord runs about the stomodeum and gives off the rostral nerve forward and a recurrent nerve backward; these serve the pharyngeal and oesophageal musculature respectively. A paired sympathetic nerve trunk with small ganglia near the ambulatory nerves 2, 3, and 4 was also found. Behind the brain and lateral to the subesophageal nervous mass there are two retrocerebral glands called G1 and G2. The gland G1 has a nervous sympathetic structure. The second gland (G2) mostly contains glandular cells and receives innervation from gland G1. The nerve ramifations issued from the glands G1 and G2 run to the musculature of the midgut.

Opiliones, harvestmen. The anatomy and the histology of the nervous system of the harvestmen have been well studied. In the protocerebral part of the brain, Gabe (1954b, 1955, 1966), Herlant-Meewis and Naisse (1957), Naisse (1959), Juberthie (1964, 1965) and Streble (1966) have described a neurosecretory system (eight groups of cells) which is connected to a neurohemal organ (paraganglionic plate) in the typical arachnid position behind the brain. Neurosecretory cells were also described in the subesophageal nerve mass (Juberthie 1964) and studied with the electron microscope (Juberthie and Juberthie-Jupeau 1974). The paired sympathetic system is formed by the esophageal nerves which originate from the rostral region of the brain, run backward along the esophagus and finally enter the esophageal ganglia. Each of these ganglia consists of few neurons and four or five glandular cells embedded in glial cells. According to their ultrastructure, they are homologous to the second organ of Schneider of spiders (Juberthie and Bonaric 1980).

Acari, mites. In the world of the mites and ticks only two taxa have been well investigated for neurosecretion and the sympathetic nervous system: the ixodids and the argasids. Gabe (1955) first discovered the neurosecretory cells and the paraganglionic plates. More recent investigators (Obenchain 1974; Gabbay and Warburg 1977) recognize as many as 10 to 18 neurosecretory centers in the condensed nervous system of the tick, with 13 distinct neurosecretory cell types (according to their staining and cytological properties): this is the highest number of neurosecretory cell types described in arachnids.

The neurosecretory product is stored in the paraganglionic plates localized at the posterior part of the brain and the dorsal part of the subesophageal nervous mass, close to the esophagus. From here the material is released into the blood sinus through the neural lamella.

Laterosegmental organs of neural nature and similar to the axillary ganglia of spiders lie in the hemocoel between the pedal nerve trunks and along the sympathetic lateral nerves. The retrocerebral organ complex described by Obenchain and Oliver (1975) in *Dermacentor variabilis* may be part of this lateral paired sympathetic system; it consists of two pairs of organs, closely related with each other, on either side of the esophagus. On each side, the organs fuse to form a compound organ with (1) latero-ventral lobes and (2) a dorsal lobe. (1) The latero-ventral lobes are ganglia and consist of few unipolar neurons embedded in a glial matrix. (2) The dorsal lobe consists of two fused ganglia supplied by a pair of nerves running on each side of the esophagus. The origin of this paired nerve lies in the cheliceral ganglion and the rostral ganglion. The dorsal lobe is also supplied by a paired accessory nerve originating from the protocerebrum. The compound nerve (rostrocheliceral and protocerebral) passes through the dorsal lobe and enters the lateroventral lobe. It terminates in front of the midgut in a proventricular plexus, which can be regarded as a neurohemal organ.

In *Hydrachna* and *Thrombidium*, Streble (1966) described two protocerebral nerves running backward along the esophagus to enter a ganglionic mass near the brain. Wright (1969) in *Dermacentor albipictus* and Sannasi and Subramoniam (1972) in *Rhipicephalus sanguineus* have shown that applications of ecdysteroids induce an interruption of larval diapause.

4 Conclusion

The morphological variations of the basic plan of the stomatogastric nervous system found in different orders of arachnids are an expression of evolutionary trends. These, however, cannot be outlined at the present state of our knowledge. The sample of species – out of more than 50,000 arachnids recently – studied anatomically, histologically, and physiologically is as yet too small to permit definitive statements.

Acknowledgments: I wish to express my particular thanks to my friend and colleague Dr. Ch. Dondale (Ottawa) who kindly read and criticized the manuscript. I thank Dr. J. C. Bonaric for his help during the preparation of this chapter.

References

Publications with important bibliographies are marked with an asterisk.

*Babu KS (1965) Anatomy of the central nervous system of arachnids. Zool Jahrb Anat 82:1–154

Babu KS (1973) Histology of the neurosecretory system and neurohemal organs of the spider *Argiope aurantia* (Lucas). J Morphol 141:77–93

Boissin L, Cazal M (1969) Étude du système nerveux et des glandes endocrines céphaliques de l'adulte femelle d'*Hysterochelifer meridianus* (L Koch) (Arachnides, Pseudoscorpiones, Cheliferidae). Bull Soc Zool Fr 94:263–268

*Bonaric JC (1980) Contribution à l'étude de la biologie du développement chez l'araignée *Pisaura mirabilis* (Clerck, 1758). Approche physiologique des phénomènes de mue et de diapause hivernale. Thèse Etat Univ Montpellier

Bonaric JC (1981) La régulation hormonale des phénomènes de mue et de diapause hivernale chez les araignées (C R VIth Coll Arachnol Exp Fr, Modena-Pisa). Atti Soc Tosc Sci Nat 88:120–131

Bonaric JC, Juberthie C (1980) La glande de mue des araignées. Étude structurale et ultrastructurale de cette formation endocrine chez *Pisaura mirabilis* Cl. (Araneae-Pisauridae). C R 5th Coll Exp Fr, Barcelone. Ed Univ Barcelona, pp 21–30

Bonaric JC, Juberthie C (1983) Ultrastructure of the neuroendocrine complex in *Pisaura mirabilis* (Cl.) (Araneae, Pisauridae). Zool Jahrb Physiol 87:55–64

Bonaric JC, Juberthie C, Legendre R (1980) Le complexe neuroendocrine rétrocérébral et les formations neurohémales des araignées. Bull Soc Zool Fr 105:101–108

*Bullock TH, Horridge GA (1965) Structure and function in the nervous systems of invertebrates. Freeman, San Francisco London

Chaudonneret J (1978) La phylogénèse du système nerveux annélido-arthropodien. Bull Soc Zool Fr 103:69–95

Eichenberger G (1970) Das Zentralnervensystem von *Ornithodoros moubata* (Murray), Ixodoidea, Argasidae und seine postembryonale Entwicklung. Acta Trop 27:15–53

Gabbay S, Warburg MR (1977) The diversity of neurosecretory cell types in the cave tick *Ornithodorus tholozani.* J Morphol 153:371–386

Gabe M (1954a) Emplacements et connexions des cellules neurosécrétrices chez quelques aranéides. C R Acad Sci 238:1265–1267

Gabe M (1954b) Situations et connexions des cellules neurosécrétrices chez *Phalangium opilio* L. C R Acad Sci 238:2450–2452

Gabe M (1955) Données histologiques sur la neurosécrétion chez les arachnides. Arch Anat Microsc Morphol Exp 44:351–383

*Gabe M (1966) Neurosecretion. Pergamon Press, Oxford

Habibulla M (1961) Secretory structure associated with the neurosecretory system of the immature scorpion *Heterometrus swammerdami.* Q J Microsc Sci 102:475–479

Habibulla M (1971) Neurosecretion in the brain of a scorpion, *Heterometrus swammerdami.* A histochemical study. Gen Comp Endocrinol 17:253–255

Henry LM (1949) The nervous system and segmentation of the head in a scorpion (Arachnida). Microentomology 14:120–125

Henry LM (1954) The cephalic nervous system and segmentation in the Pedipalpida and the Solpugida (Arachnida). Microentomology 19:2–13

Herlant-Meewis H, Naisse J (1957) Phénomènes neurosécrétoires et glandes endocrines chez les opilions. C R Acad Sci 245:858–860

Heurtault J (1969) Neurosécrétion et glandes endocrines chez *Neobisium caporiaccoi* (arachnides, pseudoscorpions). C R Acad Sci 266:1105–1108

Heurtault J (1971) Données complémentaires sur le complexe neuroendocrine rétrocérébral des pseudoscorpions. C R Acad Sci 272:1981–1983

*Heurtault J (1973) Contribution à la connaissance biologique et anatomo-physiologique des pseudoscorpions. Bull Mus Hist Nat Paris Zool 96:561–670

Heurtault J (1978) Système sympathique, structures glandulaires et neurohémales prosomatiques chez *Phrynus whitei* et *Damon* sp. (arachnides, amblypyges). Symp Zool Soc London 42:389–397

*Juberthie C (1964) Recherches sur la biologie des opilions. Ann Speleol 19:1–238

Juberthie C (1965) Données sur l'écologie, le développement et la reproduction des opilions. Rev Ecol Biol Sol 3:377–396

*Juberthie C (1983) Neurosecretory systems and neurohemal organs of terrestrial Chelicerata (Arachnida). In: Gupta AP (ed) Neurohemal organs of Arthropods. Thomas, Springfield, pp 149–203

Juberthie C, Bonaric JC (1980) Sur l'ultrastructure des organes oesophagiens des opilions et des organes de Schneider 2 des araignées. C R 5th Coll Arachnol. Exp Fr, Barcelone. Ed Univ Barcelona, pp 99–109

Juberthie C, Heurtault J (1975) Ultrastructure des plaques paraganglionnaires d'un pseudoscorpion souterrain *Neobisium cavernarum* (L. K.). Ann Speleol 30:433–439

Juberthie C, Juberthie-Jupeau L (1963) Sur la neurosécrétion et la reproduction d'un palpigrade souterrain. Spelunca Mem 3:185–189

Juberthie C, Juberthie-Jupeau L (1974) Ultrastructure of neurohemal organs (paraganglionic plates) of *Trogulus nepaeformis* (Scopoli) (Opiliones, Trogulidae) and release of neurosecretory material. Cell Tissue Res 150:67–78

*Junqua C (1966) Recherches biologiques et histophysiologiques sur un solifuge saharien. Mem Mus Hist Nat Paris (A) 43:1–124

*Kühne H (1959) Die neurosekretorischen Zellen und der retrocerebrale neuroendokrine Komplex von Spinnen (Araneae, Labidognatha) unter Berücksichtigung einiger histologisch erkennbaren Veränderungen während des postembryonalen Lebenslaufes. Zool Jahrb Anat 77:527–600

Kwartinikow MP (1980) Neurosekretorische und endokrinologische Verhältnisse beim europäischen Skorpion (*Euscorpius carpathicus* L.) I. Morphologische Beschreibung. Acta Zool Bulg 14:83–87

Legendre R (1953) Le système sympathique stomatogastrique (organe de Schneider) des araignées du genre *Tegenaria*. C R Acad Sci 237:1283–1285

Legendre R (1954a) Sur la présence de cellules neurosécrétrices dans le système nerveux central des Aranéides. C R Acad Sci 238:1267–1268

Legendre R (1954b) Données anatomiques sur le complexe neuroendocrine rétrocérébral des Aranéides. Ann Sci Nat Zool 16:419–426

Legendre R (1956) Sur la genèse du pont stomodéal chez les araignées. Bull Soc Zool Fr 81:318–323

Legendre R (1958) Sur la structure du complexe endocrine rétrocérébral de l'araignée mygalomorphe *Scodra* (=*Stromatopelma*) *calceata* Fabr. (Theraphosidae). C R Acad Sci 246:3671–3674

*Legendre R (1959) Contribution à l'étude du système nerveux des Aranéides. Ann Sci Nat Zool (12) 1:339–473

Legendre R (1971) État actuel de nos connaissances sur le développement embryonnaire et la croissance des araignées. Bull Soc Zool Fr 96:93–114

Millot J (1930) Le tissu réticulé du céphalothorax des Aranéides et ses dérivés: néphrocytes et cellules endocrines. Arch Anat Microsc 26:43–81

Naisse J (1959) Neurosécrétion et glandes endocrines chez les opilions. Arch Biol 70:217–264

Obenchain FD (1974) Neurosecretory system of the American dog tick *Dermacentor variabilis* (Acari, Ixodidae) I. J Morphol 142:433–446

Obenchain FD, Oliver JH (1975) Neurosecretory system of the American dog tick *Dermacentor variabilis* (Acari, Ixodidae) II. J Morphol 145:269–293

Patten W (1890) On the origin of Vertebrates from Arachnids. Q J Microsc Sci 31:317–378

Police G (1903) Sul sistema nervoso stomatogastrico dello scorpione. Arch Zool Napoli 1:179–200

Sadana GL (1975) Neurosecretory elements in the nervous mass of *Lycosa chaperi* (Araneida, Lycosidae). Proc 6th Int Arachnol Congr, Amsterdam 1974, pp 158–160

Sannasi A, Subramoniam T (1972) Hormonal rupture of larval diapause in the tick *Rhipicephalus sanguineus* Latr. Experientia 28:666–667

Schneider A (1892) Système stomatogastrique des Aranéides. Tabl Zool Poitiers 2:87–94

Serna de Esteban CJ de la (1973) Histologie du système neuroendocrine rétrocérébral de *Polybetes pythagoricus* (Holmberg, 1874) (Araneae, Labidognatha). Ann Sci Nat Zool 15:595–606

Serna de Esteban CJ de la (1981) Los órganos neuroendocrinos y neurohemales retrocerebrales en la hembra de *Latrodectus mirabilis* (Holmberg, 1876) (Araneae, Theridiidae). Physis 40:55–62

*Streble H (1966) Untersuchungen über das hormonale System der Spinnentiere (Chelicerata) unter besonderer Berücksichtigung des „endokrinen Gewebes" der Spinnen (Araneae). Zool Jahrb Physiol 72:157–234

*Weygoldt P (1975) Untersuchungen zur Embryologie und Morphologie der Geißelspinne *Tarantula marginemaculata* C. L. K. (Arachnida, Amblypygi, Tarantulidae). Zoomorphologie 82:137–199

Wright JE (1969) Hormonal termination of larval diapause in *Dermacentor albipictus*. Sciences 163:390–391

Yoshikura M, Takano S (1972) Neurosecretory system of the purse web spider *Atypus karschi* Doenitz. Kumamoto J Sci Biol 2:29–36

B Structure and Function of Sensory Systems

IV The Morphology and Optics of Spider Eyes

M. F. LAND

CONTENTS

1 Introduction

The modern arachnids are the only group of arthropods in which the main organs of sight are camera-type eyes, not unlike our own, rather than compound eyes. The copepod crustaceans also lack compound eyes, but their nauplius eyes are rarely more than a trio of simple eye-cups, with a handful of receptors each. By contrast, spider eyes at their best have retinae with 10^3 to 10^4 receptors, and in the salticid *Portia* the inter-receptor angles may be as small as 2.4 min of arc (Williams and McIntyre 1980), which is only six times greater than in man (cone spacing 0.42 min), and is six times smaller than in the most acute insect eye (the dragonfly *Aeschna*, minimum inter-ommatidial angle 14.4 min; Sherk 1978). Thus, in some spiders, but by no means all, vision is excellent, and rivalled amongst invertebrates only by the cephalopod molluscs.

The probable ancestors of the spiders, relatives of the xiphosurans like *Limulus* and the extinct eurypterids, had both simple and compound eyes. Re-

School of Biological Sciences, University of Sussex, Brighton BN1 9QG, Great Britain

cent opinions about the origins of the eight eyes of most spiders suggests that one pair, the principal eyes, is derived from the ancestral pair of simple eyes, and that the other three pairs (the secondary eyes) came from the splitting up of the ancestral compound eyes (Paulus 1979). In the scorpions, the secondary eyes do seem to have a relic of the ommatidial form of the compound eye, in that the retina contains star-shaped rhabdoms reminiscent of those of *Limulus* (Scheuring 1913). Whatever the initial reasons for the ancestors of spiders abandoning compound eyes in favour of camera-type eyes, it was a fortunate development. Compound eye resolution is limited by diffraction at each small facet, whereas the single large lens of a simple eye does not restrict resolution in this way. The acuity of a jumping spider eye would be unattainable by a compound eye of a size which would fit onto a spider's head (Kirschfeld 1976). The camera-type eye is basically the better optical design.

Although good anatomical descriptions of spider eyes have been available for over a century (Grenacher 1879), the spiders remain a neglected group. Of the earlier anatomical papers, those of Scheuring (1914) and Widmann (1908) are important, and more recently there has been the excellent review by Homann (1971), which deals in detail with the diversity of spider eyes, and is largely based on his own earlier studies (Homann 1951, 1952). Information about the optics of spider eyes is much more scarce. We know a great deal about the optics of jumping spider eyes (Homann 1928; Land 1969a; Williams and McIntyre 1980), and a little about lycosids (Homann 1931), thomisids (Homann 1934), pisaurids (Williams 1979) and dinopids (Blest and Land 1977). These, however, are to a large extent visual hunters, and the majority of spiders are web-builders whose principal senses are mechanoreceptive. They have eyes, but of a type which is not similar optically to those of the hunters, and which has yet to be properly studied.

Partly because of this lack of basic information I shall spend part of this chapter outlining methods appropriate to the study of spiders' eyes, in the hope that others may take up the subject. There is more at this stage to learn than there is to review.

2 Morphology of the Eyes

2.1 Principal and Secondary Eyes

Most spiders have eight eyes, arranged roughly in two rows (Fig. 1a). The first row contains the antero-median (AM) and antero-lateral (AL) eyes, and the second the postero-median (PM) and the postero-lateral (PL) eyes. The exact positions and sizes of the eyes vary greatly from family to family, and are important taxonomic guides. One pair, always the antero-medians, is different from the others. They are usually referred to as the principal eyes, although this does not necessarily indicate their size or importance (in the lycosids, for example, the postero-lateral and postero-median eyes are much larger than the antero-medians). The principal eyes differ from the remaining secondary eyes (Nebenaugen) in four ways. (1) The principal eyes develop from a thickening

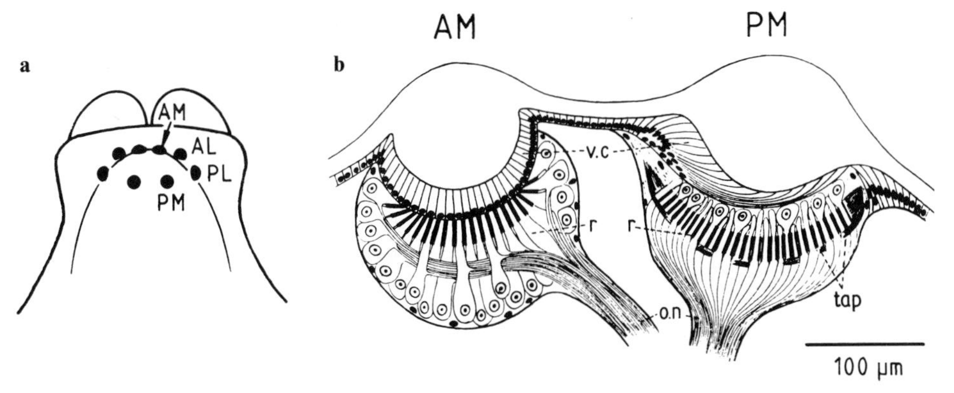

Fig. 1. a Layout of the eight eyes on the prosoma of the house spider *Tegenaria* (Agelenidae). (After Bristowe 1958). **b** Principal (*AM*) and secondary (*AL*) eyes of *Tegenaria*, from two figures of Widmann (1908). Note that the rhabdoms (dark region of the receptors) are distal to the nuclei in the principal eye and proximal to them in the secondary eye. *o.n* optic nerve; *r* receptors; *tap* tapetum; *v.c* vitreous cells

of the prosoma which folds over to produce three layers, and invaginates from below; the secondary eyes are formed from a simple invagination of the surface (Homann 1971). (2) The nuclei of the receptors in the principal eyes lie in the centre of the cell, with the rhabdoms distal to them; in the secondary eyes the nuclei lie distal to the rhabdoms (Fig. 1b). (3) The principal eyes lack a tapetum, a reflecting layer behind the receptors which is usually present in the other eyes. (4) The principal eyes are moveable. The retina, but not the lens, can be moved by a set of muscles varying in number from one to six. The retinae of the secondary eyes are always immobile. In some families (Dysderidae, Sicariidae and Oonopidae) there are only six eyes, and in these it is the principal eyes that are missing.

The secondary eyes usually have a tapetum. This consists of several layers of thin (0.1 μm) crystals, probably of guanine, which act as a colour-selective interference reflector (Land 1972a). The tapetum usually reflects green light best. Its function is probably the same as vertebrate tapeta, at least in hunting spiders, that is to reflect light back through the receptors, giving them a second opportunity to absorb photons. Homann (1971) distinguishes three types of secondary eye on the basis of tapetal morphology, although by no means all secondary eyes fit this classification (Table 1). The simplest kind of tapetum – the "primitive type" – is simply a reflecting sheet with holes in it where the axons of the receptors penetrate. This is found in the Mesothelae, Orthognatha and Haplogynae and does seem to be the ancestral form of the tapetum. The commonest kind of tapetum is the "canoe-shaped" type. This consists of two lateral walls which enclose the rhabdomeric regions of the receptors. At the base of the canoe, where its keel should be, there is a slit through which the axons of the receptors leave the eye. In all the eyes of this type that I have been able to examine, it appears that the lens forms an image well below the retina and tapetum. This implies that if the retina is to receive a focused image of any kind,

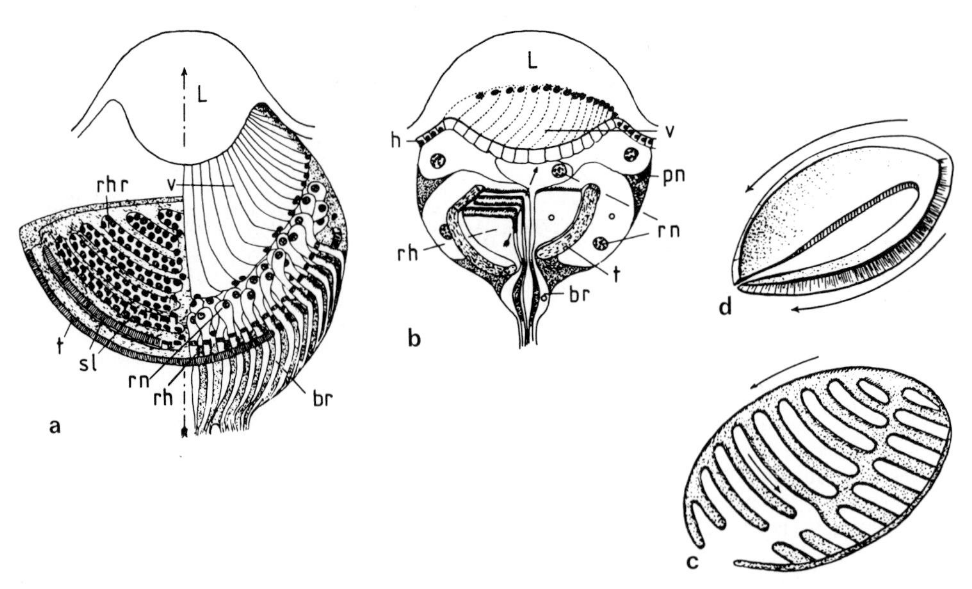

Fig. 2a–c. "Grate"- (**a, c**) and "canoe"- (**b, d**) type tapeta. *Left sides* of **a, b** show a surface view of the retinae, cut off at the level of the rhabdoms. *Right sides* show transverse sections. *Arrows* indicate the plane of symmetry of the eyes. **b** and **d** show the two types of tapeta on their own before development is complete; *arrows* indicate the direction in which development proceeds. The diagram in **b** is a section at right angles to that of the section of the secondary eye of *Tegenaria* in Fig. 1 b. *br* basal segments of receptors, *h* hypodermis, *L* lens, *pn* pigment cell nuclei, *rh* rhabdoms, *rhr* rhabdom rows, *rn* receptor nuclei, *sl* slits in the tapetum, *t* tapetum, *v* vitreous cells. (Compiled and modified from figures in Homann 1951, 1971)

the tapetum must be acting as a concave focusing mirror in conjunction with the lens; I shall return to this point later. Canoe-shaped tapeta are found in the eyes of most spider families except the hunting spiders. The third kind of tapetum – the "grate" type – has the reflecting material arranged in strips which resemble the grill of an oven. The rhabdomeric portion of each receptor sits on the tapetal strip as though on a chair, with the axon bending round and under the strip before leaving the eye (Baccetti and Bedini 1964). Grate-type tapeta are found in the hunting spiders (Lycosidae and related families) and in a modified form in crab spiders (Thomisidae). In hunting spiders the receptors and tapetum lie at the focus of the lens. The canoe and grate tapetal types are shown in Fig. 2.

It is likely that the principal and secondary eyes have different roles in the spider's behaviour, but this is only known with certainty in the case of jumping spiders (see also Forster, Chapt. XIII, this Vol.). In that family the lateral eyes detect movement, and initiate leg movements which turn the animal to face the source of the movement (Land 1971; Duelli 1978). From that point the principal eyes take over, and in particular they identify the nature of the stimulus as potential prey or a potential mate (Homann 1928; Drees 1952). During this identification period the retinae of the principal eyes move in a complicated scanning pattern (Land 1969b), which seems to be a search procedure con-

Table 1. Secondary eye types in different families of spiders (Homann 1971)

1. Tapetum of the primitive type
   Mesothelae, Orthognatha, Haplogynae, Pholcidae, Urocteidae, Filistatidae
2. Eyes with canoe-shaped tapetum
   Zodariidae, Palpimanidae, Theridiidae, Hadrotarsidae, Nesticidae, many Linyphiidae, Micryphantidae, many Araneidae, Symphytognathidae, Mimetidae, Textricellidae, Agelenidae, Argyronetidae, Prodidomidae, Gnaphosidae, Platoridae, Ammoxenidae, Homalonychidae, Cithaeronidae, Clubionidae, Amaurobiidae (*Zorocrates*), Tengellidae
3. Eyes with grate-shaped tapetum
   Lycosidae, Oxyopidae, Senoculidae, Zoropsidae; Thomisidae[a]
4. Other types
   No tapetum in: Philodromidae, Salticidae, Lyssomanidae, Uloboridae, Dinopidae, Eresidae, Tetragnathidae (*Tetragnatha*)
   Eyes with a simple tapetum but an advanced eye structure: Sparassidae (= Heteropodidae), Selenopidae
   Different types of tapetum in different eyes: some Araneidae, Tetragnathidae (*Pachygnatha*), Oecobiidae, Toxopidae, Hersiliidae, Psechridae, Acanthoctenidae
   Urocteidae have eyes between the primitive and canoe-shaped types; Hypochilidae have a thick tapetum but no lens

[a] Homann regards the Thomisidae as having an eye structure similar to that of the Lycosidae, but not closely related to the other families with grate-shaped tapeta, which he believes are monophyletic

cerned with the detection of legs or leg-like contours in the stimulus. No other spiders turn to face moving objects in the same way as jumping spiders, so it is not clear whether this division of behaviour applies in other families. It would seem likely, though, that the fixed secondary eyes will generally be concerned with the detection of motion relative to the spider, and the movable principal eyes with the examination of objects which do not necessarily move themselves.

2.2 Layout and Fields of View of the Eyes

The sketch of *Tegenaria* in Fig. 1a shows the typical double row of eyes found in most spider families. With the exception of some of the visual hunters, the eyes are usually of similar size, with the principal eyes slightly smaller than the others. Where the eyes are used in hunting, however, specialization often occurs, and interestingly it is not always the same pair or pairs of eyes that are enlarged – and by a fairly safe inference more important. In the Salticidae the principal (AM) eyes are by far the largest, and the PM eyes are vestigial in most species. In the Lycosidae, in contrast, the PM and PL eyes are enlarged, and in the nocturnal hunter *Dinopis* (Dinopidae) the PM eyes are enormous (Fig. 9). In the crab spiders (Thomisidae) the AL eyes are somewhat larger than the rest.

This specialization is reflected to some extent in the fields of view of the eyes, some of which are shown in Fig. 4. Before discussing them, however, I

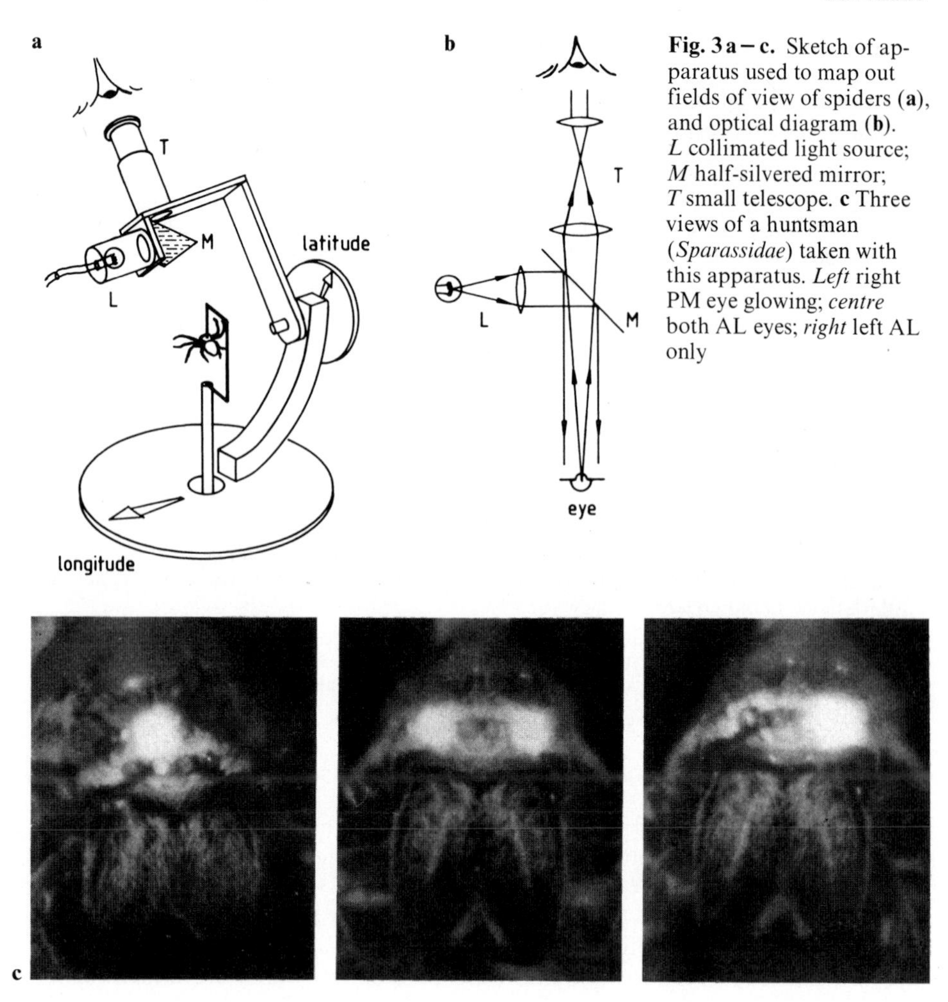

Fig. 3a – c. Sketch of apparatus used to map out fields of view of spiders (**a**), and optical diagram (**b**). *L* collimated light source; *M* half-silvered mirror; *T* small telescope. **c** Three views of a huntsman (*Sparassidae*) taken with this apparatus. *Left* right PM eye glowing; *centre* both AL eyes; *right* left AL only

should mention how they were obtained. The method makes use of the fact that in a well-focused eye with a reflecting tapetum (e.g. in a cat or a lycosid spider), light entering the eye is reflected out almost exactly along its original direction of incidence. Thus by rotating around the eye with a small telescope and a co-axial light source it is possible to establish the angular extent of the retina (strictly, the tapetum, but retina and tapetum are usually co-extensive). If a suitable Cardan arm mount is used for the telescope, the co-ordinates of the field of view can be obtained directly as latitude and longitude (Fig. 3). This method, originally devised by Homann (1928), works well with most of the hunting spiders, but less well for spiders with eyes of the canoe-shaped tapetum type, which unfortunately are the majority. The reason for this is that in these eyes the tapetum is not in the focal plane, and light emerges from the eye over an angle of 30° or more, which means that the fields of view cannot be sharply defined. Jumping spider (salticid) eyes can be mapped in this way, too, because

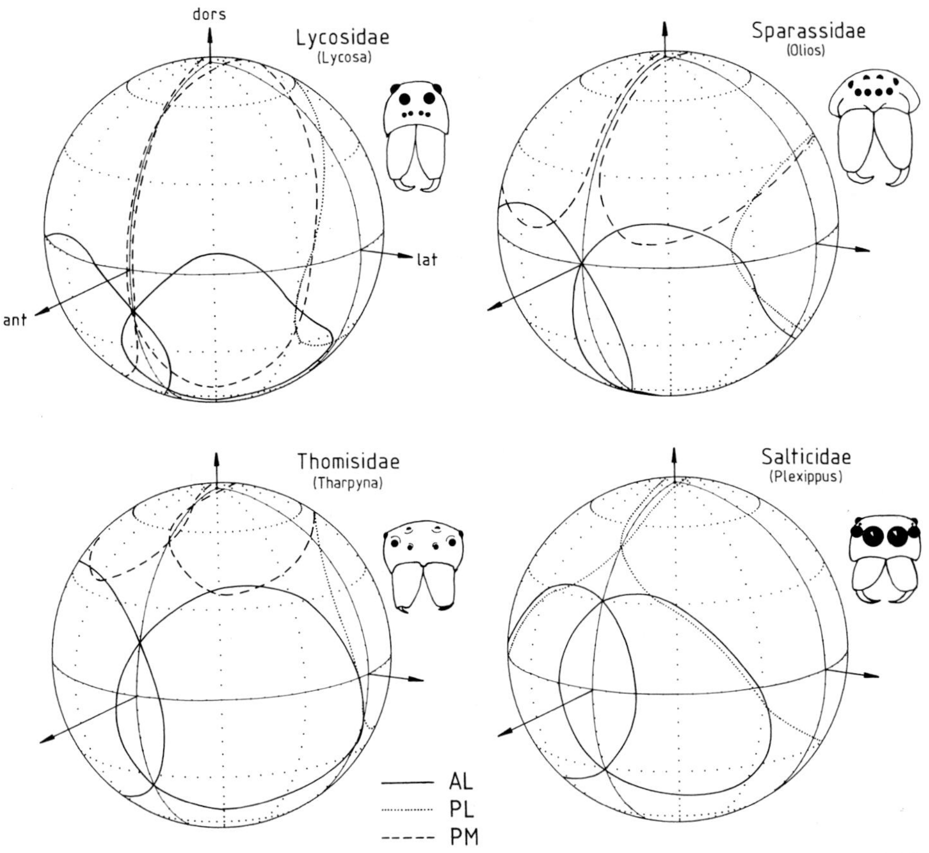

Fig. 4. Fields of view of the secondary eyes of hunting spiders from four families. The fields are plotted onto a globe with the spider at the centre. The globe is viewed from 15° above the equator, and 30° to the spider's longitudinal axis. The appearance of the retinae of the same spiders is given in Fig. 8. *ant* anterior; *dors* dorsal; *lat* lateral

although they lack a tapetum the retina reflects a characteristic pink colour. Principal eyes never have a tapetum, and are very difficult to map; they are not included in Fig. 4.

The frontal fields of view of four families of hunting spiders are shown in Fig. 4. In the huntsman *Olios* the three secondary eyes are of similar size, and the whole field of view is divided fairly equally between them: the AL eyes look forwards and down, the PM eyes look upwards and the PL eyes look laterally with a field of view extending all the way to the rear. The fields are contiguous or slightly overlapping, and meet 50° from the front and 10° above the equator. In the Lycosidae the PM eyes have moved forwards and their field has extended downwards, taking over the whole frontal region; the fields of left and right PM eye are exactly contiguous in the midline. The AL fields are still present in the normal position, but their resolution is poor (see below) and their function may well have been usurped by the PM eyes. The PL eyes are also large, and take over where the PM eyes leave off, between 60° and 80° from the front. In the thomisid crab spiders, the situation is almost opposite to that in the lycosids. The fields of the PM eyes have contracted upwards, and those of the AL eyes have extended into the dorso-frontal quadrant. In the salticids the PM eyes are usually rudimentary, and they do not have a field of view that can be mapped (this is not the case in some primitive salticids like *Portia*, where the PM eyes are still functional, and their field is an oblique 20°-wide strip between the fields of the AL and PL eyes, which in other salticids are contiguous). It is interesting that in all four families the fields of the AL eyes overlap the midline, so that there is a region of binocularity. This suggests that one of their functions may be distance judgment, and there is some evidence in favour of this in salticids (Homann 1928; Forster 1982, and Chap. XIII, this Vol.).

I realise that in the last paragraph I have implied an evolutionary progression from the sparassid type of visual field division towards the lycosid type on the one hand and the thomisid on the other. This is a possibility, and it is hard to avoid making up evolutionary stories, but one must remember that the fields of view of the eyes are only one of many characters to be considered in developing an accurate phylogeny of the spiders.

3 Basic Optics of Simple Eyes

Before proceeding to review the capabilities of spider eyes, in terms of their resolution and sensitivity, it will be helpful to give a brief summary of the theoretical performance of eyes of the type found in spiders, and also to outline the ways in which the actual performance can be discovered. A full discussion is given in Land (1981).

3.1 Resolution

The ability of an eye to resolve detail depends on the fineness of the retinal mosaic: the coarser the mosaic, in angular terms, the worse the resolution. One measure of resolution is the inter-receptor angle ($\Delta\phi$) subtended at the nodal point of the eye by an adjacent pair of receptors (Fig. 5). Thus:

$$\Delta\phi = \frac{d_{cc}}{f} \quad \text{(radians)}. \qquad (1)$$

where d_{cc} is the centre-to-centre spacing of the retinal receptors, and f is the focal length (specifically the posterior nodal distance) of the eye. Spider receptors are usually contiguous, so that $d_{cc} \simeq d_r$, the receptor width. An alternative measure of resolution is the highest resolvable spatial frequency for a grate-like image. This is also known as the sampling frequency of the retina, ν_s, and this is given by:

$$\nu_s = \frac{1}{2\Delta\phi} = \frac{f}{2\,d_{cc}} \quad (\text{radians}^{-1}). \qquad (2)$$

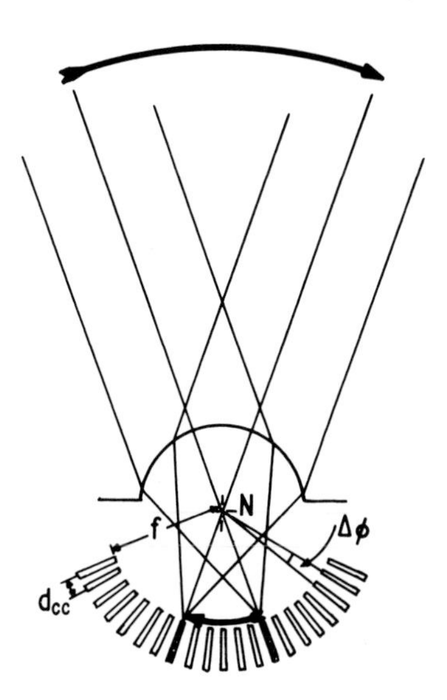

Fig. 5. Image formation in a simple eye. The inter-receptor angle ($\Delta\phi$) is given by d_{cc}/f, where d_{cc} is the centre-to-centre separation of the receptors, and f is the focal length measured from the nodal point (N) to the image. A simple method for obtaining f is shown in Fig. 6a

This has the advantage that it increases as the eye's resolution gets better. The factor of 2 in the expression simply indicates that two receptors are required to resolve each line-pair in a grating. (Note that 1 radian=$180/\pi$=57.3°).

There is also a theoretical limit to optical performance set by diffraction at the aperture of the eye; the bigger that aperture the better, theoretically, is the performance. Diffraction blurs the image of a point source into a circle of confusion known as the Airy disc, which has a roughly Gaussian distribution of light with a half-width of λ/D radians, where λ is the wavelength of light (~0.5 µm) and D is the lens diameter. The effect of this blurring on the capacity of the eye to resolve stripes in a grating is to impose an absolute limit to the fineness of gratings that can be resolved. This is known as the cut-off frequency, ν_{co}:

$$\nu_{co} = \frac{D}{\lambda} \quad (\text{radians}^{-1}). \tag{3}$$

Whether an eye's performance is limited by the receptor grain [Eq. (2)] or diffraction [Eq. (3)] clearly depends on the dimensions of the eye and the receptors. If we take the case of the PM eye of a medium sized wolf spider (Lycosidae), this has a lens diameter of about 200 µm, a focal length that is very similar, and receptors about 5 µm across. This gives ν_s equal to 20 cycles per radian, and ν_{co} equal to 400. Clearly it is the retinal mosaic that limits the eye's resolution, and not diffraction. Only one instance is known in which the retinal mosaic approches the diffraction limit, and this is in the principal eyes of the jumping spider *Portia* (Williams and McIntyre 1980), which also holds the spider resolution record (see below).

3.2 Focal Length

A crucial measurement in establishing an eye's performance is the focal length of the optics (f), the distance from the eye's nodal point to the image. The reason for choosing this definition of focal length is that the nodal point is that point in the eye through which rays pass undeviated, which means that the angle subtended by an object outside the eye is the same as the angle subtended at the nodal point by the conjugate image inside the eye (Fig. 5). Thus if O and I are the linear dimensions of the object and image, and the object is at a known large distance u from the eye, then:

$$\frac{O}{I}=\frac{u}{f}. \tag{4}$$

If f is known, it is easy to work out the size of the retinal image of any object, and equally the projection of retinal structures (especially receptors) onto object space can be determined from Eq. (4).

The location of the nodal point can often be guessed with reasonable accuracy from good central sections through an eye. In the eyes of many spiders, the cornea is almost a hemisphere and the lens structures inside it tend to be concentric with the corneal surface. Under these circumstances the nodal point must lie at the common centre of curvature, because rays directed to that point will strike all refracting surfaces at 90°, and will not be deviated. If only sectioned material is available, then the distance from the estimated nodal point to the retina is the focal length (f), provided there is good reason for thinking that the retina is in focus (again, any eye with a canoe-shaped tapetum should be treated with caution). It is much better to measure f directly, and this can be done by a simple and elegant method devised once again by Homann (1928). The cornea and lens are dissected out and placed in a hanging drop of saline so that the cornea has its outer face in air, as in life (Fig. 6a). An object of known size is placed beneath the slide with the hanging drop, at a distance of about 10 cm, and with the condenser of the microscope removed. The size of the image formed by the cornea and lens is then measured in the hanging drop, using a suitable microscope objective. With this method, O, I and u in Eq. (4) are known or measured, and f is just u I/O.

It is instructive to compare the focal lengths of spider eyes with those expected from refraction at the cornea alone. For a single spherical surface with its convex face in air the posterior nodal distance is given by:

$$f=\frac{r}{(n-1)}, \tag{5}$$

where r is the corneal radius of curvature, and n is the refractive index inside the eye. If one uses this formula with $n \simeq 1.4$ (between that of water, 1.333, and chitin, 1.52) it usually gives an answer for f which is much too long. This means that the lens behind the cornea must actually be quite powerful. An extreme example of this is provided by the nocturnal ambusher *Dinopis*, where r=660 µm, suggesting a focal length of 1650 µm. In fact the focal length measured by Homann's hanging drop method is less than half this, 771 µm. The lens in *Dinopis* does indeed have a very short focal length, and a structure reminiscent of the powerful and optically inhomogeneous lenses of fish eyes (Blest and Land 1977). On the other hand, the AM eyes of salticids have focal lengths not much shorter than that of the cornea alone, and thus have relatively weak lenses (Land 1969a).

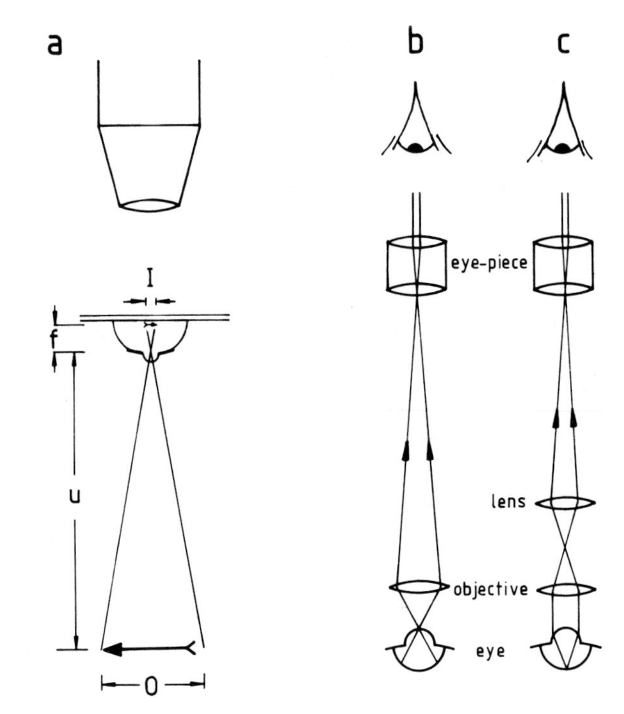

Fig. 6. **a** Homann's hanging-drop method for measuring focal length. The lens is suspended from a drop of Ringer, and the size of the image of an object at a known distance is measured. Then if u is large, f=u I/O (see text). **b, c** Conversion of a microscope (**b**) into an ophthalmoscope (**c**) by addition of a lens which focuses the *back* focal plane of the objective onto the eye-piece. This plane is conjugate with infinity, and hence with the retina. An ophthalmoscope must also contain an arrangement for illuminating the retina

3.3 Ophthalmoscopy

A useful device for studying both the anatomy and resolving power of eyes is an ophthalmoscope. The basis of all ophthalmoscopy is that one uses the optics of the subject's own eye to examine structures in or near the retina. It has the advantage over histological methods that it can be used on living animals. The optical arrangement for looking at the retina is shown in Fig. 6c, and detailed descriptions of practical instruments are given in Land (1969b, 1984). The most important feature of these instruments is that they are basically telescopes rather than microscopes; light from a point on the retina emerges from the eye as a parallel beam, which means that any instrument used to image the retina must be focused on infinity. There is a way of using a microscope as a telescope by introducing an extra lens into the path which images the *back* focal plane of the objective onto the eyepiece: the back focal plane is conjugate with infinity (Fig. 6c).

Some examples of spider retinae photographed through an ophthalmoscope are shown in Figs. 7 and 8. Figure 7 (left) shows the face of a wolf spider viewed with a microscope in its normal configuration, whereas the pictures on the right are taken with the extra lens in place, and correspond to a plane at infinity, and hence the retina (of the PM eyes). These photographs permit several important conclusions. Firstly, the grate-type tapetum is clearly visible, as are small striations across the bars of the grate: these correspond to the rhab-

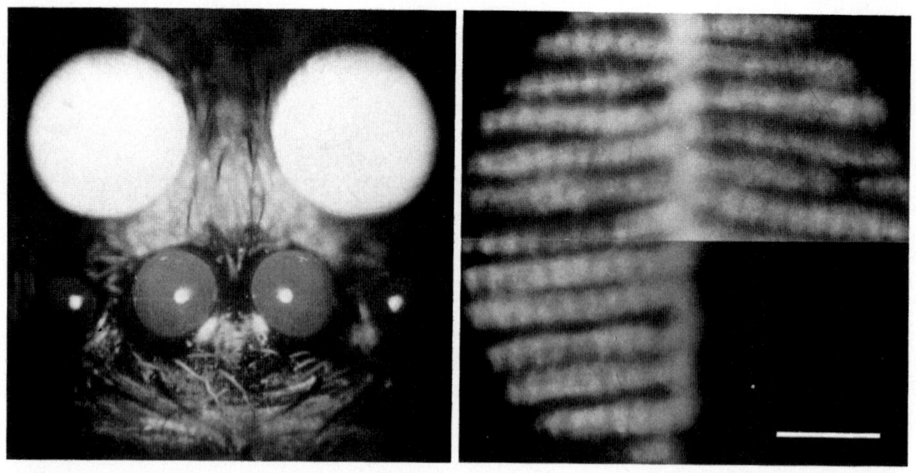

Fig. 7. Eyes of *Lycosa* sp. illuminated from in front, and viewed by ordinary microscopy (*left*) as in Fig. 6b. The largest eyes are the PM's, with a diameter of 490 μm. *Right* ophthalmoscopic view of the retinae of the PM eyes taken with the same instrument, but incorporating the extra lens of Fig. 6c. The "grate" structure of the tapetum is clearly visible, as are the receptor boundaries crossing the stripes. *Upper part* medial edges of the fields of view of both PM eyes; *lower part* that of the right PM eye alone. Note the way the two fields of view touch without overlapping (cf. with the lycosid in Fig. 4). *Dark circular surround* is the limit of the field of view of the instrument (about 25°), not of the eye. Bar=5°

doms of the receptors themselves. The fact that these are visible means that the retina is in focus, and also that the optics of the eye are very good – better in fact than the resolving power of the retinal mosaic, because if the optical blur circle were larger than the receptors they would not be visible. Secondly, the resolution of the eye can be measured directly. The ophthalmoscopic image can be calibrated in angular terms (the simplest way to do this is to remove the spider and photograph a 1-cm scale at 57.3 cm. One division on the film then equals 1°). The receptor spacing in the lycosid in Fig. 8 is about 2°, and in the centre of the AL eye of a jumping spider (Fig. 8) it is about ½°. Notice that ophthalmoscopy gives $\Delta\phi$ directly [Eq. (1)], but cannot give the receptor spacing d_{cc} or the focal length f. However, if either d_{cc} or f is known, then the other can be obtained from equation 1 and the ophthalmoscopic measurement (see Fig. 5).

The spiders which give the best ophthalmoscopic images, and hence have the best optics, are all hunting spiders. Of the families I have examined, the Salticidae, Lycosidae, Thomisidae and Sparassidae all give good images, both of retinal structures and of objects in outside space. So far I have not found an eye with a canoe-shaped tapetum in which there is any sign of an in-focus image, and this is true not only of the web-building families, but of the nocturnal hunters (Gnaphosidae and Clubionidae) as well. This is further strong evidence for thinking that in this kind of eye the image, if one is present, is not at the focus of the lens (see Fig. 11).

3.4 Sensitivity

The amount of light that an eye is able to capture in its receptors is just as important as its ability to resolve detailed images. We refer to this as the eye's sensitivity (S). The reason it is important is that light is quantal in nature, which means that at low light levels receptors may be receiving very small numbers of photons in a given time period. Small photon samples, like all other randomly distributed small samples, are unreliable, or to use the electronic analogy, they are noisy. This means that as the number of available photons decreases, shades of grey cease to be distinguishable, and ultimately black cannot be told from white. It is this fundamental statistical limitation that imposes the absolute threshold for vision in the dark, in humans and in all other animals (Pirenne 1967; Land 1981). In animals which are active at night, which probably includes the majority of spiders, it is of great importance to ensure that as many photons as possible are caught by the photopigment of the receptors.

It is possible to calculate the amount of light the receptors receive from an extended field whose luminance is known. The sensitivity (S) is given by the formula:

$$S = \frac{F_{p(abs)}}{L} = \left(\frac{\pi}{4}\right)^2 \cdot \left(\frac{D}{f}\right)^2 \cdot d_r^2 \cdot (1 - e^{-kl}), \tag{6}$$

where $F_{p(abs)}$ is the flux absorbed by a receptor, L is the luminance of the extended source, D is the lens diameter, f its focal length, d_r is the receptor diameter, l the receptor length and k is the extinction coefficient of the photopigment in the receptors (in rhabdomeric receptors the best available value for k is 0.67% per μm). A full derivation of this formula is given in Land (1981). Notice that all the quantities on the right-hand side (except k for which we have an estimate) can be determined from observation or histology, so that it is possible to determine an eye's ability to see in dim light from anatomical studies alone. Although Eq. (6) looks daunting, it is actually quite straightforward. If we ignore the $(\pi/4)^2$ term which only arises because apertures and receptors are round rather than square, the other three terms are respectively the relative aperture of the eye which determines the light flux passing through the retina, the cross-sectional area of a receptor, and the proportion of the light entering a receptor that is actually absorbed by it. Values for S can be compared between species (see below) to determine how well each would fare under low light conditions, but S can also be used to find the absolute number of photons that receptors receive. Suppose L=1 candela per m^2, the luminance of a white card at sunset. This is approximately equal to 4×10^{15} photons s^{-1} m^{-2} steradian^{-1}. Suppose, too, that anatomical studies show that S=10 (μm^2, if d is measured in μm) then the number of photons caught per receptor per second will be $L \cdot S \times 10^{-12}$ (10^{-12} is the number of μm^2 in 1 m^2). With the values of L and S chosen, this comes to 4×10^2 photons per second. This is certainly enough to see with, but had the illumination been starlight (10^{-4} cd m^{-2}) then this figure would reduce to 4 per second, which does not permit much discrimination.

Resolution and sensitivity are in competition in the design of eyes. Resolution improves as the ratio of receptor diameter to focal length decreases [Eq.

(2)], but sensitivity improves as the same ratio increases [Eq. (6)]. In an eye of a given size there is, in a sense, a choice between the two conflicting demands. If greater resolution is required without sacrificing sensitivity, then the eye must increase in size to accommodate more receptors, and if increased sensitivity is required without loss of resolution, then the eye must become larger because each receptor must be wider. *Good resolution and high sensitivity both require a large eye.* We see the former in the AM eyes of jumping spiders (Fig. 9) and the latter in the extraordinary PM eyes of *Dinopis* (Fig. 9).

4 The Optical Performance of Spider Eyes

4.1 Resolution

We know very little about the resolving power of the eyes of web-building spiders, but the various groups of hunting spiders have been comparatively well studied. A summary of available information is given in Table 2. The only family in which the inter-receptor angles ($\Delta\phi$) are less than 1 degree is the Salticidae; in the others $\Delta\phi$ is in the range $1° - 5°$, which is very much the same range as that found in the compound eyes of insects. The salticids really are much better than the rest, and in the case of *Portia* the resolution approaches that of primates. The structural basis for this fine resolution can be seen from Table 3. Recall that $\Delta\phi = d/f$, so that the smaller the receptor diameter d and the longer the focal length f, the smaller the value of $\Delta\phi$. In *Portia* the focal length of the AM eyes is nearly 2 mm, and the receptor diameter (1.4 µm) is close to

Table 2. Inter-receptor angles ($\Delta\phi$) in hunting spiders

Family (genus)	Eye				Reference
	AM	AL	PM	PL	
Salticidae					
(*Portia*)	2.4′	–	–	–	(1)
(*Evarca*)	12–39′	35′–2.2°	–	–	(2)
(*Epiblemum*)	–	–	11°	2°	(2)
(*Metaphidippus*)	11–39′	27′–1.5°	–	1°	(3)
Lycosidae					
(*Lycosa*)	4– 7°	7°	1.8–2.3°	1.8°	(4)
Thomisidae					
(*Xysticus*)	3.6°	1.8–2.6°	4.2°	3°	(5)
Sparassidae					
(*Isopeda*)	–	1.7°	3.2°	1.8°	(6)
Dinopidae					
(*Dinopis*)	–	–	1.5°	–	(7)

References. (1) Williams and McIntyre (1980); (2) Homann (1928); (3) Land (1969a); (4) Homann (1931); (5) Homann (1934); (6) present study; (7) Blest and Land (1977).
A dash (–) indicates that no data are available.

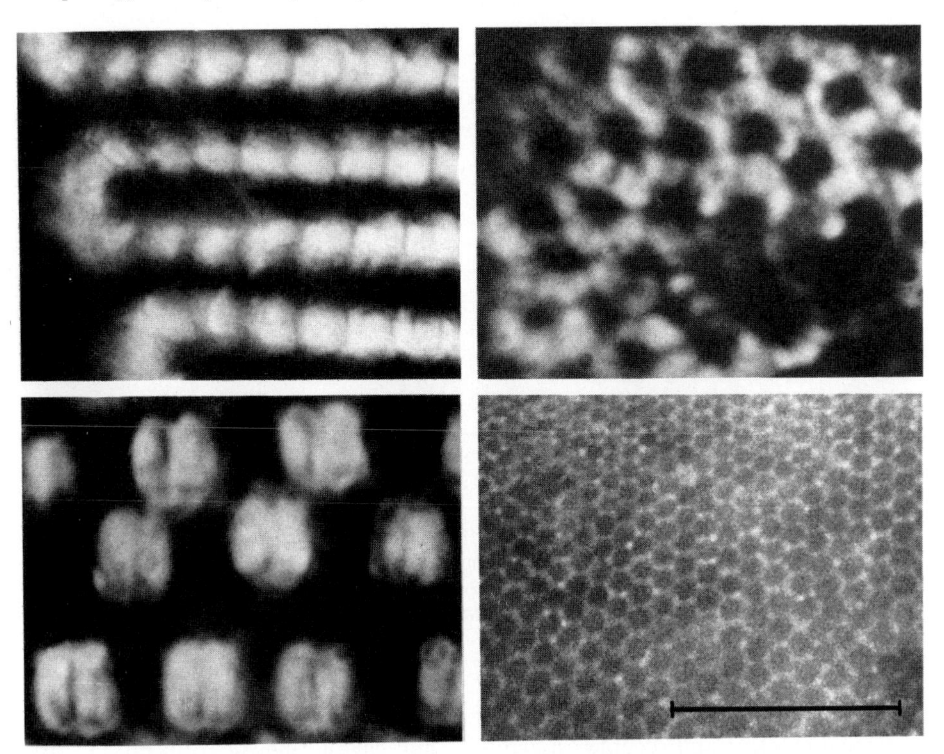

Fig. 8. Ophthalmoscope photographs of the tapeta and retinae of the eyes of the same four spiders whose fields of view are given in Fig. 4. *Lycosa* (*top left* PM eye) has a grate tapetum, and receptors arranged as in Fig. 2a. The thomisid *Tharpyna* (*lower left* AL eye) is similar, except that each receptor is surrounded by black pigment. In the sparassid *Olios* (*top right* PL eye) the tapetum is a continuous sheet (similar to Homann's "primitive type") punctuated by holes for the receptor axons. The salticid *Plexippus* has no tapetum (*lower right* AL eye) but the receptors are visible because the pigment granules in the surrounding cells reflect some light. The scale bar (5°) applies to all four figures. It is evident that the salticid has better resolution than the others. Note: "*Lycosa*" may not be the correct genus; the tapetum corresponds to Homann's description of *Trochosa*

the theoretical minimum (if receptors are narrower than about 1 μm they become "leaky", with much of the light travelling outside them; see Snyder 1975). The long focal length is achieved in an interesting way, too. The lens itself has a focal length of 1273 μm, but just above the retina there is a concave refracting interface which behaves as a second, negative, lens, and the effect of this is to extend the focal length of the combination to 1980 μm (Williams and McIntyre 1980). This combination of separated positive and negative lenses is the way that photographic telephoto lenses are designed, and in nature the same kind of arrangement is found in the eyes of hawks, with the concave foveal pit acting as the second lens (Snyder and Miller 1978).

In the secondary eyes of salticids and the other hunting spider families the values for $\Delta\phi$ given in Table 2 can be compared directly with the ophthalmoscope photographs in Fig. 8. What emerges is that the 50-year-old

Fig. 9. Photographs of the champions. *Portia* (*left*) has an inter-receptor angle of only 2.4 arc min in its large AM eyes. In the wholly nocturnal *Dinopis* (*right*) the largest eyes are secondary PM's. The inter-receptor angle (1.5°) is typical of hunting spiders, but the eyes are more than 3000 times as sensitive as the AM eyes of *Portia* (details in Table 3). (*Portia* photograph courtesy of David Blest)

measurements of Homann (1928, 1931, 1934), made by measuring separately the focal length f and receptor spacing d_{cc} and calculating $\Delta\phi$ from Eq. (1), are almost identical to those obtained directly from the retinal image. It is a pity that Homann's optical measurements were confined to only three spider families.

At the other end of the scale are the secondary eyes of web-building spiders. These usually have canoe-shaped tapeta, and several pieces of evidence point to them having rather unconventional optics, and probably very poor resolution. If we consider the secondary eyes of *Tegenaria,* which were described in detail by Widmann (1908), we find that the retina consists of just two rows of receptors, with each row containing about 40 cells. The slab-like rhabdoms are at right angles to the receptor rows and hence lie across the axis of the tapetum (Fig. 1b and 2c). Although *Tegenaria* itself has not been studied optically, we know from other spiders with canoe-shaped tapeta, for example *Storena* (Zodariidae), that the focus of the lens is deeper than the tapetum (Fig. 10). This makes the slit in the tapetum visible in the eye without ophthalmoscopy, and means that in contrast to the situation in the hunting spiders (Fig. 8) the retina is seriously out of focus. However, the tapetum is a concave mirror, and must therefore behave as part of the optical system to produce an image in the receptor layer in front of the tapetum (Fig. 11). This image will be of poor quality because the tapetum has different curvatures along and across its long axis: thus the focal lengths are different for rays in different planes, and the image will be very astigmatic. From Widmann's plates it appears that the shorter radius of curvature of the tapetum is 20 µm, and the longer 100 µm. Depending on the exact position of the focus of the lens, this would put the final image near the

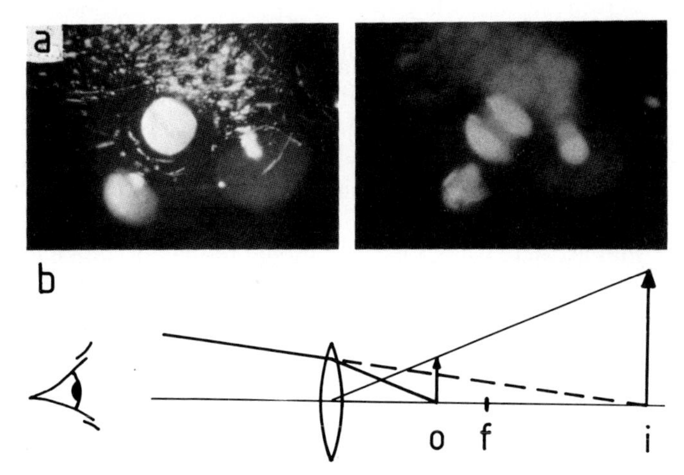

Fig. 10 a, b. Photographs of the PM eye of *Storena* (Zodariidae), which has a canoe-type tapetum. *Left* focused on the cornea; *right* focused about 0.6 mm below cornea. At this level the enlarged image of the slit in the tapetum (Fig. 2c, d) is in focus and clearly visible. **b** Explanation of **a:** if an object at *o* lies inside the focus of the lens (*f*), then an enlarged, upright virtual image of the object will be formed at *i*, beyond the focus (*f*). This means that in *Storena* (and probably most other eyes with canoe-type tapeta) the focus of the lens is behind the retina and tapetum

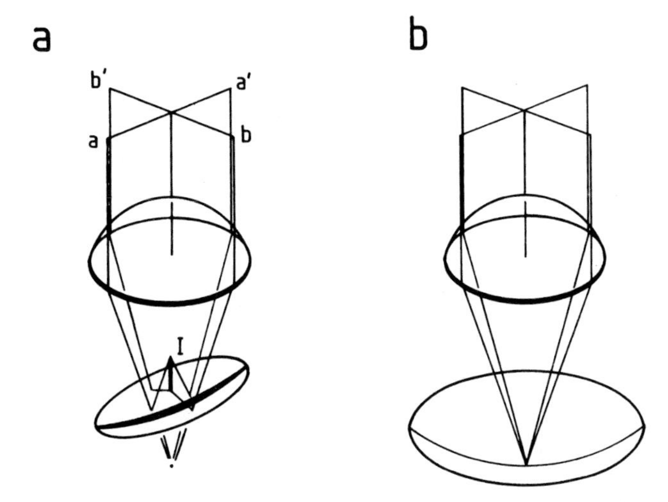

Fig. 11 a, b. An attempt to represent the optics of eyes with canoe-type tapeta. **a** Light entering the lens is focused behind the reflector. The reflector intercepts this light and brings it to a focus in front of the reflector. Because the reflector has different radii of curvature along and across its long axis (indicated by the *black slit*), the resulting image is not a point but an astigmatic line (*I*). Rays in the plane of the reflector's axis (*a, a'*) are focused nearer the lens than those at right angles to that plane (*b, b'*). **b** Where the reflector is in the focal plane of the lens, as in the eyes of cats and hunting spiders (Fig. 8), it has a purely passive role, and reflects the entering beam out of the eye back along its original course

bottom of the rhabdoms for light reflected from the sides of the tapetum, and above the rhabdoms for light reflected along the length of the tapetum. The effect of this is probably to fill the receptors efficiently with the available light, but it is hard to see how the quality of the resulting messy image can approach the potential resolution provided by the rhabdom spacing, which is about 5 μm or close to 3° in angular terms.

In certain of the secondary eyes of web-building argiopid spiders, *Epeira* for example, there is both a conventional retina not unlike that of a hunting spider, and also a canoe-shaped tapetum occupying about a quarter of the retinal area, and with the transverse slab rhabdom arrangement typical of canoe-type retinae (Widmann 1908; Uehara et al. 1978). The importance of these eyes is that they imply that both the ordinary "in-focus" optics and the less comprehensible canoe-tapetum arrangement have different but useful functions. Possibly the former resolves better, but the latter – by virtue of its shorter focal length and hence lower F-number – has higher sensitivity in dim conditions. However, until we have a clearer idea of the way eyes with canoe-shaped tapeta work optically any further speculation is premature.

4.2 Sensitivity

As pointed out in the theoretical section above, it is possible from anatomical measurements alone to arrive at a figure which indicates how much light a receptor receives when the eye is viewing a scene of known brightness. The higher this number (S) the more sensitive the eye. In practical terms, if S is 100 times greater in one eye than in another, then the first eye is capable of a similar performance to the second (in terms of its ability to resolve contrast) at light intensities 100 times lower.

The impressive feature of the sensitivity values in Table 3 is that there is an excellent agreement with habitat. At one extreme the salticids are diurnal, and at the other *Dinopis* is a strictly nocturnal hunter, remaining in a stick-like cryptic posture during the day. The sensitivity difference between the AM eyes of salticids and the PM eyes of *Dinopis* is more than 3 log units, which incidentally is much the same as that between the diurnal eye of a bee and the nocturnal eye of a moth (Kirschfeld 1974). This seems an enormous difference, and it is achieved mainly by the much larger diameter of the receptors in the nocturnal spider, and to a lesser extent by the larger ratio of lens diameter to focal length. However, impressive as these differences are, they are only part of the way towards meeting the difference between day and night, which is not three log units, but closer to six. A particularly interesting comparison is between the two dinopid species in Table 3. *Menneus* is crepuscular, building its catching net about half an hour after sunset, when the light intensity is 2–3 log units lower than daylight, whereas *Dinopis* builds its snare about 1 h after sunset when it is practically night, about 5 log units dimmer than daylight (Austin and Blest 1979). It seems that at crepuscular light levels no spectacular modifications of the eyes are required: *Menneus* PM eyes are similar in most respects to the PL eyes of the small lycosid listed in Table 3. However, to stretch the

Table 3. Optical data and sensitivities of the eyes of hunting spiders

Family (genus) Eye	f (μm)	D (μm)	d (μm)	l (μm)	Δφ (′ or °)	S (μm^2)	Reference
Salticidae							
(*Portia*)							
AM	1980	810	1.4	90	2.4′	0.09	(1)
(*Phidippus*)							
AM	767	380	2	23	9′	0.09	(2)
Lycosidae							
(*Lycosa*)							
PL	135	214	4	30[a]	1.8°	8.2	(3, 4)
Thomisidae							
(*Xysticus*)							
AL	115	150	6	30[a]	3°	12.5	(5, 6)
Sparassidae							
(*Olios*)							
AL	450	450	14	30[a]	1.8°	40	(6, 7)
Dinopidae							
(*Dinopis*)							
PM	771	1325	20	113	1.5°	387	(8, 9)
(*Menneus*)							
PM	170	235	4	36	1.4°	6.2	(10)

References. (1) Williams and McIntyre (1980); (2) Land (1969a); (3) Homann (1931); (4) Widmann (1908); (5) Homann (1934); (6) Homann (1951); (7) present study; (8) Blest and Land (1977); (9) Blest (1978); (10) Blest et al. (1980).
Notes. The sensitivity S is calculated from Eq. (6). f, focal length; D, lens diameter; d, receptor diameter; l, receptor length.
[a] Indicates a tapetum and 2×l is used to calculate S; Δφ is the inter-receptor angle.

eyes' sensitivity another 2 log units does require a huge enlargement of the receptors (20 μm in *Dinopis*, 4 μm in *Menneus*), and since the interreceptor angle ($\Delta\phi$) is the same in the two animals, this means that the whole eye must increase in size by the same amount. The result is the spectacular appearance of *Dinopis*, the "ogre-faced spider" (Fig. 9).

In Table 3 it has been assumed that photopigment-bearing microvilli occupy the whole cross-section of the receptors. In the dark this is often nearly true, but many spiders have the ability to remove most of the photoreceptor membrane during the day, leaving only a narrow "rind" of short microvilli around the cell (Blest 1978). This reduces the amount of photopigment by at least a factor of 10, and so, presumably the sensitivity as well (see Blest, Chap. V, this Vol.). In fact this seems to be the only known method of dark/light adaptation in spiders: there are no reports of mobile irises around the lenses of the eyes, nor of movement of screening pigment in the vicinity of the receptors.

In calculating sensitivities using Eq. (6), we have to assume that the receptors do effectively transduce all the photons that they receive. It is now possible to measure the absorption of photons by receptors directly, by microelectrode recording. Laughlin et al. (1980) recorded the 10 mV "bumps" which result from photon captures in the receptors of the PM eye of *Dinopis*, and compared their results with a similar study made earlier on the PL eyes of a jumping spider, *Plexippus*, by Hardie and Duelli (1978). They found that with the receptors in their depleted diurnal state, about 7% of photons were captured, which is roughly what would be expected from the amount of membrane available, and implies that in the nocturnal state a very much higher percentage would be caught. The cells gave half-maximal responses at intensities which were "midway between starlight and moonlight under clear sky conditions". Laughlin et al. found that to produce the same response in *Plexippus* required 2×10^5 times more light. Of this difference about 3 log units could be explained by structural differences [Eq. (6)] and the other 2 log units by the much greater electrical amplification in *Dinopis*, which has very much larger photon "bumps" than *Plexippus*. The electrical amplification does not necessarily improve the eye's performance, however, because it is the *number* of photon absorptions that matters, rather than the size of the response they produce; a noisy signal remains noisy when amplified. It is of great value to have this direct electrophysiological confirmation of the sensitivity estimates based solely on the eyes' anatomy.

4.3 The Tiered Retina in Salticids

The retina of the AM eyes of jumping spiders is so different from that of any other spider that it deserves separate consideration. Although the eyes had been

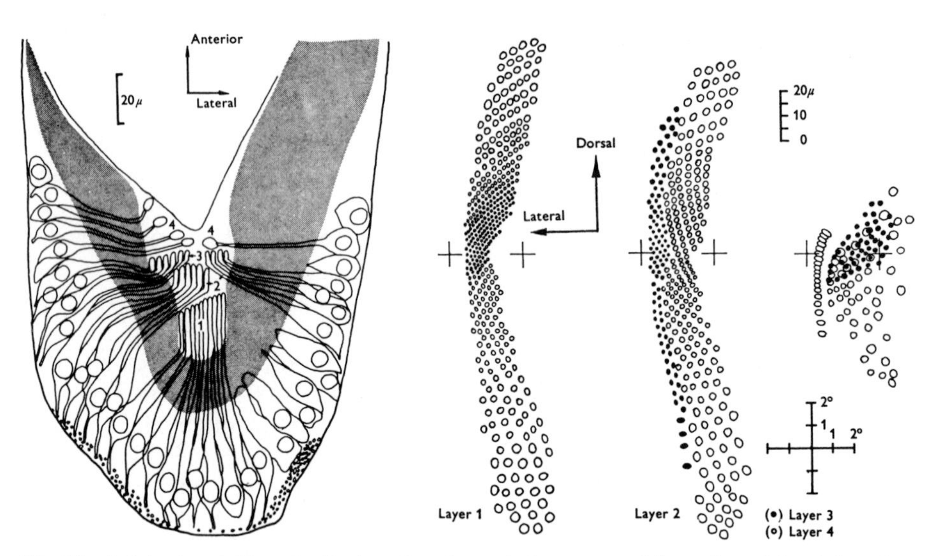

Fig. 12. *Left* horizontal longitudinal section through the centre of the retina of the right principal (AM) eye of a jumping spider (*Metaphidippus*) showing the four retinal layers. *Right* frontal (transverse) view of the four retinal layers, showing that layer 1 has the highest receptor density, and that layers 3 and 4 are confined to the centre of the retina. (Land 1969 a)

described originally by Grenacher (1879), their most intriguing feature was not found until nearly a century later. This is the fact that the retina has four layers, one behind the other in the light path (Land 1969a; Eakin and Brandenburger 1971; Blest et al. 1981). The arrangement of the layers is shown in Fig. 12. The deepest layers (1 and 2) occupy the whole boomerang-shaped area of the retina, whereas the more superficial layers (3 and 4) are present in the central region of the retina only. Layer 1 has the densest receptor packing, and hence the highest acuity, and layers 2–4 have progressively poorer acuity. There are distinct differences in the shape of the receptors of the different layers, as well as ultrastructural differences. In particular, layer 1 receptors are long and thin, and undoubtedly function as light-guides rather like insect rhabdomes. This is usually not the case in other spider eyes, mainly because the lens supplies a cone of light that is much wider than the receptors can accept. When receptors can behave as light-guides they can receive a sharply focused image at their distal tips, rather than a somewhat blurred image whose light passes obliquely through the length of the receptor, so this is an daptation to fine resolution (see Blest, Chap. V, this Vol.).

It is natural to ask why the receptors should be layered in this way. Land (1969a) considered two possibilities. The first was that the layering acted to increase the eye's depth of focus, with nearby objects focused deep in the retina, and distant objects in the more superficial layers. Although the dimensions are roughly compatible with this idea, it is unattractive because salticids have, and need, their finest resolution at large distances, so that infinity should be focused on layer 1, which would mean that the other layers are not conjugate with planes in front of the spider. Blest et al. (1981) point out that layer 1 itself has a substantial "staircase", with the distal ends of lateral receptors lying about 20 μm closer to the lens than the medial receptors. This distance is sufficient for different parts of layer 1 to receive in-focus images from 3 cm to infinity, as the eye scans across the scene (see Land 1969b), and this is surely adequate for the animal's predatory needs, since it jumps on objects closer than about 3 cm. This still leaves open the question of the function of the other three layers. The second suggestion (Land 1969a) was that the different layers are situated in the best image planes for light of different wavelengths. The lens of *Plexippus* does have substantial chromatic aberration (Blest et al. 1981), amounting to 5% of the focal length between the UV and red, so that there would be a gain in image quality to be had by situating receptors with different photopigments in different layers. One would predict that long wavelengths would be focused on layer 1, and short wavelengths on layer 4. There have been a number of studies implicating colour vision in jumping spiders (rev. Forster 1982 and Chap. XIII, this Vol.), and there have also been three intracellular studies (see also Yamashita, Chap. VI, this Vol.). DeVoe (1975) found green and UV receptors in *Phidippus regius*, and also some cells with peaks in both parts of the spectrum. Yamashita and Tateda (1976) found blue and yellow receptors in addition to UV and green, in *Menemerus*, and it would certainly be convenient to suppose that their four colour types coincided with the four layers of the salticid retina. However, they only characterised a total of seven cells, and in neither of the preceding studies were the receptors marked. In the study of Blest et al. (1981) the receptors were dye-marked. They found green receptors (λ_{max} 520 nm) in layers 1 and 2, and UV receptors (λ_{max} 360 nm) in layer 4. No layer 3 cells were marked. The situation thus is that these eyes are *at least* dichromatic, and possibly have more than two colour types; the green and UV cells that have been found are in appropriate regions of the retina in relation to the eye's chromatic aberration. The remaining problems are that layers 1 and 2 seem to have overlapping functions, which would seem to make layer 2, with its coarser mosaic, redundant, and that no function has been found for layer 3.

4.4 Eye Movements in Salticids

The very narrow and elongated structure of the AM retinae only begins to make sense when the movements of the eyes are taken into account. A fairly complete description of these is given by Land (1969b), although Homann (1928) and Dzimirski (1959) had observed them earlier. Unlike vertebrate eyes, in which the whole eye rotates in its socket, it is only the retina which moves in salticids:

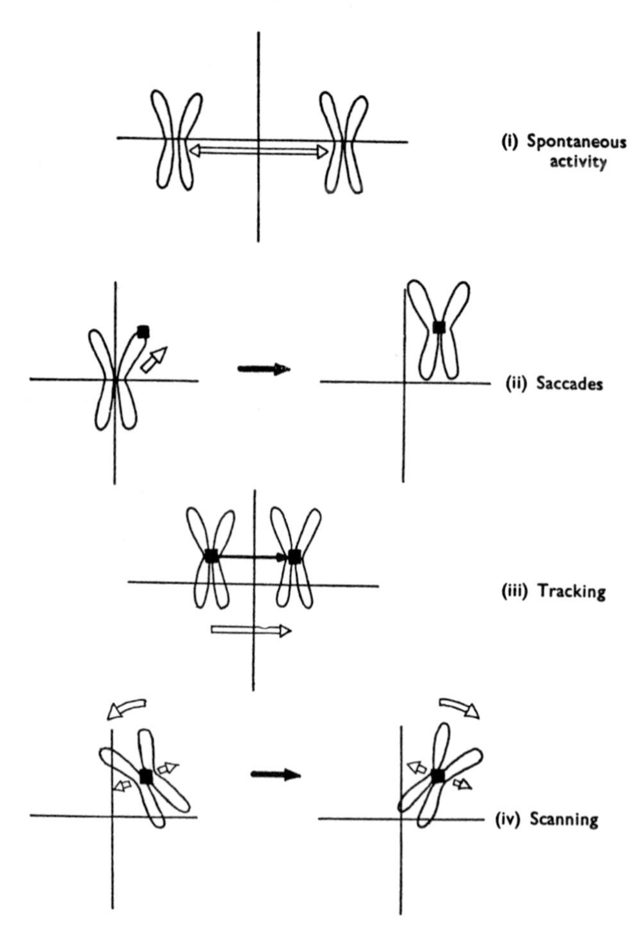

Fig. 13. Types of eye movement shown by the principal eyes of jumping spiders. The two boomerang-shaped retinae (Fig. 12) together have a field of view which is X-shaped. Movements of the retinae, brought about by the operation of six muscles attached to each eye, are indicated by the *open arrows*. Explanation in the text. (Land 1969b)

it scans the fixed image produced by the lens. The movements are of four types (Fig. 13). (1) Wide angle spontaneous scanning movements, in which the centres of the AL retinae wander at varying speed over a large vertical and horizontal field, probably co-extensive with the fields of view of the AL eyes (Fig. 4). These movements probably serve to locate objects of interest that do not themselves move. (2) Saccades: these are rapid movements which centre the cross made by the fields of view of the two retinae onto objects which have just moved, either in the field of view of the AM eyes themselves, or that of the AL eyes. (3) Tracking: if an object acquired by a saccade moves in the field of view, then the retinae keep it centred by tracking movements. (4) Scanning: these movements are remarkable and distinctive and occur whenever the AM retinae have acquired a new target. The two retinae oscillate laterally across the target at a rate of 0.5 to 1 Hz, and at the same time rotate about the eyes' long axes through about 50° at a much slower rate – about 0.1 Hz. We know from the earlier work of Heil (1936) and Drees (1952) that the important features that salticids use to distinguish prey from conspecifics are the presence of appropriately disposed "legs" or leg-like contours. It seems likely that the scanning movements which follow target acquisition are a stereotyped procedure designed to determine the presence or absence of appropriate contours in the target (Land 1969b, 1972b). Usually, both retinae move together (conjugate eye movements), but they can also move independently. Eye movements of the complexity described here are only found in salticids; in other families the AM retinae may show movements that amount to little more than a tremor, although in crab spiders (Thomisidae) there is a suggestion of movements that look more like tracking (Homann 1934).

4.5 Colour and Polarization Detection Outside the Salticidae

The few studies that have been made of the electrophysiology of spider retinae suggest that the secondary eyes are colour blind, with only one type of receptor, but that colour vision in the principal eyes may be fairly widespread. DeVoe (1972), recording from single receptors in wolf spiders, found that cells in the AL eyes had a single wavelength of maximal sensitivity in the blue-green at about 510 nm. Hardie and Duelli (1978) similarly found a single type of receptor in salticid PL eyes, with a peak at 535 nm. By comparison, DeVoe found that the AM eyes of wolf spiders had cells with two sensitivity maxima, one in the green (510 nm) and one in the UV at about 365 nm. In different cells the sensitivity ratio between green and UV varied considerably, suggesting that the cells contained varying amounts of two different photopigments. Yamashita and Tateda (1978) found cells with three different spectral sensitivities in the AM eyes of orb-web spiders (*Argiope*). One class peaked in the UV at 360 nm, one in the blue (480–500 nm) and one in the green at 540 nm. It seems that salticids are not unique in having colour vision mediated by the AM eyes (see also Yamashita, Chap. VI, this Vol.).

The pattern of polarization in skylight is a cue that all animals with microvillous receptors are potentially able to exploit for the purpose of navigation (review by Waterman 1981). This is because the geometry of the microvillus ensures that more photo-pigment molecules are aligned along the microvillar axis than at right-angles to it, and this in turn means that the microvillus will absorb preferentially polarized light with its E-vector parallel to the axis (Moody and Pariss 1961). Most or all spiders have a retinal organization that would permit the analysis of polarized light, but there seem to be only two convincing reports of spiders making use of this facility. These are in the web-building spider *Agelena labyrinthica* (Görner 1962) and the lycosid *Arctosa variana* (Magni et al., 1964, 1965). In both spiders the eyes involved appear to be the principal eyes. In *Arctosa*, rotating a sheet of polaroid above a fleeing animal changes its direction of flight, and this seems to be part of a more general mechanism for finding the home direction, based either on the direction of the sun, or when it is not visible, on the pattern of polarization in the sky (Magni et al. 1964). The same authors (1965) found that the ERG's from both principal and secondary eyes showed polarization maxima and minima, but apparently only the AM eyes are involved in the behaviour. Single receptors in the PL eyes of a jumping spider have been shown to have a modest polarization sensitivity (Hardie and Duelli 1978), but it is not known whether these spiders use polarization information.

5 Conclusions

Despite their small size and similar appearance, the simple eyes of spiders show at least as great a functional diversity as the compound eyes of insects. Specializations for extreme acuity are found in the principal (AM) eyes of jumping spiders, and for extreme sensitivity in the secondary (PM) eyes of dinopid spiders (Fig. 9). The optical performance of the eyes is only known with certainty in the hunting spiders where the lens produces a focused image on the retina; in most web-building spiders the eyes have a canoe-shaped tapetum which is not at the focus of the lens, and the way these eyes work is not properly understood. Spider photoreceptors have been studied electrophysiologically, and they are essentially similar to those of insects. However, nothing at all is known about the processing of visual information by the nervous system. In jumping spiders the secondary eyes are concerned with the detection of movement, and the moveable principal eyes with the identification of objects. Whether or not this division of labour applies in families other than the Salticidae is again not known. At all levels of analysis – optical, neurophysiological and behavioural – the study of spider vision still offers a wide range of problems for future research.

References

Austin AD, Blest AD (1979) The biology of two Australian species of Dinopid spider. J Zool (London) 189:145–156

Baccetti B, Bedini C (1964) Research on the structure and physiology of the eyes of a lycosid spider. I. Microscopic and ultramicroscopic structure. Arch Ital Biol 102:97–122

Blest AD (1978) The rapid synthesis and destruction of photoreceptor membrane by a dinopid spider: a daily cycle. Proc R Soc London Ser B 200:463–483

Blest AD, Land MF (1977) The physiological optics of *Dinopis subrufus* L. Koch: a fish lens in a spider. Proc R Soc London Ser B 196:198–222

Blest AD, Hardie RC, McIntyre P, Williams DS (1981) The spectral sensitivities of identified receptors and the function of retinal tiering in the principal eyes of a jumping spider. J Comp Physiol 145:227–239

Blest AD, Williams DS, Kao L (1980) The posterior median eyes of the dinopid spider *Menneus*. Cell Tissue Res 211:391–403

Bristowe WS (1958) The world of spiders. Collins, London

DeVos RD (1972) Dual sensitivities of cells in wolf spider eyes at ultraviolet and visible wavelengths of light. J Gen Physiol 59:247–269

DeVoe RD (1975) Ultraviolet and green receptors in principal eyes of jumping spiders. J Gen Physiol 66:193–208

Drees O (1952) Untersuchungen über die angeborenen Verhaltensweisen bei Springspinnen (Salticidae). Z Tierpsychol 9:169–207

Duelli P (1978) Movement detection in the posterolateral eyes of jumping spiders (*Evarca arcuata, Salticidae*). J Comp Physiol 124:15–26

Dzimirski I (1959) Untersuchungen über Bewegungssehen und Optomotorik bei Springspinnen (Salticidae). Z Tierspsychol 16:385–402

Eakin RM, Brandenburger J (1971) Fine structure of the eyes of jumping spiders. J Ultrastruct Res 37:618–663

Forster L (1982) Visual communication in jumping spiders (Salticidae). In: Witt PN, Rovner JR (eds) Spider communication. Princeton Univ Press, Princeton, pp 161–212

Görner P (1962) Orientierung der Trichterspinne nach polarisiertem Licht. Z Vergl Physiol 45:307–314

Grenacher H (1879) Untersuchungen über das Sehorgan der Arthropoden, insbesondere der Spinnen, Insekten und Crustaceen. Vandenhoek & Ruprecht, Göttingen

Hardie RC, Duelli P (1978) Properties of single cells in posterior lateral eyes of jumping spiders. Z Naturforsch 33c:156–158

Heil KH (1936) Beiträge zur Physiologie und Psychologie der Springspinnen. Z Vergl Physiol 23:1–25

Homann H (1928) Beiträge zur Physiologie der Spinnenaugen. I. Untersuchungsmethoden. II. Das Sehvermögen der Salticiden. Z Vergl Physiol 7:201–269

Homann H (1931) Beiträge zur Physiologie der Spinnenaugen. III. Das Sehvermögen der Lycosiden. Z Vergl Physiol 14:40–67

Homann H (1934) Beiträge zur Physiologie der Spinnenaugen. IV. Das Sehvermögen der Thomisiden. Z Vergl Physiol 20:420–429

Homann H (1951) Die Nebenaugen der Araneen. Zool Jahrb Anat 71:56–144

Homann H (1952) Die Nebenaugen der Araneen. 2. Mitteilung. Zool Jahrb Anat 72:345–364

Homann H (1971) Die Augen der *Araneae*. Anatomie, Ontogenie und Bedeutung für die Systematik. Z Morphol Tiere 69:201–272

Kirschfeld K (1974) The absolute sensitivity of lens and compound eyes. Z Naturforsch 29c:592–596

Kirschfeld K (1976) The resolution of lens and compound eyes. In: Zettler F, Weiler R (eds) Neural principles in vision. Springer, Berlin Heidelberg New York, pp 354–370

Land MF (1969a) Structure of the principal eyes of jumping spiders (Salticidae: Dendryphantinae) in relation to visual optics. J Exp Biol 51:443–470

Land MF (1969b) Movements of the retinae of jumping spiders (Salticidae: Dendryphantinae) in response to visual stimuli. J Exp Biol 51:471–493

Land MF (1971) Orientation by jumping spiders in the absence of visual feedback. J Exp Biol 54:119–139

Land MF (1972a) The physics and biology of animal reflectors. Prog Biophys Mol Biol 24:75–106

Land MF (1972b) Mechanisms of orientation and pattern recognition in jumping spiders (Salticidae). In Wehner R (ed) Information processing in the visual systems of arthropods. Springer, Berlin Heidelberg New York, pp 231–247

Land MF (1981) Optics and vision in invertebrates. In: Antrum H (ed) Handbook of sensory physiology, vol VII/6B. Springer, Berlin Heidelberg New York, pp 471–592
Land MF (1984) The resolving power of diurnal superposition eyes measured with an ophthalmoscope. J Comp Physiol 154:515–533
Laughlin S, Blest AD, Stowe S (1980) The sensitivity of receptors in the posterior median eye of the nocturnal spider, *Dinopis*. J Comp Physiol 141:53–65
Magni F, Papi F, Savely HE, Tongiorgi P (1964) Research on the structure and physiology of the eyes of a lycosid spider II. The role of different pairs of eyes in astronomical orientation. Arch Ital Biol 102:123–136
Magni F, Papi F, Savely HE, Tongiorgi P (1965) Research on the structure and physiology of the eyes of a lycosid spider III. Electroretinographic responses to polarized light. Arch Ital Biol 103:146–158
Moody MF, Pariss JR (1961) The discrimination of polarised light by *Octopus:* a behavioural and morphological study. Z Vergl Physiol 44:268–291
Paulus HF (1979) Eye structure and the monophyly of the arthropoda. In: Gupta AP (ed) Arthropod phylogeny. Van Nostrand Reinhold, New York, pp 299–383
Pirenne MH (1967) Vision and the eye. Chapman and Hall, London
Scheuring L (1913) Die Augen der Arachnoideen. I. Zool Jahrb Anat 33:553–636
Scheuring L (1913) Die Augen der Arachnoideen. II. Zool Jahrb Anat 37:369–464
Sherk TE (1978) Development of the compound eyes of dragonflies (Odonata). III. Adult compound eyes. J Exp Zool 203:61–80
Snyder AW (1975) Photoreceptor optics – theoretical principles. In: Snyder AW, Menzel R (eds) Photoreceptor optics. Springer, Berlin Heidelberg New York, pp 38–55
Snyder AW, Miller WH (1978) Telephoto lens system of falconiform eyes. Nature (London) 275:127–129
Uehara A, Toh Y, Tateda H (1978) Fine structure of the eyes of orbweavers, *Argiope amoena* L. Koch (Aranea: Argiopidae). Cell Tissue Res 186:435–452
Waterman TH (1981) Polarization sensitivity. In: Autrum H (ed) Handbook of sensory physiology, vol VII/6B. Springer, Berlin Heidelberg New York, pp 281–469
Widmann E (1908) Über den feineren Bau der Augen einiger Spinnen. Z Wiss Zool 90:258–312
Williams DS (1979) The physiological optics of a nocturnal semi-aquatic spider, *Dolomedes aquaticus* (Pisauridae). Z Naturforsch 34c:463–469
Williams DS, McIntyre P (1980) The principal eyes of a jumping spider have a telephoto component. Nature (London) 288:578–580
Yamashita S, Tateda H (1976) Spectral sensitivities of jumping spider eyes. J Comp Physiol 105:1–8
Yamashita S, Tateda H (1978) Spectral sensitivities of the anterior median eyes of the orb web spiders *Argiope bruennichii* and *A. amoena*. J Exp Biol 74:47–57

V The Fine Structure of Spider Photoreceptors in Relation to Function

A. David Blest

CONTENTS

1 Introduction

The eyes of spiders are ocelli, and it is natural to compare their performance with that of compound eyes and of the ocelli of insects. The latter, which are underfocused and usually possess receptor mosaics of indifferent quality (Wilson 1978), are better known than those of arachnids, and have not encouraged workers to examine those of spiders in much detail. Nevertheless, the ocelli of spiders range from the principal eyes of jumping spiders whose sophisticated organisation sustains high visual acuities (Land 1969a; Eakin and Brandenburger 1971; Jackson and Blest 1982a; Blest and Price 1984) to many, perhaps the majority, that can hardly be supposed to sustain much in the way of image analysis at all.

Department of Neurobiology, Research School of Biological Sciences, The Australian National University, Canberra, A.C.T. 2601, Australia

Widmann (1908) examined the eyes of a number of spiders by light microscopy in an important pioneer study, and Homann (1951, 1971) extended his work by a series of massive comparative surveys. It is a tribute to their diligence that so much of what has subsequently been revealed by electron microscopy does little more than elaborate on the foundations that they laid.

With the notable exception of the Salticidae, spiders are not primarily vision-dependent, although vision plays a significant role in the behaviour of many families. All spiders, however, must discriminate between different classes of prey, mates and territorial rivals. For those reliant to whatever degree on vision, fine spatial resolution can be accepted as an evolutionary goal, and one must ask how it has been achieved.

The present chapter summarises briefly those features of photoreceptor organisation that are common to all arthropods as an introduction to the special aspects that characterise the photoreceptors of spiders. The optics of spider ocelli are reviewed in Land, Chapter IV, this Volume. Given that images of good or poor quality are presented to spider retinae, one may ask how receptors are disposed to make the best use of them? We shall see that in jumping spiders (Salticidae) a number of devices are employed in the retinae of the principal eyes to optimise fine spatial analysis, but that the ultrastructural strategies unique to this family are not significantly developed in any other.

Spiders are not tractable material for studies of the basic biophysics of vision, and the relevant theoretical background derives from work on insects and *Limulus*. Reviews of the optics of invertebrate vision are given by Land (1980) and Snyder (1979), and of the electrophysiology of photoreceptors with particular reference to information coding by Laughlin (1980). These treatments will seldom be specifically cited in the following account, but should be regarded as necessary theoretical background.

Four questions of broad interest will be addressed:

1. How are the photoreceptors of spiders organised and with what optical consequences?
2. What is the quality achieved by the receptor mosaics, and what special features have evolved to serve the requirements of high visual acuity where it has been achieved?
3. How do strategies for the turnover of phototransductive membrane relate to functional demands?
4. To what extent can retinal organisation be employed as a taxonomic character?

2 Organisation of Spider Photoreceptors

2.1 Cytology and Function

Photons whose energy is transduced by photoreceptors to provide electrical signals are captured by integral membrane pigments which in arthropods are proteins termed *rhodopsins*. Photon capture is relatively inefficient, so that large amounts of pigment must be interposed in the light path. Thus, the membrane containing it is amplified: in arthropods, specialised domains of the photoreceptor plasma membrane are organised as *rhabdoms* composed of arrays of slender microvilli (Figs. 1–3). Microspectrophotometry of rhabdoms indicates high concentrations of rhodopsin which together with other integral membrane proteins are seen as densely packed particle arrays on the P-face of freeze-etch replicas (Fig. 3). In photoreceptors of flies, at least, the particle density of microvillar membrane matches that of nearby regions of flat plasmalemma (Schinz et al. 1982; Blest and Eddey 1984), and genetic dissection has shown the particles of both largely to consist of rhodopsin: the major specialisation of the rhabdom lies in the amplification of the membrane. Photoreceptor microvilli contain cytoskeletons (Blest et al. 1982a, b; Blest et al. 1983) and recent evidence suggests that some cytoskeletal components may serve to constrain translational diffusion of the photopigments (De Couet et al. 1984). Cytoskeletal architectures have not been described in the microvilli of spiders, but our own preliminary observations suggest that they may be rather diverse.

The receptive segment of a spider photoreceptor lies beneath and usually near to a corneal lens, and is associated with a cell soma containing the nucleus. There is a trivial difference between the dispositions of the somata in the principal and secondary eyes: in the former, somata lie proximally beneath the receptive segments, while in the latter they lie distally between the receptive segments and the "glass cells" that secrete the corneal lens. Usually, there is a pronounced "intermediate segment" near to the receptive segment, consisting of a swollen region filled with endoplasmic reticulum, mitochondria, and the lysosomal systems concerned with the turnover of rhabdomeres (see below). In the latter context, receptive segments often contain numerous pinocytotic vesicles (Fig. 5) and multivesicular bodies generated by the internalisation of rhabdomeral membrane. The photoreceptors narrow abruptly as their axons leave the retina for their final destinations in the first optic neuropil of the protocerebrum, where they form synapses with second-order visual interneurons. The basic organisation of spider photoreceptors is schematised in Fig. 8.

2.2 Glial Components of the Retina

The photoreceptors of spiders are to a greater or lesser extent ensheathed by the processes of glial cells. Two kinds of glia can be distinguished, and are deployed in a variety of ways. Somata of the *non-pigmented glia* lie distal to the receptive segments; in secondary eyes they are interspersed with the somata of the receptors. Their processes invade the layer of receptive segments and may, as in the Salticidae, achieve great architectural regularity (Eakin and Brandenburger 1971; Blest 1983, 1984; Blest and Sigmund 1984). Somata of *pigmented glial cells* lie proximal to the receptive segments. In secondary eyes they are intercalated with the intermediate segments of the receptors; in principal eyes they lie between the proximal receptor somata. In either case, a large number of receptive segments is ensheathed by processes derived from a much smaller number of glial cells.

The role of the non-pigmented glia is not clear. In insects, there is evidence to suggest that analogous glial components are concerned with ionic exchanges

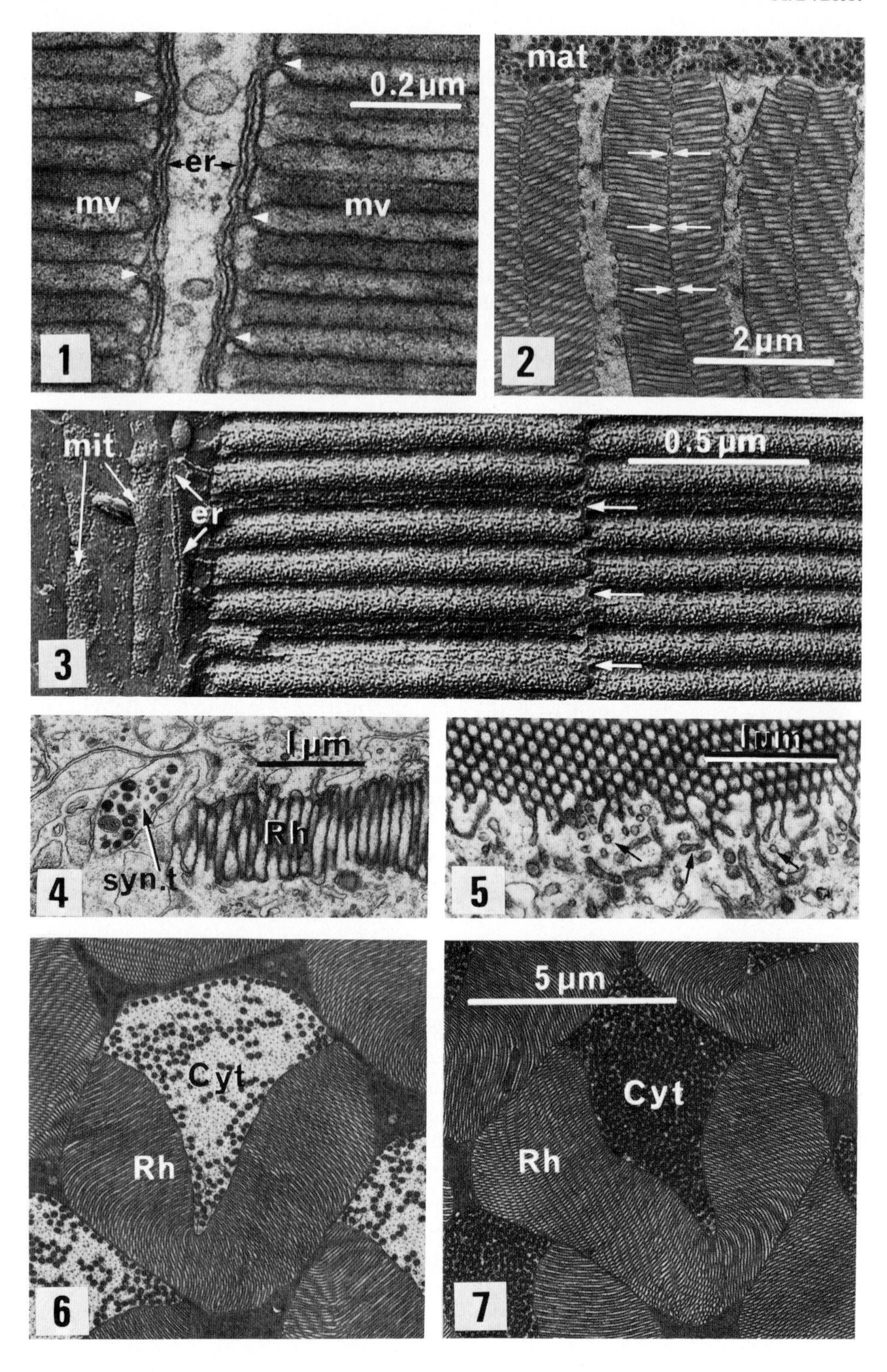
0.2 µm
er
mv
mv
1
mat
2 µm
2
mit
er
0.5 µm
3
1 µm
Rh
syn.t
4
1 µm
5
Cyt
Rh
6
5 µm
Cyt
Rh
7

Fig. 1. Microvillar bases of the paired rhabdoms of a secondary eye of *Plexippus* (Salticidae). *mv* microvilli; *er* sheets of smooth endoplasmic reticulum closely adpressed to the bases of the microvilli

Fig. 2. Receptors of layer IV of a principal eye of *Portia* (Salticidae) to illustrate the close juxtaposition of rhabdomeres of adjacent receptors. A junction between two receptors is indicated by *arrows. mat* retinal matrix

Fig. 3. A freeze-etch replica of a junction shown in Fig. 2 (*arrowed*), to show that the P-face of the microvillar membranes is densely particulate, the particles consisting of rhodopsin molecules or aggregates. *mit* tubular mitochondria; *er* a flat saccule of smooth endoplasmic reticulum

Fig. 4. A rhabdom (*Rh*) of the posterior median eye of *Lampona* (Gnaphosidae), to show a nearby synaptoid terminal (*syn.t*) embedded in non-pigmented glia

Fig. 5. A rhabdom of a posterior median eye of *Lampona* (Gnaphosidae) to show pinocytotic vesicles and other endocytotic profiles (*arrowed*) shed from the microvilli during the breakdown phase of turnover

Figs. 6, 7. Rhabdoms of peripheral layer I (**6**) and layer II (**7**) of a principal eye of *Lagnus* (Salticidae), to show that in layer I the cytoplasm is almost empty of organelles, whereas in layer II it is densely packed with tubular mitochondria that have been shown to equilibrate the refractive indices of rhabdomeres and their cytoplasmic surrounds. *Rh* rhabdom; *Cyt* cytoplasm. Scale bar on **7**

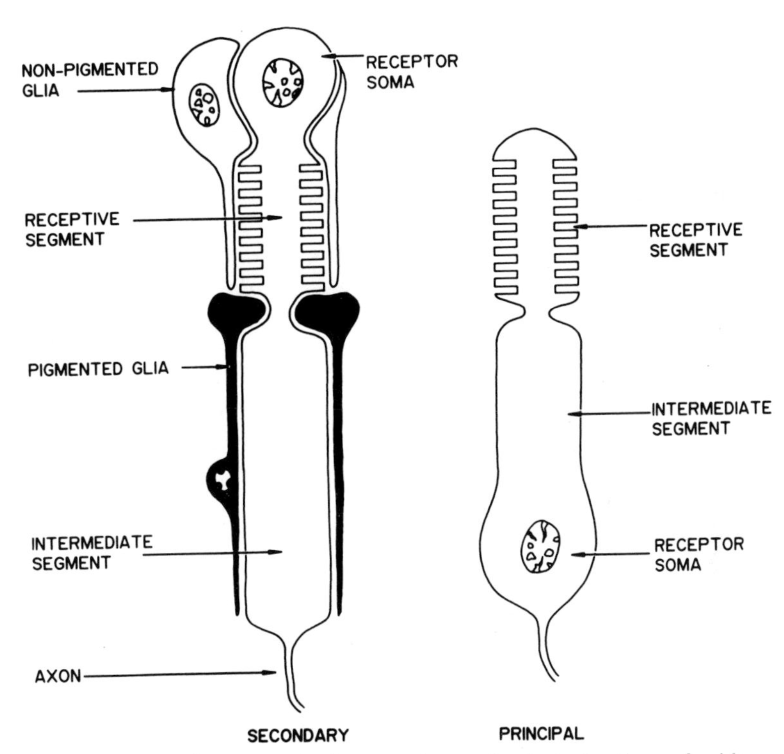

Fig. 8. The basic organisation of receptors in the secondary and principal retinae of spiders. The dioptrics lie beyond the confines of the diagram at the *top*. Associated glial components are not shown for the receptor of the principal retina, at *right*

with the receptors and with K^+ regulation in particular (e.g. Tsacopoulos et al. 1983). Whilst this may prove to be true in some spiders, in many species the non-pigmented glia is reduced at the level of the receptive segments to columns so slender as to render such a function implausible (e.g. Fig. 10).

The pigmented glia in both principal and secondary eyes shields the receptive segments from stray light transmitted antidromically through the cuticle of the cephalothorax. Although pigmented glial processes usually penetrate the layer of receptive segments, they do not always contain pigment granules at that level (cf. Blest 1983, 1984). When they do, as in the secondary eyes of diurnal Lycosids (Bacetti and Bedini 1964; Melamed and Trujillo-Cenóz 1966) and advanced Salticids (Eakin and Brandenburger 1971; Blest 1983), they are also able to shield receptors from stray light scattered from their neighbours.

There is no evidence for any substantial movement of pigment granules in the pigmented glia during light or dark adaptation. A minor invasion of the receptive layer by pigment during dark adaptation has been noted in the Dinopid *Menneus* (Blest et al. 1980), but it is too slight to be of optical significance. Even in retinae that sustain high visual acuities there are no pigment granules within the receptors themselves, so that equivalents of the longitudinal pupil mechanisms found in various insect photoreceptors (e.g. Kirschfeld and Franceschini 1969) are not present in those of spiders.

2.3 Efferent Terminals in the Retinae

The retinae of some spiders contain the efferent terminals of central neurons. Such terminals were observed by Bacetti and Bedini (1964) and Melamed and Trujillo-Cenóz (1966) in retinae of Lycosids; by Blest and Day (unpublished) in the Pisaurid, *Dolomedes;* by Blest (unpublished) in all retinae of *Clubiona* (Clubionidae), *Lampona* (Gnaphosidae: Fig. 4) and the mygalomorph *Darcyops;* and by Dr. A. Uehara in *Argiope* (Argiopidae) (unpublished: Professor H. Tateda, personal communication). The terminals are synaptoid endings located in the retinal glia, and also invade the receptors. They contain electron-dense vesicles and small mitochondria (Fig. 4).

Synaptoid terminals have not been found despite intensive search in either *Menneus* or *Dinopis* (Dinopidae), primitive or advanced Salticidae, or in a number of genera of Australian Sparassidae (Blest, unpublished observations), so that they cannot be a universal feature of the retinae of spiders. With respect to other arachnids, they have been noted in the retinae of a scorpion (Fleissner and Schliwa 1977; see also Fleissner and Fleissner, Chap. XVIII, this Vol.) and are well-known in *Limulus* to contribute to the circadian modulation of sensitivity (Barlow et al. 1977) and the control of transductive membrane turnover (Chamberlain and Barlow 1979), although in *Limulus* the exact nature of the latter phenomenon is not yet understood (cf. Stowe 1981).

The presence or absence of efferent retinal terminals amongst the families of spiders is so distributed as to indicate that although they may, in some species, be concerned with the regulation of membrane turnover, they cannot be necessary for it: Dinopids, Sparassids, and some Clubionids and Gnaphosids use membrane turnover to achieve massive geometrical changes to their rhabdoms in the course of a daily cycle along the lines described by Blest (1978) for *Dinopis*, yet in the first two of these families efferent terminals are, apparently, absent.

Similar arguments apply to the control of sensitivity. Yamashita and Tateda (1981, 1983) have shown that the sensitivity of photoreceptors of *Argiope* is modulated by efferent inputs, including some that stem from photosensitive neurons in the protocerebrum (see also Yamashita, Chap. VI, this Vol.). Nevertheless, secondary eyes of a Salticid adapted for diurnal vision (Hardie and Duelli 1978) and the posterior median eyes of the nocturnal *Dinopis* which traverse 2 log units of sensitivity during dark adaptation (Laughlin et al. 1980) both lack efferent innervation.

3 Physiological Aspects of Retinal Ultrastructure

3.1 Diurnal Adjustments of Rhabdom Size in Spiders, and the Turnover of Rhabdomeral Membranes

The rhabdoms of invertebrate photoreceptors are not stable structures: the phototransductive membrane is in a state of flux, and is continuously removed and replaced by a variety of strategems. The turnover of rhabdomeral membrane was initially disclosed by Eguchi and Waterman (1967, 1976), White (1968) and White and Lord (1975). Blest (1980) has reviewed breakdown processes on a comparative basis, and Blest et al. (1984) provide a broad overview of the physiological and molecular correlates of turnover.

The nocturnal spider *Dinopis*, the posterior median eyes of which are remarkably specialised to achieve high sensitivity (Blest and Land 1977; Laughlin et al. 1980) offered the first demonstration that changes in the dimensions of some rhabdoms follow natural daily cycles of illuminance, and are triggered by the transitions between light and darkness (Blest 1978). At dawn, membrane is removed from the rhabdoms, so that during the day they are small, whilst at nightfall, new membrane is added, so that they become large. Later studies found that similar events take place in the photoreceptors of certain crabs (Nassel and Waterman 1979; Stowe 1980, 1981) and locusts (Williams 1982, 1983). Our own unpublished results demonstrate that identical changes take place in the rhabdoms of Clubionid, Gnaphosid and Sparassid spiders. The results of Blest and Day (1978) indicate that they are also found in retinae of nocturnal Pisaurids, although in that study the changes observed were not unfortunately related to real daily cycles of illuminance. The ultrastructure of the eyes of Argiopid spiders described by Uehara et al. (1977, 1978) implies that daily adjustments of rhabdom size may occur in that family, too.

These phenomena pose two obvious questions:

1. What happens to the membrane that is removed during the breakdown phase? With one exception, in secondary eyes of the salticid spider *Plexippus* (Blest and Maples 1979), membrane is shed from the rhabdomeral microvilli by basal pinocytosis and enters a lysosomal system (Blest et al. 1978a, b). Acid

phosphatase ultrastructural cytochemistry (Blest et al. 1979) indicates that the membrane and its associated rhodopsins are probably degraded rather than merely being recycled, an inference supported but not yet proved by more detailed work on analogous crustacean systems (Blest et al. 1980; De Couet and Blest 1982).

2. What effects do increments in rhabdom size have upon the sensitivity and spatial resolution of the receptors? Laughlin et al. (1980) attempted to answer this question for the posterior median eyes of *Dinopis* using electroretinograms and intracellular recordings. They failed, because test flashes of light presented to the retinae were sufficient to disrupt the process of rhabdom assembly at dusk, so that recordings from rhabdoms in the night state could not be obtained. Thus, although ca. 2 log units increase of sensitivity were observed during the first 2 h of dark adaptation at the normal time of nightfall, it took place in the absence of significant morphological changes to the rhabdoms. Calculations derived solely from the dimensions of the receptors and their rhabdoms suggest that a day rhabdom of *Dinopis* can be expected to catch some 6% of photons incident upon a single receptor, and a night rhabdom 74% (Blest and Land 1977; Blest 1978). Intracellular recordings by Laughlin et al. (1980) support the first value, so that the second may be reasonable. An increase of ca 1 log unit of sensitivity attributable to morphological adjustments to the rhabdom is unremarkable compared to the 2 log units apparently contributed by other processes. The consequences of changes to the dimensions of rhabdoms during the daily cycle of illuminance were measured for photoreceptors of the compound eye of a locust by Williams (1983); the lack of any increase of sensitivity to photons delivered to an ommatidium on-axis, coupled with identical particle densities revealed by freeze-etch preparations on the P-faces of day and night rhabdomeral membranes (Williams 1982) implies that the replacement of day rhabdoms by new membrane at night is not a device to augment the concentration of photopigment in the membrane.

How, precisely, new rhabdomeral membrane is differentiated within a photoreceptor, transported to the region occupied by a rhabdom, and reorganised as microvilli is still, in detail, unresolved. In *Dinopis* and many other spiders with massive turnover routines, intermediate segments are occupied by large amounts of endoplasmic reticulum, a feature common to many photoreceptors (Whittle 1976). As in a crab (Stowe 1980), it appears to be the immediate precursor of rhabdomeral membrane, but despite the current ultrastructural evidence, such a supposition implies many difficulties that remain to be resolved (Blest et al. 1984).

3.2 Scotopic and Photopic Receptors

Blest (1980) noted that arthropod photoreceptors with low absolute sensitivities appear also to exhibit a sluggish turnover of rhabdomeral membrane. Receptors of species whose ecology requires them to function over the whole daily cycle (e.g. locusts, the crab *Leptograpsus*, and the spider *Dinopis*) perform massive exchanges of microvillar membrane in the course of a day. These differences (which have yet to be more widely surveyed, and must be examined more precisely in terms of exact rates of turnover, and of absolute sensitivities measured as voltage gains of transduction) invite an obvious speculation: do any retinae of spiders contain receptors of both kinds, so that they can respond effectively at both high and low levels of illuminance?

Two unusual eyes suggest that retinae with both photopic and scotopic receptors may exist. Widmann (1908) described a posterior median eye of an Argiopid, *Epeira*, which is apparently similar to the posterior and anterior lateral eyes of *Argiope* whose ultrastructure was figured by Uehara et al. (1978). Two different kinds of receptor are disposed at different distances from the corneal lens. The receptive segments have not been observed at different points in a daily cycle, but the populations of secondary lysosomes within them imply that one class of photoreceptor may engage in substantial turnover of rhabdomeral membrane, whilst the other is relatively stable.

A similar possibility is suggested by the extraordinary posterior median eye of a Gnaphosid, *Lampona* (Land and Blest, in preparation). A major class of receptors engages in massive turnover and reorganisation during a daily cycle, whilst a minor population does not.

4 Functional Implications of Retinal Organisation

4.1 Receptor Mosaics

Receptors in the retinae of spiders display mosaics of very variable quality. Although there is much diversity of detail, receptor mosaics can be broadly classified as follows:

4.1.1 Rhabdomeral Networks

Rhabdoms of adjacent receptive segments are contiguous and microvilli derived from contiguous rhabdoms may wholly or partially interdigitate (Fig. 9). Because the receptive segments are cylindrical or roughly hexagonal in transverse profile and their entire surfaces are composed of microvilli, the rhabdoms are arranged rather like the wax in a honeycomb, so that transverse sections of the receptors present the appearance of a network (Fig. 10).

Contiguity and interdigitation of the microvilli of adjacent cells mean that there is optical pooling between receptors. There may also be electrical coupling, a possibility discussed in more detail by Laughlin et al. (1980).

Rhabdomeral networks are typical of the retinae of many nocturnal spiders, and the eyes to which they belong are necessarily of limited spatial acuity. For receptors of the posterior median eye of *Dinopis* in the "day" state, the mean acceptance angle, calculated from the half widths of angular sensitivity functions measured from intracellular recordings, was 2.3° (Laughlin et al. 1980). The angular periodicity of the receptors is 1.5°, and the corneal lens, which is substantially corrected for spherical aberration and has an f-ratio of 0.58, forms an image of a point source with a diameter of 1.8° (Blest and Land 1977).

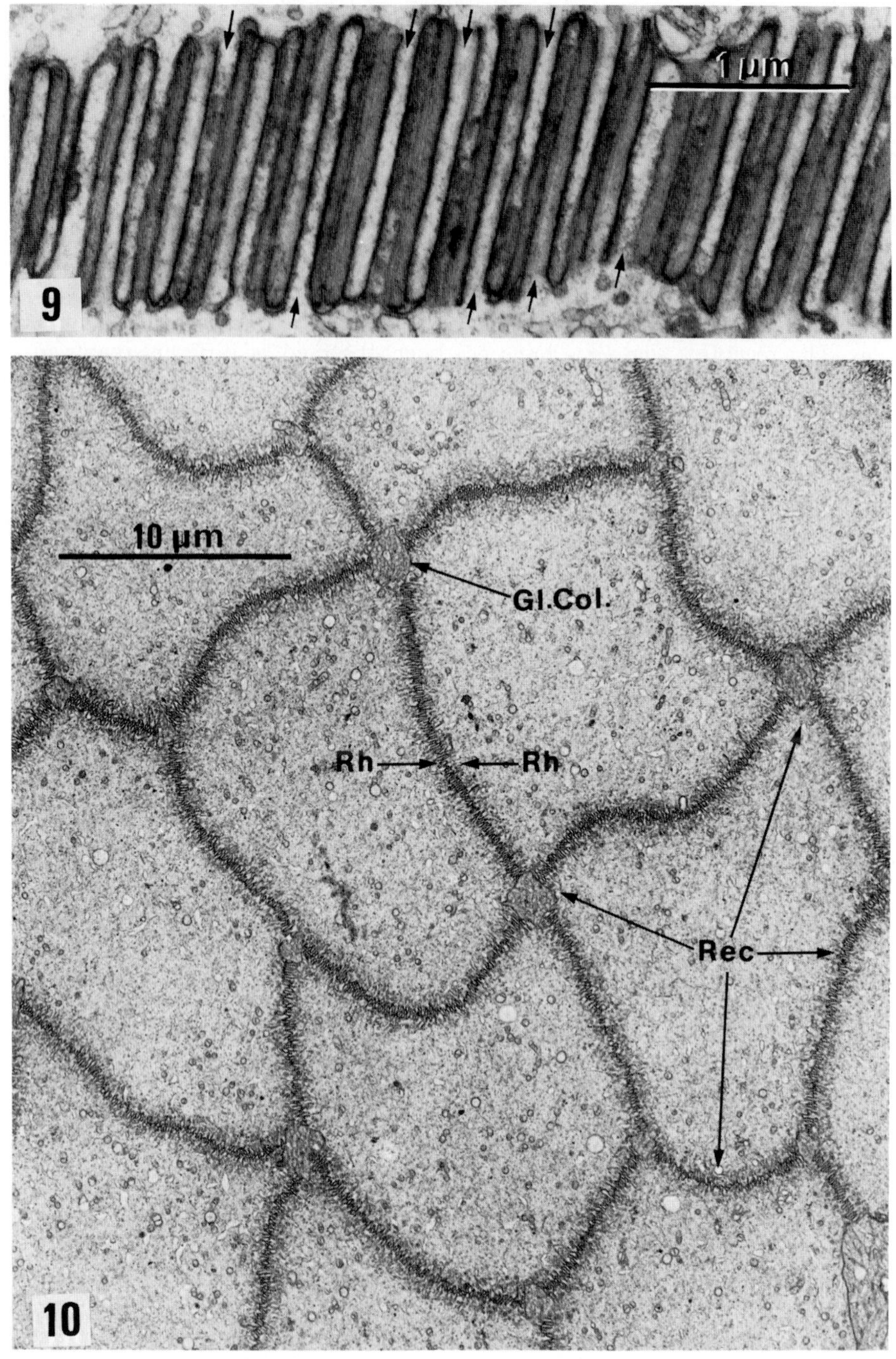

Fig. 9. Microvilli of the adjacent rhabdomeres of a posterior median retina of *Lampona* (Gnaphosidae) to illustrate total interdigitation. Open bases of microvilli are indicated by *arrows*

Fig. 10. Transverse section through the rhabdomeral network of an anterior median eye of *Isopeda* (Sparassidae). *Rh* rhabdomeres; *Rec* limits of a receptor; *Gl.Col.* glial column

Laughlin et al. (1980) note that the inferior angular resolution of the receptors implied by their measurements could have resulted, in part, from quite small mechanical deformations of the retinae imposed by the recording procedure. Even if that were so, coupling between receptors must severely prejudice spatial resolution, so that the architecture of the rhabdoms makes it impossible for a receptor to attain the spatial resolving power suggested by its own angular diameter.

In the "night" state, spatial acuity can only be worse: additional phototransductive membrane is synthesised and incorporated into microvilli; the transverse profiles of the receptive segments become more complex and irregular, and the receptive segments lengthen (Blest 1978) increasing the scatter of a focused point source into adjacent receptors (Laughlin et al. 1980).

It is not known to what extent the performance of eyes with rhabdomeral networks belonging to other families is inferior to that of *Dinopis.* In Sparassidae and probably in the anterior median (AM) eye of the Gnaphosid *Lampona* there is much to suggest that the retinae receive focused images. Accessory eyes with boatshaped tapeta, however, are underfocused (see also Land, Chap. IV, this Vol.), and their spatial resolution can only be poor. It is possible that rhabdomeral networks that receive good images from superior dioptrics are late evolutionary products that represent the best that could be achieved from unpromising precursors. Comparison between the two Australian genera of Dinopidae, *Dinopis* and *Menneus,* discussed below (Sect. 5.1) makes this simplistic view questionable.

4.1.2 Punctate Mosaics Without Refractive Index Barriers

Retinae of all eyes in many phylogenetically advanced spiders present mosaics that are, in some sense punctate: rhabdomeres are closely juxtaposed and the resultant rhabdoms are arranged in regular arrays (Fig. 11). In terms of the image analysis that they can sustain, such mosaics are a considerable advance on the rhabdomeral networks of more primitive spiders, but they are often suboptimal in two respects:

a) The receptors usually contain two rhabdomeres situated on opposite sides of a receptive segment. A rhabdom, however, may be composed of the closely contiguous rhabdomeres of two adjacent cells. Thus, in optical terms, such a receptive unit pools the information received by two receptors and the ideal solution in which one rhabdomere can correspond to a single point in object space is not achieved. Spatial resolution can only be less than would be implied simply by the intervals between photoreceptors. This unsatisfactory anatomy is found in the retinae of Pisauridae (Blest and Day 1978) and in some nocturnal Lycosidae (e.g. *Geolycosa:* Land (1984) and unpublished data), but is resolved in at least one diurnal hunting spider in the Oxyopidae, where the two rhabdomeres of each receptive segment are reasonably close together, and are well separated from those of adjacent receptors (Fig. 11B). In some, presumably diurnal, Lycosidae, receptive segments are isolated from each other by pigmented glia (Bacetti and Bedini 1964; Melamed and Trujillo-Cenóz 1966).

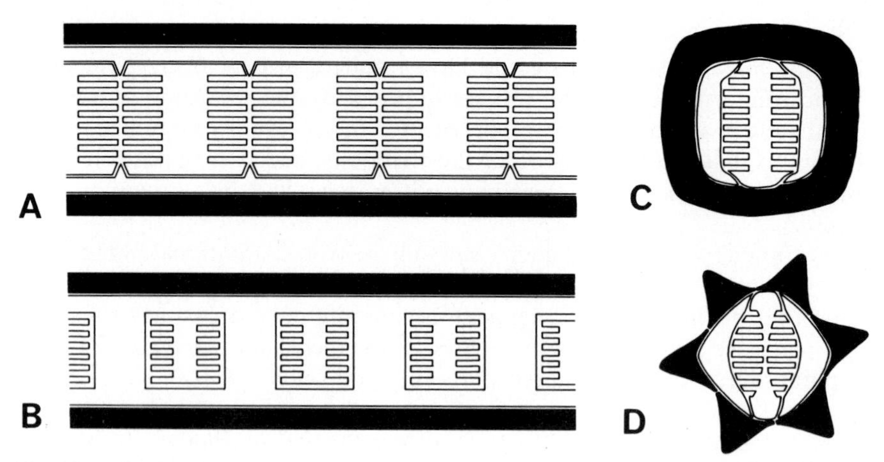

Fig. 11 A – D. Trends in the evolution of punctate receptor mosaics. **A** Rhabdoms constructed from the contiguous rhabdomeres of adjacent receptors, implying optical pooling along linear strips (e.g. Pisauridae). **B** Receptive segments isolated but not optically screened from each other (e.g. Oxyopidae). **C** Receptive segments optically isolated by pigmented glia to form a true punctate mosaic (e.g. some diurnal Lycosidae). **D** Receptive segment optically isolated by (a) a refractive index barrier, and (b) screening pigment (secondary eyes of advanced Salticidae)

b) The absence of a refractive index barrier between rhabdoms and their surrounds must be interpreted in the light of two considerations: firstly, if the rhabdoms are short, little light will be scattered from them if images are accurately focused within the receptive layer, *but* absorption of photons will be inefficient (see Land 1984) and signal-to-noise ratios will be poor. If, on the contrary, rhabdoms are long, scatter, particularly for the case of a corneal lens of wide aperture, will be significant, but both absorption and signal-to-noise ratios will be improved.

It is remarkable and so far unexplained that these two serious problems for the design of ocelli have only once been solved in spiders, by all eyes of the advanced Salticidae.

4.1.3 Punctate Mosaics with Refractive Index Barriers

A substantial refractive index difference between a rhabdom and its immediate surround allows light focused on the tips of a rhabdom to be confined within it, throughout its length, by total internal reflection (rev. Snyder 1979). This means that an image of superior quality received by a punctate mosaic can be conserved even through the rhabdoms that register and transduce it are long: absorption of photons, and signal-to-noise ratios can be optimised.

Two classes of receptor in the advanced Salticidae exploit this design: (a) photoreceptors and their glial surrounds in the secondary eyes, and (b) Layer I photoreceptors of the principal eyes, whose anatomy and *modus operandi* are described in Land, Chapter IV, this Volume.

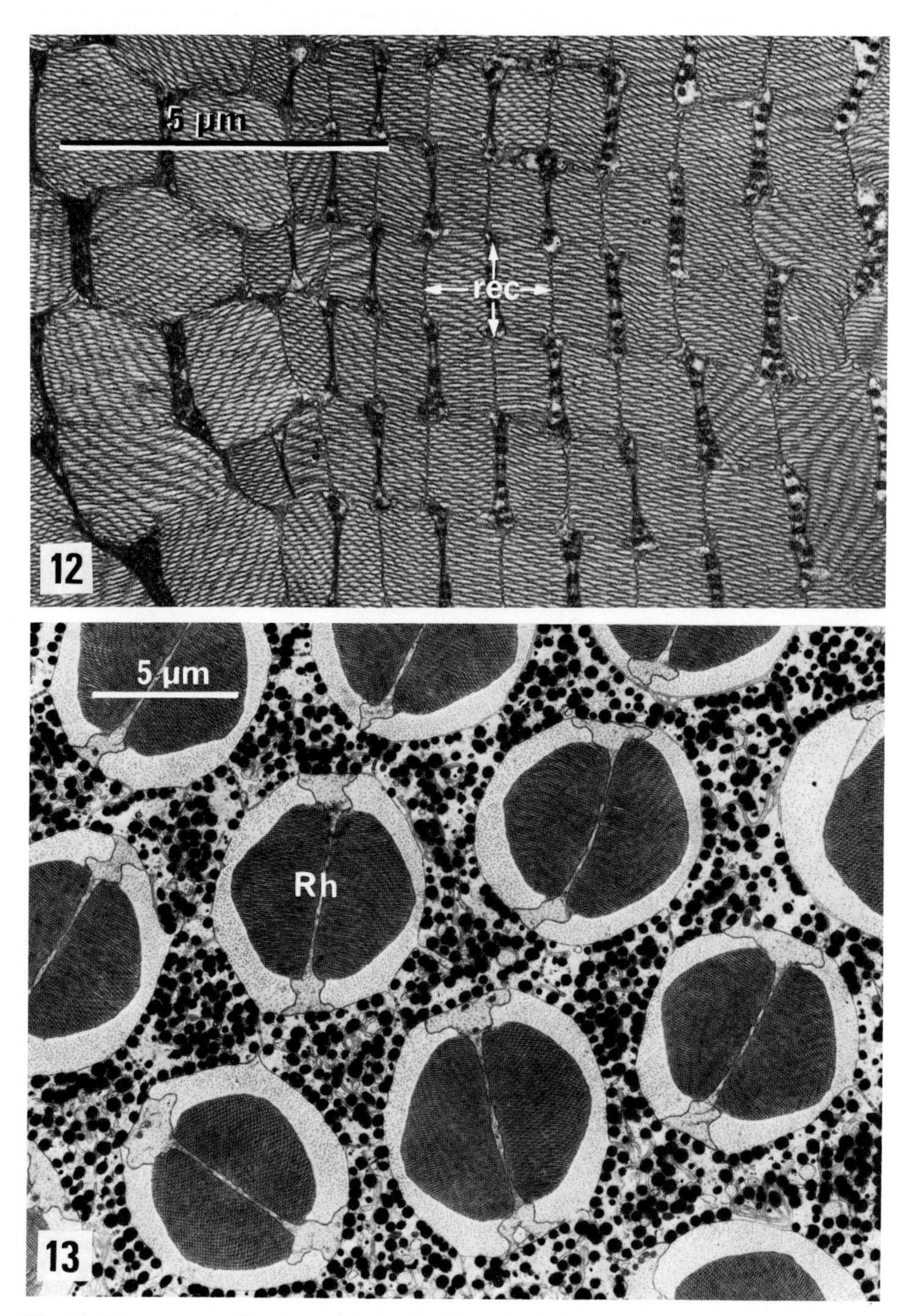

Fig. 12. Transverse section through layer II of the principal retina of *Lyssomanes* (Salticidae). *rec* limits of a single receptive segment with twin rhabdomeres

Fig. 13. Transverse section through an anterior lateral (secondary) eye of *Plexippus* (Salticidae). Each rhabdom (*Rh*) is surrounded by a sleeve composed of non-pigmented glia of lower refractive index, and the complex is ensheathed by six processes of the pigmented glia, densely packed with pigment granules

4.2 The Design and Evolution of Photoreceptors in the Salticidae

4.2.1 The Secondary Eyes

The secondary eyes of advanced Salticidae determine extremely precise orientation responses (Land 1969b). Rhabdoms consist of two rhabdomeres with hemicircular transverse profiles, separated from each other by insignificant cytoplasmic clefts (Fig. 13). The amount of material interposed between lens and receptive segment is reduced in the anterior lateral eyes by a displacement of the receptor somata to the margins of the retinal cup, and the bulk of the cytoplasm of the receptive segment and its organelles is situated proximally, below the tapered rhabdom (Eakin and Brandenburger 1971). Each rhabdom is isolated from its neighbours by two components: a refractive index barrier is created by two processes of the non-pigmented glia that contain almost no cytoplasmic elements other than microtubules and parallel the cylindrical shape of the rhabdom as a sleeve; surrounding these processes in turn are six processes of the pigmented glia which protect the rhabdoms against stray light (Eakin and Brandenburger 1971). Hardie and Duelli (1978) showed by intracellular recording that these receptors are of low absolute sensitivity, possess spectral sensitivities with maxima around 530–550 nm which contrast with the 510–520 nm recorded for the nocturnal *Dinopis* (Laughlin et al. 1980), and achieve excellent angular resolutions (see also Land, Chap. IV, this Vol.).

The posterior median eyes of advanced salticids are vestigial (Eakin and Brandenburger 1971), incapable of forming images, and make no detectable contribution to the spiders' visual fields (Land, Chap. IV, this Vol.). Their conservation as functional ocelli in the Spartaeinae is a major reason for supposing the subfamily to be primitive (Wanless 1984). Although the Lyssomaninae are usually regarded as primitive on other grounds, they, too, have vestigial posterior median eyes (Blest 1983). The secondary retinae of *Lyssomanes*, two genera of the Spartaeinae (*Yaginumanis* and *Portia*), and the advanced Salticidae present an interesting series: rhabdoms of *Lyssomanes* are not shielded by a refractive index barrier (Fig. 14) and for much of their lengths are virtually contiguous. Rhabdoms of *Yaginumanis* and *Portia* have rectangular transverse profiles, and the processes of the non-pigmented glia that surround them,

Fig. 14. Transverse section through two receptive segments of an anterior lateral (secondary) eye of *Lyssomanes* (Salticidae), to illustrate their close contiguity. *Rh* rhabdom; *gl* glial partitions

Fig. 15. Transverse section through a posterior median (secondary eye) of *Yaginumanis* (Salticidae). *Rh* rhabdom; *npg* non-pigmented glial process; *1–4* four processes of the pigmented glia, which in this species does not have pigment granules at the level of the receptive segments

Fig. 16. Transverse section through an anterior lateral (secondary) eye of *Portia* (Salticidae), to show that the arrangement of components is identical to that of *Yaginumanis*, except for the content of pigment granules in the four processes of the pigmented glia that surround each rhabdom

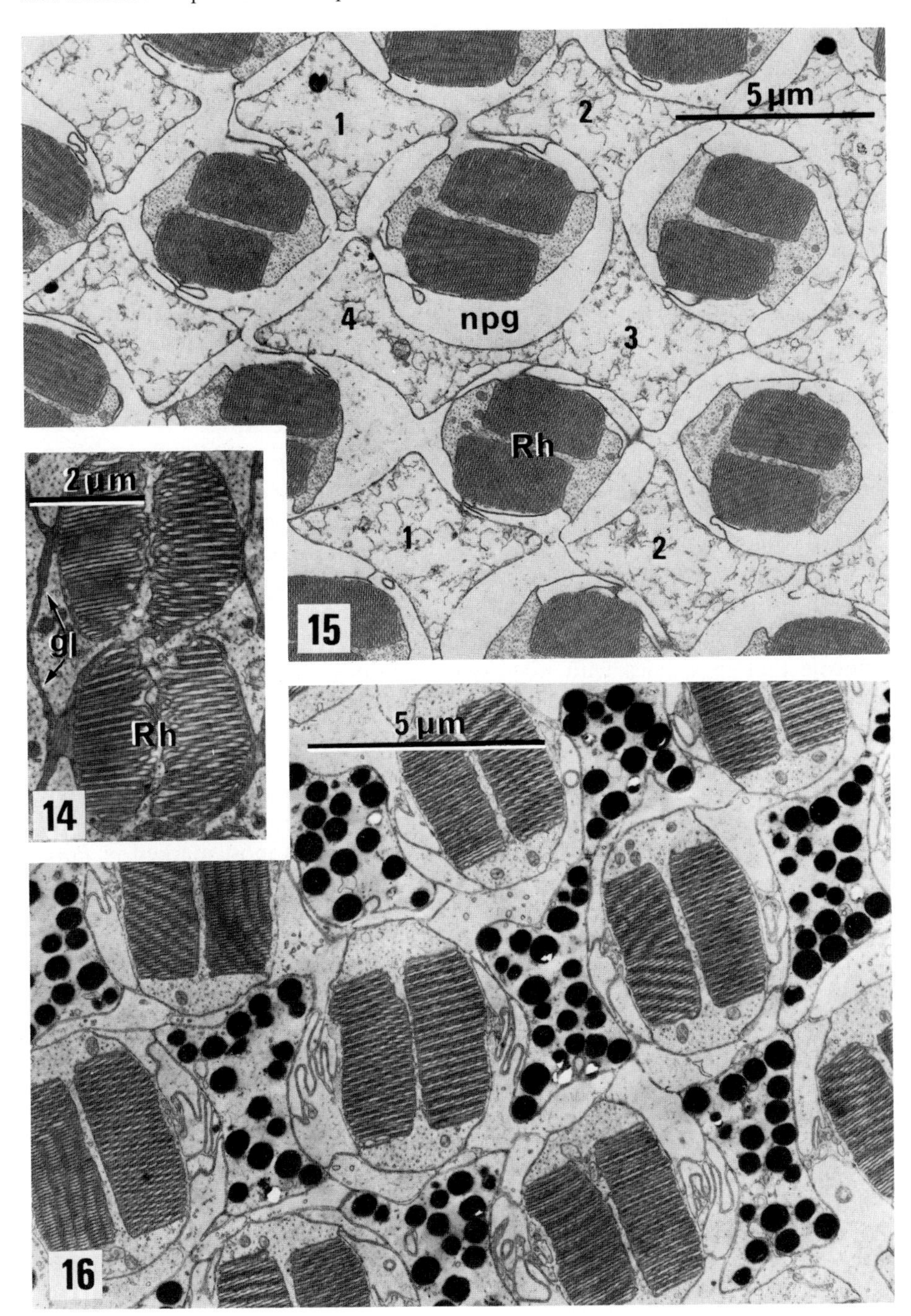
5 µm
1
2
4
npg
3
Rh
1
2
15
2 µm
gl
Rh
14
5 µm
16

although devoid of microtubules, are much divided distally, and would seem to provide a relatively inefficient refractive index barrier. In both species, an outer sleeve is provided by four processes of the pigmented glia: that of *Yaginumanis* is not equipped with pigment at the level of the receptive segments (Fig. 15); in *Portia*, each process contains large, rather scattered pigment granules (Fig. 16). Optically, the arrangement can only be suboptimal.

This small series can be regarded as representative of stages in the evolution of the sophisticated secondary eyes of advanced jumping spiders, although it must be emphasised that the taxonomic relationships between the three presumptively primitive genera are controversial and incompletely understood.

4.2.2 The Principal Eyes

Despite the diversity of the secondary eyes, the complex anatomy of the principal eyes is, in essence, fully developed and largely conserved throughout the Salticidae (Blest and Price 1984; Blest and Sigmund 1984). The tiering of the retinae to provide four layers, and its optical implications are described and illustrated in Land, Chapter IV, this Volume. Briefly: the telephoto optics of the dioptric system analysed by Williams and McIntyre (1980) and Blest et al. (1981) in conjunction with the ultrastructural anatomy (Blest and Price 1984) indicate that in species with high visual acuities (Jackson and Blest 1982a) layer I, farthest from the dioptrics, alone possesses a mosaic quality sufficient to sustain them (Fig. 17). Foveal layer I receptive segments each bear a single rhabdom on one face of a cell whose cytoplasm lacks organelles. Because rhabdoms are long and are isolated by a refractive index barrier, they both can and must act as light guides (Williams and McIntyre 1980; Blest and Price 1984).

Specialisation for high visual acuity at the fovea is most marked in *Portia* (Fig. 18), whose unusual ethology places a premium on the making of fine visual discriminations at substantial distances (Jackson and Blest 1982b; see also Forster, Chap. XIII, this Vol.). Microtubules have been eliminated from the receptor cytoplasm, perhaps as a device marginally to enhance the refractive index difference between rhabdom and surround. The 1.4-μm spacing between receptors is close to the physical limit below which there would be significant optical cross-talk between receptors. The long foveal rhabdoms can be supposed to optimise signal-noise ratios. Perhaps it is not surprising that these detailed adaptations are found in the receptors of a mosaic that in combination with its associated dioptrics achieves a minimal visual angle of 2.4 min of arc

Fig. 17. Transverse section through a principal retina of *Plexippus* (Salticidae) at the level of distal layer I, to indicate the region illustrated for *Portia* in Fig. 18. The anatomical outer side of the retina is at the *left*, and the mid-point is indicated by *arrows*. Note the "boomerang" conformation

Fig. 18. Transverse section through distal layer I of a principal retina of *Portia* (Salticidae). *Large arrows* indicate the foveal region of highest acuity. *RET* transverse limits of the field of receptors; *MAT* retinal matrix. Note the absence of microtubules in the cytoplasm of the receptive segments

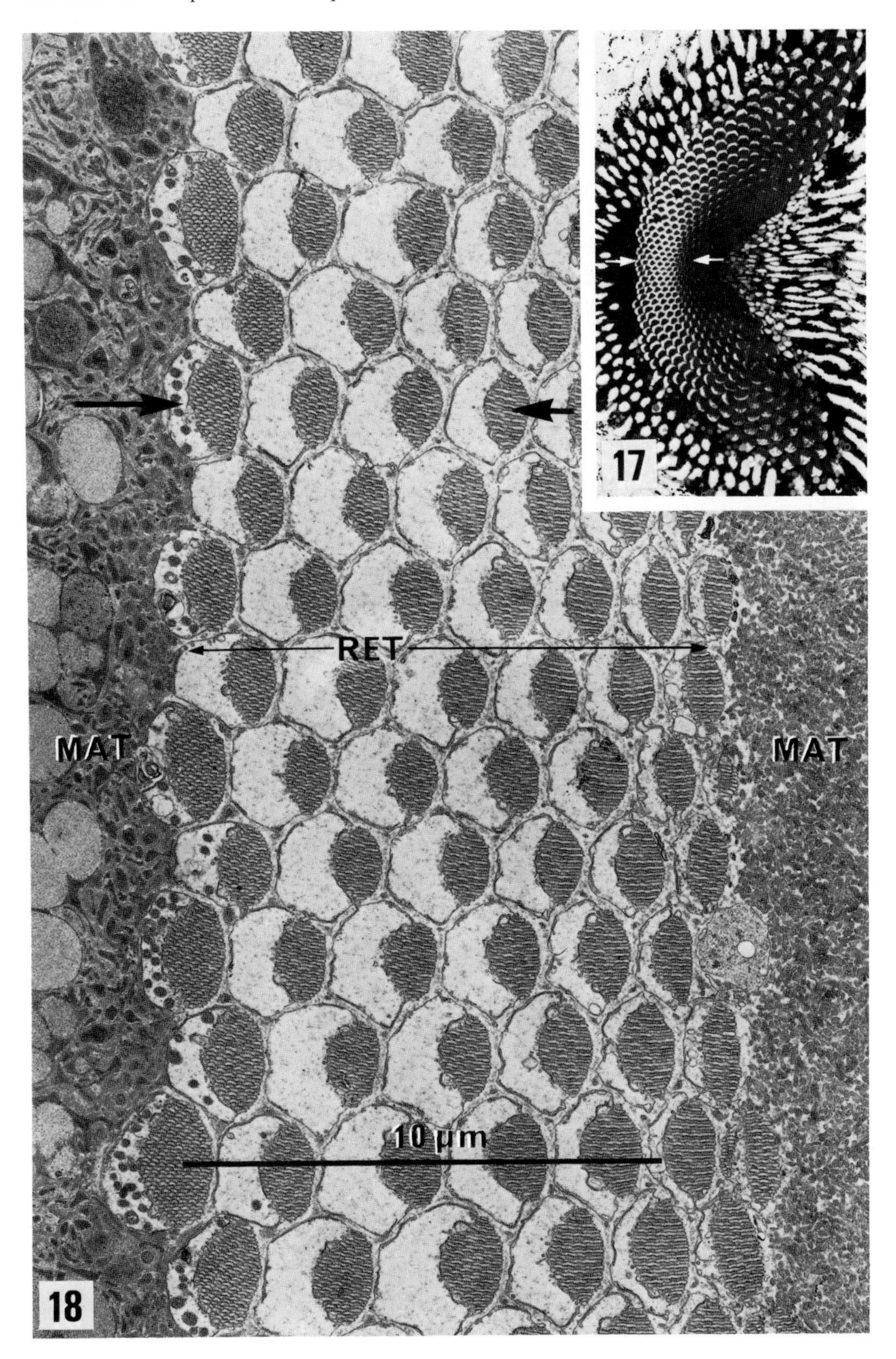
17
RET
MAT
MAT
10 µm
18

(Williams and McIntyre 1980), the highest spatial resolution that has been estimated for any invertebrate ocellus.

Space does not permit an account of layers II–IV: they are, however, of inferior mosaic quality, and present an interesting cellular adaptation. Degradation of images received by layer I would be significant if they were required to pass through three tiers of receptors with large refractive index differences between rhabdomeres and cytoplasm. The cytoplasm of layers II–IV is packed with long, tubular mitochondria (Figs. 6 and 7) which have been shown to equilibrate the cytoplasmic and rhabdomeral refractive indices (Williams and McIntyre 1980) and appear not to be metabolically active (Blest and Price 1984).

Portia, although primitive in many respects, is ethologically very specialised (Jackson and Blest 1982b) and cannot be taken as a paradigm of the Spartaeinae within which it is currently placed (Wanless 1984). Less specialised Spartaeinae (e.g. *Yaginumanis* and *Spartaeus*) and *Lyssomanes* have principal eyes whose organisation gives some hint as to the pathway by which high acuity retinae may have been evolved. A speculative model is discussed by Blest and Sigmund (1984). Here, discussion must be confined to the difficult matter of layers I and II which in *Plexippus* have been shown to have identical spectral peaks in the green at ca. 520 nm (Blest et al. 1981). Optical models that account for the properties of layer I fail to offer a convincing role for layer II, which must receive blurred images. Trivially, this can be held to account for the mostly poor mosaics of layer II, but evades an important question: why should there be two spectrally identical tiers in the first place?

In *Yaginumanis* and *Lyssomanes* foveal layer I mosaics are of poor quality: each receptive segment bears two rhabdomeres, and rhabdomers of adjacent receptors are closely contiguous, so that there must be substantial optical pooling. Layer II receptive segments are organised in the same way and are of equivalent mosaic quality (cf. Fig. 12). If we assume that in these forms layer I and II receptors have identical spectral sensitivities as they do in *Plexippus*, tiering of the two layers can be interpreted as Land (1969a) originally proposed as an alternative to colour vision extending into the red: focused images will be received by the two layers from objects at different distances so that tiering compensates for the inability of the visual system to accommodate. Blest and Sigmund (1984) model the evolution of the principal eye by assuming that this situation is phylogenetically primitive and was subsequently modified to allow the evolution of a mosaic in layer I able to sustain high visual acuities. The principal retina of *Spartaeus* (Blest and Sigmund, 1985) represents a putative intermediate stage in this posited transformation.

Interpretations of Salticid principal retinae are constrained by the paucity of data concerning the spectral sensitivities of the receptors. Using different species, De Voe (1975) and Blest et al. (1981) found only green and ultraviolet receptors, the latter belonging to marked cells in layer IV. Yamashita and Tateda (1976) found green, yellow, blue and ultraviolet receptors in *Menemerus* which they did not localise to layers. The resolution of these ambiguities is critical for optical models that seek to explain how Salticid retinae work, and how they may have evolved (see also Yamashita, Chap. VI, this Volume).

5 Taxonomic and Phylogenetic Implications of Retinal Ultrastructure

5.1 Taxonomic Implications of Retinal Structure

Homann's (1951, 1971) survey of retinal organisation throughout the spiders was conducted by light microscopy. The major phylogenetic conclusions of his seminal study remain unimpaired by subsequent work. At a more detailed anatomical level, ultrastructural analysis poses some interesting and so far unresolved questions:

1. To what extent should it be supposed that retinal organisations are conservative, and can be used as taxonomic characters?
2. Alternatively, to what extent can it be supposed that eyes are so subject to stringent selection pressures that they will always be modified and optimised to a degree that renders them useless as taxonomic characters?

As far as we know at present, basic retinal organisations are well-conserved within taxonomic groupings, forms conventionally regarded as phylogenetically primitive retaining receptors arranged as optically inefficient rhabdomeral networks; in these groups adaptive radiation is probably achieved primarily by modifications to the dioptrics, and by manipulations of the exact dispositions and functional roles of the tapeta underlying the receptors (not dealt with in this account).

The two following cases introduce a note of caution and suggest that the ultrastructure of spider photoreceptors can be subject to radical modification by rapid selection.

5.2 The Phylogenetic Lability of Photoreceptor Mosaics

The posterior median eyes of Dinopid spiders in the genera *Dinopis* and *Menneus* provide a pertinent example of the dangers of inferring too much from differences of retinal ultrastructure alone in a systematic context. The anatomy of the receptive segments is given in Figs. 19 and 20.

In *Dinopis*, the big posterior median eyes are specialised for nocturnal vision (Blest and Land 1977; Blest 1978; Laughlin et al. 1980). The receptors present a rhabdomeral network with large (20 – 22 μm) centre-to-centre spacings between receptive segments, each of which in the day state bears rhabdomeres disposed as a thin peripheral rind whose microvilli interdigitate with those of the photoreceptors that flank it. In the night state, the receptive segments lengthen by the addition of more rhabdomeral membrane throughout: the optical consequences are noted under Section 3.1 above.

A closely related genus, *Menneus*, possesses small posterior median eyes whose anatomy and optics are discussed by Blest et al. (1980). They have f-numbers of 0.72 (compared to 0.58 for *Dinopis*); the retinal illuminances of *Menneus*:*Dinopis* have a ratio of 1:1.54. However, the receptive segments of *Menneus* are disproportionately small, so that the relative illuminance of single receptors becomes a ratio of ca. 1:38. If this ratio is extrapolated by including

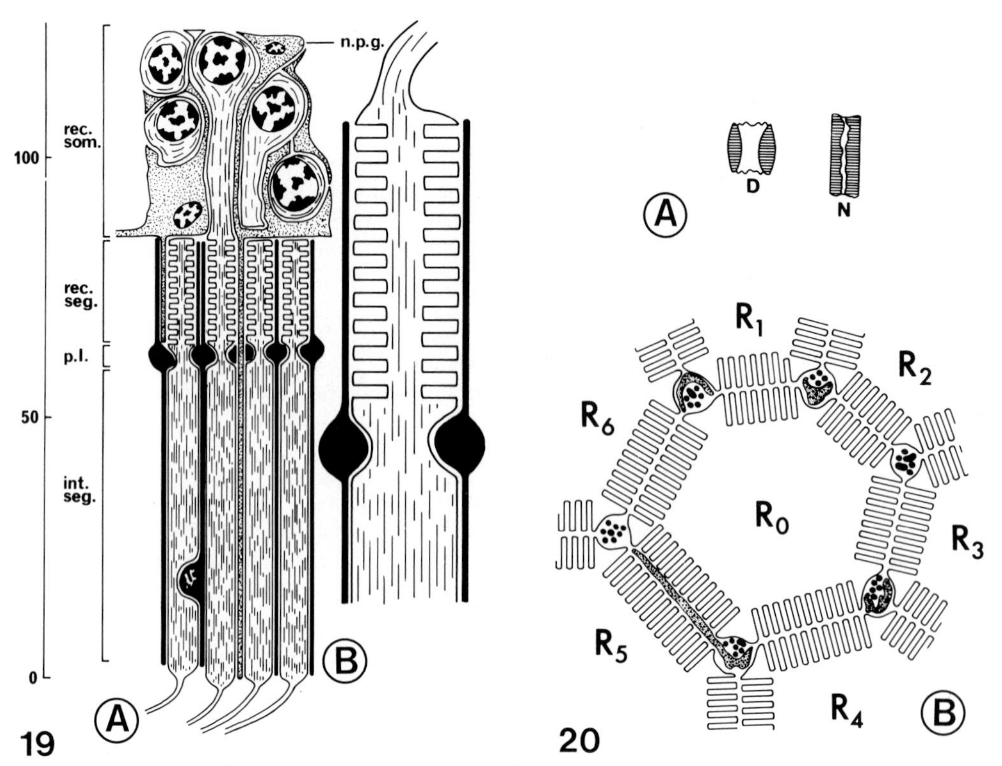

Figs. 19, 20. Comparison of the retinal architectures of the posterior median eyes of *Menneus* (**A**) and *Dinopis* (**B**), displayed by schematic longitudinal (**19**) and transverse (**20**) sections, the latter taken at the level of the receptive segments. The smaller components are those of *Menneus*, structures being represented in the day state, with the exception of one of the transverse profiles of a *Menneus* receptive segment in the night state given in **20 A** for comparison. *n.p.g.* non-pigmented glia; *rec. som.* receptor somata; *rec. seg.* receptive segments; *p.l.* pigment layer; *int. seg.* intermediate segment; *D* day state; *N* night state; $R_0 - R_6$ receptors. Scale at *left* of **19** in microns

the relative lengths of the receptors in the calculation, it seems probable that the relative photon capture efficiency for *Menneus:Dinopis* amounts to a ratio of ca. 1:100.

This difference in presumptive photon capture efficiency (which necessarily relates to sensitivity), amounts to some two orders of magnitude derived from differences in optical design alone, and can be related to the ecology of the two genera. Austin and Blest (1979) found that in the same habitat, *Dinopis* only becomes active and constructs a snare after nightfall; *Menneus* does so before dusk.

None of this accounts for the radical distinction between the anatomy of the photoreceptive segments in the two genera: *Dinopis* has a rhabdomeral network, whilst *Menneus* has receptive segments each with two rhabdomeres that are not contiguous with those of adjacent receptors (Figs. 19 and 20). Were it not for the difference in size between the posterior median eyes of the two genera, *Dinopis* and *Menneus* might well be treated by taxonomists as congeneric, yet one has a receptor organisation associated with relatively primitive families with poor vision, the other receptors which resemble those of partially vision-dependent families such as the Lycosidae.

A second example concerns *Supunna,* an Australian genus of diurnally active spiders currently placed in the Clubionidae. Our unpublished data show that the secondary eyes possess vestigial and non-functional tapeta; they are overlain by receptor mosaics that bear no relationship to the rhabdomeral networks of the Clubionidae and the related Gnaphosidae, but seem, instead, to be irregular equivalents of the gridiron receptor mosaics of phylogenetically more advanced spiders such as the Lycosidae and Thomisidae. There is little to indicate whether this circumstance means that retinal architectures are more labile than has been supposed, or whether *Supunna* has been assigned to an inappropriate family. Further examination of the eyes of spiders will doubtless reveal other anomalies.

6 Conclusions

The foregoing account demonstrates that the ocelli of spiders contain a range of retinal types: on the one hand, optically inefficient rhabdomeral networks; or, at another extreme, precise mosaics serving superior dioptrics in such a way that the eyes of some salticids can plausibly be supposed to be diffraction-limited (Williams and McIntyre 1980; Blest et al. 1981). The latter ideal state has not been achieved by any compound eye known to us. We are left with the problem posed by Laughlin et al. (1980): what advantage does the compound eye present over the spider ocellus, given what the latter can be shown to have achieved? Why have so few spiders developed ocelli to their best advantage? There would appear to be many reasons, and it is beyond the scope of a short review to discuss them. The diversity of spider ocelli in terms of their functional designs should persuade us that they are worthy of more detailed physiological analysis, despite the formidable technical difficulties of doing so.

Acknowledgements. I am indebted to successive technical assistants for the skillful preparation of thin sections: Ling Kao, Karen Powell, Joanne Maples, Dean Price, Wendy Eddey and Claudia Sigmund. Dr. Sally Stowe prepared the freeze-etch replica illustrated in Fig. 3, and with Drs. M. F. Land, P. McIntyre, S. B. Laughlin and D. S. Williams provided many stimulating discussions. Gary Brown prepared the line drawings.

References

Austin AD, Blest AD (1979) The biology of two species of Australian dinopid spider. J Zool (London) 189:145–156

Bacetti B, Bedini C (1964) Research on the structure and physiology of eyes of a lycosid spider. I. Microscopic and ultramicroscopic structure. Arch Ital Biol 102:97–122

Barlow RB, Bolanowski SJ, Brachman ML (1977) Efferent optic nerve fibres mediate circadian rhythms in the *Limulus* eye. Science 197:86–89

Blest AD (1978) The rapid synthesis and destruction of photoreceptor membrane by a dinopid spider. Proc R Soc London Ser B 200:463–483

Blest AD (1980) Photoreceptor membrane turnover in arthropods: comparative studies of breakdown processes and their implications. In: Williams TP, Baker BN (eds) The effects of constant light on visual processes. Plenum Press, New York, pp 217–245

Blest AD (1983) Ultrastructure of secondary retinae of primitive and advanced jumping spiders (Araneae, Salticidae). Zoomorphology 102:125–141

Blest AD (1984) Ultrastructure of the secondary eyes of a primitive jumping spider, *Yaginumanis* (Araneae, Salticidae, Spartaeinae). Zoomorphology 104:223–225

Blest AD, Day WA (1978) The rhabdomere organisation of some nocturnal pisaurid spiders in light and darkness. Philos Trans R Soc London Ser B 283:1–23

Blest AD, De Couet HG, Sigmund C (1983) The microvillar cytoskeleton of leech photoreceptors: a stable bundle of actin microfilaments. Cell Tissue Res 234:9–16

Blest AD, Eddey W (1984) The extrarhabdomeral cytoskeleton in photoreceptors of Diptera: II. Plasmalemmal undercoats. Proc R Soc London Ser B 220:353–359

Blest AD, Hardie RC, McIntyre P, Williams DS (1981) The spectral sensitivities of identified receptors and the function of retinal tiering in the principal eyes of a jumping spider. J Comp Physiol 145:227–239

Blest AD, Kao L, Powell K (1978) Photoreceptor membrane breakdown in the spider *Dinopis:* the fate of rhabdomere products. Cell Tissue Res 195:425–444

Blest AD, Land MF (1977) The physiological optics of *Dinopis subrufus:* a fish lens in a spider. Proc R Soc London Ser B 196:197–222

Blest AD, Maples J (1979) Exocytotic shedding and glial uptake of photoreceptor membrane by a salticid spider. Proc R Soc London Ser B 204:105–112

Blest AD, Powell K, Kao L (1978) Photoreceptor membrane breakdown in the spider *Dinopis:* GERL differentiation in the intermediate segments. Cell Tissue Res 195:277–297

Blest AD, Price GD (1984) Retinal mosaics of the principal eyes of some jumping spiders (Salticidae: Araneae): adaptations for high visual acuity. Protoplasma 120:172–184

Blest AD, Price GD, Maples J (1979) Photoreceptor membrane breakdown in the spider *Dinopis:* localisation of acid phosphatases. Cell Tissue Res 199:455–472

Blest AD, Sigmund C (1984) Retinal mosaics of two primitive jumping spiders, *Yaginumanis* and *Lyssomanes* (Araneae, Salticidae): clues to the evolution of Salticid vision. Proc R Soc London Ser B 221:111–125

Blest AD, Sigmund C (1985) Retinal mosaics of a primitive jumping spider, *Spartaeus* (Araneae: Salticidae: Spartaeinae): a phylogenetic transition between high and low visual acuities. Protoplasma (in press)

Blest AD, Stowe S, De Couet HG (1984) Turnover of photoreceptor membranes in arthropods. Science Prog, Oxford 69:83–100

Blest AD, Stowe S, Eddey W (1982) A labile, Ca^{2+}-dependent cytoskeleton in rhabdomeral microvilli of blowflies. Cell Tissue Res 223:553–573

Blest AD, Stowe S, Eddey W, Williams DS (1982) The local deletion of a microvillar cytoskeleton from photoreceptors of tipulid flies during membrane turnover. Proc R Soc London Ser B 215:469–479

Blest AD, Williams DS, Kao L (1980) The posterior median eyes of the dinopid spider *Menneus.* Cell Tissue Res 211:391–403

Chamberlain SC, Barlow RB (1979) Light and efferent activity control rhabdom turnover in *Limulus* photoreceptors. Science 206:361–363

De Couet HG, Blest AD (1982) The retinal acid phosphatase of a crab, *Leptograpsus:* characterisation, and relation to the cyclical turnover of photoreceptor membrane. J Comp Physiol 149:353–362

De Couet HG, Stowe S, Blest AD (1984) Membrane-associated actin in the rhabdomeral microvilli of crayfish photoreceptors. J Cell Biol 98:834–846

DeVoe RD (1975) Ultraviolet and green receptors in principal eyes of jumping spiders. J Gen Physiol 66:193–208

Eakin RW, Brandenburger JL (1971) Fine structure of the eyes of jumping spiders. J Ultrastruct Res 37:618–663

Eguchi E, Waterman TH (1967) Changes in retina fine structure induced in the crab *Libinia* by light and dark adaptation. Z Zellforsch 79:202–229

Eguchi E, Waterman TH (1976) Freeze-etch and histochemical evidence for cycling in crayfish photoreceptor membranes. Cell Tissue Res 169:419–434

Fleissner G, Schliwa M (1977) Neurosecretory fibres in the median eyes of the scorpion, *Androctonus australis* L. Cell Tissue Res 178:189–198

Hardie RC, Duelli P (1978) Properties of single cells in posterior lateral eyes of jumping spiders. Z Naturforsch 33c:156–158

Homann H (1951) Die Nebenaugen der Araneen. Zool Jahrb Anat 71:56–144

Homann H (1971) Die Augen der Araneae. Z Morph Tiere 69:201–272

Jackson RR, Blest AD (1982a) The distances at which a primitive jumping spider makes visual discriminations. J Exp Biol 97:441–445

Jackson RR, Blest AD (1982b) The biology of *Portia fimbriata*, a web-building jumping spider (Araneae: Salticidae) from Queensland: utilisation of webs and predatory versatility. J Zool (London) 196:255–293

Kirschfeld K, Franceschini N (1969) Ein Mechanismus zur Steuerung des Lichtflusses in den Rhabdomeren des Komplexauges von *Musca*. Kybernetik 6:13–22

Land MF (1969a) Structure of the retinae of the eyes of jumping spiders (Salticidae: Dendryphantinae) in relation to visual optics. J Exp Biol 51:443–470

Land MF (1969b) Movements of the retinae of jumping spiders (Salticidae: Dendryphantinae) in response to visual stimuli. J Exp Biol 51:471–493

Land MF (1980) Optics and vision in invertebrates. In: Autrum H (ed) Handbook of sensory physiology, vol VII/6B. Springer, Berlin Heidelberg New York, pp 471–592

Land MF (1984) The resolving power of diurnal superposition eyes measured with an ophthalmoscope. J Comp Physiol 154:515–533

Laughlin SB (1980) Neural principles in the visual system. In: Autrum H (ed) Handbook of sensory physiology, vol VII/6B. Springer, Berlin Heidelberg New York, pp 133–280

Laughlin SB, Blest AD, Stowe S (1980) The sensitivity of receptors in the posterior median eye of the nocturnal spider *Dinopis*. J Comp Physiol 141:53–66

Melamed J, Trujillo-Cenóz O (1966) On the fine structure of the visual system of *Lycosa* (Araneae, Lycosidae). I. Retina and optic nerve. Z Zellforsch Anat 74:12–31

Nassel DR, Waterman TH (1979) Massive, diurnally modulated photoreceptor membrane turnover in crab light and dark adaptation. J Comp Physiol 131:205–216

Schinz RH, Lo M-V, Larrivee DC, Pak W (1982) Freeze-fracture study of the *Drosophila* photoreceptor membrane. J Cell Biol 93:961–969

Snyder AW (1979) The physics of vision in compound eyes. In: Autrum H (ed) Handbook of sensory physiology, vol VII/6A. Springer, Berlin Heidelberg New York, pp 225–313

Stowe S (1980) Rapid synthesis of photoreceptor membrane and assembly of new microvilli in a crab at dusk. Cell Tissue Res 211:419–440

Stowe S (1981) Effects of illumination changes on rhabdom synthesis in a crab. J Comp Physiol 142:19–25

Tsacopoulos M, Orkand RK, Coles JA, Levy S, Poitry S (1983) Oxygen uptake occurs faster than sodium pumping in bee retina after a light flash. Nature (London) 301:604–606

Uehara A, Toh Y, Tateda H (1978) Fine structure of the eyes of orb-weavers, *Argiope amoena* L. Koch (Araneae: Argiopidae). 2. The anterolateral, posterolateral and posteromedial eyes. Cell Tissue Res 186:435–452

Wanless FR (1984) A review of the spider subfamily Spartaeinae nom. nov. (Araneae: Salticidae) with descriptions of six new genera. Bull Br Nat Hist (Zool) 46 (2):135–205

White RH (1968) The effect of light deprivation upon the ultrastructure of the larval mosquito eye. II. The rhabdom. J Exp Zool 166:405–425

White RH, Lord E (1975) Diminution and enlargement of the mosquito rhabdom in light and darkness. J Gen Physiol 65:583–598

Widmann E (1908) Über den feineren Bau der Augen einiger Spinnen. Z Wiss Zool 90:258–312

Williams DS (1979) The physiological optics of a nocturnal, semi-aquatic spider, *Dolomedes aquaticus* (Pisauridae). Naturforscher 34c:463–469

Williams DS (1982) Ommatidial structure in relation to turnover of photoreceptor membrane in the locust. Cell Tissue Res 225:595–617

Williams DS (1983) Changes of photoreceptor performance associated with the daily turnover of photoreceptor membrane in locusts. J Comp Physiol 150:509–515

Williams DS, McIntyre P (1980) The principal eyes of a jumping spider have a telephoto component. Nature (London) 288:578–580

Wilson M (1978) The functional organisation of locust ocelli. J Comp Physiol 124:297–316

Whittle AC (1976) Reticular specialisations in photoreceptors: a review. Zool Scr 5:191–206

Yamashita S, Tateda H (1976) Spectral sensitivities of jumping spider eyes. J Comp Physiol 105:1–8

Yamashita S, Tateda H (1981) Efferent neural control in the eyes of orb-weaving spiders. J Comp Physiol 143:477–483

Yamashita S, Tateda H (1983) Cerebral photosensitive neurons in the orb-weaving spiders, *Argiope bruennichii* and A. *amoena*. J Comp Physiol 150:467–462

VI Photoreceptor Cells in the Spider Eye: Spectral Sensitivity and Efferent Control

SHIGEKI YAMASHITA

CONTENTS

1 Introduction

Most species of spiders have four pairs of simple eyes arranged in two rows, anterior and posterior, in the frontal part on the prosoma. These eyes are referred to as the anterior median, anterior lateral, posterior median, and posterior lateral eyes. The anterior median eyes are generally referred to as principal eyes and the other three pairs as secondary eyes.

The light responses of the eyes have been recorded electrophysiologically in wolf spiders (DeVoe 1962, 1967a, b, 1972; DeVoe et al. 1969), jumping spiders (DeVoe 1975; Yamashita and Tateda 1976a, b, 1982; Hardie and Duelli 1978; Blest et al. 1981), orb-weaving spiders (Yamashita and Tateda 1978, 1981, 1983) and net-casting spiders (Laughlin et al. 1980). These studies have revealed that spider eyes possess more complex functions than hitherto assumed.

Biological Laboratory, Kyushu Institute of Design, Shiobaru, Fukuoka 815, Japan

Yamashita and Tateda (1981) showed that efferent axons in the optic nerve control the photoreceptor cells in orb-weaving spiders. They (1983) found that the efferent neurons are directly sensitive to light, and that the responses of these neurons are affected significantly by illumination of the eyes. In this chapter, attention is directed to the spectral sensitivities of the eyes, the cerebral photosensitive neurons (the efferent neurons), and the neural interaction between the photoreceptors and the cerebral photosensitive neurons of orb-weaving spiders.

2 Spectral Sensitivities

2.1 Anterior Median Eyes

The spectral sensitivities of the receptor cells in the anterior median eyes of the orb-weaving spiders, *Argiope bruennichii* and *A. amoena* were examined by intracellular recordings of receptor potentials. Light stimulation elicited a depolarizing receptor potential in all cases. The anterior median eyes had three types of receptor cells, UV cells with maximum sensitivities at about 360 nm, blue cells at 480 – 500 nm, and green cells at 540 nm (Fig. 1 A). The average intensity-response curves of these cells in the dark-adapted state are shown in Fig. 1 B. In the dark-adapted state, the sensitivity of the blue cell is higher than that of the green and the UV cells. On the other hand, in the light-adapted state, the sensitivity of the blue cell was lower than that of the green and UV cells (Fig. 2).

As can be seen in Fig. 1 A, each curve has a small secondary peak in the ultraviolet or visible region. Furthermore, spectral sensitivity curves which had two or three peaks were sometimes obtained (Fig. 3). In such cases, however, the spectral sensitivities often changed with time following impalement, i.e., the spectral sensitivity curves obtained immediately following impalement had single narrow peaks, but those obtained several minutes later had two or three broad peaks. These observations suggest that a certain amount of damage occurs to the photoreceptor cell at the site of microelectrode insertion, and that the high secondary spectral sensitivity peak may be the result of some sort of artificial coupling. Thus, in orb-weaving spiders, it probably cannot be concluded that there are electrical interactions among different receptor cells or a mixture of two or even three different visual pigments within a cell.

The recovery processes following illumination of the less sensitive cells (UV and green cells) and those of the more sensitive cells (blue cells) differed. Figure 2 shows two examples, one obtained from a green and the other from a blue cell. Initially, the control light, serving also as the test light, was presented. This was followed by the conditioning light and after various time intervals by the test light. The amplitude of the response of the green cell to the control light was about 7 mV. Ten seconds after the conditioning stimulus, the response to the test light increased (about 8 mV). This increase in response to the test stimulus reflects an increase in sensitivity of the photoreceptor cells following illumination rather than during a complete adaptation to the dark (Yamashita

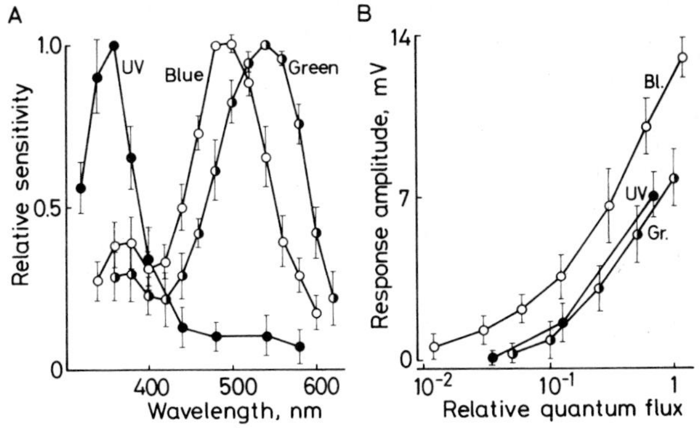

Fig. 1 A, B. The average spectral sensitivity curves (**A**) and the average intensity response curves (**B**) of three UV, two blue, and four green cells in the dark-adapted anterior median eyes of *Argiope bruennichii* and *A. amoena*. The intensity response curves were determined at 360 nm for the UV, at 480 nm for the blue and at 540 nm for the green cells. *Vertical lines* indicate the standard deviation. (Yamashita and Tateda 1978)

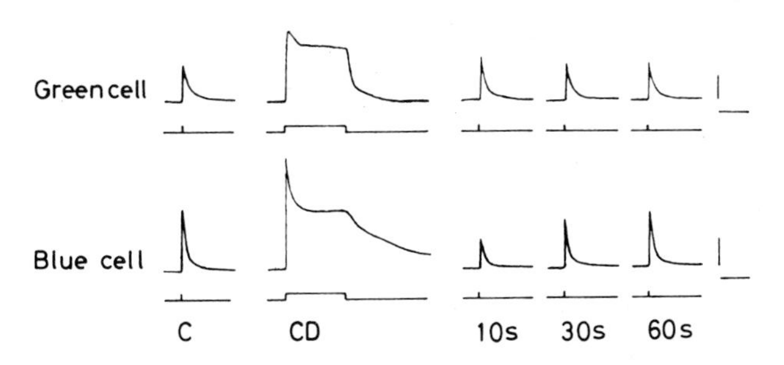

Fig. 2. Increased (green cell) and decreased (blue cell) responses following a light stimulus of 1 s duration (*Argiope*). The control light (*C*) of 2.5 ms duration was presented to the dark-adapted eye. The test lights were presented to the eye 10 s, 30 s and 60 s following the conditioning stimulus (*CD*). Calibrations: 5 mV, 0.5 s (Yamashita and Tateda 1978)

and Tateda 1976b, 1978, 1982). The response of the blue cell to the control light was about 10 mV. Ten seconds after the conditioning stimulus, the response to the test light was very small, and recovery to the dark-adapted level was gradual.

2.2 Posterior Lateral Eyes

All recorded receptor cells, except for one cell in the posterior lateral eyes, showed peak sensitivities at about 480–500 nm and/or at about 540 nm. The exceptional cell had a peak sensitivity at about 360 nm and a secondary peak at about 500 nm. Figure 3 shows the average spectral sensitivity curves for single-peaked blue and green cells, and the double-peaked "UV" cell. The peak sensitivities of these cells are similar to those of the UV, blue, and green cells in the anterior median eyes. Thus, it is reasonable to conclude that the posterior lateral eyes also have three types of visual cell, as do the anterior median eyes.

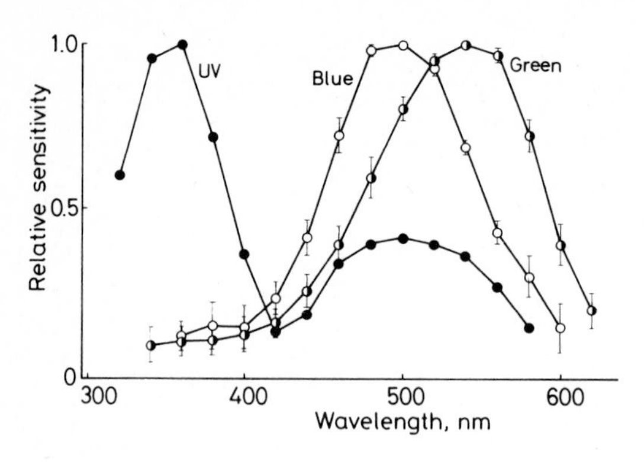

Fig. 3. The average spectral sensitivity curves of three blue cells, five green cells, and one "UV" cell in the dark-adapted posterior lateral eyes of *Argiope*

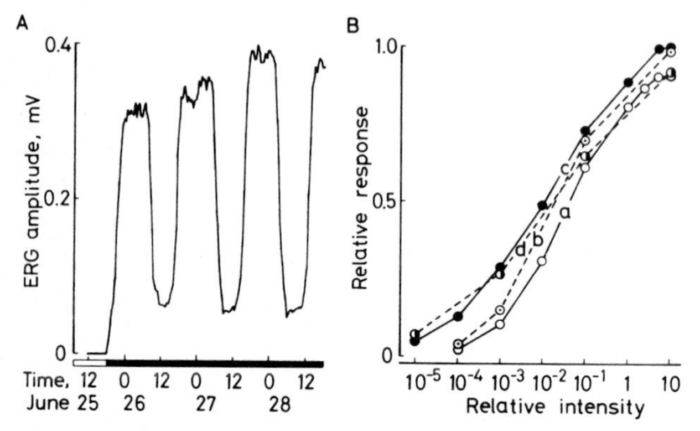

Fig. 4. **A** Circadian changes in ERG amplitude for the anterior median eye (*Argiope*). The spider was collected from the open field in the morning and recording was begun at 11:30 h under background illumination. After cessation of background illumination at 18:00 h, ERG's were recorded under constant darkness for 3 days. Throughout these periods, a light flash of 10 ms duration was automatically presented every 2 min. (Yamashita and Tateda 1978). **B** Circadian changes in ERG intensity-response curves determined at about 18:00 h (curve *a*), 19:30 h (*b*), 22:30 h (*c*) and 7:30 h (*d*). (Yamashita and Tateda 1981)

3 Efferent Signals in the Optic Nerve

3.1 Circadian Rhythm

The eyes of the intact orb-weaving spiders show a circadian oscillation of sensitivity (Fig. 4A and B). In Fig. 4A, the amplitude of the ERG from the anterior median eye changes with a period of approximately 22 h, under conditions of constant darkness. ERG intensity-response curves of the anterior median eye under constant darkness were determined at various times during "low" and "high" sensitive states (Fig. 4B). In this figure, four series of responses obtained at about 18:00 h (curve a in Fig. 4B), 19:30 h (b), 22:30 h (c) and 07:30 h (d) are plotted. The intensity-response curve determined at about 18:00 h is typical for the "low" sensitive state. At about 19:30 h just after the threshold begins to increase, the saturated value markedly increases, i.e. the slope of the intensity-response curve is the steepest at that time. At about 22:30 h, the threshold is more than 10 times lower and the saturated value is about 1.1 times higher than during the "low" sensitive state. At about 07:30 h, while the threshold is kept at a low level, the saturated value markedly decreases, i.e. the slope of the intensity-response curve is the slightest at that time. Subsequently, there was a reversion of intensity-response curves for the "low" sensitive state. These results show that changes in the saturated value of the light response which suggest changes in the gain of the photoreceptors are rapid and transient, and that changes in the threshold which suggest changes in the quantum catch are slow and sustained.

The spectral sensitivities of the anterior median eyes during the *high* and *low* sensitive states were studied by recording ERG's from intact eye. Figure 5 A shows spectral response curves for the high and low sensitive states, under conditions of constant darkness, and a spectral response curve for the high sensitive state during dim white light adaptation. During the "high" sensitive state under dark conditions, the peak responses are observed at 360 nm in the ultraviolet and at 480 – 540 nm in the visible. During the low sensitive state, the ERG amplitude to each monochromatic light decreases, as compared with the ERG's for the high sensitive state. The decrease in the ERG amplitude is large in the blue, but small in the ultraviolet and the red. Consequently, peak responses for the low sensitive state are observed at 360 nm and at 540 nm. These results suggest that the blue cells have a circadian oscillation of sensitivity. If such is indeed the case, the same spectral response curve for the low sensitive state under constant darkness should be obtained for the high sensitive state by a moderate adaptation of the blue cells. To test this hypothesis, ERG's were recorded during the high sensitive state under various intensities of dim light. As can be seen in Fig. 5 A, the spectral response curve for the "high" sensitive state during dim light adaptation was all but coincident with that for the low sensitive state in the dark. It is, therefore, concluded that the blue cells show a circadian oscillation of sensitivity.

The spectral response curves for the posterior lateral eye during the high and low sensitive states under constant darkness were similar to those for the anterior median eye, except that the curves for the posterior lateral eye showed no obvious peak in the ultraviolet. Thus, it is likely that the blue cells in the posterior lateral eye also show a circadian oscillation of sensitivity.

The circadian sensitivity change of the eyes may be mediated by efferent optic nerve signals. If this is true, the circadian change should be lost after cutting the optic nerve. Figure 5 B shows the ERG spectral response curves for the anterior median eye with intact and cut optic nerves in the same preparation in situ. During the most sensitive state, peak responses for the eye with the intact optic nerve (right AM) are observed at 360 nm in the ultraviolet and at 480 – 540 nm in the visible. For the eye with the cut optic nerve (left AM), the ERG amplitude to each monochromatic light is small, as compared to the ERG for the eye with the intact optic nerve. The difference of the ERG amplitude is large in the blue, but small in the ultraviolet and the red. The spectral response curves for the eyes with intact and cut optic nerves were very similar to those

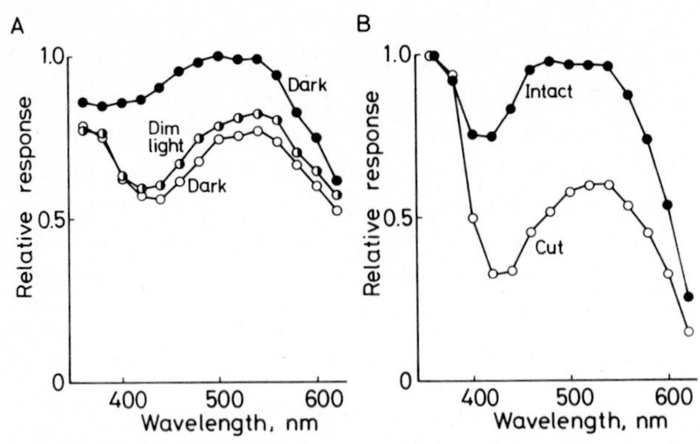

Fig. 5. A The spectral response curves (ERG) for the circadian states of high (*closed circles*) and low (*open circles*) sensitivity under constant darkness, and that for the "high" sensitive state during dim light adaptation (*half-closed circles*) (*Argiope*). The relative amplitude of response to a monochromatic light stimulus of equal quanta is plotted against the stimulus wavelength. The magnitude of the response to a 480 nm flash for the "high" sensitive state in the dark is referred to as 1.0. (Yamashita and Tateda 1978.) **B** Effects of cutting the optic nerve on the spectral response curves (ERG). The optic nerve for the left anterior median eye (AM) was cut (*open circles*) and that for the right AM was intact (*closed circles*). Both curves are normalized to a relative response of 1.0 at 360 nm. (Yamashita and Tateda 1981)

for the circadian high and low sensitive states, respectively (Fig. 5A). Electrical stimulation applied to the distal end of the cut optic nerve produced effects on the ERG spectral responses of the anterior median eye similar to endogenous efferent activity, i.e., the spectral response curves determined before and 80–100 min after the onset of electrical stimulation were also similar to those for the circadian high and low sensitive states, respectively. Thus, it is concluded that efferent signals in the optic nerve mediate the circadian sensitivity change of the photoreceptor cells, especially of the blue cells.

The presence of efferent optic nerve fibers which control the circadian sensitivity change in the eyes has been also reported for *Limulus* (Barlow et al. 1977; Chamberlain and Barlow 1977; Kaplan and Barlow 1980) and for scorpions (Fleissner and Schliwa 1977; Fleissner and Fleissner 1978; see Fleissner, Chap. XVIII, this Vol.). In both *Limulus* (Fahrenbach 1969, 1973) and scorpions (Fleissner and Schliwa 1978; Fleissner and Heinrichs 1982), the efferent fibers are neurosecretory. It is reasonable to consider that efferent fibers in the optic nerves of spiders are neurosecretory as well, since they have several morphological and physiological similarities to neurosecretory fibers (Melamed and Trujillo-Cenóz 1966; Yamashita and Tateda 1981, Uehara, personal communication).

3.2 Illumination of the Brain

Efferent optic nerve activity in response to light stimulation of the brain was studied in isolated preparations consisting of the brain and optic nerve. In most

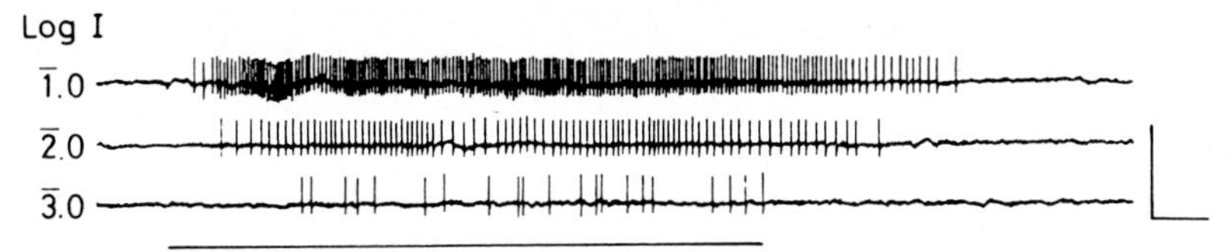

Fig. 6. Efferent optic nerve impulses in response to white light stimulation of the brain (10 s duration at various intensities) obtained from an isolated brain-optic nerve preparation after blocking the synaptic action by the addition of 2 mM Co^{2+} to the saline (*Argiope*). *Horizontal bar* duration of illumination. Calibrations: 0.2 mV, 1 s. (Yamashita and Tateda 1983)

cases, recording from the optic nerve in the dark gave rhythmic bursts of impulses (Fig. 8). However, impulse frequencies in the dark varied from preparation to preparation. For example, while no impulse was observed in one case about 10 imp s^{-1} was observed in another. Frequencies of over 20 imp s^{-1} were obtained within the bursts. The efferent optic nerve activity increased during illumination of the brain, showing that certain photosensitive neurons are present in the brain. To determine whether or not the efferent neurons were directly photosensitive, efferent impulses were recorded after blocking chemical transmission by adding 2 mM Co^{2+} to the saline bath. Studies have shown these cations to be potent blockers of transmission at chemical synapses (e.g., Weakly 1973). In Co^{2+}-containing saline, rhythmic bursts of impulses characteristic of the dark-adapted state disappeared completely. Illumination of the brain, however, elicited efferent activity consisting of continuous discharges of impulses (Fig. 6). The impulse frequency increased with increasing light intensity. These observations show that the efferent neurons are directly sensitive to light and receive inputs from other neurons involved in the rhythmic bursts of impulses.

The spectral sensitivity of the efferent neurons was determined by recording efferent impulses from the optic nerve after blocking the synaptic action. As shown in Fig. 7, each spectral sensitivity curve has a single peak at 420–440 nm, suggesting that the efferent neurons contain a single photopigment.

3.3 Illumination of the Eyes

The discharge rate of efferent optic nerve impulses decreased during illumination of the eyes, in the case of the isolated eyes–optic nerve–brain preparation (Fig. 8). The impulse frequency during each stimulation period decreased with an increase in the intensity of illumination. This observation shows that receptor cells of the eyes are inhibitory to the efferent neurons. The spectral sensitivity for the inhibition by ocular illumination is shown in Fig. 9. The spectral sensitivity curve has a broad peak at 480–540 nm in the visible region and an additional peak at 360–380 nm in the ultraviolet one. These peaks were similar to those of the UV, blue, and green cells in the eyes (Figs. 1A and 3), suggesting that all three kinds of photoreceptor cell may have inhibitory effects on the efferent neurons.

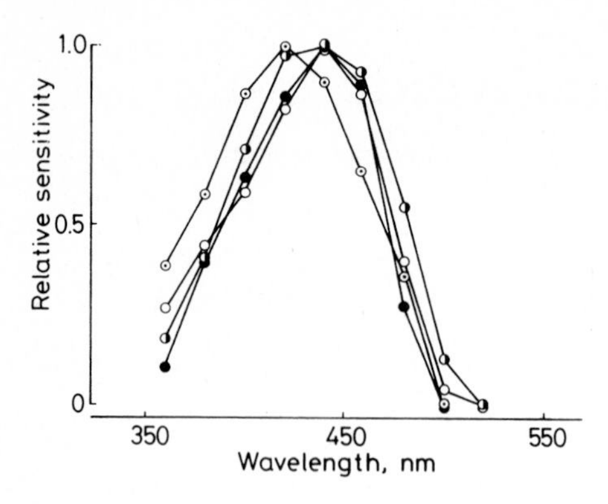

Fig. 7. Spectral sensitivities of the cerebral photosensitive neurons obtained from four preparations by recording efferent impulses (*Argiope*). (Yamashita and Tateda 1983)

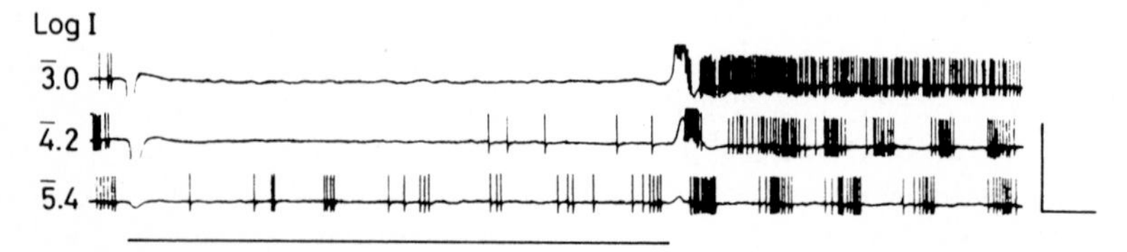

Fig. 8. Efferent optic nerve impulses in response to white light stimuli of the eyes (10 s duration at various intensities) obtained from an isolated brain-optic nerve-eyes preparation (*Argiope*). An opaque chamber divided into two was used for recording. Selective illumination of the eyes was achieved by placing an opaque sheet over the part of the recording chamber that contained the brain. The slow transients at onset and at offset of illumination are ERG's of the eyes. Calibrations: 0.2 mV, 1 s. (Yamashita and Tateda 1983)

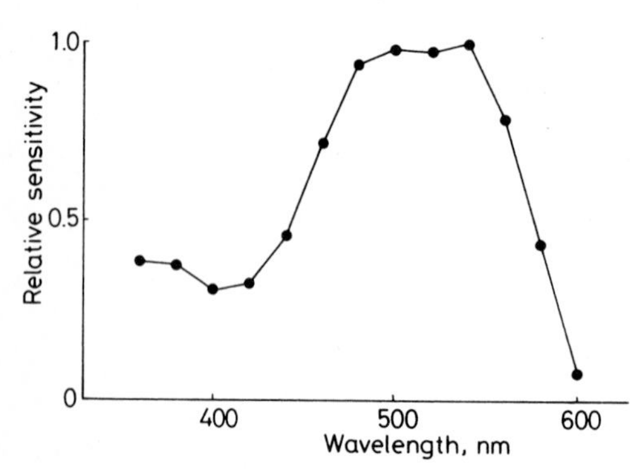

Fig. 9. Spectral sensitivity curve (ERG) for inhibition of the efferent optic nerve impulses during illumination of the eyes (*Argiope*). The relative sensitivity was defined, for convenience, as the reciprocal of the relative number of quanta required to elicit five impulses during 10 s of ocular illumination, and calculated from intensity-response curve for each wave-length. (Yamashita and Tateda 1983)

3.4 Dimming

The discharge rate of efferent optic nerve impulses increased transiently following diminution of the light intensity striking the eyes (Fig. 10A and cf. Fig. 8). The response to the dimming of light is hereafter called the "dimming response". The minimum decrement of light intensity necessary to elicit the smallest dimming response was about 0.04 log units or less. The dimming re-

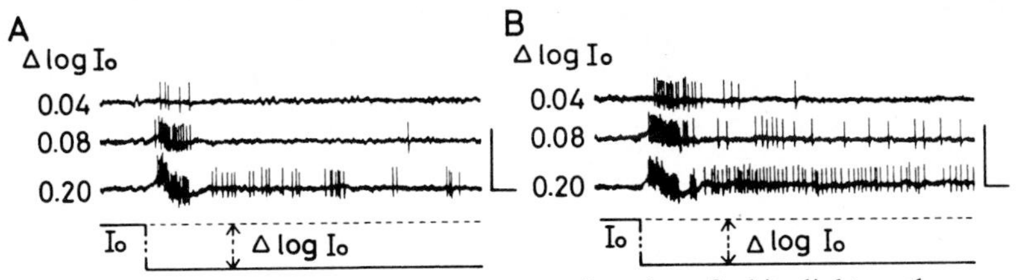

Fig. 10 A, B. Efferent optic nerve impulses in response to dimming of white light on the eyes (**A**) and those to both the eyes and the brain (**B**) (*Argiope*). Decrement of light intensity (Δlog I_0) is indicated for each trace in log units. **A** and **B** are the same preparation. Calibrations: 0.1 mV, 0.5 s. (Yamashita and Tateda 1983)

sponse increased as the decrement in intensity increased. Saturation occurred at a decrement of 0.6 – 1.0 log units.

When the lights striking the eyes and the brain were dimmed simultaneously, the dimming response was enhanced, as compared to the response obtained to a dimming of the light striking the eyes alone (Fig. 10 B). The difference was particularly obvious following small decrements in intensity. For example in Fig. 10, five impulses followed a 0.04 log units decrement of ocular illumination (Fig. 10 A), whereas 25 impulses were recorded following simultaneous ocular and cerebral illumination (Fig. 10 B). This observation suggests that the cerebral photosensitive neurons play a role in increasing the dimming response, i.e., short-lasting excitability of the cerebral photosensitive neurons (the efferent neurons) following diminution of cerebral illumination (Fig. 6) should strengthen transient increase in excitability following diminution of ocular illumination (Fig. 10 A).

In addition, efferent effects on the receptor cells of the eyes also play a role in increasing the dimming response. When the efferent activity increases, the resting membrane potential of the photoreceptor cell hyperpolarizes (Yamashita and Tateda 1981). Similar hyperpolarization is also directly elicited by diminution in the intensity of the light striking the eyes and resulting in an increase in efferent activity. This means that there is a positive feedback loop for the dimming response between the cerebral photosensitive neurons and the eyes. The interaction between the efferent neurons and the eyes may well enable these spiders to detect even a slight diminution in light intensity.

Prominent responses to the offset and the dimming of light have been shown for the second order cells (I-cells) and the third order cells (A-cells) in the barnacle visual pathways (Stuart and Oertel 1978; Oertel and Stuart 1981) and for the pallial nerve of *Spisula solidissima* (Kennedy 1960). The impulses generated in the A-cells of barnacles in response to dimming cause a withdrawal of the cirri and closure of the opercular plates (Gwilliam 1963, 1965, 1976); impulses in the afferent axon of the pallial nerve of *Spisula* cause withdrawal of the siphon (Kennedy 1960). In orb-weaving spiders, impulses in the cerebral photosensitive neurons produced as a result of the interaction of the cerebral photosensitive neurons and the eyes seem to play a role in shadow reaction behavior.

4 Color Vision and Behavior

4.1 Salticidae

A number of behavioral activities of the jumping spiders, a hunter species, are initiated by visual stimuli (see also Forster, Chap. XIII, this Vol.). The anterior median eyes play a significant role in the initiation of various behavioral events and are aptly named the principal eyes. Homann (1928) and Crane (1949) showed that blinding the principal eyes results in a cessation of prey-catching as well as courtship behavior, while blinding two pairs of the lateral eyes affects only the initial turn toward an object moving outside the visual field of the principal eyes. The anterior median eyes of jumping spiders are capable of a certain extent of color recognition. The majority of jumping spiders have specific colorful adornments. Peckham and Peckham (1894) showed that painting various parts of live female jumping spiders in atypical colors greatly reduces male sexual display. Crane (1949) found that the yellow patches on the clypeus of *Corythalia xanthopa* are essential for the release of threat display. A crude two-dimensional model with similar yellow patches initiated threat display, while a model with white patches did not. Such differences, according to Crane, cannot be attributed to differences in intensity. Kästner (1950) observed that jumping spiders, *Evarcha falcata*, leap upon and climb up a striped wall. Black and white stripes elicited the behavior, most effectively, while black and gray stripes were only weakly effective. On the other hand, colored (blue or orange) and gray stripes were equally as effective as black and white stripes. Kästner concluded that the anterior median eyes are capable of color discrimination.

Land (1969) observed that the retina of the anterior median eyes of the jumping spiders, *Phidippus johnsoni* and *Metaphidippus aeneolus*, consists of four layers of receptor cells (1, 2, 3 and 4 from the deepest layer forward). He showed that red light from infinity focused on layer 1, blue-green on 2, violet or near ultraviolet on 3, and ultraviolet on 4. Land proposed two hypotheses to explain the layering of the receptors in the anterior median eye. (1) The several receptor layers act in lieu of a focusing system, the animal using one or another layer to examine objects at different distances. (2) Each receptor layer contains a different photopigment situated in a plane for maximum absorption of light for that pigment, and thus for best image formation (see also Land, Chap. IV, this Vol.).

Yamashita and Tateda (1976a) found that the anterior median eye of the jumping spider, *Menemerus confusus* has UV cells with a maximum sensitivity at about 360 nm, blue cell at 480 – 500 nm, green cell at 540 nm and yellow cell at 580 nm (see Table 1). Possession of four types of receptor cells suggests that the anterior median eyes are capable of color discrimination. The spectral sensitivity curves determined by ERG's had a very broad peak between 490 nm and 590 nm in the visible, and an additional peak at 360 nm in the ultraviolet. The ERG's recorded from the optic nerve side (proximal part of the retina) were affected considerably by long wave chromatic light and those on the corneal side (distal part of the retina) by short wave chromatic light. These results support the latter hypothesis proposed by Land.

Table 1. Spectral sensitivities of the principal eyes

Family	Species	Receptor type (λ_{max})	Futher observation	Reference
Salticidae	*Phidippus regius*	UV (370), green (532) UV-green (370 and 525)	Ratios of UV-to-green sensitivities for the UV-green cells varied greatly from cell to cell	DeVoe, 1975
	Menemerus confusus	UV (360), blue (480–500), green (520–540), yellow (580)	ERGs from proximal part of retina were affected greatly by long wavelength and those from distal part by short wavelength	Yamashita and Tateda, 1976 a
	Plexippus validus	UV (360), green (520)	Marking. Peripheral layer 1 and peripheral and central layer 2 receptors: green cells; layer 4: UV cells	Blest et al. 1981
Lycosidae	*Lycosa baltimoriana* *L. miami* *L. lenta*	UV-green (360–370 and 510)	Each cell contains variable amounts of both UV and green pigments	DeVoe, 1972
Argiopidae	*Argiope bruennichii* *A. amoena*	UV (360), blue (480–500), green (540)	The blue cells are the most sensitive and have a circadian oscillation of sensitivity	Yamashita and Tateda, 1978

On the other hand, DeVoe (1975) found UV cells (maximum sensitivity at 360 nm), green cells (532 nm) and UV-green cells (both 370 nm and 525 nm) in the retinae of *Phidippus regius*, but not blue and yellow cells. Blest et al. (1981) also found only UV (maximum is 360 nm) and green (520 nm) cells in the retinae of *Plexippus validus*. In both species, the spectral sensitivity curves determined by ERG's have a single narrow peak in the visible at about 530 nm for *Phidippus* (DeVoe et al. 1969) and at about 500 – 520 nm for *Plexippus* (Blest et al. 1981).

Blest et al. (1981) marked spectrally characterized cells by the injection of Lucifer Yellow, and showed that peripheral layer 1 and peripheral and central layer 2 of the retina contain green cells, and layer 4 contains UV cells. Cells in the central layer 1 and layer 3 were not marked. They also found that the spacing between receptor layers 1 and 4 is matched to the chromatic aberration of the eye; if green light from an object in front of the spider is focused on layer 1, UV light will be focused on layers 4 and 3, and that a staircase of the distal ends of layer 1 receptors enable the spider to receive in-focus images from objects at distances between 3 cm–∞ in front of it. Blest et al. concluded that tiering of the receptors and the staircase of layer 1 compensate both for the chromatic aberration of the dioptrics of the eye and for its inability to accommodate. The discrepancy between the results of Yamashita and Tateda (1976 a) and those of Blest et al. (1981) remains to be elucidated (see also Blest, Chap. V, this Vol.).

Table 2. Spectral sensitivities of the secondary eyes (AL, PM and PL) and cerebral photosensitive neurons (CP). AL, anterior lateral eye; PM, posterior median eye; PL, posterior lateral eye

Family	Species	Eyes	Receptor type (λ_{max})	Method	Reference
Salticidae	*Menemerus confusus*	AL, PL	Green (535–540)	ERG	Yamashita and Tateda, 1976a
	Plexippus validus	PL	Green (535)	Intracellular recording	Hardie and Duelli, 1978
Lycosidae	*Lycosa baltimoriana* *L. miami* *L. carolinensis*	AL, PM PL	Green (505–510)	ERG	DeVoe et al., 1969
	L. baltimoriana *L. miami* *L. lenta*	AL	Green (510)	Intracellular recording	DeVoe, 1972
Argiopidae	*Argiope bruennichii* *A. amoena*	PL	UV (360), blue (480–500), green (540)	Intracellular recording	This chapter
	A. bruennichii *A. amoena*	CP	blue (420–440)	Extracellular recording of impulses	Yamashita and Tateda, 1983
Dinopidae	*Dinopis subrufus*	PM	Green (517)	Intracellular recording	Laughlin et al. 1980

The anterior and posterior lateral eyes of the jumping spiders are movement detectors (Homann 1928; Crane 1949; Land 1971; Duelli 1978). Land (1971) showed that when any small object is moved within a short distance in the visual field of the lateral eyes, the spider executes a single rapid turn to face the moving object. For detection of moving objects, it is not necessary for the lateral eyes to discriminate color. The lateral eyes indeed have only one type of visual cell with a maximum sensitivity at about 535–540 nm for *Menemerus* (Yamashita and Tateda 1976a) and at 535 nm for *Plexippus* (Hardie and Duelli 1978) (see Table 2).

4.2 Lycosidae

In studies on wolf spider eyes, DeVoe (1972) showed that visual cells in the anterior lateral eye have a maximum sensitivity in the visible at 510 nm and a secondary maximum in the near ultraviolet at 380 nm. Cells in the anterior median eyes all responded maximally both in the visible at 510 nm and in the ultraviolet at 360–370 nm. The ratio of ultraviolet to visible sensitivities varied from cell to cell. DeVoe concluded that visual cells of the anterior median eye contain variable amounts of both a visual pigment absorbing maximally in the visible light region and one absorbing in the ultraviolet.

DeVoe et al. (1969) studied the spectral sensitivities of wolf spider eyes by recording ERG's from intact animals, and found that when the spectral sensitivities of the anterior median eyes were once again obtained from the same eyes after 7–10 weeks, the spectral sensitivities differed from those obtained at the first recording. The difference was large in the visible, but small in the ultraviolet. The difference is similar to that between the circadian high and low sensitive states of the orb-weaving spiders. Therefore, the wolf spider eyes may have a similar circadian change in sensitivity, although DeVoe et al. (1969) thought such changes were unlikely in the case of wolf spider eyes.

4.3 Dinopidae

The posterior median eye of the net-casting spider, *Dinopis subrufus*, a nocturnal predator, has a single type of visual cell with a maximum sensitivity at 517 nm (Laughlin et al. 1980).

4.4 Argiopidae

In orb-weaving spiders, both the anterior median eyes and the posterior lateral eyes have three types of visual cell, suggesting that there is no remarkable difference in function among the four pairs of eyes (cf. Figs. 1 A and 3). Possession of three types of visual cell suggests the possibility for color vision, although there is no behavioral evidence to support this idea. The sensitivity of the blue cell was higher than that of the green and UV cells. The orb-weaving spiders build their webs before sunrise. It is reasonable to consider that in this species there are mechanisms which permit adjustment of the visual system over a wide range of light intensities. This can be explained by the duplicity theory of vision. The blue cells may correspond functionally to the rods of the vertebrate retina while the green and UV cells may correspond to the cones.

4.5 Comparison with Other Arthropods

Many insects, though belonging to another class of Arthropoda, can discriminate color by means of compound eyes which have different types of visual cells, UV and blue and/or green cells (see review by Menzel 1979). In most cases, maximum sensitivities of the UV, blue, and green cells are around 330–380 nm, 420–460 nm and 490–540 nm, respectively. In addition to the UV, blue and green cells, long wavelength receptors, red cells, have been detected photochemically and physiologically in 9 out of 17 species of butterflies tested (maximum sensitivity is 610 nm) (Bernard 1979), electrophysiologically in the butterfly, *Papilio aegeus* (610 nm) (Horridge et al. 1983; Matič 1983) and the dragonfly, *Sympetrum rubicundulum* (620 nm) (Meinertzhagen et al. 1983). The peak sensitivities of UV and green cells in spider eyes correspond to those of UV and green cells in insect eyes. The peak sensitivities of blue cells

(480–500 nm) in spiders, *Menemerus* and *Argiope* are similar to those of green cells rather than blue cells in insects. Yellow cells in *Menemerus* may correspond to red cells in butterflies and dragonflies.

References

Barlow RB Jr, Bolanowski SJ Jr, Brachman ML (1977) Efferent optic nerve fibers mediate circadian rhythms in the *Limulus* eye. Science 197:86–89

Bernard GD (1979) Red-absorbing visual pigment of butterflies. Science 203:1125–1127

Blest AD, Hardie RC, McIntyre P, Williams DS (1981) The spectral sensitivities of identified receptors and the function of retinal tiering in the principal eyes of a jumping spider. J Comp Physiol 145:227–239

Chamberlain SC, Barlow RB Jr (1977) Morphological correlates of efferent circadian activity and light adaptation in the *Limulus* lateral eye. Biol Bull 153:418–419

Crane J (1949) Comparative biology of salticid spiders at Rancho Grande, Venezuela. Part IV. An analysis of display. Zoologica 34:159–214

DeVoe RD (1962) Linear superposition of retinal action potentials to predict electrical flicker responses from the eye of the wolf spider, *Lycosa baltimoriana* (Keyserling) J Gen Physiol 46:75–96

DeVoe RD (1967a) Nonlinear transient responses from light-adapted wolf spider eyes to changes in background illumination. J Gen Physiol 50:1961–1991

DeVoe RD (1967b) A nonlinear model for transient responses from light-adapted wolf spider eyes. J Gen Physiol 50:1993–2030

DeVoe RD (1972) Dual sensitivities of cells in wolf spider eyes at ultraviolet and visible wavelengths of light. J Gen Physiol 59:247–269

DeVoe RD (1975) Ultraviolet and green receptors in principal eyes of jumping spiders. J Gen Physiol 66:193–207

DeVoe RD, Small RJW, Zvargulis JE (1969) Spectral sensitivities of wolf spider eyes. J Gen Physiol 54:1–32

Duelli P (1978) Movement detection in the posterolateral eyes of jumping spiders (*Evarcha arcuata, Salticidae*). J Comp Physiol 124:15–26

Fahrenbach WH (1969) The morphology of the eyes of *Limulus*. II. Ommatidia of the compound eye. Z Zellforsch 93:451–483

Fahrenbach WH (1973) The morphology of the *Limulus* visual system. V. Protocerebral neurosecretion and ocular innervation. Z Zellforsch 144:153–166

Fleissner G, Fleissner G (1978) The optic nerve mediates the circadian pigment migration in the median eyes of the scorpion. Comp Biochem Physiol (A) 61:69–71

Fleissner G, Heinrichs S (1982) Neurosecretory cells in the circadian-clock system of the scorpion, *Androctonus australis*. Cell Tissue Res 224:233–238

Fleissner G, Schliwa M (1977) Neurosecretory fibers in the median eyes of the scorpion, *Androctonus australis* L. Cell Tissue Res 178:189–198

Gwilliam GF (1963) The mechanism of the shadow reflex in Cirripedia. I. Electrical activity in the supraesophageal ganglion and ocellar nerve. Biol Bull 125:470–485

Gwilliam GF (1965) The mechanism of the shadow reflex in Cirripedia. II. Photoreceptor cell response, second-order responses, and motor cell output. Biol Bull 129:244–256

Gwilliam GF (1976) The mechanism of the shadow reflex in Cirripedia. III. Rhythmical patterned activity in central neurons and its modulation by shadows. Biol Bull 151:141–160

Hardie RC, Duelli P (1978) Properties of single cells in posterior lateral eyes of jumping spiders. Z Naturforsch 33c:156–158

Homann H (1928) Beiträge zur Physiologie der Spinnenaugen. I. Untersuchungsmethoden. II. Das Sehvermögen der Salticiden. Z Vergl Physiol 7:201–268

Horridge GA, Marčelja L, Jahnke R, Matič T (1983) Single electrode studies on the retina of the butterfly *Papilio*. J Comp Physiol 150:271–294

Kaplan E, Barlow RB Jr (1980) Circadian clock in *Limulus* brain increases response and decreases noise of retinal photoreceptors. Nature (London) 286:393–395

Kästner A (1950) Reaktionen der Hüpfspinnen (Salticidae) auf unbewegte farblose und farbige Gesichtsreize. Zool Beitr 1:12–50

Kennedy D (1960) Neural photoreception in a lamellibranch mollusc. J Gen Physiol 44:277–299

Land MF (1969) Structure of the retinae of the principal eyes of jumping spiders (Salticidae: Dendryphantinae) in relation to visual optics. J Exp Biol 51:443–470

Land MF (1971) Orientation by jumping spiders in the absence of visual feedback. J Exp Biol 54:119–139

Laughlin S, Blest AD, Stowe S (1980) The sensitivity of receptors in the posterior median eye of the nocturnal spider, *Dinopis*. J Comp Physiol 141:53–65

Matič T (1983) Electrical inhibition in the retina of the butterfly *Papilio*. I. Four spectral types of photoreceptors. J Comp Physiol 152:169–182

Meinertzhagen IA, Menzel R, Kahle G (1983) The identification of spectral receptor types in the retina and lamina of the dragonfly *Sympetrum rubicundulum*. J Comp Physiol 151:295–310

Melamed J, Trujillo-Cenóz O (1966) The fine structure of the visual system of *Lycosa* (Araneae: Lycosidae). Part I. Retina and optic nerve. Z Zellforsch 74:12–31

Menzel R (1979) Spectral sensitivity and colour vision in invertebrates. In: Autrum H (ed) Comparative physiology and evolution of vision in invertebrates. Handbook of sensory physiology, vol VII, 6A. Springer, Berlin Heidelberg New York, pp 503–580

Oertel D, Stuart AE (1981) Transformation of signals by interneurons in the barnacle's visual pathway. J Physiol (London) 311:127–146

Peckham GW, Peckham EG (1894) The sense of sight in spiders with some observations of the color sense. Trans Wiss Acad Sci Arts Lett 10:231–261

Stuart AE, Oertel D (1978) Neuronal properties underlying processing of visual information in the barnacle. Nature (London) 275:287–290

Weakly JN (1973) The action of cobalt ions on neuromuscular transmission in the frog. J Physiol (London) 234:597–612

Yamashita S, Tateda H (1976a) Spectral sensitivities of jumping spider eyes. J Comp Physiol 105:29–41

Yamashita S, Tateda H (1976b) Hypersensitivity in the anterior median eye of a jumping spider. J Exp Biol 65:507–516

Yamashita S, Tateda H (1978) Spectral sensitivities of the anterior median eyes of the orb web spiders, *Argiope bruennichii* and *A. amoena*. J Exp Biol 74:47–57

Yamashita S, Tateda H (1981) Efferent neural control in the eyes of orb weaving spiders. J Comp Physiol 143:477–483

Yamashita S, Tateda H (1982) Importance of calcium and magnesium ions for postexcitatory hypersensitivity in the jumping spider (*Menemerus*) eye. J Exp Biol 97:187–195

Yamashita S, Tateda H (1983) Cerebral photosensitive neurons in the orb weaving spiders, *Argiope bruennichii* and *A. amoena*. J Comp Physiol 150:467–472

VII Mechano- and Chemoreceptive Sensilla

R. F. FOELIX

CONTENTS

1 Introduction

Arachnids do not have antennae, but bear most of their sensory organs on their extremities. In particular the palps and first two pairs of legs carry a variety of mechano- and chemoreceptors. The basic receptor form is represented by the sensory hair or *hair sensillum.* According to the mode of innervation we can distinguish two broad categories: (1) hair sensilla with dendrites ending at the hair base (mechanoreceptors), and (2) hair sensilla with dendrites that enter the hair shaft and communicate with the outside through pores in the hair wall (chemoreceptors). Other sensilla, e.g., for thermo- and hygroreception, certainly occur in arachnids but have not yet been studied systematically.

2 Mechanoreceptors: Tactile Hairs, Spines, Scopula Hairs

Mechanoreceptive sensilla of arachnids comprise (1) *tactile hairs,* (2) *trichobothria,* (3) *slit sensilla,* and (4) *joint receptors.* Only the first type of receptor will be discussed here in detail, the other sensory organs are dealt with in Chapters VIII, IX, and XII, this Volume.

Université de Fribourg, Institut d'Anatomie, 1, Rue Gockel, CH-1700 Fribourg, Switzerland

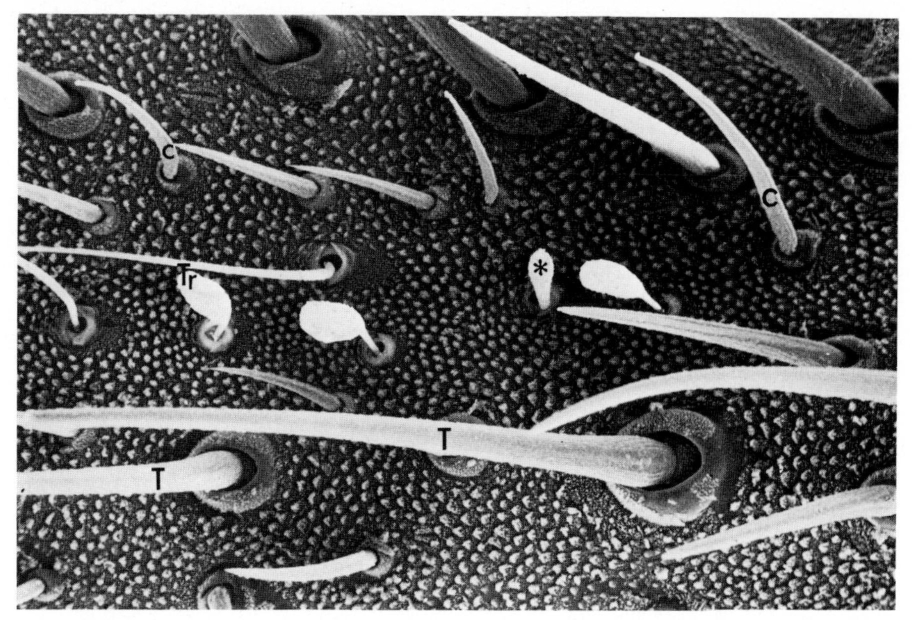

Fig. 1. Different types of hair sensilla on the tarsus of the spider *Liphistius*. Irregular rows of tactile hairs (*T*), taste hairs (*c*), and trichobothria (T_r, two types here) are identified. A small, club-shaped sensillum of unknown significance is marked by an *asterisk*. ×350

Most arachnids are active at night and thus do not primarily rely on vision. Their most important sensory input comes from mechanical stimuli such as touch, vibrations and air currents. Even among the hunting spiders, which are supposedly visually oriented (e.g., the Lycosidae), we find many species which lie quietly in ambush and react mainly to mechanical cues. Direct contact is perceived by movable tactile hairs, air currents by slender trichobothria, and substrate vibrations by special slit sensilla embedded in the exoskeleton (see also Reißland and Görner, Chap. VIII and Barth, Chap. XI, this Vol.).

2.1 Tactile Hairs

Despite a large variety in size and shape (Fig. 1), the mechanosensitive hair sensilla of arachnids share the following characteristics: (1) A hollow cuticular shaft, suspended movably in a socket via an articulating membrane, and (2) several sensory cells whose dendrites are attached to the base of the hair shaft (Fig. 2). A multiple innervation is typical for arachnid mechanoreceptors, whereas tactile hairs in insects are singly innervated (for review see McIver 1975). Table 1 gives the number of neurons associated with tactile hairs in three different arachnid orders.

The dendritic terminals always end at the proximal side of the hair base and contain a characteristic *tubular body* (Thurm 1964). This structure consists of

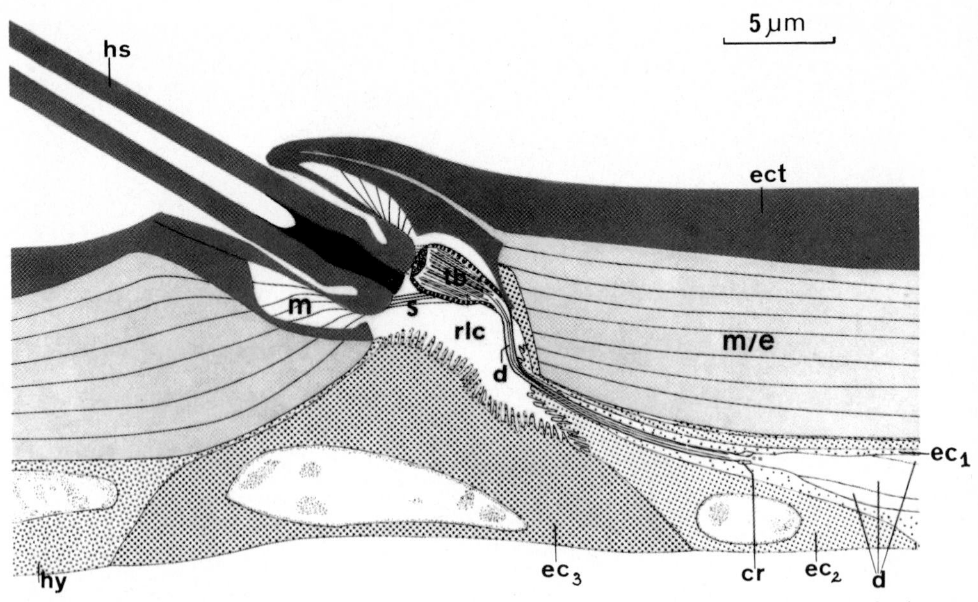

Fig. 2. Tactile hair of a spider, longitudinal section. (Henksmeyer 1983.) The hair shaft (*hs*) is suspended in a narrow socket via an articulating membrane (*m*). Three dendrites (*d*) with tubular bodies (*tb*) terminate at the hair base. *cr* ciliary region; ec_{1-3} enveloping cells; *ect* exocuticle; *hy* Hypodermis; *m/e* meso/endocuticle; *rlc* receptor lymph cavity; *s* socket septum

Table 1. Tactile hair sensilla in arachnids

Order	Number of sensory cells	References
Acari	2	Chu-Wang and Axtell 1973 Haupt and Coineau 1975 Hess and Vlimant 1983
Araneae	3	Eckweiler 1983 Foelix and Chu-Wang 1973a Harris and Mill 1973
Scorpiones	7	Brownell and Farley 1979 Foelix and Schabronath 1983

tightly packed microtubules which are usually interconnected by an electron-dense substance; furthermore, the peripheral microtubules form short bridges to the dendritic membrane (Fig. 3a). The tubular body of insect sensilla is considered to represent the site of sensory transduction (Gnatzy and Tautz 1980; Moran et al. 1976; Thurm 1965, 1982). Since the structure of the tubular body of arachnid sensilla corresponds closely to that of insect mechanoreceptors, it seems reasonable to assume the same function. The occurrence of a tubular

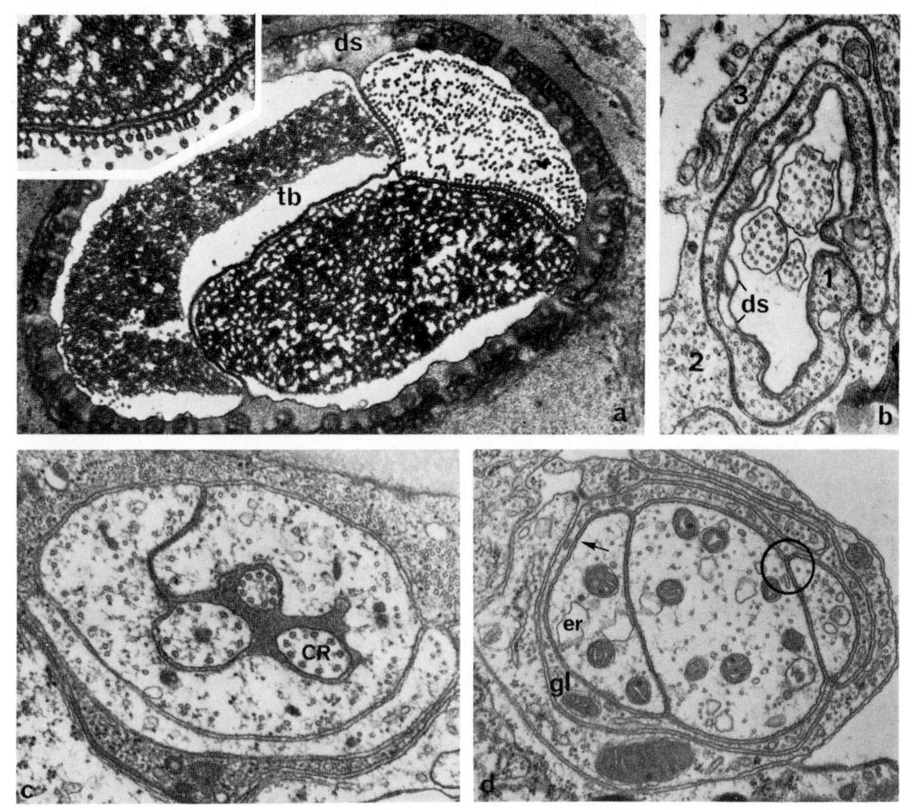

Fig. 3a–d. Innervation of tactile hairs in spiders (cross-sections). **a** *Lycosa punctulata.* Three dendritic endings with tubular bodies (*tb*), surrounded by a dendritic sheath (*ds*) and a layer of cuticular granules. ×21,000. *Inset* the peripheral microtubules of each tubular body connect to the dendritic membrane. ×77,000. **b, c** *Amaurobius ferox.* **b** The three outer dendritic segments, enclosed by a thin dendritic sheath (*ds*) and three enveloping cells (1–3). ×19,000. **c** Three ciliary regions (*CR*) with the typical 9 × 2 + 0 arrangement of microtubules. ×28,000. **d** *Zygiella x-notata.* Three sensory axons, wrapped by a glial cell (*gl*). Axons show septate (*circle*) and gap junctions (*arrow*). *er* smooth endoplasmic reticulum. ×20,000

body in a dendritic terminal is now generally accepted to be a diagnostic feature of an arthropod cuticular mechanoreceptor, comparable to the characteristic microvillar cell border in photoreceptor cells.

Dendrites with tubular bodies are not restricted to purely mechanoreceptive sensilla but also occur in many contact chemoreceptors, thus providing a dual modality for that sensillum (see below). In fact, some arachnid orders (e.g., Amblypygi, Uropygi) seem to have only such combined receptors and lack purely tactile hairs.

Another common feature of arthropod sensilla is a narrow *ciliary region* which divides the dendritic process into an inner and an outer segment (Fig. 3c). This ciliary region consists of nine peripheral doublets of microtubules which arise from triplets in a basal body (cf. Fig. 7). A central pair of microtubules, as is typical for motile cilia, is lacking in the sensory cell's dendrite, yet single ("free") microtubules may be seen to traverse the central region. The ciliary region of arthropod sensilla is thus characterized by the formula $9 \times 2 + 0$, and this arrangement of microtubules holds generally true for most arachnids as well. Some tick species, however, show consistent deviations from this pattern, namely $10 \times 2 + 0$, $11 \times 2 + 0$, and $12 \times 2 + 0$ (for summary see Hess and Vlimant 1982). The significance of the ciliary structure is not yet understood, but perhaps it plays an active role in sensory transduction (Moran et al. 1977).

The dendrites of the sensory neurons are enclosed by several enveloping cells (Fig. 3b), which probably correspond to the tormogen cell, the trichogen cell and the dendritic sheath cell of insect sensilla (Schmidt and Gnatzy 1971; Ernst 1972; Gaffal 1976; Steinbrecht and Müller 1976; Keil 1978). A clear homology cannot be drawn so far, because there are only very few and incomplete ontogenetic studies on arachnid sensilla (Harris 1977; Haupt 1982). It is likely, however, that the innermost enveloping cell produces the dendritic sheath that surrounds the outer dendritic segments. The following sheath cell (trichogen cell) is probably responsible for the formation of the cuticular hair shaft, while the outer sheath cell (tormogen cell) surrounds the receptor lymph cavity underneath the hair base (Fig. 2).

The dendritic sheath is attached to the hair base and also to the inside of the hair socket. Its inner surface is studded with peculiar granules which may indent the dendritic membrane (Fig. 3a). The outer dendritic segments are surrounded by an additional extracellular sheath, the so-called socket septum. This structure is well known from insect sensilla (Gaffal et al. 1975; Gaffal and Theiß 1978), where it apparently transmits a compression onto the dendritic terminal, when the hair is bent.

The somata of the bipolar sensory cells lie 100–200 μm proximal of the hair base and measure 20–80 μm in length (Foelix 1970a; Eckweiler 1983). They always have a rather large, round nucleus and relatively little cytoplasm. The axons exit proximally from the soma and are engulfed by a glial cell (Fig. 3d). The axons from one sensory hair may run separately within the hypodermis or they may join axons of other sensilla to form larger bundles. Eventually they enter the hemolymph-filled lumen of the leg in the form of a small sensory nerve.

Only few studies have dealt with the physiological aspects of tactile hairs in arachnids (Harris 1977; Sanjeeva-Reddy 1971). Moving a single hair while recording from it electrophysiologically has yielded several classes of spikes. Since it was known from histological studies that these sensilla are multiply innervated, this was to be expected. However, it is puzzling that of the three sensory cells in a spider's tactile hair only two cells responded and likewise only four to five units of the seven neurons in a scorpion sensillum reacted. In the spider tactile hair the two identified receptor cells are mainly phasic units which respond primarily to a downward deflection (Harris and Mill 1977a). Movements of the hair shaft in other directions produces a reduced response in both units; thus there is little evidence for a directional sensitivity, where each sensory cell would react preferentially to a certain direction of bending. Such a multidirectional sensitivity has, however, been demonstrated for spider tri-

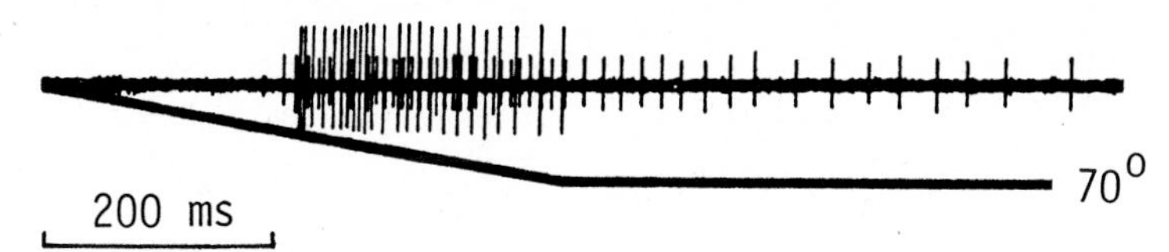

Fig. 4. Recording from a large spine during full elevation (70°). *Lower trace* mechanical displacement of spine. Note response of one phasic (*large spike*) and one phasic-tonic unit (*small spike*). (Harris 1974, unpublished)

chobothria (Görner 1965; Harris and Mill 1977a; see also Reißland and Görner, Chap. VIII, this Vol.).

2.2 Spines

Another interesting mechanosensitive hair that has been studied in spiders refers to the erectile bristles or spines (Harris and Mill 1977a). These large hairs normally lie rather flat along the leg cuticle but can be moved hydraulically (i.e., by increased hemolymph pressure) to an almost vertical position. Electron microscope examination showed that the spines have the same triple innervation as the smaller tactile hairs. Electrophysiologically, however, spikes could be elicited only during erection of the spine but not during deflection. Two units could be distinguished, namely a fast-adapting phasic unit and a slowly adapting phasic-tonic unit (Fig. 4). Since the spines lie flat when the spider is at rest and respond only during erection, it seems questionable whether they can be classified as tactile hairs. Instead, the authors suggested that the function of the erectile spines is to act as hemostatic pressure receptors.

2.3 Scopula Hairs in Spiders

The scopula hairs represent a type of adhesive organ that enables some ground spiders to walk sure-footedly on vertical or even on overhanging smooth surfaces. Each hair resembles a miniature brush with about 1000 cuticular extensions ("endfeet") that act as points of adhesion with the substrate (Homann 1957; Foelix and Chu-Wang 1975; Hill 1977). True scopula hairs are restricted to the very tip of a leg where they form the so-called *claw tufts* (Fig. 5a). In many wandering spiders very similar *scopulate* hairs extend onto the ventral surface of the tarsus and sometimes even the metatarsus. These scopulate hairs (Fig. 5c) are apparently specialized for prey capture, providing an effective "power grip" for large, struggling prey (Rovner 1978; Foelix et al. 1984). Immediately before grasping prey, the scopulate hairs are activated hydraulically to move out of their flat resting position, in a way similar to that described for the erectile spines.

Until recently it was thought that these adhesive hairs were not innervated (Foelix and Chu-Wang 1973a). A closer examination of the claw tufts in a

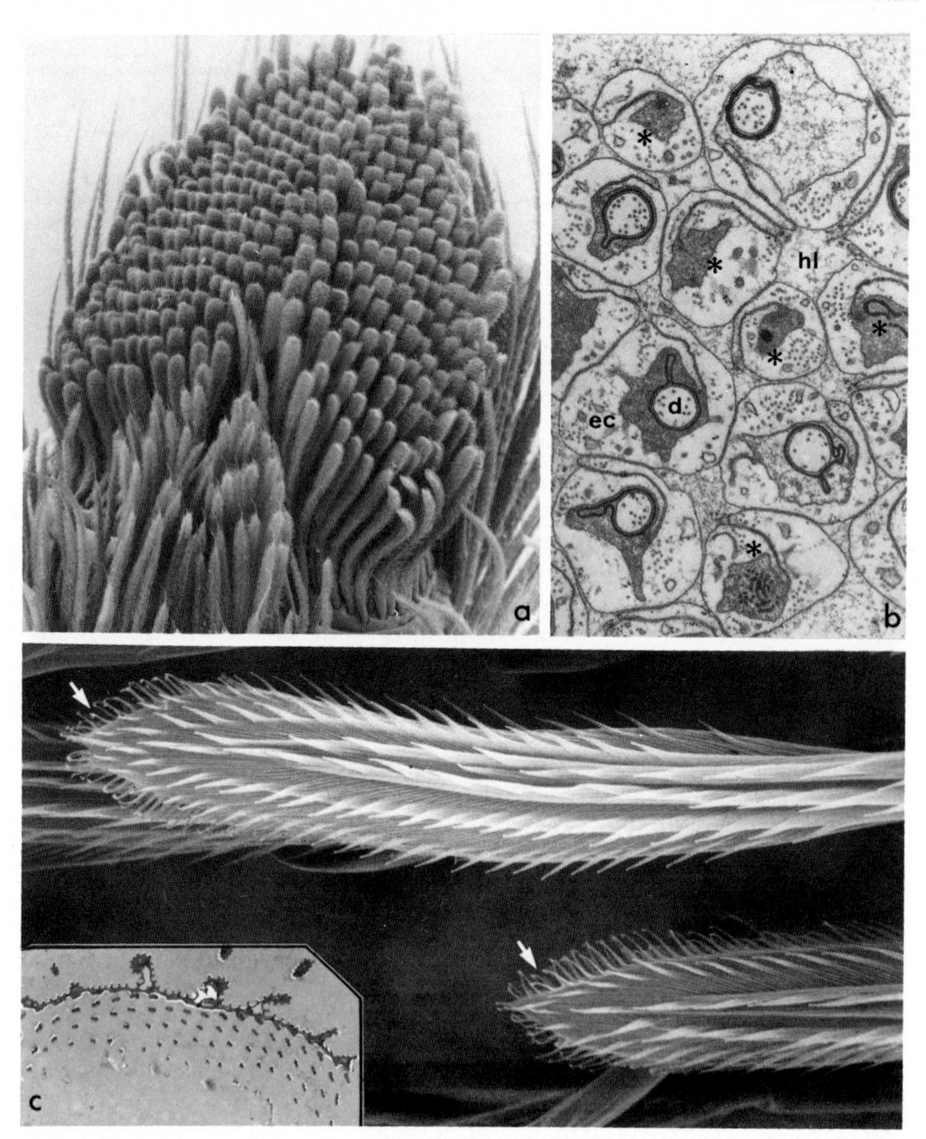

Fig. 5. **a** Claw tuft of *Phidippus audax*, ventral view. About 200 scopula hairs are arranged symmetrically beneath the claws. (Hill 1977.) ×250. **b** Cross-section of scopula hairs of *Portia schultzi* below the hair base. Most hairs are provided by one dendrite (*d*) and one enveloping cell (*ec*), but several sensilla (*asterisk*) lack innervation. (Henksmeyer 1983.) *hl* hemolymph ×15,000. **c** Two scopulate hairs from the metatarsus of *Portia*. Tiny adhesive endfeet (*arrows*) are protruding from the underside of each hair. ×2800. *Insect* cross-section of scopulate hair showing leaf-like structure of hair shaft and regularly emerging stalks of endfeet. (Henksmeyer 1983.) ×7000

jumping spider showed, however, that most of the scopula hairs are supplied by one sensory neuron (Fig. 5b; Henksmeyer 1983). This implies that these hairs not only serve as an adhesive device, but also provide a sensory feedback after contact has been made.

The tactile sense certainly plays an important role in arachnid behavior, be it when inspecting friend or foe or when attacking prey. The primary stimulus will be a depression of the hair sensillum on contact with the substrate. The sensitivity of this system is perhaps best exemplified by the observation that stimulating a single hair suffices to trigger a behavioral reaction of the animal, i.e., either escape or attack (Foelix 1982; Foelix and Schabronath 1983).

3 Chemoreceptors: Contact Chemoreceptors, Pore Hairs, Tarsal Organs

An exchange of chemical signals may have been one of the oldest forms of communication to evolve among arachnids (Weygoldt 1977). Long-standing and numerous observations have shown that arachnids react to odors as well as to direct contact with certain "chemicals" (taste). It took, however, almost a century to identify the corresponding receptors.

Basically, we can distinguish three types of receptors: (1) contact chemoreceptors (taste hairs), (2) pore hairs, and (3) tarsal organs. The first type is characterized by a rather thick-walled hair with many dendrites inside and an opening at the hair tip. Pore hairs have thin or thick walls which are perforated by numerous small pores. Tarsal organs are usually pits or invaginations on the tarsus with several dendrites ending near tiny pores.

3.1 Contact Chemoreceptors

Practically all arachnids probe their environment with the tips of their legs and apparently can perceive the chemical properties of a substrate by simply touching it. This ability of "taste-by-touch" (Bristowe 1941) is also well known from many insects: flies and butterflies, for example, extend their proboscis after their tarsi have contacted a sugar solution (Dethier 1963). Their tarsal chemoreceptors (taste hairs) have been extensively studied, both morphologically (for reviews see Altner and Prillinger 1980; Zacharuk 1980) and physiologically. Arachnids have very similar chemosensitive hairs (Foelix 1970b, 1976). Although they may vary in their outer appearance, these sensilla are always socketed and have a relatively thick wall with a double lumen (Fig. 6). The inner, circular lumen contains many dendrites which run through the entire shaft to an apical pore opening. This pore provides for a communication with the environment and is therefore a crucial prerequisite for a chemoreceptor.

It is rather difficult, however, to demonstrate that small orifice (diameter less than 0.5 μm) under the microscope. An indirect method calls for submerging the taste hairs in dye solutions (crystal violet, methylene blue) and observing the gradual diffusion of the dye into the hair tip (Slifer 1970; Foelix 1970b). Even under the scanning electron microscope, the pore opening may not be seen, because it is often clogged with receptor lymph. In the transmission electron microscope the pore is, however, clearly visible, provided one can achieve sections of the very

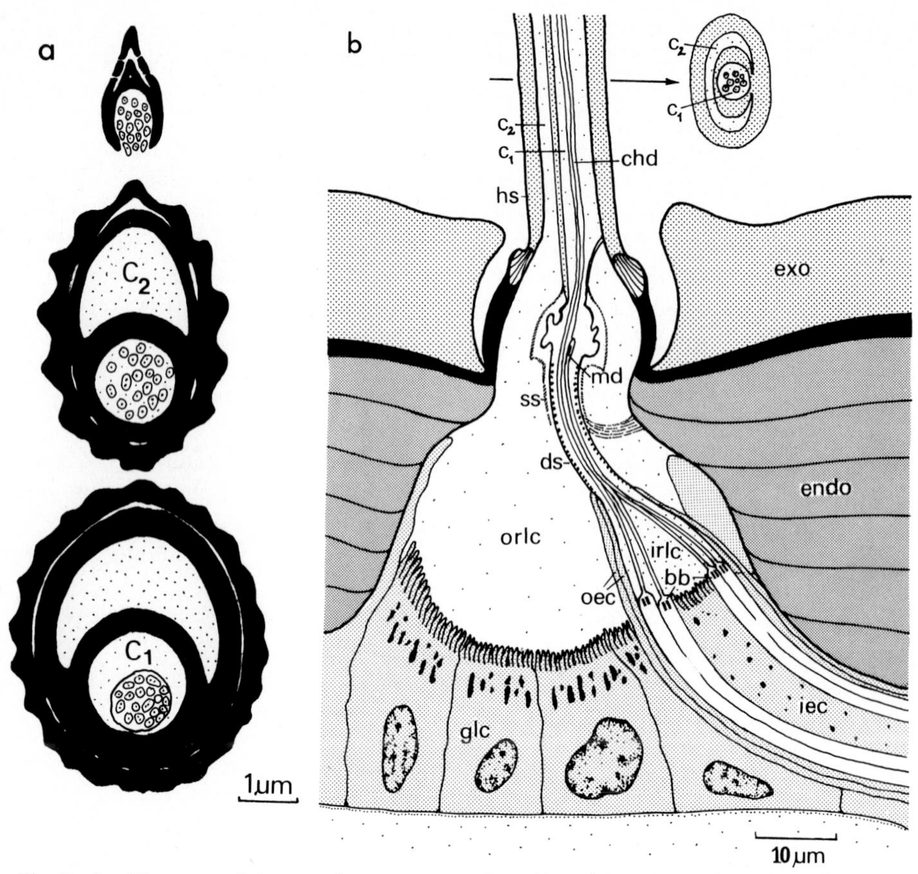

Fig. 6 a, b. Diagrams of contact chemoreceptors in spiders (**a**) and scorpions (**b**). **a** Series of cross-sections of the hair shaft. *Bottom* in the lower part the dendrites are still surrounded by the dendritic sheath while ascending the circular lumen C_1. The crescent-shaped lumen C_2 is filled with receptor lymph. *Middle* the dendritic sheath vanishes soon after the dendrites enter the C_1-lumen. *Top* just below the hair tip the lumen C_1 opens to the outside. (After Foelix and Chu-Wang 1973b.) **b** Longitudinal section of hair base. Two mechanosensitive dendrites (*md*) end inside the socket septum (*ss*), while the chemosensitive dendrites (*chd*) enter the C_1-lumen of the hair shaft (*hs*). The large outer receptor lymph cavity (*orlc*) extends into the C_2-lumen, the inner receptor lymph cavity (*irlc*) into the C_1-lumen. *Inset* shows cross-section of hair shaft. (Foelix and Schabronath 1983.) *d* inner dendritic segments with basal bodies (*bb*); *ds* dendritic sheath; *exo, endo* exo- and endocuticle; *glc* gland cell; *iec, oec* inner and outer enveloping cell

hair tip (Foelix and Chu-Wang 1973b) or if one looks at whole mounts of exuviae (Foelix 1974; Harris 1977).

Most interesting is the fact that all dendrites actually reach the pore opening and are thus directly exposed to the environment. In electrophysiological experiments one could stimulate single taste hairs by slipping a tiny pipette containing salt solutions over the hair tip (Drewes and Bernard 1976; Egan 1976; Harris and Mill 1977b). Several classes of spikes could be recorded, but it was not

Table 2. Contact chemoreceptors in arachnids

Order	Number of chemoreceptive dendrites	Number of mechanoreceptive dendrites	References
Acari	3–8	2	Chu-Wang and Axtell 1973 Foelix and Chu-Wang 1972 Hess and Vlimant 1982
Amblypygi	ca. 10	2	Foelix et al. 1975
Araneae	ca. 20	2	Foelix and Chu-Wang 1973 b Harris and Mill 1973
Opiliones	ca. 16	?	Foelix 1976
Pseudoscorpiones	3–5	?	Foelix unpubl.
Scorpiones	ca. 20	4	Foelix and Schabronath 1983
Solifugae	12	4	Haupt 1982

possible to identify the sensory cells individually as in the taste hairs of the blowfly (sugar cell, water cell, two salt cells). This is understandable, because the taste hairs of arachnids have up to 20 chemosensitive neurons per sensillum, compared to only 4 in flies.

Another characteristic of taste hairs in arachnids is the presence of "accessory" mechanoreceptors at the hair base. These dendrites terminate with typical tubular bodies as in the purely tactile hair sensilla. In contrast to insects, which have only one accessory mechanoreceptor in their taste hairs, we find usually several among arachnids (Table 2). In most cases these mechanoreceptors simply join the chemosensitive dendrites and are enclosed by the same dendritic sheath and enveloping cells. In ticks, however, the mechanoreceptive dendrites are separated by their own sheath (Foelix and Axtell 1972; Foelix and Chu-Wang 1972).

The presence of mechanoreceptive dendrites in taste hairs implies that they actually have a dual function, namely first to register the mechanical contact and then to test the chemical properties. Both functions could be proven in electrophysiological recordings, although only one mechanoreceptive unit responded consistently (Harris and Mill 1977 b).

The fine structure of the chemosensitive hairs corresponds largely to that outlined for tactile hair sensilla (e.g., several bipolar neurons with dendritic ciliary regions, three enveloping cells; Figs. 6, 7). The number of sensory cells per taste hair in various arachnid orders is given in Table 2 (see also Fig. 8).

Distribution. Contact chemoreceptors occur on all extremities of arachnids and are particularly concentrated on the distal segments (tarsus, metatarsus) of the front legs (Foelix 1970 b; Foelix and Schabronath 1983), where the first contact with the environment is made. However, taste hairs are also found on the mouth parts and chelicerae, and even on the spinnerets of spiders (Zimmermann 1975). Most arachnids have at least several hundred contact chemo-

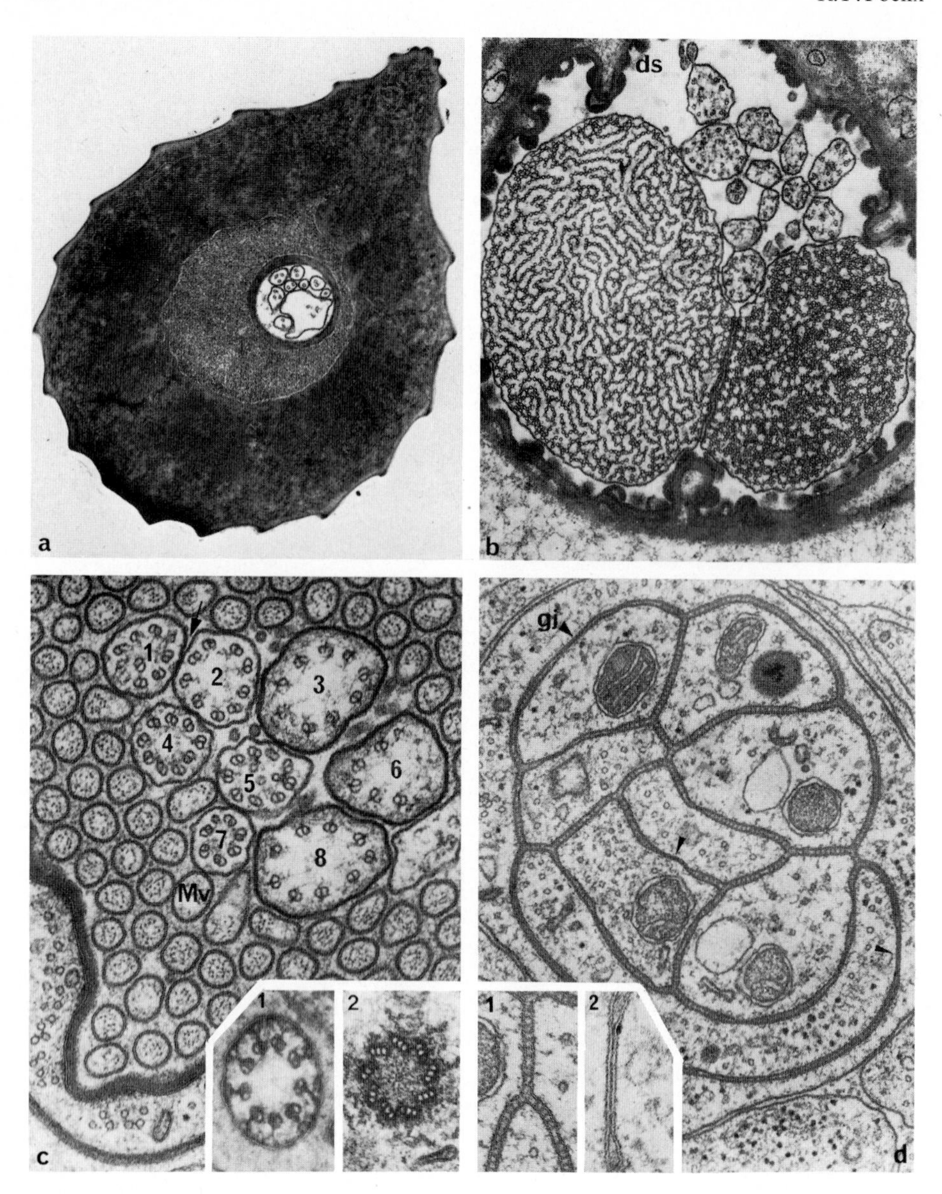

Fig. 7 a – d. Contact chemoreceptors in consecutive cross-sections (**a, b** in the whip spider *Heterophrynus;* **c, d** in the scorpion *Androctonus*). **a** Hair shaft with typical double lumen; the dendrites lie only in the circular lumen ×43,500. **b** Below the hair base the dendrites are surrounded by a granular dendritic sheath (*ds*). The two large dendrites exhibit tubular bodies typical of mechanoreceptors, the 12 small dendrites are chemosensitive ×21,000. **c** Ciliary regions of eight dendrites (*1–8*), surrounded by microvilli (*Mv*) of the inner enveloping cell ×45,000. *Inset 1* ciliary structure after tannin-uranyl-acetate treatment. The nine doublets are connected to the dendritic membrane by an electron-dense substance ×60,000. *Inset 2* the ciliary region arises from a basal body with a triplet structure ×45,000. **d** Many septate junctions (cf. Inset 1, ×48,000) and gap junctions (*gj, arrows;* cf. Inset 2, ×82,000) occur between inner dendritic segments and also toward the inner enveloping cell. ×34,000

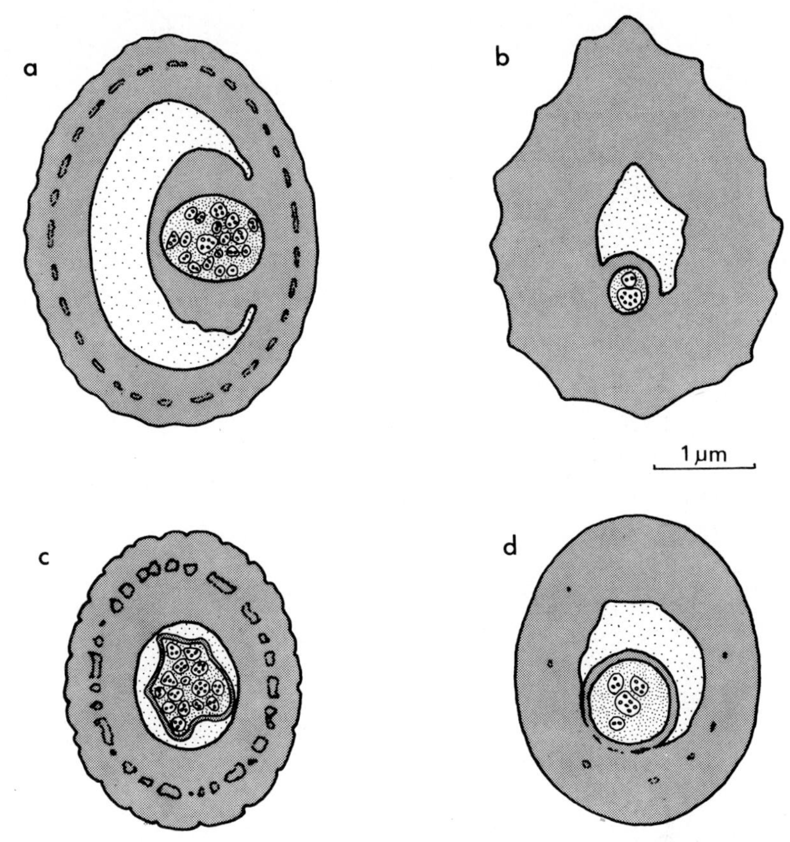

Fig. 8a–d. Cross-sections of chemosensitive hairs in various orders of arachnids: **a** Scorpiones (*Euscorpius*); **b** Pseudoscorpiones (*Neobisium*); **c** Opiliones (*Siro*); **d** Acari (*Amblyomma*). Note the relatively thick hair wall and the characteristic double lumen

receptors covering their legs, but some orders (e.g., Acari) have relatively few – although these are "strategically" located.

Function. In addition to the obvious gustatory function, contact chemoreceptors also play a role in courtship. Male spiders, for instance, follow the trail of a conspecific female and perform typical courtship rituals in response to female threads (Tietjen 1979; Tietjen and Rovner 1980, 1982). It is likely that some kind of pheromone is associated with the female drag lines. Ticks and certain parasitic mites choose their hosts on the basis of chemical information gathered by tarsal contact chemoreceptors (Egan 1976).

3.2 Pore Hairs

Among insects, pore hairs represent the typical olfactory receptors. Pore hairs are usually nonsocketed and lack any mechanoreceptive innervation. In arach-

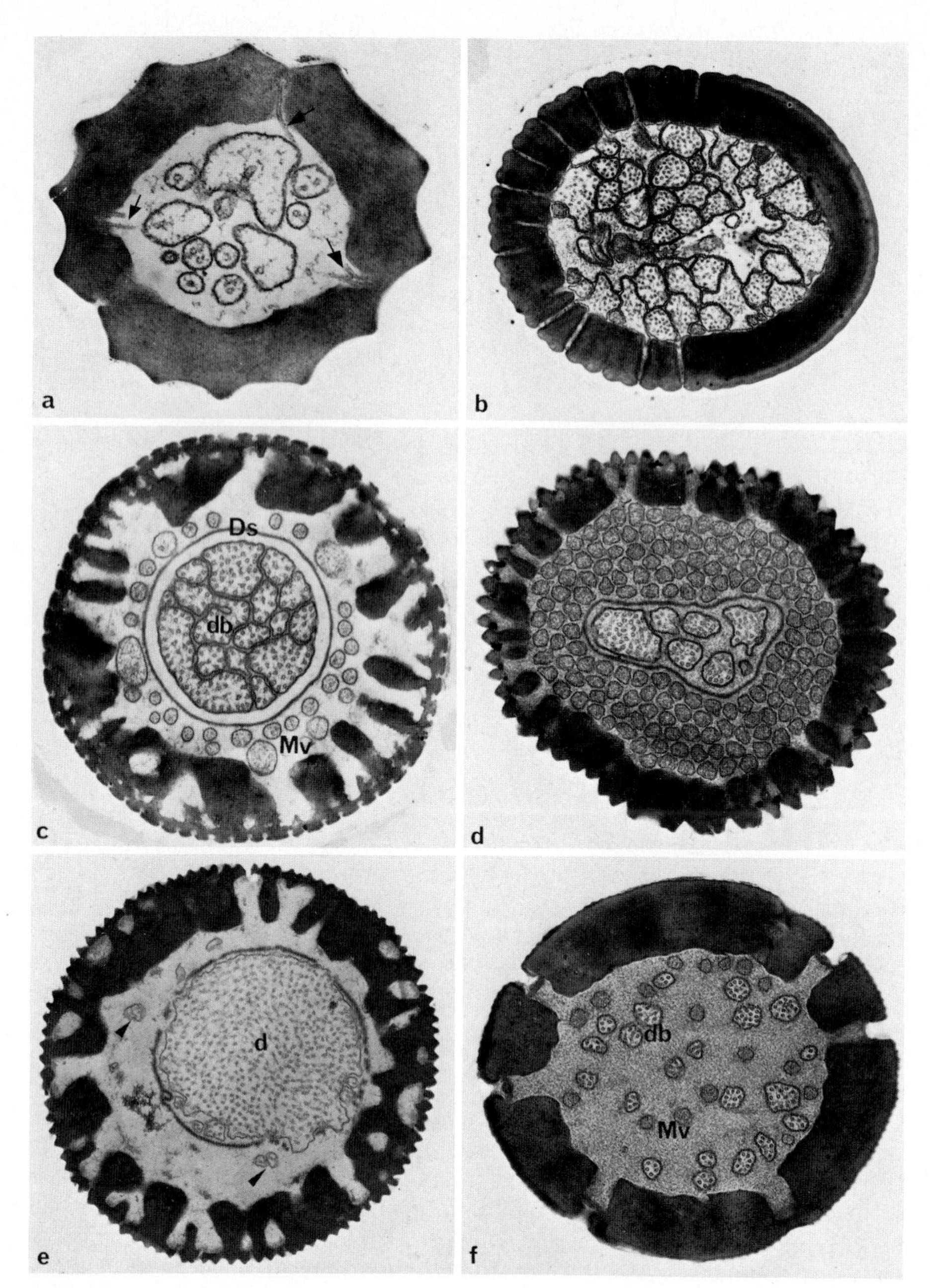

Fig. 9a–f. Various types of pore hairs in cross-sections. **a–d** *Heterophrynus* (Amblypygi); **e** *Gradungula* (Araneae); **f** *Amblyomma* (Acari). **a** Typical pore tubules (*arrows*) connect pores and dendritic membranes ×43,500. **b** Dendritic processes enter the pores, yet pore tubules are absent ×13,000. **c** Thin-walled rod hair with one branching dendrite (*db*) inside a dendritic sheath (*Ds*), surrounded by microvilli (*Mv*) of sheath cell ×18,500. **d** Thick-walled rod hair with six dendrites, dendritic sheath and many microvilli ×18,500. **e** Pore hair with a single dendrite (*d*) that gives off small branches (*arrows*) ×13,500. **f** Thick-walled pore hair with lentil-shaped plugs in each pore. The lumen of the hair shaft contains many dendritic branches (*db*) and microvilli (*Mv*). ×18,000

nids we find some orders (Acari, Amblypygi) which have almost identical pore sensilla, yet in some others (Araneae, Scorpiones) pore hairs are very rare or absent. The largest variety of pore sensilla occurs on the antenna-like front legs of whip spiders (Fig. 9a–d; Foelix et al. 1975; Beck et al. 1977). One type has relatively thick, perforated walls with typical pore tubules, which are considered as a stimulus-conducting system for odor molecules (Steinbrecht 1973; Keil 1982). The dendrites inside the hair shaft are unbranched and belong to 20–30 sensory cells. A similar type of pore hair has even more neurons (40–45) but lacks pore tubules. Another type of pore sensillum, the rod hair, occurs only in clusters at the tip of the first leg; here only a single sensory cell with branching dendrites enters the hair lumen. Although the function of these pore hairs has not been tested electrophysiologically, olfaction seems most probable. Behaviorally, olfaction can easily be demonstrated in whip spiders: exposure of the leg tip to strong odors leads to a quick withdrawal of the entire leg.

In ticks we find also several kinds of pore sensilla (Foelix and Axtell 1971, 1972; Leonovich 1977; Hess and Vlimant 1982): (1) Single-walled hairs with large, *plugged* pores and pore tubules (Fig. 9f), and (2) double-walled sensilla with narrow slit pores. In the first category – which is typical for all Acari – the walls can be either thick or thin and the hair shaft contains dendritic branches of 3–5 neurons. In the second category the dendrites of the sensory cells (1–7) remain unbranched. From a few electrophysiological studies it seems evident that the plugged pore sensilla are olfactory (Sinitzina 1974; Hess and Vlimant 1980), but the function of the double-walled pore hairs remains to be determined.

In spiders, opilionids and scorpions, pore hairs seem to be rare or even absent. Among the many spider species investigated with the electron microscope, I have found only one specimen, namely the New Zealand spider *Gradungula*, which exhibited a single pore hair (Fig. 9e). It was located on the dorsal side of the tarsus, distal of the tarsal organ (cf. Fig. 11). Some pore hairs were noticed on the legs of opilionids (Foelix 1976; Gnatzy and Dumpert 1978, personal communication), but none have been found in scorpions so far. Although it is quite likely that new types of pore hairs will be discovered among the arachnids, it still seems safe to say that they are generally less developed than in insects.

3.3 Tarsal Organs

Originally, tarsal organs have been described only for spiders (Blumenthal 1935). They represent small depressions or invaginations on the dorsal side of each tarsus (cf. Fig. 10). From the bottom of such a pit a small cuticular cone projects, which is perforated by 6–7 canals. Each canal contains 3–4 dendrites which end in small pores at the tip of the cone (Foelix and Chu-Wang 1973b). It has long been postulated that the tarsal organ serves for olfaction and possibly humidity reception (Blumenthal 1935). Only recently did we obtain proof for the olfactory function of this organ by recording electrophysiologically from it (Dumpert 1978); however, hygroreception is still not ruled out.

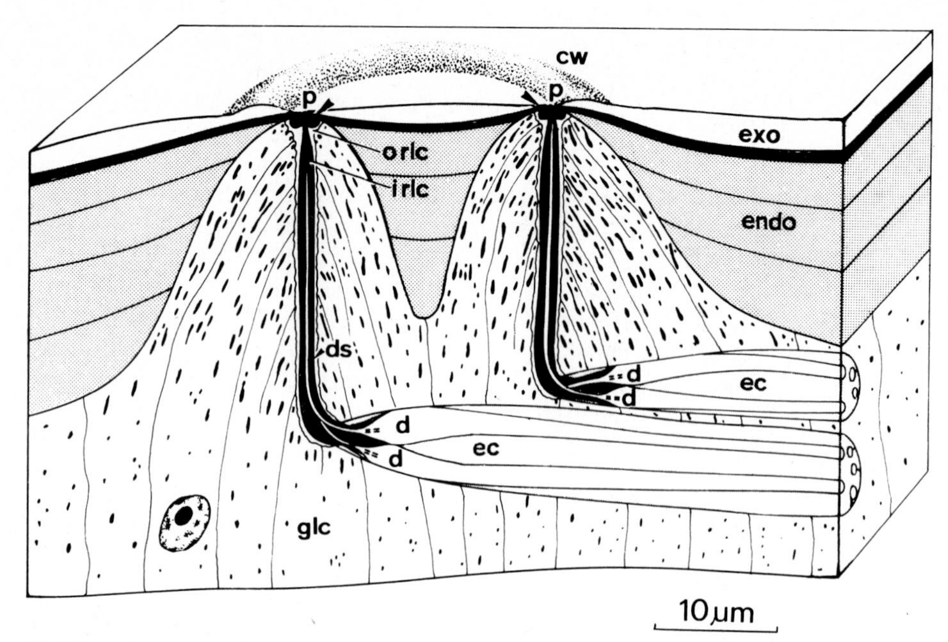

Fig. 10. Diagram of the tarsal organ in a scorpion. (Foelix and Schabronath 1983.) Two small pore openings (*p*) lie inside a slight cuticular wall (*cw*) and several dendrites (*d*) are exposed there to the outside. *exo, endo* exo- and endocuticle; *irlc, orlc* inner and outer receptor lymph cavity; *ds* dendritic sheath; *ec* enveloping cells; *glc* gland cell

In a few cases the tarsal organs form cuticular extensions rather than invaginations (Forster 1980). Although they look superficially like hair sensilla, their internal structure is typically that of tarsal organs (Fig. 11c). An interesting question is which form represents the ancient type – whether the hair-like tarsal organs are plesiomorphic and the invaginated form is to be considered derived, or vice versa.

Recently we found tarsal organs also in scorpions (Foelix and Schabronath 1983) and probably also in whip spiders (Foelix et al. 1975). Perhaps Haller's organ in ticks is a homologous structure, although it is restricted to the first pair of legs, whereas the tarsal organs occur on all legs of spiders and scorpions. The olfactory function of Haller's organ and its role in host detection is well-known from the classical work of Lees (1948). In web spiders olfaction seems to be involved in courtship, i.e., male spiders are attracted by the scent of the mature female (Blanke 1973). Most likely the tarsal organs are the receptors for the pheromone, as was shown electrophysiologically for the ctenid spider *Cupiennius* (Dumpert 1978).

4 Other Receptors

Several specialized sensilla occur among the arachnids which could not be discussed in this brief review. I only want to point out the pecten sensilla of scor-

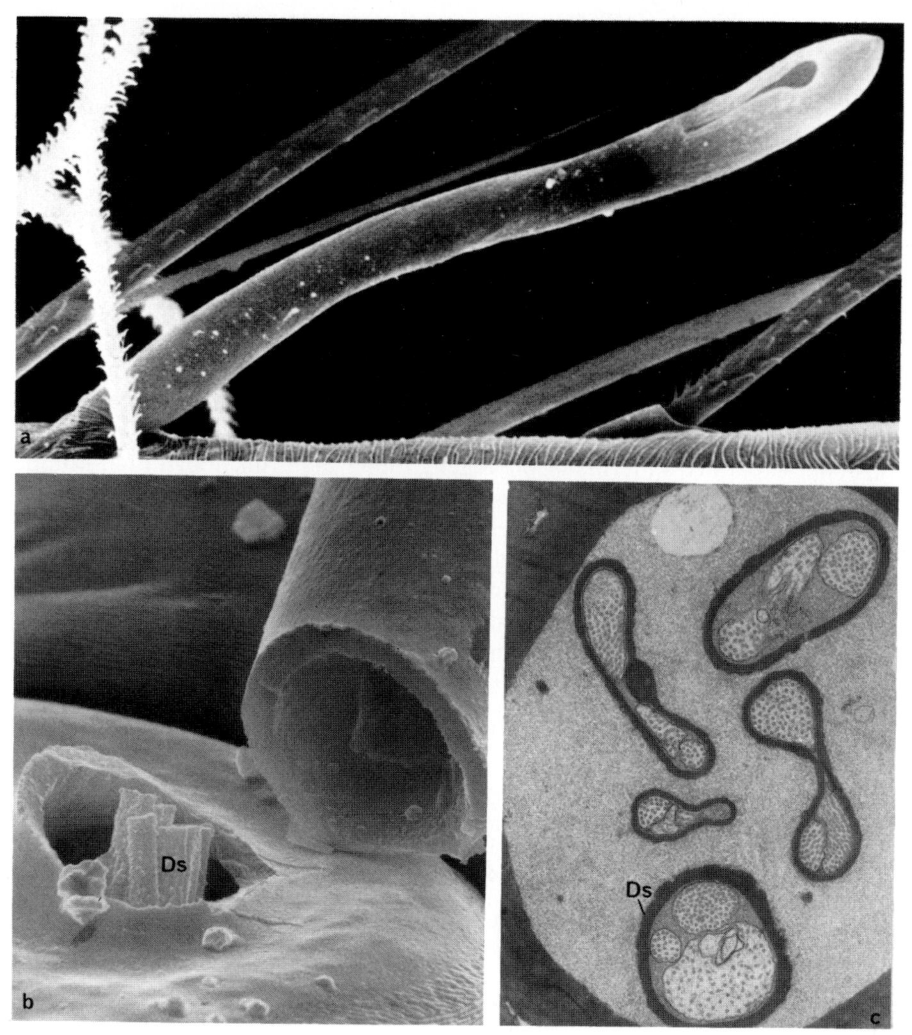

Fig. 11 a–c. Hair-shaped tarsal organs from the New Zealand spiders *Otira* (**a**) and *Gradungula* (**b, c**). **a** A slit near the tip of the hair shaft connects the outside and the dendrites inside ×570. **b** The broken-off hair shaft reveals several dendritic sheaths (*Ds*) entering the hair base ×9600. (Photos **a, b** Forster.) **c** Cross- section below hair base shows five dendritic sheaths (*Ds*) containing three dendrites each. ×15,000

pions, which now seem to be contact chemoreceptors (Foelix and Müller-Vorholt 1983; Ivanov 1981), and the malleolar organs in solpugids, which may be olfactory (Brownell and Farley 1974). Even less clear is the function of the hundreds of "sensilla ampullacea" which lie inside the tarsus of the pedipalps in solpugids (Bauchhenss 1983). Furthermore, new types of hair sensilla are being discovered now in spiders (Pulz 1983, personal communication) and ticks (Hess and Loftus 1984) which act as thermoreceptors.

5 Concluding Remarks

As this survey has amply shown, hair sensilla of arachnids closely resemble those of other arthropods, especially those of insects. A remarkable difference lies in the number of neurons per sensillum, i.e., most arachnid sensilla have relatively more sensory cells than insect sensilla. This is particularly true for mechanoreceptors, which are singly innervated in insects, but multiply innervated in arachnids. Whether this "surplus" of sensory neurons is of any physiological advantage seems questionable, because we have hardly any evidence that the additional sensory cells provide any additional information. The tactile hairs of spiders, for example, did not show a directional sensitivity despite their triple innervation. The same is true for the contact chemoreceptors, where one can only wonder about some 20 neurons per sensillum, when flies do apparently quite well with only four (Dethier 1971). Perhaps this difference can be better explained in terms of phylogeny rather than function: a high number of neurons per sensillum may simply be an ancient feature that was retained in arachnids. In this respect it is interesting that the scorpions, which belong to the oldest (and most "primitive") arachnids, also have the highest number of sensory cells, both in their mechano- and chemoreceptors. As we shall see in Chapter X, this Volume, a similar reasoning can be applied to the organization of the peripheral nervous system in arachnids.

References

Altner H, Prillinger L (1980) Ultrastructure of invertebrate chemo-, thermo-, and hygroreceptors and its functional significance. Int Rev Cytol 67:69–151

Bauchhenss E (1983) Morphology and ultrastructure of sensilla ampullacea in Solifugae (Chelicerata: Arachnida). Int J Insect Morphol Embryol 12:129–138

Beck L, Foelix R, Gödecke E, Kaiser R (1977) Morphologie, Larvalentwicklung und Haarsensillen des Tastbeinpaares der Geißelspinne *Heterophrynus longicornus* Butler (Arach., Amblypygi). Zoomorphologie 88:259–276

Blanke R (1973) Nachweis von Pheromonen bei Netzspinnen. Naturwissenschaften 10:481

Blumenthal H (1935) Untersuchungen über das 'Tarsalorgan' bei Netzspinnen. Z Morphol Oekol Tiere 29:667–719

Bristowe WS (1941) The comity of spiders, vol II. Ray Soc No 128, London

Brownell PH, Farley RD (1974) The organization of the malleolar sensory system in the solpugid, *Chanbria sp.* Tissue Cell 6:471–485

Brownell PH, Farley RD (1979) Detection of vibrations in sand by tarsal sense organs of the nocturnal scorpion *Paruroctonus mesaensis.* J Comp Physiol 131:23–30

Chu-Wang IW, Axtell RC (1973) Comparative fine structure of the claw sensilla of a soft tick, *Argas (Persicargas) arboreus* Kaiser, Hoogstraal, and Kohls, and a hard tick, *Amblyomma americanum* (L.). J Parasitol 59:545–555

Dethier VG (1963) The physiology of insect senses. Methuen, London

Dethier VG (1971) A surfeit of stimuli: a paucity of receptors. Am Sci 59:706–715

Drewes CD, Bernard RA (1976) Electrophysiological responses of chemosensitive sensilla in the wolf spider. J Exp Zool 198:423–428

Dumpert K (1978) Spider odor receptor: Electrophysiological proof. Experientia 34:754–755

Eckweiler W (1983) Topographie von Proprioreceptoren, Muskeln und Nerven im Patella-Tibia- und Metatarsus-Tarsus-Gelenk des Spinnenbeins. Diplomarbeit, Univ Frankfurt

Egan ME (1976) The chemosensory bases of host discrimination in a parasitic mite. J Comp Physiol 109:69–89

Ernst KD (1972) Die Ontogenie der basiconischen Riechsensillen auf der Antenne des Aaskäfers *Necrophorus* (Coleoptera). Z Zellforsch 129:217–236

Foelix RF (1970a) Structure and function of tarsal sensilla in the spider *Araneus diadematus*. J Exp Zool 175:99–124

Foelix RF (1970b) Chemosensitive hairs in spiders. J Morphol 132:313–334

Foelix RF (1974) Application of the transmission electron microscope to the examination of spider exuviae and silk. Psyche 81:507–509

Foelix RF (1976) Rezeptoren und periphere synaptische Verschaltungen bei verschiedenen Arachnida. Entomol Germ 3:83–87

Foelix RF (1982) Biology of Spiders. Harvard Univ Press, Cambridge, Mass, p 71

Foelix RF, Axtell RC (1971) Fine structure of tarsal sensilla in the tick *Amblyomma americanum* L. Z Zellforsch 114:22–37

Foelix RF, Axtell RC (1972) Ultrastructure of Haller's organ in the tick *Amblyomma americanum* L. Z Zellforsch 124:275–292

Foelix RF, Chu-Wang IW (1972) Fine structural analysis of palpal receptors in the tick *Amblyomma americanum* L. Z Zellforsch 129:548–560

Foelix RF, Chu-Wang IW (1973a) The morphology of spider sensilla. I. Mechanoreceptors. Tissue Cell 5:451–460

Foelix RF, Chu-Wang IW (1973b) The morphology of spider sensilla. II. Chemoreceptors. Tissue Cell 5:461–478

Foelix RF, Chu-Wang IW (1975) The structure of scopula hairs in spiders. Proc 6th Int Congr Arachnol, Amsterdam, pp 56–58

Foelix RF, Müller-Vorholt G (1983) The fine structure of scorpion sensory organs. II. Pecten sensilla. Bull Br Arachnol Soc 6:68–74

Foelix RF, Schabronath J (1983) The fine structure of scorpion sensory organs. I. Tarsal sensilla. Bull Br Arachnol Soc 6:53–67

Foelix RF, Chu-Wang IW, Beck L (1975) Fine structure of tarsal sensory organs in the whip spider *Admetus pumilio* (Amblypygi, Arachnida). Tissue Cell 7:331–346

Foelix R, Jackson R, Henksmeyer A, Hallas A (1984) Tarsal hairs specialized for prey capture in the salticid *Portia*. Rev Arachnol 5:329–334

Forster RR (1980) Evolution of the tarsal organ, the respiratory system and the female genitalia in spiders. Proc 8th Int Congr Arachnol, Vienna, pp 269–284

Gaffal KP (1976) Die Feinstruktur der Sinnes- und Hüllzellen in den antennalen Schmecksensillen von *Dysdercus intermedius* Dist. (Pyrrhocoridae, Heteroptera). Protoplasma 88:101–115

Gaffal KP, Theiß J (1978) The tibial thread-hairs of *Acheta domesticus* (L.) (Saltatoria, Gryllidae). The dependence of stimulus transmission and mechanical properties on the anatomical characteristics of the socket apparatus. Zoomorphologie 90:41–51

Gaffal KP, Tichy H, Theiß J, Seelinger G (1975) Structural polarities in mechanosensitive sensilla and their influence on stimulus transmission (Arthropoda). Zoomorphologie 82:79–103

Gnatzy W, Tautz J (1980) Ultrastructure and mechanical properties of an insect mechanoreceptor: Stimulus-transmitting structures and sensory apparatus of the cercal filiform hairs of *Gryllus*. Cell Tissue Res 213:441–463

Görner P (1965) A proposed transducing mechanism for a multiply innervated mechanoreceptor (trichobothrium) in spiders. Cold Spring Harb Symp Quant Biol 30:69–73

Harris DJ (1977) Hair regeneration during moulting in the spider *Ciniflo similis* (Araneae, Dictynidae). Zoomorphologie 88:37–63

Harris DJ, Mill PJ (1973) The ultrastructure of chemoreceptor sensilla in *Ciniflo* (Araneida, Arachnida). Tissue Cell 5:679–689

Harris DJ, Mill PJ (1977a) Observations on the leg receptors of *Ciniflo* (Araneida, Dictynidae). I. External mechanoreceptors. J Comp Physiol 119:37–54

Harris DJ, Mill PJ (1977b) Observations on the leg receptors of *Ciniflo* (Araneida, Dictynidae). II. Chemoreceptors. J Comp Physiol 119:55–62

Haupt J (1982) Hair regeneration in a solpugid chemotactile sensillum during moulting (Arachnida: Solifugae). Wilhelm Roux' Arch Entwicklungsmech Org 191:137–142

Haupt J, Coineau Y (1975) Trichobothrien und Tastborsten der Milbe *Microcaeculus* (Acari, Prostigmata, Caeculidae). Z Morphol Tiere 81:305–322

Henksmeyer A (1983) Funktionelle Anatomie der hydraulisch aufrichtbaren Haare am Spinnenbein. Examensarbeit, Ruhr-Univ Bochum

Hess E, Loftus R (1984) Warm and cold receptors of two sensilla on the fore leg tarsi of the tropical bont tick *Amblyomma variegatum*. J Comp Physiol A 155:187–195

Hess E, Vlimant M (1980) Morphology and fine structure of the tick *Amblyomma variegatum* (Acarina, Ixodidae, Metastriata), including preliminary electrophysiological results. Proc Olfaction and Taste, vol VII, Paris, p 190

Hess E, Vlimant M (1982) The tarsal sensory system of *Amblyomma variegatum* Fabricius (Ixodidae, Metastriata). I. Wall pore and terminal pore sensilla. Rev Suisse Zool 89:713–729

Hess E, Vlimant M (1983) The tarsal sensory system of *Amblyomma variegatum* Fabricius (Ixodidae, Metastriata). II. No pore sensilla. Rev Suisse Zool 90:157–167

Hill DE (1977) The pretarsus of salticid spiders. Zool J Linn Soc 60:319–338

Homann H (1957) Haften Spinnen an einer Wasserhaut? Naturwissenschaften 44:318–319

Ivanov VP (1981) Sense organs of scorpions. Acad Sci USSR Proc Zool Inst 106:4–33 (in Russian)

Keil T (1978) Die Makrochaeten auf dem Thorax von *Calliphora vicina* Robineau-Desvoidy (Calliphoridae, Diptera). Zoomorphologie 90:151–180

Keil T (1982) Contacts of pore tubules and sensory dendrites in antennal chemosensilla of a silkmoth: Demonstration of a possible pathway for olfactory molecules. Tissue Cell 14:451–462

Leonovich SA (1977) Electron microscopy of Haller's organ of the tick *Ixodes persulcatus* (Ixodidae). Parazitologya 11:340–347 (in Russian)

Lees AD (1948) The sensory physiology of the sheep tick, *Ixodes ricinus* L. J Exp Biol 25:145–207

McIver SB (1975) Structure of cuticular mechanoreceptors of arthropods. Annu Rev Entomol 20:381–397

Moran DT, Rowley JC, Zill SN, Varela FG (1976) The mechanism of sensory transduction in a mechanoreceptor. Functional stages in campaniform sensilla during the molting cycle. J Cell Biol 71:832–847

Moran DT, Varela FJ, Rowley JC (1977) Evidence for active role of cilia in sensory transduction. Proc Natl Acad Sci USA 74:793–797

Rovner JS (1978) Adhesive hairs in spiders: Behavioral functions and hydraulically mediated movement. Symp Zool Soc London 42:99–108

Sanjeeva-Reddy P (1971) Function of the supernumerary sense cells and the relationship between modality of adequate stimulus and innervation pattern of the scorpion hair sensillum. J Exp Biol 54:233–238

Schmidt K, Gnatzy W (1971) Die Feinstruktur der Sinneshaare auf den Cerci von *Gryllus bimaculatus* Deg. (Saltatoria, Gryllidae). II. Die Häutung der Faden- und Keulenhaare. Z Zellforsch 122:210–226

Sinitzina EE (1974) Electrophysiological reactions of the neurons of the Haller's organ to the odour stimuli in the tick *Hyalomma asiaticum*. Parazitologiya 8:223–226 (in Russian)

Slifer EH (1970) The structure of arthropod chemoreceptors. Annu Rev Entomol 15:121–142

Steinbrecht RA (1973) Der Feinbau olfaktorischer Sensillen des Seidenspinners (Insecta, Lepidoptera). Rezeptorfortsätze und reizleitender Apparat. Z Zellforsch 139:533–565

Steinbrecht RA, Müller B (1976) Fine structure of the antennal receptors of the bed bug, *Cimex lectularis* L. Tissue Cell 8:615–636

Thurm U (1964) Mechanoreceptors in the cuticle of the honey bee: Fine structure and stimulus mechanism. Science 145:1063–1065

Thurm U (1965) An insect mechanoreceptor, part I. Fine structure and adequate stimulus. Cold Spring Harbor Symp Quant Biol 30:75–82

Thurm U (1982) Biophysik der Mechanorezeption. In: Hoppe W, Lohmann W, Markl H, Ziegler H (eds) Biophysik, 2nd edn. Springer, Berlin Heidelberg New York, pp 691–696

Tietjen WJ (1979) Is the sex pheromone of *Lycosa rabida* (Araneae, Lycosidae) deposited on a substratum? J Arachnol 6:207–212

Tietjen WJ, Rovner JS (1980) Physico-chemical trail-following behaviour in two species of wolf spiders: sensory and eco-ethological concomitants. Anim Behav 28:735–741

Tietjen WJ, Rovner JS (1982) Chemical communication in lycosids and other spiders. In: Witt PN, Rovner JS (eds) Spider communication, mechanism and ecological significance. Princeton Univ Press, Princeton, NJ

Weygoldt P (1977) Communication in crustaceans and arachnids. In: Sebeok TA (ed) How animals communicate. Indiana Univ Press, Bloomington, Indiana, pp 303–333

Zacharuk RY (1980) Ultrastructure and function of insect chemosensilla. Annu Rev Entomol 25:27–47

Zimmermann W (1975) Biologische und rasterelektronenmikroskopische Feststellungen an Oecobiinae, Uroecobiinae und Urocteinae. Dissertation, Univ Bonn

VIII Trichobothria

ANDREAS REISSLAND and PETER GÖRNER

CONTENTS

1 Introduction

Trichobothria are hair sensilla on the integument of various terrestrial arthropods. They are characterized by a cup-shaped cuticular structure (Latin *bothrium*, cup) out of which a hair (Greek ϑϱίξ, τϱιχόσ, hair) protrudes into the air. This hair is either long and thin (therefore in insects the sensillum is usually referred to as thread-hair) or it has the shape of a racket or a club. In spiders trichobothria have been known for a century. In 1883 Dahl observed that they were deflected by the sound of a violin and therefore called them Hörhaare (hairs of hearing). The racket-shaped trichobothria were described by Simon in 1892. The respective literature on spider trichobothria has been reviewed by Chrysanthus (1953), Schuh (1975), Krafft and Leborgne (1979) and Barth (1982).

Here we will give a survey on where trichobothria occur, briefly describe the morphology of these sensilla in spiders and then deal in detail with their physiology and behavioral significance. Nothing is known yet on information processing in the central nervous system that links data from the receptor with behavior.

Universität Bielefeld, Fakultät für Biologie, Postfach 8640, D-4800 Bielefeld 1, Federal Republic of Germany

2 Occurrence and Definition

Trichobothria are found on the walking legs and pedipalps of all true spiders (Millot 1968), but not in Solifugae (wind scorpions), Ricinulei and Opiliones (harvestmen).

Not only the distinct location of the trichobothria in different arthropods (Table 1) suggests that they are apomorphal structures. Indications as to the different origin of insect and arachnid sensilla are the following: (1) Trichobothria (thread-hairs) in insects are innervated by one mechanosensitive receptor cell (see, e.g., Altner 1977; Juberthie and Piquemal 1977), while those in arachnids possess several such cells (Christian 1971, 1972; Ignatiev et al. 1976; Meßlinger 1981). (2) The mechanosensitive dendrite of the insect sensilla ends in the distally open molting channel of the hair, whereas the dendrites of spiders, mites, and myriapods have only indirect contact with the hair base (Gossel 1936; Görner 1965; Christian 1971, 1972; Meßlinger 1981). Whether the trichobothria of scorpions and spiders are homologous structures is not clear (Weygoldt and Paulus 1978).

Despite the different phylogenetic origins of the trichobothria it is worthwhile comparing them from a functional point of view: the sensory system may have similar biological tasks, and there must be common reasons for the convergence of trichobothrial morphology. Let us first have a look at what is known about biological utilization of the organs.

Scorpions use their trichobothria in anemotactic orientation (Linsenmair 1968) and for the location of prey (Krapf on *Androctonus* and *Buthus*, personal communication). In cockroaches it has been shown that the "normal" trichobothria serve for detecting puffs of air which signal an approaching predator (Camhi and Tom 1978 a, b; Camhi 1980; Tautz and Markl 1978). Trichobothria with a club-shaped hair on the cerci of crickets are gravity receivers Nicklaus 1969; Bischof 1974, 1975; Horn and Bischof 1983) and possibly ground-vibration receivers. Some trap-door spiders have spoon- or racket-shaped trichobothria on their front leg tarsi, which they probably use for the detection of substrate vibrations that indicate nearby prey (Buchli 1969). The

Table 1. Occurrence of trichobothria in arachnids other than spiders

Animals	Location	Reference
Pseudoscorpions	Pedipalps	Weygoldt 1966 Gabbut 1969, 1972 v. Helversen 1966 Mahnert 1976
Scorpions	Pedipalps	Vachon 1973 Hoffmann 1965, 1967 Meßlinger 1981
Mites	Body and tarsi	Pauly 1956 Tarman 1961 Haupt and Coineau 1975

spectrum of biological functions is obviously wide. Surprisingly, one of the oldest ideas, i.e., that trichobothria might be tactile organs (Hansen 1917) has been confirmed recently: Bauer (1982) showed that carabid larvae detect collemboles, on which they feed, by touching them with their antennal trichobothria. The touch is slight enough for the springtails not to notice, so that the larvae have a chance of catching them.

It is obvious that not only the structure, but also the biological function of trichobothria appears to be diverse. The reason for this is the morphological definition which refers mainly to the cup-like socket of these hair sensilla. We would arrive at a more functional definition if we precluded such sensilla that have a broadened hair. Since most "normal" trichobothria share the property of being air movement receivers (cf. Sects. 4.2 and 4.3), we could redefine trichobothria as sensilla that are receivers of weak air currents and sound near-fields. However, in this case we would be excluding the tactile sensilla. To avoid such difficulties we will leave the term as it is and specify it when necessary.

Experiments on the biological function of the trichobothria in spiders will be dealt with in detail at the end of this chapter. First we will examine their morphology and their physiology.

3 Morphology of Spider Trichobothria

Distribution. Spider trichobothria are located on the tibia and the tarsal segments of all walking legs and the pedipalps. They stand together in groups or rows which are innervated by a common nerve (see below). This order is characteristic of each family and, in some cases, even of the species. Trichobothria have therefore been used as taxonomic characters (Dahl 1911; Lehtinen 1967, 1975; Emerit 1976). Scorpions and pseudoscorpions can be classified by this criterion (Vachon 1973; Mahnert 1976).

The trichobothria of spiders develop and grow gradually during ontogeny (Emerit 1970), the first ones appearing during the first molting. Their development has been investigated in Liphistiidae (Vachon 1958), Pholcidae (daddy-long-legs spiders) (Vachon 1965), Argiopidae (orbweavers) (Emerit 1972) and Pisauridae (nursery-web spiders) (Emerit 1974; Bonaric 1975; Emerit and Bonaric 1975). The first hairs that occur in *Agelena labyrinthica* – a funnel-web spider – are those that eventually become the most distal ones. Their topographic position on each leg segment moves distally with each molt while new sensilla emerge (unpublished work). The final number of hairs in the tarsal and the metatarsal rows and the tibial groups (cf. Görner 1965) varies somewhat with the eventual size of the adult animal (Fig. 1).

Length and Surface Structure. Agelenidae have about 10 trichobothria on each pedipalp and about 25 trichobothria on every walking leg (Palmgren 1936). Their lengths diminish within most fields from distal to proximal. The longest hair in *Agelena labyrinthica* is found on the metatarsus, measuring up to 755 μm. This is similar to the hair length in the caterpillar of the cabbage moth

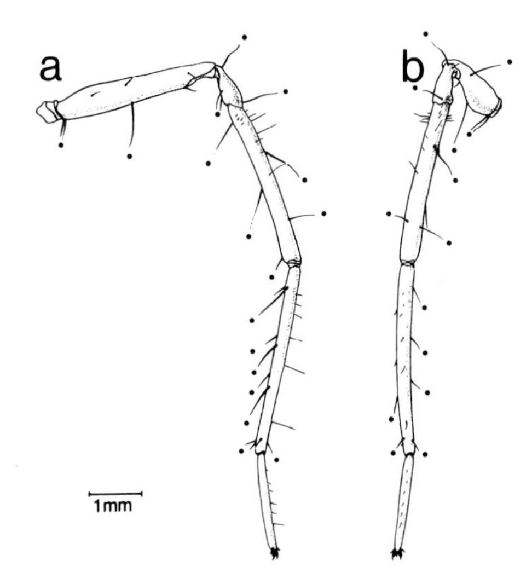

Fig. 1 a, b. Topography of trichobothria of first right leg of *Agelena labyrinthica*. **a** caudal, **b** lateral view. Bristles are also drawn and indicated by *dots*. (Görner and Andrews 1969)

Barathra brassica (up to 650 μm) (Tautz 1977). In the wandering spider *Cupiennius* the maximal length is 1.3 mm (Barth 1982), in the bird spider *Sericopelma* it is 2.5 mm (Den Otter 1974). In other arthropods the figures are similar, e.g., 1.8 mm in scorpions (Meßlinger 1981) and 3 mm in the cricket *Gryllus* (Gnatzy and Schmidt 1971).

The hair bears many fine emergences twisted around its axis (Figs. 2a, b, 3) so that the viscous forces that deflect the hair are increased. Scanning electron microscope (SEM) pictures of scorpion trichobothria (Meßlinger 1981) show that in some families the hair is flat and in some cases also twisted probably to increase its surface. In both scorpions and crickets the hair can be longitudinally ribbed.

Lumina. All trichobothrial hairs are hollow. In *Agelena* the cavity is subdivided into three lumina of varying width (Christian 1971). All are connected to the epithelial lumen but do not contain dendrites. The dendrites end in a helmet-like cuticular structure (Figs. 2c, 3) (Görner 1965). In scorpions, the dendrites end in a short separate basal lumen (Meßlinger 1981) and in insects in an open tube, the molting channel (Gnatzy and Schmidt 1971; Schmidt and Gnatzy 1971; Gnatzy 1976).

Hair Base. The hair base is suspended in its socket by a corona of radial fibrils that are covered distally by a membrane (Fig. 3) terminating the hair socket (Fig. 2c). Such a socket or inner cup is typical of arachnid trichobothria. Fibrillar hair suspension also occurs in scorpions (Ignatiev et al. 1976; Meßlinger 1981), in mites (Haupt and Coineau 1975) and in some insects (Juberthie and Piquemal 1977), while in other insects the hair is hinged (Gnatzy and Schmidt 1971; Gnatzy 1972, 1976; Gnatzy and Tautz 1977, 1980). Beneath the membrane the hair is connected with the helmet (Fig. 3). By microscope observation

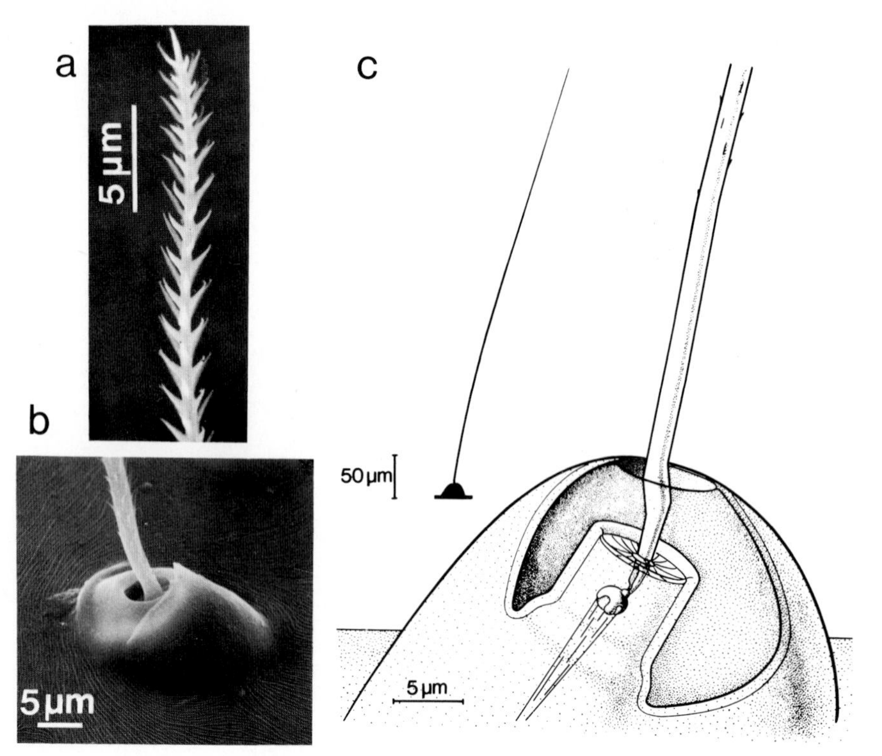

Fig. 2a–c. Spider trichobothrium. **a** hair tip (*Tarantula*); **b** cup and basal hair shaft (*Tegenaria*); **c** microscopic view (*Tegenaria*). (**a, b** courtesy H. Müller; **c** after Görner 1965)

Görner (1965) found that the helmet appears to be tri-lobed in most tibial sensilla. This has not been confirmed by SEM investigation so far (cf. Christian 1972).

Receptor Cells. In *Tegenaria* the trichobothrium receptor cells are situated close to the cup just beneath the hypodermis (Gossel 1936). Four ciliary dendrites insert into the helmet (Christian 1971, 1972). Here at least one of them possesses a tubular body (tarsi and metatarsi of legs); in some groups a tubular body is found in three or all four dendrites (on all tibiae). If the hair is deflected in a given direction, the appropriate dendrite is bent over the rim of the helmet and most probably the tubular body is compressed (Görner 1965). Dendrites with tubular bodies have been found in all cuticular mechano-sensilla of arthropods (Thurm 1964; Barth 1971, 1981; Gaffal and Hansen 1972; McIver 1975 and others), and it is agreed upon that it is their deformation that causes the receptor potential (Thurm 1964, 1982).

Extracellular electrophysiological recordings from spiders upon hair deflections in different directions (Görner 1965; Harris and Mill 1977; Reißland 1978) have shown that either two or three receptors respond. The function of the fourth dendrite which, like the other dendrites, probably belongs to a sepa-

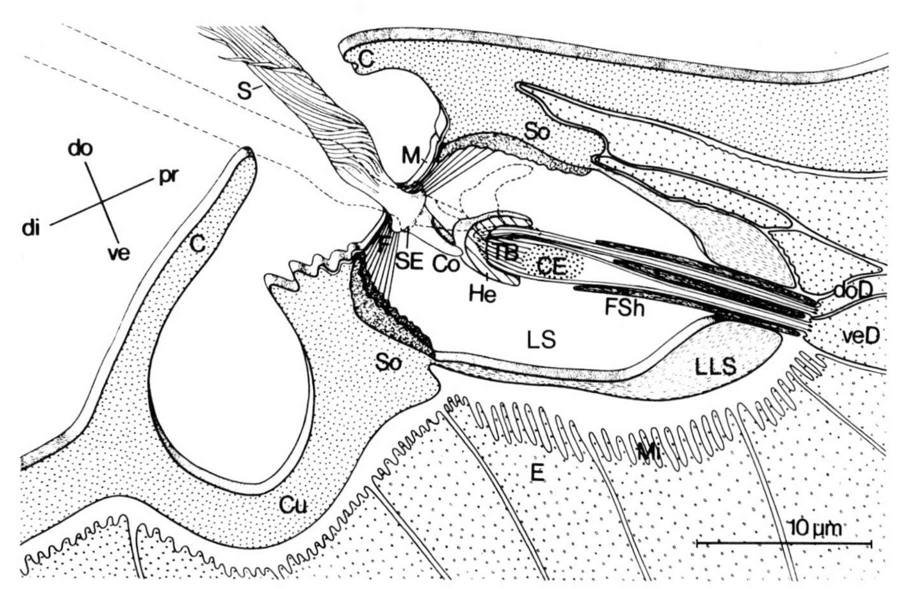

Fig. 3. Fine structure of trichobothrium (*Tegenaria*). *C* cup; *Cu* cuticula; *S* hair shaft; *SE* extension of hair shaft; *F* fibrils; *M* membrane; *So* cuticular socket; *Co* connection of hair with helmet; *He* helmet; *TB* tubular body; *CE* extension of cilia; *FSh* fibrillar sheath; *LS* receptor lymph space; *LLS* inner lining of lymph space; *E* epidermis; *Mi* microvilli; *ve* ventral; *do* dorsal; *di* distal; *pr* proximal; *veD* ventral, *doD* dorsal dendrite. (Christian 1972)

rate bipolar neuron, is unknown. Similarly, recordings from scorpion trichobothria revealed that fewer cells respond (one in *Euscorpius carpathicus,* Hoffmann 1967; two to three in *Buthus eupeus,* Ignatiev et al. 1976) than can be found by EM methods (probably four in *Euscorpius,* Meßlinger, personal communication; six to seven in *Buthus*). It appears that supernumerary dendrites are characteristic for cuticular mechanoreceptive sensilla in arachnids. In all cases the function of the silent cells remains obscure. Peripheral synapses that occur occasionally in arachnids (Yamashita and Tateda 1981; Hayes and Barber 1982; Foelix 1975, 1976; Foelix, Chap. X, this Vol.) have not been found in these sensilla.

The Cup. The outer cuticular cup (bothrium) that surrounds the basal hair shaft is strengthened inside by lamellae of varying structure (Hoffmann 1967; Barth 1969; Meßlinger 1981). The obvious purpose of the cup is to limit the deflection of the hair, preventing damage to its suspension. In Agelenidae the cup is inclined by about 30° distally (Görner 1965) and the hair can move 23° distally, but only 13° proximally before touching the edge of the bothrium. Curiously, the stimulated receptor cell concerned stops discharging at extensive hair deflections in distal direction, while it continues to respond in proximal direction (Reißland 1978). This can be explained by asymmetries of the bothrium, the helmet, or the dendritic attachment.

The innervation of the extremities of spiders consists of three parallel nerves (Parry 1960; Rathmayer 1966; Emerit 1967, 1969; Foelix et al. 1980) of which the dorsal nerve (nerve C) supplies the trichobothria and other mechano-receptive hairs. At each of the three segment groups, trochanter-femur, tibia, and metatarsus-tarsus, this nerve bifurcates into a dorsal and a ventral root. The ramifications of the dorsal root supply the various trichobothrial groups (see also Foelix, Chap. VII, this Vol.).

4 Functional Properties

The exposure of the trichobothrial hairs to the air and their good movability suggest at least one of the following functions: (1) the reception of transient or stationary air currents; (2) the reception of air vibrations; (3) the reception of substrate vibrations that result in motion of the hair base relative to the hair. Experiments and calculations that were carried out to test these possibilities will be described in the following. There is, however, the further possibility, that the trichobothrium is a tactile organ as found by Bauer (1982) for carabid larvae. This alternative can be discarded for spiders as most of the trichobothria are shorter than the majority of the other mechano-receptive hairs and bristles (Foelix 1976; McIver 1975; Foelix, Chap. X, this Vol.).

4.1 Detection of Transient Air Movements

We shall consider now whether airstreams caused by prey animals can be detected by the trichobothria. Generally these currents are weak.

Humming midges emit a dipole field. This was shown by sound pressure measurements on the yellow-fever mosquito *Aedes aegypti* by Tischner and Schief (1955). Tautz and Markl (1978) measured the current and the vibrational field generated by a wasp (*Dolichovespula media*) by aid of a hot-wire

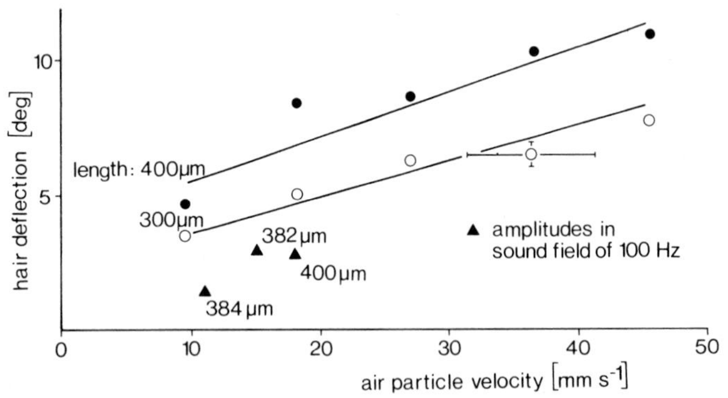

Fig. 4. Deflection of trichobothria of different lengths in laminar airstream (●, ○) and sound far-field (▲) (*Tegenaria*). *Bars* indicate estimated errors of measurement. (Reißland and Görner 1978)

anemometer. They found that as far as 70 cm behind the wasp the air movement is strong enough to deflect the trichobothria of *Barathra* caterpillars. The airstream produced by the beating wings measured at a distance of 50 cm from the wasp amounted to 9 ± 2 cm s^{-1}.

Reißland and Görner (1978) report that a buzzing fly held 1 cm away from a house spider (*Tegenaria* sp.) caused deflection of trichobothria and oscillation about the resulting position. Figure 4 shows the velocities of stationary laminar airstreams that cause hair deflections in the range of physiological efficiency (Görner 1965; Harris and Mill 1977; Reißland 1978, 1979). Their order of magnitude is 1 cm s^{-1}. This means that air currents generated by insects are strong enough to deflect the trichobothria. However, we shall see that it is not the position of the hair but its movement that stimulates the receptor cell.

4.2 The Sound Field of Buzzing Insects

Let us now consider the second candidate stimulus, viz. the airborne sound field. This has been the subject of detailed investigations during the last 20 years (Görner 1965; Görner and Andrews 1969; Markl and Tautz 1975; Tautz 1977, 1978, 1979; Reißland and Görner 1978; Fletcher 1978). We will first take a closer look at the physics of sound in a gaseous medium.

The propagation of sound is determined by the wave equation which reads in Cartesian coordinates (cf., e.g., Skudrzyk 1971)

$$\Delta p = \frac{1}{c^2}\frac{\partial^2 p}{\partial t^2}, \qquad \Delta = \frac{\partial^2}{\partial x^2} + \frac{\partial^2}{\partial y^2} + \frac{\partial^2}{\partial z^2}, \tag{1}$$

where p is the sound pressure, c the sound propagation velocity ($c = 343$ ms^{-1} at 20 °C and a static pressure of 1 bar), and t time. c refers to the propagation of the dynamic disturbance of the medium but not to the particle velocity of the air molecules that vibrate about their resting position. As every sound event can be described as a superposition of sine waves, we need only consider these. The wavelength λ of a sine wave is given by

$$\lambda = \frac{c}{f}, \tag{2}$$

where f denotes the frequency of the wave. We shall see that frequencies which trichobothrium receptors can transduce into pulse trains range from 0.1 Hz to 1 kHz, corresponding to sound wavelengths from 3430 m to 34.3 cm.

The wave equation has many solutions from which we shall pick the ones that apply to our biological problem. The most simple solution describes a sound field that is generated by a pulsating sphere (monopole). Further solutions describe an oscillating solid sphere (dipole) and other more complicated sources (Skudrzyk 1971). A whirring insect emits a dipole field, so we are interested in the solution describing a wave that travels away from a dipole.

The monopole function can be obtained if we introduce the spherical coordinates r, θ and φ into the wave equation. For spherical waves this then depends only on r (distance from sound source) and reads

$$\frac{\partial^2 (r\Phi)}{r^2} = \frac{1}{c^2}\frac{\partial^2 (r\Phi)}{\partial t^2}, \tag{3}$$

where Φ denotes the velocity potential from which the sound pressure p and the air particle velocity $\vec{v}$ can be derived:

$$p = \varrho \frac{\partial \Phi}{\partial t} \quad \text{and} \quad \vec{v} = -\operatorname{grad} \Phi. \tag{4}$$

The dipole solution can be derived by superimposing two monopole functions, generated by a positive and a negative monopole. The final solution for the sound pressure generated by a dipole reads (cf. Skudrzyk 1971)

$$p = -\frac{k^2 \varrho c D}{4\pi r} \cos\Theta \left[\cos(\omega t - kr) + \frac{1}{kr}\sin(\omega t - kr)\right], \tag{5}$$

where ϱ is the density of the air ($\varrho = 1{,}2\ \mathrm{kg\ m^{-3}}$), $k = 2\pi/\lambda$ is the wave number, D is the dipole moment, and $\cos\Theta$ describes the directionality of the dipole. The first term has, apart from the factor $\cos\Theta$, the form of the solution for a pulsating sphere and is therefore a monopole function. It decreases with $1/r$. The second term is the dipole function. It decreases with $1/r^2$.

It must be noted that p is a scalar quantity. An object that is small compared with the wavelength can be compressed, but not moved, because there is no pressure gradient between both sides of the object.

The air particle velocity, on the other hand, is a vector. It has one component (v_r) that is directed radially away from the dipole and one (v_ϑ) that is directed transversally to v_r, lying in the plane of the dipole. The radial component is (cf. Morse and Ingard 1968)

$$v_r = -\frac{k^2 D}{4\pi r}\cos\Theta\left[\cos(\omega t - kr) + \frac{2}{kr}\sin(\omega t - kr) - \frac{4}{k^2 r^2}\cos(\omega t - kr)\right]. \tag{6}$$

The first term is a monopole function. It oscillates in phase with the pressure and, consequently, is responsible for the energy transport in the sound wave. Like the pressure, it declines with $1/r$. The second term of v_r represents the particle flow that is $90°$ out of phase with the pressure and therefore does not contribute to the energy transport. It decreases with $1/r^2$ and thus is restricted to the vicinity of the sound source: doubling the distance results in a decrease by 12 dB. This term is called near-field velocity, and the monopole term the far-field velocity of the sound field. Near the dipole, the near-field velocity is proportional to k and thus is greater than the far-field velocity which is proportional to k^2. The third term decreases with $1/r^3$ and can be neglected except for very small distances from the sound source.

The transversal component is (Morse and Ingard 1968)

$$v_\vartheta = \frac{kD}{4\pi r^2}\sin\theta\left(\sin(\omega t - kr) - \frac{1}{kr}\cos(\omega t - kr)\right). \tag{7}$$

v_ϑ is zero in the direction $\vartheta = 0$, i.e., in the direction of the dipole axis. The first term is 90° out of phase with the dipole oscillation and 180° out of phase with the radial component of the near-field velocity. Figure 5a shows the result of an experiment that confirms this phase difference. The second term can be neglected except in very close vicinity to the sound source.

4.3 Dynamics of the Receiver

Let us now have a look at the physics of the air particle velocity receiver (Markl and Tautz, Reißland and Görner, Fletcher, already cited; Barth and Blickhan 1984). Trichobothria can be moved only by the near-field velocity of biological sound sources.

If the trichobothrial hair is considered to be a stiff cone that is hinged at its base (Tautz 1978) the moment of inertia I_h can be calculated as

$$I_h = \int_0^l \pi \varrho_0 r^2 y^2 dy = \pi \varrho_0 r_0^2 l^3/30 \tag{8}$$

(Fletcher 1978), where l is the length and r_0 is the basal radius of the hair and $\varrho_0 \approx 1100 \text{ kg m}^{-3}$ is its mass density. We will assume $l = 500 \,\mu\text{m}$ and $r_0 = 1 \,\mu\text{m}$, which are dimensions found for spider trichobothria.

Because of the small hair diameter the force acting on the hair in a sound field is caused almost entirely by the viscosity of the air (Stokes 1851, Fletcher 1978). This force can be separated into two components. One (F_1) is the damping force caused by the air that streams past the hair. It is 180° out of phase with the hair velocity. The second component (F_2) arises from the air that entrains the hair by viscous forces. This is in phase with the hair movement and appears as a mass added to the mass of the hair itself. The moments $L_{1,2}$ exerted on the hair by the two forces are

$$L_1 = \int_0^l F_1 y \, dy \quad \text{and} \quad L_2 = \int_0^l F_2 y \, dy. \tag{9}$$

If the deflection angle of the hair is denoted by α, the equation of motion reads

$$I_a \ddot{\alpha} + k \alpha = L_1 + L_2, \tag{10}$$

where k α describes the restoring torque caused by the elastic suspension at the hair base. Calculating L_1 and L_2 and introducing them into the equation, we obtain (Fletcher 1978)

$$(I_h + I_a) \ddot{\alpha} + \frac{2}{3} \varrho \nu G l^3 \dot{\alpha} + k \alpha = \varrho \nu G l^2 \left(v - \frac{\pi \dot{v}}{4 g f} \right), \tag{11}$$

where $\nu = 1.5 \cdot 10^{-5} \text{ m}^2 \text{ s}^{-1}$ is the dynamic viscosity of air under normal conditions; $g = 0.58 + \ln \frac{r}{2} \sqrt{\frac{2 \pi f}{\nu}}$; $G = -\frac{g}{g^2 + \left(\frac{\pi}{4}\right)^2}$; v is the sound particle velocity. The equation describes a damped harmonic oscillator driven by the force that appears on the right side.

$$I_a = -\frac{\pi \varrho \nu G l^3}{6 g f} \tag{12}$$

is the moment of inertia of the entrained air particles. As I_a is positive and varies with $1/f$ it contributes increasingly more to the overall moment of inertia with decreasing frequency.

The ratio of I_h and I_a for a spider trichobothrium amounts to 2.7×10^{-3} at 1 Hz, 1.1×10^{-2} at 10 Hz, 7.5×10^{-2} at 100 Hz, 0.46 at 1 kHz and to 1 at 1.45 kHz. This means that the moment of inertia of the hair itself can be neglected in the whole range of frequencies that affect the trichobothria.

Measurements in an intense sound field did not reveal resonances of trichobothria in spiders and the same counts for the deflection by a weak alternating electric field in a vacuum (Reißland and Görner 1978).

This was doubted by Tautz (1979), who found a resonance of approximately 150 Hz for *Barathra* caterpillar trichobothria if driven by vibrating air and not by a strong electric field. To settle the discrepancy we observed the free movement of a spider trichobothrial hair from a deflected position back to its resting position. It appeared that it returned to its resting position in about 1 s without overshoot and was therefore hypercritically damped. Resonances cannot be observed in this case.

For the equation of motion [Eq. (11)], the strong damping of spider trichobothra has the consequence that the first term on the left side can be neglected. Solving the simplified equation for the case of a resting leg and quiet air, we obtain an exponentially decreasing function α (t). In the case of the harmonic sound particle velocity $v = v_0 \sin 2\pi f t$ the equation reads (cf. Fletcher 1978)

$$\frac{2}{3} \varrho \nu G l^3 \dot{\alpha} + k \alpha = \varrho \nu G l^2 v_0 \sqrt{\left(1 - \frac{\pi^2}{2 g}\right)} \sin (2 \pi f t - \delta_0), \tag{13}$$

where $\tan \delta_0 = \pi^2/2\,g$. For the present case we get $\delta_0 = -38.1°$ at 10 Hz and $-43.8°$ at 100 Hz. The equation has no compact solution. There is a phase lead of about 30° in the frequency range from 10 to 120 Hz (Fig. 5a). Therefore the system differentiates the input signal or, in other words, has high-pass properties.

Obviously, the trichobothria are deflected by airborne vibrations if the substrate on which the spider sits is at rest. However, Hergenröder and Barth (1983a) have shown that substrate-borne vibrations may also cause deflections (Fig. 5b). These are substantially smaller than deflections caused by sound propagation perpendicular to the hair.

The movability of the trichobothria in spiders is about the same in all deflection directions as was shown by measurements in the sound fields at various frequencies (Reißland and Görner 1978).

For other trichobothria the directional mechanical sensitivity is anisotropic (Nicklaus 1965, 1967; Hoffmann 1967; Bischof 1974; Gnatzy and Schmidt 1971; Gnatzy and Tautz 1980; Tautz 1977). In *Barathra* caterpillars sensilla can be found with both isotropic and anisotropic

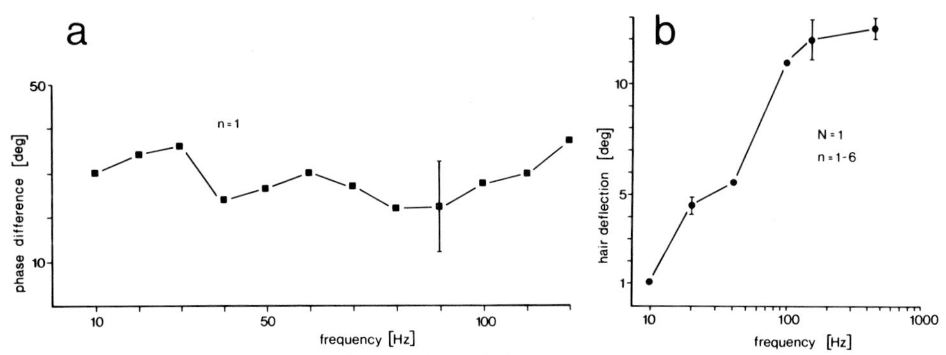

Fig. 5. **a** Phase-frequency response in sound near-field (*Tegenaria*); **b** amplitude-frequency re-response upon substrate vibration (*Cupiennius*). (**a** Reißland and Görner 1978, **b** Hergenröder and Barth 1983a)

directional sensitivity. This can be explained by the bend of the hair which results in direction-dependent viscous forces.

4.4 Transfer Properties of the Receptor Cell

Extracellular electrophysiological investigations of trichobothrial receptor cells have been made in Dictynidae (Harris and Mill 1977), in Agelenidae (Görner 1965; Reißland 1978, 1979), in Argiopidae (Frings and Frings 1966), in scorpions (Hoffmann 1967; Ignatiev et al. 1976), and in insects (Nicklaus 1965; Camhi 1969; Drašlar 1973; Bischof 1975; Tautz 1978; Buño et al. 1981; Tobias and Murphey 1979; Westin 1979). In spiders the receptors are not spontaneously active as they are in cockroaches (Buño et al. 1981).

The signal transfer properties of the receptor cell typical for *Tegenaria* trichobothria has been summarized in a heuristic model (Fig. 6). The input signal is a deflection of the hair by which the dendritic tubular body is compressed.

Rectifier. The rectifier symbolizes that the majority of the cells only respond to deflections away from the resting position (Fig. 7a). However, in *Tegenaria*

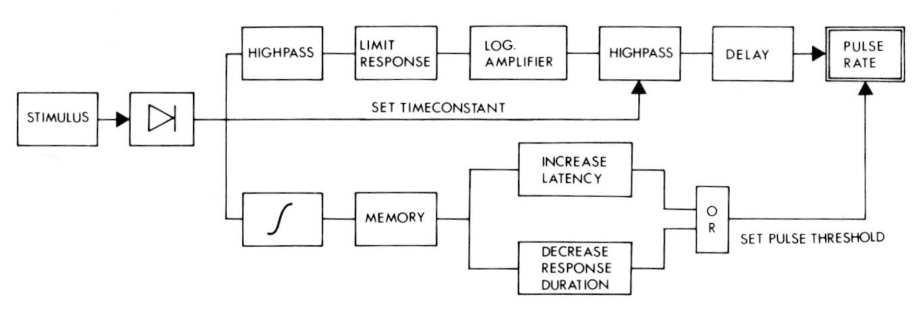

Fig. 6. Model of the *Tegenaria* trichobothrium receptor. Explanation in the text

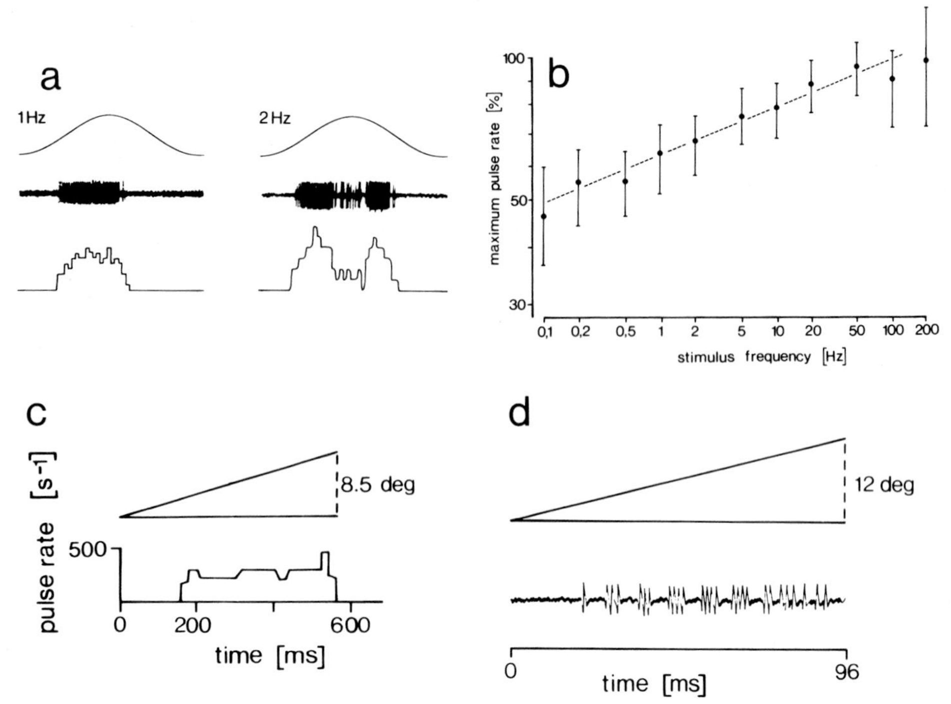

Fig. 7a–d. Response of *Tegenaria* trichobothrium receptors to hair deflections (direct coupling to vibrator) from resting position to one side; **a** *left* normal receptor, responding to cosine deflection; *right* receptor responding to return movement as well; **b** amplitude-frequency response (mean of 13 receptors, *bars* indicate standard deviation; hair deflection amplitude approximately 8.5°). **c** response to ramp stimulus; **d** volley discharge. Note that the first differentiation of natural stimuli which results from the transformation of air movement into hair movement is not to be seen in the figure

30% of the receptor cells responded to the return movement as well (Reißland 1979). The responses were generally weaker and only occurred in the medium working range.

High-passes. The upper branch of the model contains two high-passes. The first high-pass means that the input signal is differentiated. Figure 7a demonstrates this for a cosine stimulus that causes a pulse rate of the same shape with a phase lead of approximately 90°. As the pulse rate is constant at constant hair deflection velocity, the receptor is a velocity receiver (Fig. 7c). This is typical at least for the upper half of the working range (deflection velocity from 1 to 100° s^{-1}). Half of the receptors investigated are purely phasic for all hair deflection velocities.

In scorpions, trichobothria are phasic in the lower working range (Hoffmann 1967; unpublished work of the authors), however, some have an initial peak in the pulse rate curve if a ramp stimulus of higher velocity is applied. In cockroach and cricket trichobothria such a peak occurs regularly (Buño et al. 1981; Bischof 1975).

The second high-pass refers to the slow exponential decrease of the receptor response to step stimuli. Contrary to the maximal pulse rate, these depend on the hair deflection angle which therefore is responsible for the "memory" of the receptor.

Response Limiter. The response limiter in the model means that the pulse rate cannot exceed a maximal value, which can be close to 1000 impulses per second. In scorpions the figures are similar. The maximal pulse rate does not depend on the hair deflection angle (unpublished work).

The exponential decline of the response is in discordance with the power-law dynamics (proportionality to t^{-k}) of the adaptation course of many other sensory cells (Thorson and Biederman-Thorson 1974; Barth 1967, 1978; Tautz 1978). We may suppose that in the case of trichobothria, adaptation is caused by a single relaxation process, while in most other cases several such processes distributed in time and space are superimposed (Thorson and Biederman-Thorson 1974).

Amplifier. The *logarithmic amplifier* in the model symbolizes the fact that the amplitude-frequency response is a logarithmic curve (Fig. 7b). Above 200 Hz the response consists of only one pulse per cycle.

Delay. The delay is a constant period of approximately 2 ms (Reißland 1979) that is probably mainly due to pulse generation and propagation to the site of pulse recording around 2 mm from the trichobothrium.

Adaptation. The lower branch of the model describes the mechanism of receptor adaptation. The *integration symbol* means that adaptation takes place on repeated stimulation. The *memory* symbolizes the experimental finding that the effect of hair deflection decreases with time; the decrease has been assumed to be exponential. In trichobothrial receptors, time-dependent processes can include transient behavior of strain-sensitive conductances in the transducer membrane producing the generator current and accommodation of the axonal encoder determining the impulse frequency (Mann and Chapman 1975). As accommodation appears to be negligible except for the lowest deflection velocities, we suppose that adaptation is caused mainly by the properties of the transducer membrane. The maximum instantaneous pulse rate is hardly affected by adaptation. Therefore it is likely that the spider evaluates this parameter.

The effect of adaptation can be measured by the *increase of latency* and the *decrease of response duration.* The pulse train length increases with the hair deflection amplitude. Pre-stimulus static hair deflection has an additional adaptation effect (Fig. 8). Except for the higher stimulus frequencies the receptor eventually ceases to respond. In the model, adaptation is assumed to be an increase of pulse generation threshold.

The receptor model has been realized on an analog computer. Some calculated curves are shown in Fig. 9. They show the same essential features that have been found by electrophysiological recordings from original trichobothrium receptors in *Tegenaria.* A few experimental findings which have not been incorporated in the model will be described in the following.

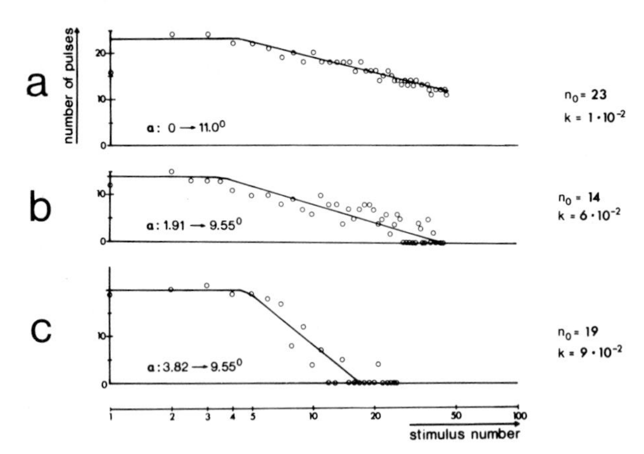

Fig. 8a–c. Receptor adaptation (*Tegenaria*). **a** adaptation due to repeated stimulation with constant deflection velocity; **b** adaptation after a static pre-stimulus hair deflection that continued for 4 min; **c** same for a greater pre-stimulus hair deflection. α deflection angles of pre-stimulus and stimulus deflection; n_0 initial number of pulses; k time constant of the adaptation. (Reißland 1978)

The threshold hair deflection angle at which the receptor responds does not depend on the deflection velocity except for very slow deflections. It amounts to 1°–2°, a value that is in agreement with that of scorpions (2°–3°, Hoffmann 1967) and of *Barathra* caterpillar larvae (2.5°, Tautz 1978). At smaller deflections the dendritic tubular body obviously is not deformed sufficiently to induce action potentials. At very slow hair deflections the threshold angle increases which is probably due to accommodation.

The lower limit of the working range of spider trichobothria is approximately 1° s^{-1}. Here the pulse train is interrupted by irregular breaks, and sometimes the receptor is silent for a whole cycle. The tip of a 500-μm-long hair then has a velocity of less than 90 μm s^{-1}, a value that is smaller by a factor 100 than the airstream velocity necessary for physiological hair deflections (Fig. 4). Assuming this velocity is reached within a time interval as long as 10 s, there is still a safety factor of 10 by which the air particle velocity exceeds the hair tip velocity necessary for evoking a receptor response. Therefore every slow natural air movement that deflects at least the longer hairs causes a response of the tri-

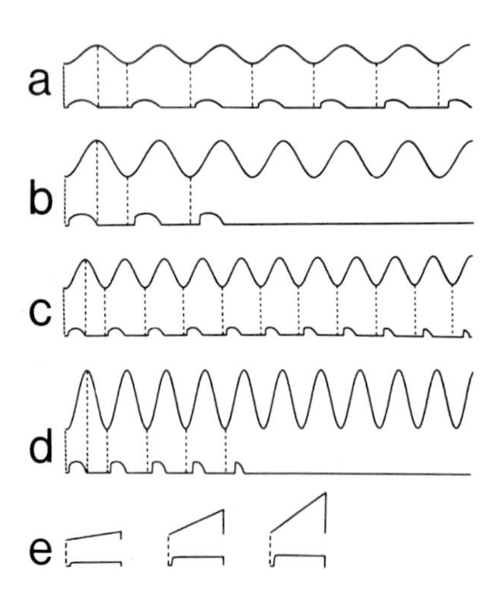

Fig. 9a–e. Input signals, equivalent to direct coupling as in Fig. 7 (*upper curves*) and output signals (*lower curves*) calculated by an analog computer according to the model of Fig. 6. **a** Latency increases upon repeated stimulation; **b** response duration decreases if input amplitude is doubled; **c** both effects diminish by doubling of frequency; **d** both effects increase by doubling of amplitude; **e** constant responses to constant velocity inputs, increasing logarithmically with slope of input curves. Note that the output curves of **a–d** are not pieces of sine curves but of logarithmically deformed sine curves

chobothrial receptors. If the hair is moved sinusoidally with an amplitude of about 10°, the receptors respond to signals of 0.1 to 0.2 Hz (Reißland 1979). Such slow signals are emitted neither by prey nor by predators. These measurements, therefore, suggest that the trichobothria in funnel-web spiders may serve as wind-detectors.

Some of the trichobothrium receptors in Agelenidae respond to constant hair deflection velocity with a volley-like pulse train rather than a constant pulse rate (Fig·7d). Therefore we suppose that the signal time course is evaluated by the superposition of responses of different receptors.

4.5 What is the Adequate Stimulus?

The question as to what is the adequate stimulus for a trichobothrium requires consideration of both the mechanical and the receptor cell transfer properties. The mechanical structure has high-pass properties, and the same applies for the receptor cell. Consequently, the natural input signals are differentiated twice. Therefore the adequate stimulus is suggested to be air particle acceleration. Behavioral experiments on cockroaches are in agreement with this assumption. They escape from puffs of air that exceed a threshold acceleration (Camhi 1980).

If we assume that the adequate signal range is where the least signal deformation by the receptor takes place we arrive at a signal frequency range around 1 Hz. Even if a straightforward conclusion cannot be drawn from this consideration we can at least suppose that the hum frequency of insects is not a candidate parameter to be evaluated by the trichobothria. If the peak instantaneous pulse rate were evaluated, a single trichobothrium would be able to encode a constant hair deflection velocity of around 100° s^{-1} or a frequency up to about 200 Hz. By summing the outputs of several sensilla, even higher frequencies could be discriminated, as in this case Wever's volley principle would apply. However, behavioral experiments appear not to support this hypothesis.

5 Behavior

Behavioral experiments with definite stimuli, that are crucial for revealing the biological function of a sensory system, have been carried out on trichobothria only during the last 15 years (Görner and Andrews 1969; Hergenröder and Barth 1983a, b; Reißland and Habigsberg, in press). From observations by Görner and by Riechert and Łuczak (1982) it is known that Agelenids are able to seize insects that fly by within a few millimeters from the spider. The plausible assumption that the trichobothria are involved in the catching reaction was tested by using a vibrating paper disc as a prey substitute (Görner and Andrews 1969). *Agelena labyrinthica* turned towards the dummy prey, and in many cases even seized it. The turning angle (for definition see legend of Fig. 10) was always somewhat smaller than the stimulus angle, a result that was also found in *Cupiennius* (Hergenröder and Barth 1983b) for turning reactions toward a buzzing fly. The turning angle as defined in both series of experiments has its vertex outside the spider and therefore is probably not the angle that is cal-

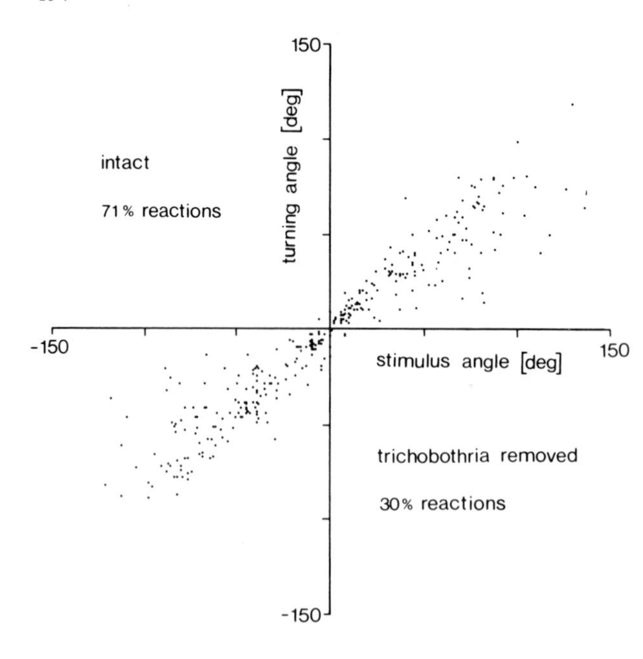

Fig. 10. Orienting reaction of *Agelena* toward a vibrating disc. Intact side of spider compared with side deprived of trichobothria. *Stimulus angle* angle between long axis of spider and axis through stimulus site and center of prosoma before reaction; *turning angle* angle between the spider's long axis before and after reaction; stimulus frequency 20 Hz, N=5

culated by the animal. Experimental data with respect to a spider-centered coordinate system are not available so far.

The reaction frequency of *Agelena labyrinthica* is substantially reduced by ablation of trichobothria (Görner and Andrews 1969). When unilaterally deprived animals were stimulated from the front they turned to the intact side. Therefore the trichobothria were made responsible for prey detection and considered to be Ferntastsinnesorgane (organs of touch at a distance, term coined by Dijkgraaf in 1947). Later experiments showed that the turning angle is the same on stimulation of either side of unilaterally deprived spiders (Fig. 10). The orienting reaction was the same in both stimulus directions even if stimulation was performed from the front of the spider, a discordance that is not yet settled. However, it was established that localization of prey is also possible without trichobothria and therefore other organs must be involved. These cannot be receivers of web vibrations induced by airborne sound, since Görner and Andrews performed many of their experiments with spiders partly sitting on solid ground. What remains as a candidate stimulus is the sound pressure field that can be received by single slit sensilla on the tarsus (Barth 1967, 1982). We will assume these to be the only such sensilla in the following.

To stimulate the trichobothria and the single slit sensilla separately different stimuli were applied: (1) a disc vibrating with a frequency of 10 Hz; (2) a bundle of fine copper wires that was periodically charged with a frequency of 200 Hz (both these stimuli deflected the trichobothria, but electrophysiological recordings showed that the single slit sensilla were not excited); (3) the disc vibrating with 200 Hz (stimulating both single slit sensilla and trichobothria) as a control stimulus. The reaction frequency upon stimulation of the trichobothria alone was significantly smaller than in the control experiments

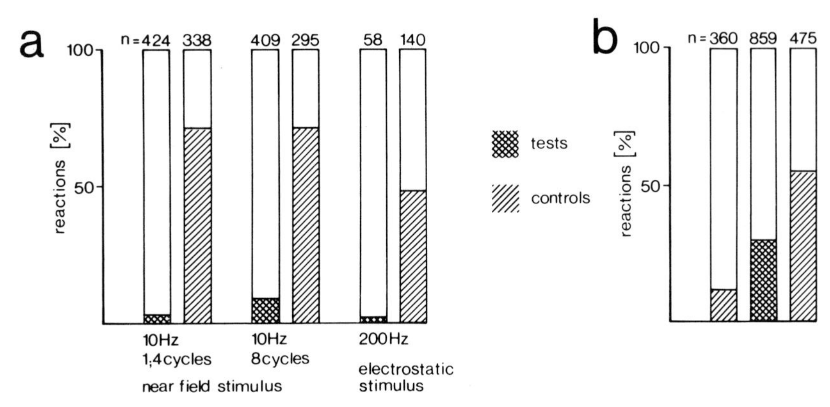

Fig. 11 a, b. Percentage of reactions of *Tegenaria* upon **a** stimulation of trichobothria by a sound near-field of 10 Hz and different duration, and by an electric field of 200 Hz; differences significant; N=4. **b** stimulation with minispeaker generating a sound far-field of 200 Hz; *right bar* control with vibrating disc; *left bar* control with silent minispeaker whether trichobothria were accidentally touched; differences significant; N = 2. (Reißland and Habigsberg, in press)

(Fig. 11 a). On the other hand, a modified miniature loudspeaker was used to generate a far-field stimulus. Excitation of the single slit sense organs alone can release a catching reaction (Fig. 11 b). Moreover, localization of the dummy does not deteriorate (unpublished work). Therefore trichobothria in agelenids are not essential for localizing prey whirring in the air. It rather appears that they usually play a role as a trigger of the catching reaction. Only if the intact side competes with the side that is deprived of trichobothria, which is the case if a stimulus is applied from the front, can the turning angle be influenced by the trichobothria.

Hergenröder and Barth (1983 a, b; see also Barth, Chap. XI, this Vol.) investigated the release of predatory and flight behavior in the wandering spider *Cupiennius salei* by vibratory stimuli. If *Cupiennius* is stimulated by substrate vibrations of a frequency lower than 100 Hz and a peak-to-peak amplitude below 140 μm, it responds in many cases with an orienting movement (jump or turn) toward the side of the stimulus, or with search movements in that direction. At higher stimulus frequencies or amplitudes increasingly more withdrawal reactions occur. The substrate vibrations also cause stimulation of the trichobothria which has the following consequences for the behavior.

(1) If the trichobothria are removed, the threshold vibration displacement for the release of withdrawal is increased (Fig. 12a), the supra-threshold probability of approach reactions is increased, and the probability of withdrawal reactions is decreased (Fig. 12b). This means that the trichobothria inhibit the readiness for approach. Moreover, Hergenröder and Barth (1983 a) have shown that the trichobothria of only one leg are sufficient to cause the full effect of inhibition (Fig. 12c, 2nd, 4th and 6th bar). A buzzing fly does not alter this result. The inhibition, however, is smaller if two adjacent legs are intact (Fig. 12c, 5th bar).

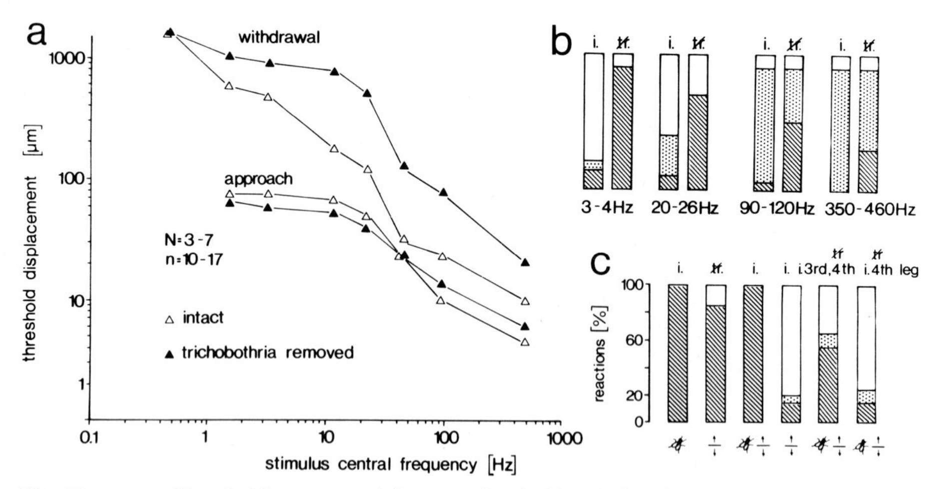

Fig. 12a–c. **a** Threshold curves and **b** supra-threshold reaction frequencies of *Cupiennius* stimulated by substrate vibration with band-limited noise centered at different frequencies. *Hatched* and *dotted areas* of histograms denote percentage of reactions of approach and of withdrawal, respectively; free areas denote no response at 140 μm stimulus amplitude. (Hergenröder and Barth, 1983a). **c** Percentage of reactions found with stimulation by substrate vibration and/or fly buzzing in air. *i* intact legs, *tr* deprived of trichobothria (5th column: all trichobothria except those on 3rd and 4th leg ablated; 6th column: all trichobothria except those on 4th leg ablated); areas of histograms as in **b**. (After Hergenröder and Barth 1983a)

(2) At stimulus frequencies above 1 Hz, the threshold for withdrawal is significantly increased by removing the trichobothria. From (1) and (2) it is evident that these sensilla reduce the readiness for predatory reactions and lower the threshold for releasing escape reactions. However, simultaneous stimulation by a fly which generates a different spatio-temporal near-field pattern in conjunction with an airborne far-field appears to compete with the inhibition of predatory behavior (Fig. 12c, 1st, 3rd and 5th bar).

(3) Moreover, the trichobothria allow discrimination between substrate-borne sine stimuli and band-limited noise centered at the corresponding frequency (10 Hz and 140 Hz tested, Hergenröder and Barth 1983a). Therefore signals from the trichobothria may well be used by the spider to recognize the frequency content of vibrational stimuli.

To summarize these results we conclude that the behavioral response of *Cupiennius* results from competition of the input signals caused by substrate-borne vibrations and by airborne vibrations and currents (cf. Hergenröder and Barth 1983a).

Hergenröder and Barth (1983b) also analyzed the turning angles of *Cupiennius* under various stimulus conditions. These were always smaller than the stimulus angle and did not differ significantly, whether intact spiders or spiders deprived of trichobothria were stimulated by substrate vibration, or whether intact spiders were stimulated by a buzzing fly. If the trichobothria were removed from all legs except the third and fourth on one side and both were stimulated by a fly and simultaneous ground vibrations, the turning angles resulting from both stimuli added up. Under the prerequisite that the new findings on the role of airborne sound receivers in *Agelena* (Reißland and Habigsberg, in press) also apply for the orienting behavior of *Cupiennius* (Klärner and Barth 1982; Hergenröder and Barth 1983a), it can be assumed that the neuronal signals from the single-slit sense organs converge with those from the substrate vibration receptors whereas the trichobothria do not contribute to the angle of orientation.

Orb weavers possibly use their trichobothria for the detection of enemies. *Zygiella* and *Nephila* often respond to airborne stimuli with defensive behavior, but never with attack behavior (Klärner and Barth 1982).

6 Conclusions

We know that scorpions use their trichobothria for different purposes. It may well be that this is also the case in spiders. Clear-cut behavioral experiments have been made on only a few species. Spiders that live on solid ground can use these sensilla both for releasing flight and predatory behavior (Hergenröder and Barth 1983a, b). In funnel-web spiders, which have a densely woven web, the significance of airborne stimuli for prey detection can still be considerable (Reißland and Habigsberg, in press). Here trichobothria play a role in predatory but perhaps not in flight behavior. Contrarily, orb weavers, living on very elastic webs, detect their prey after it has been caught in the web and here trichobothria serve probably only for the detection of enemies (Klärner and Barth 1982). Wind detection may be another purpose of spider trichobothria, but no experiments have been done on this.

References

Altner H (1977) Insect sensillum specificity and structure: an approach to a new typology. In: Le Magnen J, Mac Leod P (eds) Olfaction and Taste, vol VI, Paris 1977. Information Retrieval, London Washington DC, pp 295–303

Barth FG (1967) Ein einzelnes Spaltsinnesorgan auf dem Spinnentarsus: seine Erregung in Abhängigkeit von den Parametern des Luftschallreizes. Z Vergl Physiol 55:407–449

Barth FG (1969) Die Feinstruktur des Spinnenintegumerits I. Die Cuticula des Laufbeines adulter häutungsferner Tiere (*Cupiennius salei* Keys). Z Zellforsch 97:137–159

Barth FG (1971) Der sensorische Apparat der Spaltsinnesorgane (*Cupiennius salei* Keys, Araneae). Z Zellforsch 112:212–246

Barth FG (1978) Slit sense organs: "Strain Gauges" in the arachnid exoskeleton. Symp Zool Soc Lond 42:439–448

Barth FG (1981) Strain detection in the arthropod exoskeleton. In: Laverack MM, Cosens DJ (eds) Sense organs. Blackie, Glasgow, pp 112–141

Barth FG (1982) Spiders and vibratory signals: Sensory reception and behavioral significance. In: Witt PN, Rovner JS (eds) Spider communication. Princeton Univ Press, Princeton, pp 67–122

Barth FG, Blickhan R (1984) Mechanoreception. In: Bereiter-Hahn J, Matoltsy AG, Richards KS (eds) Biology of the integument, vol I. Invertebrates, Springer, Berlin Heidelberg New York Tokyo, pp 554–582

Bauer T (1982) Prey-capture in a ground-beetle larva. Anim Behav 30:203–308
Bischof H-J (1974) Verteilung und Bewegungsweise der keulenförmigen Sensillen von *Gryllus bimaculatus* Deg. Biol Zentralbl 93:449–457
Bischof H-J (1975) Die keulenförmigen Sensillen auf den Cerci der Grille *Gryllus bimaculatus* als Schwererezeptoren. J Comp Physiol 98:277–288
Bonaric JC (1975) Utilisation des barèmes trichobothriotaxiques comme critère d'age chez *Pisaura mirabilis* Cl. (Araneae, Pisauridae). Ann Sci Nat Zool 12 Ser 17:521–534
Buchli H (1969) Hunting behavior in the Ctenizidae. Am Zool 9:175–193
Buño W Jr, Monti-Bloch L, Mateos A, Handler P (1981) Dynamic properties of cockroach cercal "threadlike" hair sensilla. J Neurobiol 12:123–141
Camhi JM (1969) Locust wind receptors. I. Transducer mechanism and sensory response. J Exp Biol 50:335–348
Camhi JM (1980) The escape system of the cockroach. Sci Am 6, 243:144–156
Camhi JM, Tom W (1978a) The escape behavior of the cockroach *Periplaneta americana* L. I. Turning response to wind puffs. J Comp Physiol 128:193–201
Camhi JM, Tom W (1978b) The escape behavior of the cockroach *Periplaneta americana* L. II. Detection of natural predators by air displacement. J Comp Physiol 128:203–212
Christian U (1971) Zur Feinstruktur der Trichobothrien der Winkelspinne *Tegenaria derhami* (Scopoli), (Agelenidae, Araneae). Cytobiology 4:172–185
Christian UH (1972) Trichobothrien, ein Mechanorezeptor bei Spinnen. Elektronenmikroskopische Befunde bei der Winkelspinne *Tegenaria derhami* (Scopoli) (Agelenidae, Araneae). Verh Dtsch Zool Ges 66:31–36
Chrysanthus F (1953) Hearing and stridulation in spiders. Tijdschr Entomol 96:57–83
Dahl F (1883) Über die Hörhaare bei den Arachniden. Zool Anz 6:267–270
Dahl F (1911) Die Hörhaare (Trichobothrien) und das System der Spinnentiere. Zool Anz 37:522–532
Den Otter CJ (1974) Setiform sensilla and prey detection in the bird-spider *Sericopelma rubronitens* Ausserer (Araneae, Theraphosidae). Neth J Zool 24:219–235
Dijkgraaf S (1947) Über die Reizung des Ferntastsinnes bei Fischen und Amphibien. Experientia 3:206–208
Drašlar K (1973) Functional properties of trichobothria in the bug *Pyrrhocoris apterus* (L.). J Comp Physiol 84:175–184
Emerit M (1967) Innervation trichobothriale et axiale de la patte de l'Aranéide, *Gasteracantha versicolor* (Walck.) (Argiopidae). C R Acad Sci 265:1134–1137
Emerit M (1969) Contribution a l'étude des Gasteracanthes (Aranéides, Argiopides) de Madagascar et des îles voisines. Thèse Fac Sci Montpellier, AO 2888
Emerit M (1970) Nouveau apports a la théorie de l'arthrogenèse de l'appendice arachnidien. Bull Mus Hist Nat 2 Ser 41:1398–1402
Emerit M (1972) Le développement des Gasteracacanthes (Aranéida, Argiopidae). Une contribution a l'étude de l'appendice aranéidien. Ann Mus Afr Cent Sci Zool 195:1–103
Emerit M (1974) Observations sur la trichobothriotaxie des *Néphiles* (Araneae, Araneidae, Nephilinae). Bull Mus Hist Nat 3 Ser 260:1613–1628
Emerit M (1976) Quelques reflexions sur la trichobothriotaxie des Aranéides. C R Col Arachnol Fr, Les Eyzies, 1976, Académie de Paris, Stn Biol Eyzies, Paris, pp 40–51
Emerit M, Bonaric JC (1975) Contribution a l'étude du développement de l'appareil mécanorécepteur des araignées: la tricobothriotaxie de *Pisaura mirabilis* Cl (Araneae, Pisauridae). Zool Jahrb Anat 94:358–374
Fletcher NH (1978) Acoustical response of hair receptors in insects. J Comp Physiol 127:185–189
Foelix RF (1975) Occurrence of synapses in peripheral sensory nerves of arachnides. Nature (London) 254:146–148
Foelix RF (1976) Rezeptoren und synaptische Verschaltungen bei verschiedenen Arachnida. Entomol Germ 3:83–87
Foelix RF, Müller-Vorholt G, Jung H (1980) Organization of sensory leg nerve in the spider *Zygiella x-notata* (Clerck) (Araneae, Araneidae). Bull Br Arachnol Soc 5:20–28
Frings H, Frings M (1966) Reactions of orb-weaving spiders (Argiopidae) to airborne sounds. Ecology 47:578–588

Gabbutt PD (1969) Pseudoscorpions: Growth and trichobothria. Bull Mus Nat Hist Nat Paris 2 Ser 41 (Suppl 1):134–140
Gabbutt PD (1972) Differences in the disposition of trichobothria in the Chernetidae (Pseudoscorpiones). J Zool Lond 167:1–13
Gaffal KP, Hansen K (1972) Mechanorezeptive Strukturen der antennalen Haarsensillen der Baumwollwanze *Dysdercus intermedius* Dist. Z Zellforsch 132:79–94
Gnatzy W (1972) Die Feinstruktur der Fadenhaare auf den Cerci von *Periplaneta americana* L. Verh Dtsch Zool Ges 66:37–42
Gnatzy W (1976) The ultrastructure of the thread-hairs on the cerci of the cockroach *Periplaneta americana* L: The intermoult phase. J Ultrastruct Res 54:124–134
Gnatzy W (1980) Morphogenesis of mechanoreceptor and epidermal cells of crickets during the last instar and its relation to molting-hormone level. Cell Tissue Res 213:369–391
Gnatzy W, Schmidt K (1971) Die Feinstruktur der Sinneshaare auf den Cerci von *Gryllus bimaculatus* Deg (Saltatoria, Gryllidae). Z Zellforsch 122:190–209
Gnatzy W, Tautz J (1977) Sensitivity of an insect mechanoreceptor during moulting. Physiol Entomol 2:279–288
Gnatzy W, Tautz J (1980) Ultrastructure and mechanical properties of an insect mechanoreceptor: Stimulus-transmitting structures and sensory apparatus of the cercal filiform hairs of *Gryllus*. Cell Tissue Res 213:441–463
Görner P (1965) A proposed transducing mechanism for a multiply-innervated mechanoreceptor (trichobothrium) in spiders. Cold Spring Harbor Symp Quant Biol 30:69–73
Görner P, Andrews P (1969) Trichobothrien, ein Ferntastsinnesorgan bei Webespinnen (Araneen). Z Vergl Physiol 64:301–317
Gossel P (1936) Beiträge zur Kenntnis der Hautsinnesorgane und Hautdrüsen der Cheliceraten und der Augen der Ixodiden. Z Morphol Oekol Tiere 30:177–205
Hansen HJ (1917) On the trichobothria ("auditory hairs") in arachnida, myriapoda and insecta. Entomol Tidskr 38:240–259
Harris DJ, Mill PJ (1977) Observations on the leg receptors of *Ciniflo* (Araneidae: Dictynidae). I. External Mechanoreceptors. J Comp Physiol 119:37–54
Haupt J, Coineau Y (1975) Trichobothrien and Tastborsten der Milbe *Microcaeculus* (Acari, Prostigmata, Caeculidae). Z Morphol Tiere 81:305–322
Hayes WF, Barber SB (1982) Peripheral synapses in *Limulus* chemoreceptors. Comp Biochem Physiol 72A:287–293
Helversen O von (1966) Über die Homologie der Tasthaare bei Pseudoskorpionen (Arach). Senckenbergiana Biol 47:185–195
Hergenröder R, Barth FG (1983a) The release of attack and escape behavior by vibratory stimuli in a wandering spider (*Cupiennius salei* Keys). J Comp Physiol 152:347–359
Hergenröder R, Barth FG (1983b) Vibratory signals and spider behavior: How do the sensory inputs from the eight legs interact in orientation? J Comp Physiol 152:361–371
Hoffmann C (1965) Die Trichobothrien der Skorpione. Naturwissenschaften 52:436–437
Hoffmann C (1967) Bau und Funktion der Trichobothrien von *Euscorpius carpathicus* L. Z Vergl Physiol 54:290–352
Horn E, Bischof H-J (1983) Gravity reception in crickets: The influence of cercal and antennal afferences on the head position. J Comp Physiol 150:93–98
Ignatiev AM, Ivanov VP, Balashov YS (1976) The fine structure and function of the trichobothria in the scorpion *Buthus eupeus* Koch (Scorpiones, Buthidae). Entomol Rev 55:12–18
Juberthie C, Piquemal F (1977) L'équipement sensoriel de Trechinae souterrains (Coléoptères). II. Ultrastructure des trichobothries de l'élytre. Int J Speleol 9:137–152
Klärner D, Barth FG (1982) Vibratory signals and prey capture in orb-weaving spiders (*Zygiella x-notata, Nephila clavipes;* Araneidae). J Comp Physiol 148:445–455
Krafft B, Leborgne R (1979) Perception sensorielle et importance des phenomènes vibratoires chez les araignées. J Psychol 3:299–334
Lehtinen PT (1967) Classification of the cribellate spiders and some allied families, with notes on the evolution of the suborder Araneomorpha. Ann Zool Fenn 4:199–468
Lehtinen PT (1975) Notes on the phylogenic classification of Araneae. In: Vlijm L et al. (eds) Proc IVth Int Arachnol Congr, Vrije Univ Amsterdam 1976, pp 26–29

Linsenmair KE (1968) Anemotaktische Orientierung bei Skorpionen (Chelicerata, Scorpiones). Z Vergl Physiol 60:445–449

Mahnert V (1976) Etude comparative des trichobothries de pseudoscorpions au microscope électronique à balayage. CR Séanc Soc Phys Hist Nat 11:96–99

Mann WD, Chapman DM (1975) Component mechanism of sensitivity and adaptation in an insect mechanoreceptor. Brain Res 97:331–336

Markl H, Tautz J (1975) The sensitivity of hair receptors in caterpillars of *Barathra brassicae* L (Lepidoptera, Noctuidae) to particle movement in a sound field. J Comp Physiol 99:79–87

McIver SB (1975) Structure of cuticular mechanoreceptors of arthropods. Annu Rev Entomol 20:381–397

Meßlinger K (1981) Vergleichende Untersuchungen zur Feinstruktur und Funktionsmorphologie der Trichobothrien von Skorpionen. Diplomarbeit, Univ Würzburg

Millot J (1968) Classe des arachnides. I. Morphologie générale et anatomie interne. In: Grassé PP (ed) Traîté de zoologie, Tome VI. Masson, Paris, p 302

Morse PM, Ingard KU (1968) Theoretical acoustics. McGraw-Hill, New York, p 312

Nicklaus R (1965) Die Erregung einzelner Fadenhaare von *Periplaneta americana* in Abhängigkeit von der Größe und Richtung der Auslenkung. Z Vergl. Physiol 50:331–362

Nicklaus R (1967) Zur Richtcharakteristik der Fadenhaare von *Periplaneta americana.* Z Vergl Physiol 54:434–437

Nicklaus R (1969) Zur Funktion der keulenförmigen Sensillen auf den Cerci der Grillen. Verh Dtsch Zool Ges 62:393–398

Palmgren P (1936) Experimentelle Untersuchungen über die Funktion der Trichobothrien bei *Tegenaria derhami* Scop. Acta Zool Fenn 19:3–27

Parry DA (1960) The small leg-nerve of spiders and a probable mechanoreceptor. Q J Microsc Sci 101:1–8

Pauly F (1956) Zur Biologie einiger Belbiden (Oribatei, Moosmilben) und zur Funktion ihrer pseudostigmatischen Organe. Zool Jahrb Syst 84:275–328

Rathmayer W (1966) Die Innervation der Beinmuskeln einer Spinne *Eurypelma hentzi* Chamb. (Orthognatha, Aviculariidae). Verh Dtsch Zool Ges 59:505–511

Reißland A (1978) Electrophysiology of trichobothria in orb-weaving spiders (Agelenidae, Aranea). J Comp Physiol 123:71–84

Reißland A (1979) Funktion des Trichobothrienrezeptors von Webespinnen. Verh Dtsch Zool Ges 72:297

Reißland A, Görner P (1978) Mechanics of trichobothria in orb-weaving spiders (Agelenidae; Araneae). J Comp Physiol 123:59–69

Reißland A, Habigsberg A (1985) Ethology of the orientation to airborne sound in funnel-web spiders (Agelenidae) (in press)

Riechert SE, Łuczak J (1982) Spider foraging: Behavioral responses to prey. In: Witt PN, Rovner JS (eds) Spider communication. Princeton Univ Press, Princeton, pp 353–385

Schmidt K, Gnatzy W (1971) Die Feinstruktur der Sinneshaare auf den Cerci von *Gryllus bimaculatus* Deg (Saltatoria, Gryllidae). II. Die Häutung der Faden- und Keulenhaare. Z Zellforsch 122:210–226

Schuh RT (1975) The structure, distribution and taxonomic importance of trichobothria in the Miridae (Hemiptera). Am Mus Nov 2585. Am Mus Nat Hist, New York, pp 1–26

Simon E (1892) Histoire naturelle des araignées. Libr Encycl Roret, Paris

Skudrzyk E (1971) The foundations of acoustics, Springer, Wien New York, pp 270–366

Stokes GG (1851) On the effect of the internal friction of fluids on the motion of pendulums. Reprinted in: Mathematical and physical papers, vol III, Cambridge Univ Press, Cambridge 1922, pp 1–140

Tarman K (1961) Über Trichobothrien und Augen bei Oribatei. Zool Anz 167:51–58

Tautz J (1977) Reception of medium vibration by thoracal hairs of caterpillars of *Barathra brassicae* L (Lepidoptera, Noctuidae). I. Mechanical properties of the receptor hairs. J Comp Physiol 118:13–31

Tautz J (1978) Reception of medium vibration by thoracal hairs of caterpillars of *Barathra brassicae* L (Lepidoptera, Noctuidae). II. Response characteristics of the sensory cell. J Comp Physiol 125:67–77

Tautz J (1979) Reception of particle oscillation in a medium – an unorthodox sensory capacity. Naturwissenschaften 66:452–461

Tautz J, Markl H (1978) Caterpillars detect flying wasps by hairs sensitive to medium vibration. Behav Ecol Sociobiol 4:101–110

Thorson J, Biederman-Thorson M (1974) Distributed relaxation processes in sensory adaptation. Science 183:161–172

Thurm U (1964) Mechanoreceptors in the cuticle of the honeybee. Fine structure and stimulus mechanism. Science 145:161–172

Thurm U (1982) Biophysik der Mechanorezeption. In: Hoppe W, Lohmann W, Markl H, Ziegler H (eds) Biophysik, 2nd edn. Springer, Berlin Heidelberg, pp 691–696

Tischner H, Schief A (1955) Fluggeräusch und Schallwahrnehmung bei *Aedes aegypti* L. Zool Anz Suppl 18:453–460

Tobias M, Murphey RK (1979) The response of cercal receptors and identified interneurons in the cricket (*Acheta domesticus*) to airstreams. J Comp Physiol 129:51–59

Vachon M (1958) Contribution a l'étude du développement postembryonnaire des araignées. 2e note. Orthognates. Bull Soc Zool Fr 83:429–461

Vachon M (1965) Contribution a l'étude du développement postembryonnaire des araignées. 3e note. *Pholcus phalangioides* (Fussl). Bull Soc Zool Fr 90:607–620

Vachon M (1973) Etude des caractères utilisées pour classer les familles et les genres de Scorpions (Arachnides). 1. La trichobothriotaxie en arachnologie. Sigles trichobothriaux et types de trichobothriotaxie chez les Scorpions. Bull Mus Hist Nat 3 Ser 140:857–958

Westin K (1979) Responses to wind recorded from cercal nerve of the cockroach *Periplaneta americana*. I. Response properties of single sensory neurons. J Comp Physiol 133:97–102

Weygoldt PW (1966) Moos- und Bücherskorpione. Neue Brehm-Bücherei 365. Ziemen, Wittenberg Lutherstadt

Weygoldt PW, Paulus HF (1979) Untersuchungen zur Morphologie, Taxonomie und Phylogenie der Chelicerata. I. Morphologische Untersuchungen. Z Zool Syst Evolutionsforsch 17:85–116

Yamashita S, Tateda H (1981) Efferent neural control in the eyes of orb-weaving spiders. J Comp Physiol 143:477–483

IX Slit Sensilla and the Measurement of Cuticular Strains

FRIEDRICH G. BARTH

CONTENTS

1 Introduction

Information about the strains occurring in a piece of material under load is important in many design problems, where economy of material investment and optimization of mechanical properties are essential. Accordingly, strain-measuring techniques are applied in a wide range of industries, including such different fields as aircraft and bridge construction and design of pianos and skis.

The most commonly used strain-measuring device in technology is the strain gauge. Basically, this is a piece of wire or semi-conductor which changes its resistance when strained. By incorporating the strain gauge into an electrical circuit the mechanical events can be measured with high precision.

Zoologisches Institut der J. W. Goethe-Universität, Gruppe Sinnesphysiologie, Siesmayerstraße 70, D-6000 Frankfurt am Main 1, Federal Republic of Germany

Arthropods have sensilla built into their exoskeleton which can be considered a biological version of technical strain gauges. Based on their own functional principles, these sensory receptors measure the effects of loads generated in the cuticular skeleton by muscular activity, hemolymph pressure, gravitational forces, and substrate vibrations. The most intriguing version known of this unusual mechanical sense, which is unique to arthropods and intimately linked with their exoskeleton, is that of spiders. Their "slit sensilla", which are also found in the other arachnid groups, form a sensory system of considerable refinement which evidently provides its bearer with a very detailed picture of the mechanical events going on in its exoskeleton.

This chapter illustrates the functional principles involved in the activity of these sensilla and the behavioral relevance of a sense whose study has been neglected for a long time, due in part perhaps to its remoteness from human sensation.

2 Definitions and Principles

Adequate appreciation of a particular type of sense organ largely depends on an understanding of its natural input. Here we first ask what happens in a piece of material that is subjected to forces applied externally. Several parameters are relevant with respect to the slit sensilla and also their insect analogs, the campaniform sensilla.

Stress. This is the force per unit area (A) generated in the material under external forces (P) and resisting them. Normal to a given plane, stress is

$$\sigma = \frac{P}{A} \, [N/m^2 = Pa]. \tag{1}$$

Strictly speaking applied stress has to be distinguished from resulting stress. Applied stress is the ratio of load per unit-area.

Strain. Stress and strain are intimately related with each other. Inevitably stress is paralleled by a deformation of the specimen, minute as it may be. Strain (ε) is defined as relative change in length of a unit volume in a given direction,

$$\varepsilon = \frac{\Delta l}{l_0}, \tag{2}$$

l_0 being the original length and Δl the change in length. Strain then is without dimension. Negative strain describes a decrease of l_0, i.e., compression, in everyday language.

Young's Modulus. Up to the elastic limit (Hookean range) stress and strain are often proportional to each other. In the simple case of a bar under tension

$$\varepsilon = \frac{\sigma}{E} \quad \text{or} \quad E = \frac{\sigma}{\varepsilon} \, [Pa]. \tag{3}$$

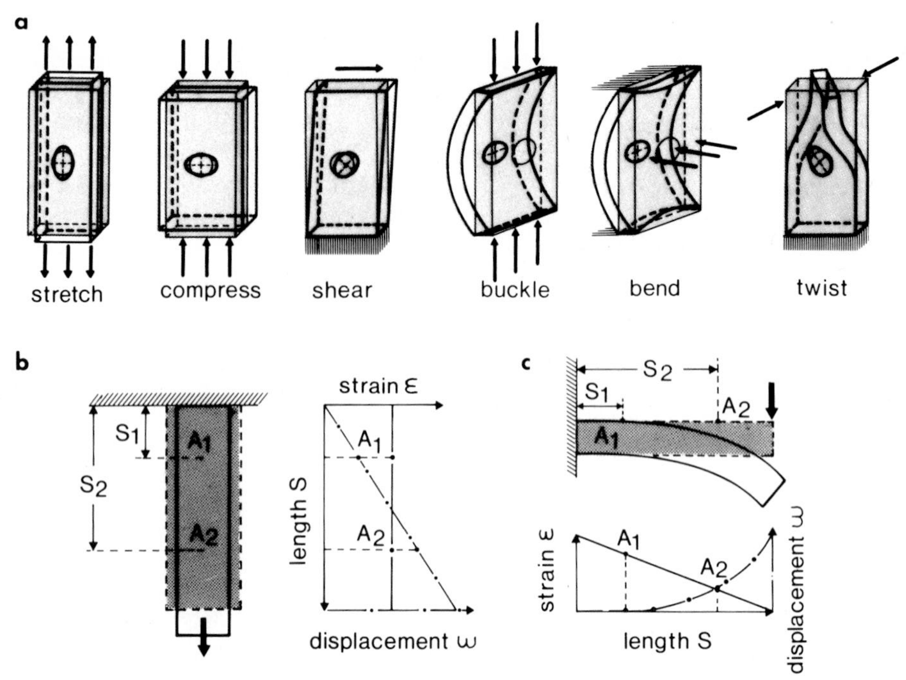

Fig. 1a–c. Some basic mechanical principles. **a** A piece of material (*shaded*) is loaded (*arrows*) in different ways and thereby deformed (*unshaded*). Basically the same phenomenon occurs in all cases at the surface. A circle drawn onto it deforms into an ellipse, indicating strains in two directions perpendicular to each other and with opposite sign. **b** A bar under tension (*arrow*) shows that displacement has to be distinguished from strain. **c** As seen from a single-ended cantilever under load (*arrow*), the sites of maximal stress or strain need not coincide with those of maximal displacement. See text. (**a** Barth 1972a; **b** and **c** Barth 1972b)

E, the ratio of stress to strain, is called the elastic modulus in tension or Young's modulus. There is also an elastic modulus in shear, giving the ratio of tangential force per unit-area to angular deformation.

Displacement. Strain of a unit volume is different from the displacement (ω) of a particular point in a piece of material under load. In the simple case of monoaxial stress, which is found in a bar under tension for instance (Fig. 1b), displacement equals the integral of strains at the site s under consideration

$$\omega = \int_0^s \varepsilon \, ds. \tag{4}$$

With multiaxial stresses the situation is more complex. Even in the simple case of the bar under tension, strain (and stress) do not change along the specimen in the same way as displacement. Whereas strain (and stress) are constant all along its axis, displacement increases with its length, due to the increase of the summed strains of the unit volumes. It is also worth noting that places of greatest stress do not coincide with those of greatest displacement. This is easily shown by a simple cantilever with a point load at its end (Fig. 1c). Stress is maximal where the coupling maximally resists displacement, whereas displacement is greatest at its free end, where strain (and stress, respectively) is minimal.

Surface. There are many different ways in which a piece of material can be loaded. It may be stretched, compressed, sheared, bent, buckled, or twisted (Fig. 1a). At the surface of a specimen the same situation applies to all these cases: at a given point there will always be tension ($+\varepsilon$) in one direction and compression ($-\varepsilon$) perpendicular to it. Unless load is applied directly to the surface, there will be only stresses and strains parallel to it. This is important to note, since the sensory endings of the strain detectors under consideration here attach to the surface of the cuticular exoskeleton. Interpretation of the mechanical effects resulting from the various ways in which loads may be applied is thereby simplified.

Definition of Receptors. According to numerous studies using the electron microscope, electrophysiological techniques and model experiments (Barth 1971, 1972a, b, 1981; Barth and Pickelmann 1975) the auxiliary structures of the slit sense organs are "designed" and arranged in a way that provides deformation even by very small forces. The most important one of the parameters listed above is displacement: the slit sense organs (and the analogous campaniform sensilla in insects) measure absolute displacement of two points or lines in the cuticle, and not strain, which denotes a *relative* change in length.

A characterization of the receptors according to these physical parameters depends on the level of consideration, however. Whereas "displacement detectors" would be the adequate terminology at the receptor level, the sensilla may well be called "strain detectors" at the level of the larger skeletal region they are built in. Like strain gauges used by technology they measure the *local* deformation of the skeleton. Taking the whole body part (mainly the legs, where the organs are most numerous) as reference, the slit sensilla of arachnids and the campaniform sensilla of insects present themselves as "force detectors", measuring the forces which lead to cuticular deformations and finally displacement at the receptors (Barth 1976, 1978, 1981; Seyfarth 1978b; Blickhan and Barth 1985).

Seen from the point of view of the animal's behavior, the main question is: which parameter is the spider or other arachnid most interested in? This is hard to say. It seems likely, however, that displacement at the receptor as such is not the main point of interest. A more reasonable assumption is that *strain* and thus also the deformation of the skeleton is one type of information needed, for instance to avoid its failure. In addition, it seems likely that the animal relies on information about the *forces* acting on its skeleton, since these are of great potential importance for its locomotor behavior (Seyfarth 1978a, b; Blickhan and Barth 1985). Since slit sensilla in arachnids vary widely with respect to their arrangement in the skeleton, it is not surprising that they have been found to be involved in quite different aspects of behavior, where different parameters such as stepping pattern, substrate vibration and muscular force are the biologically relevant input (see below). Similarly, technical strain-gauges are used to measure different physical quantities such as displacement, pressure, load, acceleration and others, depending on their arrangement in the particular type of transducer.

3 Occurrence and Distribution of Slit Sensilla

Slit sense organs, a peculiarity of arachnids, were first mentioned by Bertkau in 1878. Following early descriptions of their occurrence in spiders (Vogel 1923; Kaston 1935; Mc Indoo 1911), study of their distribution in the exoskeleton of various arachnids was taken up again, together with work on their functional design and physiology, after Pringle's (1955) pioneering experiments had shown that they are mechanoreceptive.

The best-studied case is still that of the spiders. Different species are very similar with respect to their slit sensilla. In *Cupiennius salei,* a large Central American hunting spider (Ctenidae; Lachmuth et al. 1985), about 3300 slits were counted (Fig. 2) (Barth and Libera 1970). The majority of these, i.e., 86%, lie on the legs and pedipalps, embedded in hard, sclerotized exocuticle. The 96 slits found on the opisthosoma (without petiolus) are exceptional in being surrounded by soft mesocuticle.

The slit sensilla of spiders – and in fact of all arachnids – are classified as single isolated slits, relatively loose groups of slits, and compound or lyriform organs with up to 30 slits that are closely arranged in parallel (Fig. 3a). In *Cupiennius* about half of all slits are either single slits or small groups. The rest forms a total of 144 lyriform organs which are only found on the extremities (134 on walking legs and pedipalps and 10 on spinnerets and chelicerae). While close vicinity to joints is typical of lyriform organs, isolated slits usually occur at some distance from articulations. They form rows on the anterior and posterior side of the walking legs and pedipalps; some lie close to muscle attachment sites, for instance on the ventral opisthosoma. Both isolated slits and lyriform organs are typically (with exceptions) found on the lateral aspects of the respective appendage and are oriented roughly parallel to its long axis. Typical groups occur on the walking leg trochanter, the chelicerae and petiolus (Fig. 2).

A comparison of the slit sensilla on the walking legs among representatives of five arachnid orders (spiders, scorpions, whip spiders, whip scorpions, harvestmen) shows the wide spectrum of sensillar supply and the relative share held by isolated slits, groups of slits, and lyriform organs (Barth and Stagl 1976). While *Cupiennius* has at least 325 slits on each leg, there are only 58 and 45 in case of the whip spider and harvestman. Also, lyriform organs are characteristic of spiders; there are 15 on each leg, but none or only one or two in the case of other arachnids. In all species studied, there is a concentration of slits proximally in the leg and the trochanter is the segment always most richly supplied.

The mechanical significance of these morphological findings is discussed in Sections 6 and 7.

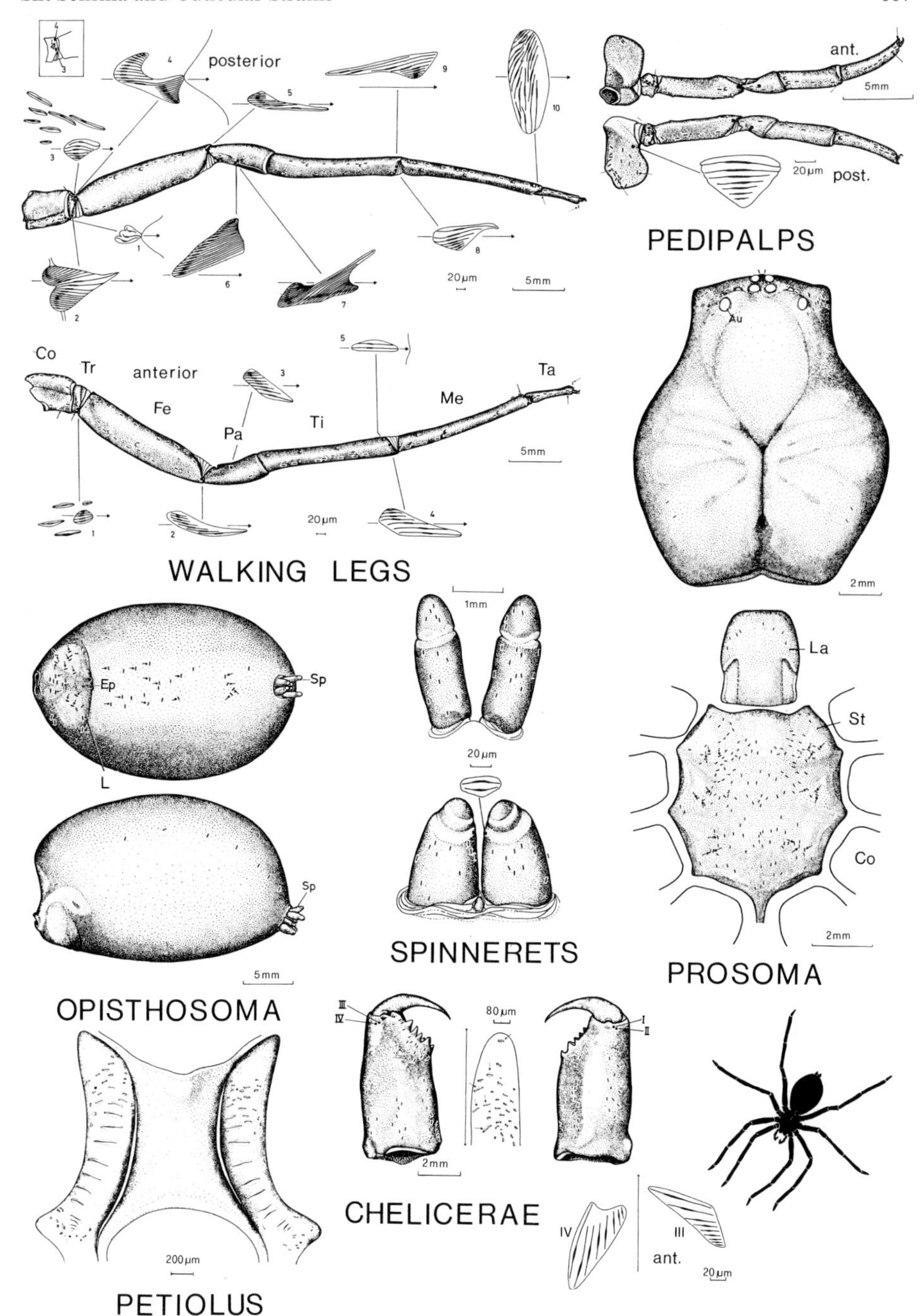

Fig. 2. Distribution of the slit sensilla in the exoskeleton of the spider *Cupiennius salei* Keys. *Small arrows* point to single slits longer than 30 µm. *Long arrows in insets* give orientation of long axis of the respective body appendage and point distally. *Co* coxa; *Tr* trochanter; *Fe* femur; *Pa* patella; *Ti* tibia; *Me* metatarsus; *Ta* tarsus; *Sp* spinneret; *L* lung slit; *Ep* epigyne; *La* labium; *St* sternum. For the walking leg and the pedipalp both the anterior and posterior aspects are shown. (After Barth and Libera 1970)

4 Functional Morphology of the Receptors

4.1 Structure

Slit sensilla are slit-shaped holes in the cuticular exoskeleton. They are from 8 to 200 µm long and 1 to 2 µm wide, and covered by a membrane only about 0.25 µm thick and made up mainly of epicuticular material. To this membrane the dendrite tip of a bipolar sensory cell attaches in an area specialized as a small "coupling cylinder". The dendrite ends in the cylinder's interior (Fig. 3d). As in other arthropod cuticular mechanoreceptors the prominent fine structural feature of the dendrite tip is a tubular body. It is one of the few modality-specific structures known in sensory receptors.

Seen in more detail the hole in the cuticle consists of two parts. The outer part is that referred to above: it lies in the exocuticle and is trough-shaped and flattened toward both ends of its longitudinal axis. The dendrite runs through a small hole in its bottom upward to the coupling cylinder, surrounded by "receptor lymph" (see below). Apart from the mechanosensory ending, all cellular components are contained in the inner, second part of the cuticular hole in the meso- and endocuticle (see Barth 1969, 1970, 1973 for details of spider cuticle). This second part extends toward the epidermis bell-shaped from the bottom of the exocuticular hole.

Most of the cellular components of the slit sensilla are those typical of arthropod cuticular sensilla in general (see Foelix, Chap. VII, this Vol.). The distal section of the dendrite is surrounded by a noncellular dendritic sheath and proximally starts with a ciliary segment. It is the continuation of the dendrite's inner segment which ends as a prominent swelling containing a rich supply of mitochondria. At least three auxiliary cells are associated with a slit sensillum and surround the dendrite as sheath cells. While the innermost secretes the dendritic sheath, the other two are considered homologous to the insect trichogen cell and tormogen cell respectively (Barth 1971).

A striking cellular peculiarity of slit sensilla is the presence of a second dendrite originating from the soma of a second bipolar sensory cell, the function of which is not understood at all (Barth 1971; Seyfarth and Pflüger 1984, Seyfarth et al. 1984). Similar "supernumerary" sense cells, from which no spike responses could be recorded so far, are also typical of other arachnid sensilla. In case of the spider slit sensilla, the dendrite of the second cell ends at the inner membrane of the exocuticular slit and a number of arguments were put forward against its significance in monitoring slit deformation (Barth 1971, 1972a, b).

The reader interested in more details of the internal structure of slit sensilla is referred to an electron microscope study (Barth 1971) and earlier reviews (Barth 1976, 1981). General aspects on the cellular organisation of arachnid cuticular sensilla are covered in Foelix, Chapter VII, this Volume.

4.2 Input Transformation and Mechanical Filtering

Conduction and transformation of the adequate mechanical input from the general cuticle toward the dendrite tip has been described in detail, together

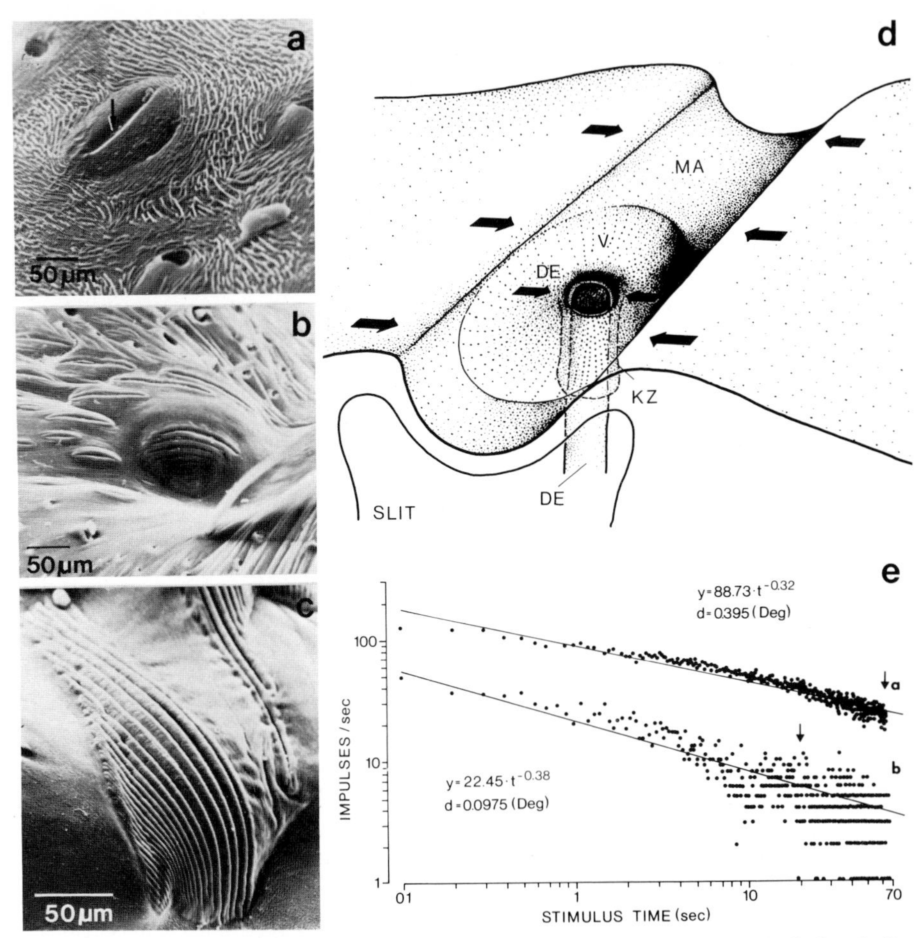

Fig. 3a – c. Types of arrangements. Slit sensilla of arachnids occur as single isolated slits (**a** showing example on opisthosoma; *arrow* points to dendrite attachment area), relatively loosely arranged groups (**b** example on trochanter), or lyriform or compound organs with a close parallel arrangement of slits (**c** example on trochanter). **d** stimulus uptake and transformation. Compression of the slit proper by forces (*arrows*) at right angles to its long axis further bends pre-bent outer membrane (*MA*), which in turn results in a deformation of the dendrite tip (*DE*) within the coupling cylinder (*KZ*) predominantly by monoaxial compressional forces. **e** step responses of slit sensilla typically follow a power-law function. Example shows response of slit 2 of lyriform organ HS-8 (Fig. 5a) to two step stimuli differing in amplitude (given as lateral displacement *d* of the metatarsus in degrees) by a factor of four. Note corresponding change in equation which implies constant gain. *Continuous lines* are fitted to measuring points (average frequency for 0.1 s intervals) to the *left of arrows* and correspond to power function. (**a – c** Barth 1981; **d** Barth 1976; **e** Bohnenberger 1981; *Cupiennius salei*, Araneae)

with the functional implications resulting from the morphology of the structures involved (Barth 1972a, b, 1976, 1981). It may therefore suffice to point out some aspects of particular importance here.

1. In principal, arthropod mechanoreceptors and in fact all mechanoreceptors can be divided into two groups: displacement receivers and force receivers. Although there is no displacement reception without the involvement of at least small forces, and no force reception without at least small displacements, the extreme cases can certainly be distinguished (Barth and Blickhan 1984): arachnid trichobothria (see Reißland and Görner, Chap. VIII, this Vol.) – as insect thread hairs – are typical displacement receivers. Their hair shaft, which is an extremely light-weight structure and articulated very flexibly in the cuticular skeleton, is deflected by the slightest *movement* of the air. The pressure or force aspect of the stimulus is of only minor importance here. Arachnid slit sensilla and insect campaniform sensilla (review: Barth 1981) on the other hand, are so designed as to detect minute deformations in the skeleton due to the *forces* set up by various internal and external loads.

2. As in all mechanoreceptors, transduction of the mechanical input energy into the electrochemical response of the sensory cell finally relies on dendrite deformation. This deformation is nonuniform and brought about by nonuniform pressure (in contrast to "hydrostatic" pressure). The resulting strains are predicted to be negative (compressive) parallel to the long axis of the dendrite and positive perpendicular to it.

3. The cuticular auxiliary structures of slit sensilla, which take up the stimulus from the surrounding cuticle, are designed and arranged in a way that provides for the dendrite's deformation even by very small forces. Thus the slit is set apart from the cuticular holes associated with other arthropod cuticular receptors, since it is its own deformation which is the adequate input to the sensillum.

4. According to fine structural, model, and electrophysiological studies (Barth 1971, 1972a, b, 1976, 1981; Barth and Pickelmann 1975) the sequence of events leading to the deformation of the dendrite tip in the coupling cylinder is the following (Fig. 3d): Compression of the slit at right angles to its long axis; as a consequence further inward bending of the covering membrane and in turn deformation of the coupling cylinder, in particular at its outer side and predominantly at a right angle to the long axis of the slit; finally deformation of the dendrite tip in the coupling cylinder by monoaxial compressive forces.

5. There are several mechanical "tricks" involved in this sequence by which the stimulus is focused from the general cuticle to less than 1 μm^2 of dendrite area (Barth 1972b). They help explain the high sensitivity (see below) of slit sensilla even when embedded in cuticle like that of the spider leg. Here the elastic (Young's) modulus of the cuticle (18 GPa, Blickhan and Barth 1979) is very similar to that of vertebrate bone (Yamada 1970). Skeletal deformations are therefore bound to be very small (also due to the hollow cylindrical shape of the leg segments which implies high second moments of inertia). These "tricks" are: (a) Due to its elongated shape, the slit is most easily deformed by

forces roughly perpendicular to its long axis[1]. (b) Due to its arched cross-section and thinness the covering membrane is most effectively further bent by lateral forces acting on it during slit compression. The trough-like shape enhances the bending moment resulting from such lateral forces and at the same time stiffens the covering membrane against bending along its long axis. (c) Finally, the coupling cylinder is attached to the bottom of the covering membrane, that is, to an area where the bending moment is greatest.

4.3 Morphological Variation

When comparing the many slit sensilla found in different places in one species (Barth and Libera 1970; Barth and Wadepuhl 1975) or at corresponding body parts in arachnids of different orders (Barth and Stagl 1976), morphological parameters such as slit length, location of dendrite tip along slit, and slit orientation relative to the main axis of the respective body part are seen to vary. In addition, the degree of association with other slits to form "loose groups" or compound lyriform organs varies to a large extent, and even the curvature of the covering membrane was found to vary. Since these variations all have a pronounced effect on the deformation of a slit in a given position and thus also on its sensitivity, they are likely to reflect important aspects of the adaptive radiation of slit sensilla.

Slit length varies from just a few microns to about 200 μm. As described earlier (Barth 1972a, b, 1981; Barth and Pickelmann 1975), both the absolute amount of deformation and its directionality increase with increasing slit length. From the directionality of the slit follows that highest sensitivity is achieved in cases with an orientation roughly perpendicular to the load direction. In a single slit (the more complex situation in lyriform organs is treated below) deformation is invariably greatest in the middle along its length, which is therefore the most sensitive position of the dendrite tip and in fact typical for single slits. Curvature of the covering membrane, which – through its impact on the bending moment – influences deformability as well, was seen to be less pronounced in the short slits of lyriform organs (presumably adding to their relative insensitivity due to short length) and more so in the long ones (Barth 1972b).

4.4 Sensitivities

Although the mechanical sensitivity of the dendrite proper must be considered crucial, the measurements of threshold sensitivities in slit sensilla seem to underline also the effectiveness of the non-neural stimulus-conducting struc-

1 Mechanically, dilatation of the slit is equivalent to compression. Electrophysiological experiments, however, demonstrate that it is an ineffective stimulus. This in turn can be explained mechanically by the arrangement and shape of the coupling cylinder, for which slit dilatation is not at all equivalent to its compression (Barth 1972b).

tures. Stimulation of slit sensilla ultimately depends on the deformation of the exoskeleton. Two examples (*Cupiennius salei*) demonstrate that sensitivity of slit sensilla is high despite the rigidity of the skeleton.

1. Lyriform organ HS-8 (Fig. 4b) can be stimulated effectively by a backward deflection of the metatarsus. When using ramp stimuli angular displacements (zero position: articular condyles just touch each other) between 0.01° and 0.92° excite most of the slits (Barth and Bohnenberger 1978). With sinusoidal stimulation at 100 Hz values are as small as 0.006° (Bohnenberger 1978, 1979). The forces inducing such a deflection are about 40 μN. Even at 0.001 Hz the gravitational forces acting on the spider leg would suffice to excite at least some of the slits in this organ (Bohnenberger 1981).

2. The metatarsal lyriform organ, an exteroceptor, is highly sensitive to substrate-borne vibrations reaching it through tarsal displacement (see Barth, Chap. XI, this Vol.). At frequencies of 1 kHz, threshold displacement is as small as 10^{-6} to 10^{-7} cm; smallest threshold acceleration values are ca. 0.02 cm s^{-2} at 1 Hz (Barth and Geethabali 1982).

5 Receptor Mechanisms and Transfer Functions

5.1 Receptor Mechanisms

Our knowledge of the mechnisms underlying mechano-electrical transduction in arachnid slit sensilla is far from complete. Questions such as the exact distribution of currents leading to the generation of spike potentials are still open for further argument, as in most other sensory receptors. From a number of quite diverse observations, however, valuable insights into some of the key problems involved have emerged. Despite the similarities in the cellular "Bauplan" of the slit sensilla with that of insect cuticular sensilla (Barth 1971, 1981), the most interesting findings may be those pointing to differences in their primary processes.

Potentials. A transepithelial standing potential can be measured at the site of many insect epithelial cuticular receptors, both mechano- and chemosensitive. It shows the receptor lymph in the extracellular space surrounding the outer dendrite segment to be 10 to 100 mV positive with respect to the hemolymph space. Upon stimulation, a negative receptor potential is recorded which decreases the transepithelial potential and sets off the spike discharge (reviews see: Thurm 1974, 1982a, b; Thurm and Küppers 1980). All three types of potential are also known from insect campaniform sensilla (on the haltere), the insect version of arthropod cuticular strain detectors. Thurm (1974) has suggested that the sensor region of the dendrite is a variable resistance whose conductivity is increased by adequate stimulation of the receptor. Stimulation in turn would increase the current flow to the sensory cell that is driven to a considerable extent by a non-neural battery of one of the auxiliary cells also responsible for the standing potential. Several findings indicate that this battery is located across the microvillous apical border of the tormogen cell and is an electrogenic K^+-pump driving K^+ into the receptor lymph (Küppers 1974; Thurm 1982a; Thurm and Wessel 1979).

There are significant differences between the above insect cases and slit sensilla. There is no standing potential. No receptor potential can be recorded and the main component of the spikes is negative and not positive as in most insect cases under similar recording conditions (Barth 1976; Thurm and Wessel 1979; Seyfarth et al. 1982).

Receptor Lymph. The lack of a standing potential in slit organs and other spider sensilla is paralleled by a difference in the chemical composition of the receptor lymph. It is rich in K^+ in the insect cases, which supports the electrogenic pump concept (Küppers 1974; Kaissling and Thorson 1980). By contrast, it is rich in Na^+ and poor in K^+ in the slit sensillum which correlates with a lack of fine-structural evidence for an electrogenic K^+-pump in the tormogen cell membrane (Barth 1971; Rick et al. 1976).

Electrical Stimulation. In a recent study Seyfarth et al. (1982) have added new aspects to the transduction processes in slit sensilla. Outward current (surface-negative) passed through the slit was found to be excitatory and inward current to be inhibitory, just the opposite from the case of insect epithelial receptors referred to above. This finding is interpreted as indicating that spike initiation occurs in the dendrite, which seems reasonable in a sensillum with the cell soma about 200 μm away from the slit. A localized-spike initiation site and transduction processes not involving the distant soma and implying relatively small current fields are also compatible with the failure of all our attempts to record receptor potentials from the cuticular surface. Finally, the high Na^+ concentration in the outer lymph space (instead of K^+) is as expected for conventional electrical excitability.

5.2 Transfer Functions and Adaptation

The transfer functions of the spider slit sensilla are similar to those of homologous insect campaniform sensilla (Chapman et al. 1979) and in fact to many receptors of even different modalities (Thorson and Biederman-Thorson 1974). With few exceptions known so far, they are slowly adapting and their step responses follow a simple power-law function (Fig. 3e)

$$y(t) = a \cdot d \cdot t^{-k}, \qquad (5)$$

where a is a constant describing gain, d stimulus amplitude, t time, and k a receptor constant which determines both the slopes of the response decline and of the frequency response. k was found to vary between 0.2 and 0.7, which implies that slit sensilla take a position in between pure displacement receptors (the response of which would be independent of stimulus frequency; $k = 0$) and pure velocity receptors, that is, first-order differentiators ($k = 1$). The frequency range of slit sensilla is very wide, like that of campaniform sensilla starting at

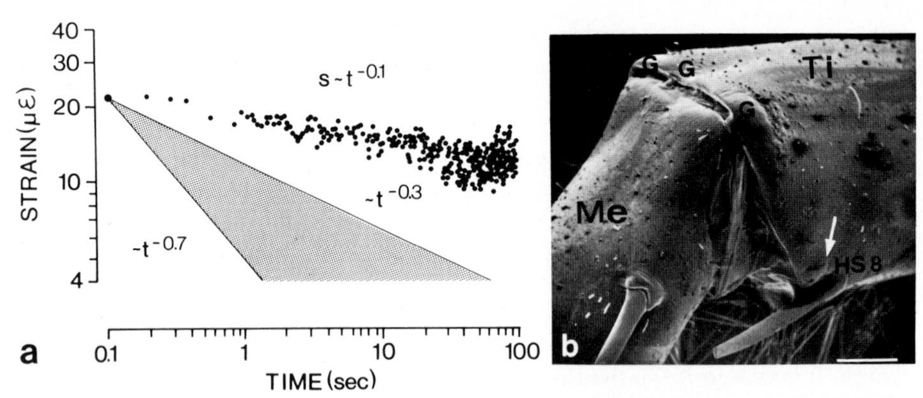

Fig. 4a, b. Joint mechanics and step response. Comparison of the relaxation of strain (*filled circles* k = −0.1) measured at the site of lyriform organ HS-8 (Fig. 4b) on the tibia (*Ti*) and the spike response (Imp s^{-1}) of slits in the same organ to ramp and hold stimuli. The starting frequency of the spike response is given the same coordinates as the initial strain value in order to make the difference in decline steepness clear. *Shaded area* of receptor responses covers the range of decline velocities (k varies between −0.3 and −0.7) found in different slits of same organ. Obviously the high-pass characteristics of the slits are only little influenced by the viscoelastic properties of the joint. *G* articular condyle; *Me* metatarsus. (After Blickhan et al. 1984)

roughly 5 mHz and covering about 5 powers of 10 (Bohnenberger 1979; Chapman et al. 1979).

Probably, the cause for the power-function decline and the mechanisms underlying adaptation can be assigned exclusively neither to the mechanical filter properties of the stimulus conducting cuticular structures nor to the spike initiation process itself. Much of the mechanical behavior of highly polymerized materials such as plastic and cuticle is commonly described by power-law dynamics (Bohnenberger 1981; Wainwright 1976). The viscosities explaining the adaptation during maintained mechanical stimuli could, for instance, be contained in the slit's covering membrane, its coupling cylinder or in the outer receptor lymph. Viscous relaxation of the forces acting in the joint region, however, could be excluded in the case of lyriform organ HS-8. This organ is located distally on the tibia and effectively stimulated by slight lateral displacement of the metatarsus. The force measured at the metatarsus as well as the cuticular strain at the site of the organ (Fig. 4) change only little during step-like stimulation (Bohnenberger 1981; Blickhan et al. 1984). Examination of the visco-elastic properties of this joint has essentially confirmed this. The k value for the force decline which follows a power-law function is only 0.04, which implies a nearly perfect elastic behavior of this joint, and would not explain the decline in the spike response (Blickhan 1983, Blickhan and Barth 1985).

The following arguments are in favor of the transduction and/or encoding processes as the main cause for the power-law dynamics in the spike response: (a) Power-law dynamics are found in receptors of many different modalities including chemoreceptors and eyes which are not subject to the mechanical filter properties of stimulus conducting structures (Biedermann-Thorson and Thorson 1971; Thorson and Biedermann-Thorson 1974); (b) power-law dynamics are common in the electric behavior of dielectric materials (Jonscher 1977); (c) in campaniform sensilla of the cockroach adaptation was found to be mainly due to the encoding processes and not to coupling and transduction (Mann and Chapman 1975).

Experiments using electrical instead of mechanical stimuli could potentially solve the question unambiguously: similar adaptation under both conditions would support the above arguments. As a matter of fact, there is no adaptation to maintained electrical stimulation. The latter result is difficult to interpret, however, because the generator current paths and spike initiation sites may differ from those under mechanical stimulation (Seyfarth et al. 1982).

6 Lyriform Organs

The most obvious morphological variation among arachnid slit sensilla is that among compound or lyriform organs where several slits (2 to ca. 30) are closely arranged in parallel. Slit deformation is drastically influenced by this type of association.

So far, the consequences of only one such arrangement have been studied in detail, using model experiments and morphological studies to investigate the main mechanical implications and electrophysiological recordings to see the physiological properties directly (Barth and Pickelmann 1975; Barth and Bohnenberger 1978; Bohnenberger 1981). The main results of these combined studies – all done with lyriform organ HS-8 distally on the spider tibia (Figs. 4b and 5a) – are the following.

1. There is a considerable difference in deformation among the seven to eight slits, which depends both on slit length and on slit position within the group. The neighboring slits reduce the deformation of a given slit, peripheral slits take up more load than intermediate ones. In a model lyriform organ HS-8 compression of the slits varied by up to 26 dB (Barth and Pickelmann 1975).

2. The differences predicted from these findings with regard to threshold sensitivity of the component slits were demonstrated by electrophysiological experiments. They amount to about 40 dB, which does not even include the most and the least sensitive slit of the group (which have so far resisted reliable recordings). There is no stimulus range fractionation among the slits, however, which implies that the stimulus amplitude ranges covered by the more sensitive slits include also those of the less sensitive ones (Barth and Bohnenberger 1979).

3. A concentration of different threshold sensitivities at one particular location in the exoskeleton is one advantage of a lyriform group of slits over a single slit. Another is the increase of the linear working range and range of high sensitivity (Bohnenberger 1981). Increment sensitivities of the individual slits drop considerably at stimulus amplitudes only 10 dB or less above threshold. The linear parts of the working ranges of the organ's slits, however, are linked with little overlap (Fig. 5c and d). This, as well as the responsiveness to low fre-

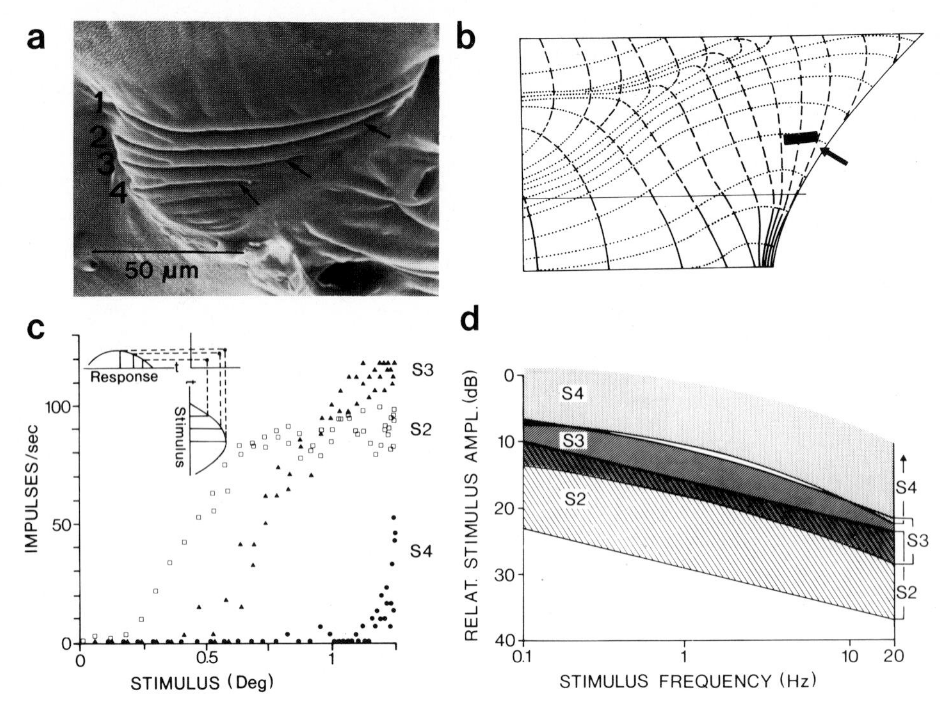

Fig. 5a–d. Functional design of lyriform organ HS-8 (*Cupiennius salei*, Araneae). **a** lyriform organ HS-8 on the posterior side of the tibia (Fig. 4b). Out of eight slits four were studied electrophysiologically (*numbers*); *arrows* point to those presented in **c**. **b** lateral aspect of distal tibia with location of lyriform organ HS-8 (*bar* indicated by *arrow*) according to tension optical experiment using an araldite model under load, equivalent to that generated by flexor muscle activity in the original leg. Organ HS-8 lies in an area where compression lines (– –) run roughly at a right angle to its long axis. Note tension lines (—) ventrally. **c** response of three slits (S2, S3, S4) to sinusoidal stimulation of organ (0.01 Hz). Instantaneous response (spike frequency) plotted versus instantaneous stimulus amplitude. **d** linearity ranges of same slits as in **c** as calculated from values in **c** (see Bohnenberger 1981). (**a**, **c** and **d** Bohnenberger 1981; **b** Barth and Pickelmann 1975)

quencies down to fractions of 1 Hz, seems appropriate for an organ that has proprioceptive functions (Barth and Seyfarth 1971; Seyfarth and Barth 1972; Seyfarth 1978a, b; see also Seyfarth, Chap. XII, this Vol.) and that is probably part of a feedback system and monitoring small deviations from the desired state.

4. According to the threshold curves (tuning curves), the different slits in organ HS-8 (Bohnenberger 1981) all behave like high-pass filters with relatively high thresholds at low frequencies up to roughly 10 Hz and steadily decreasing values toward higher frequencies.

They share this high-pass characteristic with the pretarsal ("claw") slits (Speck and Barth 1982) and with the slits of the vibration-sensitive metatarsal lyriform organ (Barth and Geethabali 1982; Bleckmann and Barth 1984). The only exception known is a tarsal slit, which responds to air-borne sound and exhibits a minimum threshold curve with highest sensitivity at frequencies between 0.3 kHz and 0.7 kHz (Barth 1967).

5. The tuning curves (0.005 Hz to 1 kHz) of the individual component slits of organ HS-8 (and also of other organs: Barth and Geethabali 1981; Speck and Barth 1982; see Barth, Chap. XI, this Vol.) show that a previous attractive hypothesis is wrong. All slits behave like high-pass filters; their differences in length do not result in differing tuning to particular small ranges of frequencies (Bohnenberger 1981). In this sense, the term "lyriform" is misleading, since there is strong evidence against the involvement of different resonance frequencies in frequency discrimination.

6. Orientation of a single slit and of lyriform organs with respect to load direction is of crucial consequence for its deformation, and thus for its sensitivity. As a rule of thumb, compression of the slits is greatest with pressure loads roughly perpendicular to the long axis of the organ. The tension-optical analysis of a model-leg tibia – subjected to forces simulating joint forces naturally generated by muscular activity – showed that the lines of principal stresses at the site of lyriform organ HS-8 are compression lines and in fact oriented almost perpendicular to the organ's long axis (Fig. 5b) (Barth and Pickelmann 1975).

From a theoretical point of view, it is hard to predict exactly what mechanical consequences arise from the various slit arrangements found in different lyriform organs (Fig. 2). In addition, the application of forces from varying directions in the original and simultaneous measurement of slit deformation and/or recording of their nervous activity meets with considerable difficulties. Therefore model studies were designed again to see what the various shapes of lyriform organs might be good for (Barth et al. 1984).

The results indicate several remarkable effects and considerable refinement in the design of this biological strain-measuring system. They suggest that the original organs are specifically adapted to meet the different requirements for strain measurement that are relevant at the particular site where they occur. Such requirements would, for instance, be an especially wide angular working range, a large range of absolute sensitivities, or directional properties providing a basis for an analysis of load direction (Fig. 6).

Five slits were cut into Araldite or Plexiglas discs which were loaded statically in their plane from various directions. Deformation of the slits was measured with a strain gauge device inserted at different points along the slits.

The wealth of shapes found in arachnid lyriform organs was reduced to three basic configurations named according to their outline shape (Fig. 6) (a) "oblique bar" with slits of equal length and a regular lateral shift between each other; (b) "triangle" with regular graduation of slit lengths and various modifications of their lateral shift resulting in isosceles, rectangular and scalene triangles, respectively; (c) "heart" with the largest slit in the middle and shorter ones on both sides. In arrangement (a) the deformation of all slits is very similar and the range of load angles leading to slit compression (i.e., the angular working range) is particularly wide (Fig. 6). Possibly a wide spread of strain directions and its measurement is typical of the sites in the exoskeleton where such arrangements occur. From a mechanical point of view, great differences in threshold sensitivity among the original slits are as unlikely as great changes in the excitation pattern within the group with load direction, which in other cases may provide a basis for the central nervous analysis of strain direction. The main feature of the triangular arrangements (b) is a great difference in deformation among the slits. This suggests stimulus amplitude range-fractionation and the occurrence of strains widely differing in amplitude at the respective sites in the skeleton. The heart-shaped configurations (c) are peculiar for the

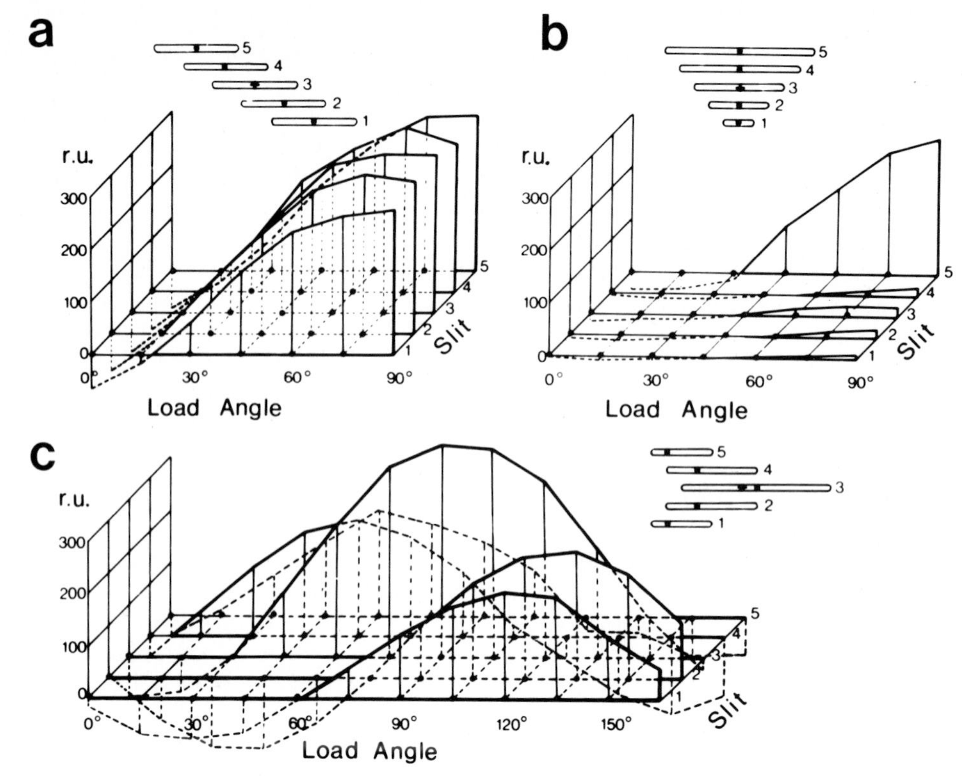

Fig. 6a–c. Deformation of model lyriform organs. Five slits (*1–5*) were cut into araldite discs in three arrangements (*insets*) typical of many original lyriform organs. Graphs give deformation of individual slits (*1–5*) at the site corresponding to dendrite attachment (*filled circles in insets*) in relative units (*r.u.*). Slit compression plotted above x-axis, slit dilatation below, as a function of load angle (0° parallel to slits). Note great differences in deformation patterns. (Barth et al. 1984)

large spread of the load angles at which deformation of the five model slits is highest. According to this finding and a considerable change in the deformation pattern within the group with changing load direction the heart configuration may be adapted to the analysis of strain direction.

7 Sources of Load and Sensitivity to Load Components

Having dealt with many aspects of the slit sensilla per se the main question is now: which are the sources of cuticular strains and how effective are they? This is a complex field of research and only recently methods have become available to do the adequate mechanical measurements with the necessary precision (Blickhan and Barth 1979; Blickhan 1983; Barth and Blickhan 1984; Blickhan et al. 1984).

In this paragraph I will first extend Section 3 on the occurrence of slit sensilla, and add a few general statements on the functional significance of topographical features typical of many of them. Then the sources of load and the mechanical sensitivity of the exoskeleton will be discussed for a representative example. Finally, in vivo measurements of cuticular strain will be presented.

7.1 General Implications of Slit Distribution

All sorts of load applications such as bending, buckling, twisting, compressing, stretching and shearing, and combinations of these are potential sources of adequate stimulation for slit sensilla (Fig. 1a). This may in part explain the wide spectrum of their topographical peculiarities. A few unifying rules pertaining at least to a large proportion of the slits nevertheless emerged from a comparison of the walking legs – where the majority of the slits occurs – of representatives of five arachnid orders (Barth and Stagl 1975).

1. In all arachnid species studied there is a concentration of slits proximally on the leg. Concentration of musculature proximally in the leg (Frank 1957; Parry 1957; Millot et Vachon 1949; Snodgrass 1965) has been suggested as the main reason, plus the fact that the main return and power-stroke components are produced here during locomotion. Since the activity of these muscles causes cuticular strains, the accumulation of slits proximally on the leg is no surprise.
2. The lyriform organs all lie close to or directly at the joints. Here forces are transferred from one segment to the next by articular condyles or hinge lines, the surface of which is much smaller than the cross-sectional area of the leg. The resulting concentration of stresses seems to be a good argument for the location of slit sensilla right behind the load-bearing surfaces. Some, however, are found clearly not in the immediate neighborhood of condyles and hinges. As argued before (Fig. 1b), displacement may well increase up to some distance from the areas of load transfer. In addition, bending of the cuticle, which is likely to occur when compressional forces are transferred at the joint, will be larger at some distance from the joint proper, where it is kept minimal due to particularly thick cuticle that is heavily sclerotized and well-placed with respect to the second moment of area.
3. In contrast to most lyriform organs, single slits on the leg (and opisthosoma of spiders) are often found far from the joints but close to muscle attachment sites, which are another source of bending forces.
4. A general finding on the legs of spiders and other arachnids are slits not only close to a joint but laterally on the leg and oriented roughly parallel to the long leg axis. As shown for a lyriform organ on the spider tibia (Fig. 5b) this particular position is in an area where strains resulting from the joint forces generated by muscular activity are not only compressive (i.e., stimulating), but in addition oriented roughly at right angles to the long axis of the slits. This latter finding implies maximal mechanical sensitivity due to the directionality of the lyriform organ (Figs. 3d, 6).

Although the above rules can serve as a guide to an understanding of the mechanical implications of the topography of many slit sensilla, they are by no means exhaustive. In many cases the particularities of an area in the skeleton have to be considered separately. An example is the vibration-sensitive metatarsal lyriform organ (see Barth, Chap. XI, this Vol.).

It should also be noted that in addition to the magnitude and orientation of strains at a particular area their time course may be another criterion for the presence of a slit (see Sect. 7.3).

7.2 Sources of Load and Mechanical Sensitivity

Recently the strains actually occurring in the exoskeleton under biologically relevant loading conditions and the force components producing them were measured (Blickhan et al. 1982; Blickhan 1983; Blickhan and Barth 1985; Blickhan et al. 1984). For that purpose miniaturized transducers were developed, which permit the precise measurement of all relevant parameters simultaneously and also in vivo. Thus, apart from minute strain gauges (active area 0.5 mm^2) for the measurement of cuticular strains, strain gauge-based pressure and force transducers were designed. Obviously these cannot be applied everywhere. The analysis was therefore restricted to the tibia of the walking leg and its distal joint with the metatarsus. This area was selected for two reasons: (1) four lyriform organs are present here, one of which is the best studied lyriform organ of all (see review Barth 1981); (2) here extensor muscles are lacking and functionally substituted by hemolymph pressure as at other joints with a dorsal hinge line and typical for spiders. Consequently the mechanical effects of hemolymph pressure and muscular activity can be separated.

Sources of Load. There are two "active" sources of load in a spider sitting on the ground, carrying its own weight and with the tibia under dorso-ventral load (Fig. 7). One is muscular activity flexing the joint, the other is the hemolymph pressure, which extends the joint and thus functions as antagonist of the flexor muscles. These two active components compensate the dorso-ventral components F_y of the ground reaction force and the torque generated by it. Lateral (F_z) and axial (F_x) forces generated extrinsically (as for instance by the body weight) are instead mainly passively absorbed by the cuticular material of the joint[2].

The momenta produced by the two active load sources are determined according to the following equations:

$$\text{i.} \quad \vec{M}_m = r_c \times \vec{F}_m . \tag{6}$$

This tells us that $\vec{M}_m$, the momentum (M) of the rotation generated by the flexor muscles (m), is a function of both the direction and the amplitude of the

2 In *Cupiennius* but not in theraphosid bird spiders, there is a third median condylus at the tibia-metatarsus joint. It may to some extent be used for active compensation by an asymmetry of the flexor activity.

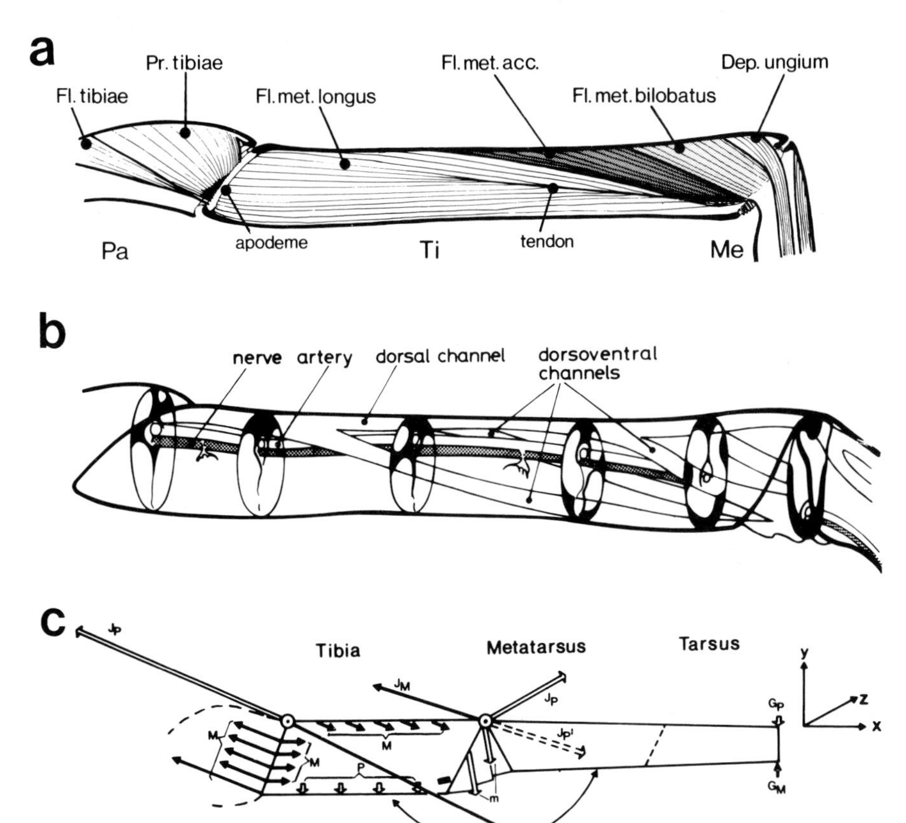

Fig. 7a–c. Active sources of load in the spider leg exoskeleton, represented by the tibia-metatarsus joint where extensor muscles are missing and functionally substituted by hemolymph pressure. **a** Flexor muscles (*Fl*) in the tibia (*Ti*); *Dep* depressor; *Pr* promotor; *Pa* patella; *Me* metatarsus. **b** blood channels ("open") in the tibia; note dorsoventral channels ending ventrally across Ti-Me joint, i.e., far away from dorsal axis of rotation, which implies a large momentum produced by the hemolymph pressure. **c** forces acting on tibia under dorsoventral load. *Filled arrows* indicate forces induced by flexor muscles (*M*); G_M ground reaction force; I_M joint force developed under load. *Open arrows* indicate forces induced by hemolymph pressure; *P* wall pressure; *m* forces induced in the articular membrane which is folded like a bellows; G_p ground reaction force, I_p joint force developed under load, I_p' joint force including *m*. ▬ lyriform organ HS-8. (**a** and **b** Blickhan and Barth 1985; **c** Barth and Blickhan 1984)

force vector ($\vec{F}_m$) and depends also on the position of the center (c) of gravity (r_c) of the muscle attachment area relative to the joint axis.

$$\text{ii.} \quad \vec{M}_h = PA(\vec{r}_s \times \vec{n}_A) \qquad (7)$$

The momentum (M) produced by the hemolymph (h) pressure is given by the pressure amplitude (P), the area (A) and orientation ($\vec{n}_A$) of the effective cross-section, and the position of its center of gravity ($\vec{r}_s$) with respect to the relevant axis. The overall joint force is not only determined by the two momenta above, but also by momentum $\vec{M}_G$, which results from the ground reaction force $\vec{F}_G$ and its moment arms of force vector $\vec{r}$:

$$\text{iii.} \quad \vec{M}_G = \vec{r} \times \vec{F}_{xyz}. \tag{8}$$

Mechanical Sensitivity to Force Components. The next question is: how sensitive are the various cuticular sites where the lyriform organs are found to the above load components. The mechanical sensitivity, which serves as a calibration constant, is defined here as

$$S = \frac{\text{strain}}{\text{force}} \, [\varepsilon/\text{N}]. \tag{9}$$

This question could be answered quantitatively by using miniaturized strain gauges glued to the cuticular surface and loading the distal metatarsus in the dorsoventral, axial, and lateral directions, respectively (Blickhan and Barth 1979; Blickhan 1983; Blickhan and Barth 1985). The measurements quoted here were done on a big theraphosid (bird) spider which has a relatively thick tibia (diameter ca 3 mm). The sites studied are those of lyriform organs HS-8 and HS-9 (posterior), VS-4 (anterior), and VS-5 (ventral) (Figs. 2, 4, 8).

1. S_y, for the active dorsoventral forces F'_y, measures up to 20 $\mu\varepsilon$/mN (that is, a piece of cuticle changes its length by a factor of 20×10^{-6} when loaded by 1 mN). While muscle forces induce negative strain (compression, i.e., adequate stimulation) at the site of organs HS-8, HS-9 and VS-4, they induce dilatation (i.e., no stimulation) of ventral organ VS-5. By contrast, organ VS-5 is compressed by the hemolymph pressure, which in turn dilates the other organs. Except for the site of organ HS-8 strains developed by hemolymph pressure are much higher than those due to muscular contraction at the other organs. At the site of organ HS-8 strains are remarkably independent of the joint angle.

2. S_z, for the passive lateral forces F'_z, measures up to 10 $\mu\varepsilon$/mN at organ HS-8, but less than 3 $\mu\varepsilon$/mN at organs HS-9 and VS-4. These values are much dependent on the joint angle α. Thus organ HS-8 is compressed both by posterior deflection of the metatarsus if $\alpha < 180°$, and by anterior deflection if $\alpha > 180°$. Similar findings apply to the other organs.

3. S_x, for the axial forces F'_x, is low with less than 0.8 $\mu\varepsilon$/mN and, like S_z, strongly depending on α. Under pressure load organs HS-8, HS-9 and VS-4 will be compressed, if $\alpha < 170°$ and under tension load, if $\alpha > 170°$. By contrast organ VS-5 is compressed at the reversed load directions. All organ sites are more or less unaffected mechanically at a joint angle of about 170°.

It is easy to see that from these data a number of predictions can be made on the strains occurring in the freely walking animal. In vivo measurements are described in the following.

7.3 In Vivo Measurements of Cuticular Strain During Locomotion

With the new techniques strains together with the leg-kinematics, ground reaction force and hemolymph pressure could be measured in the tibial exo-

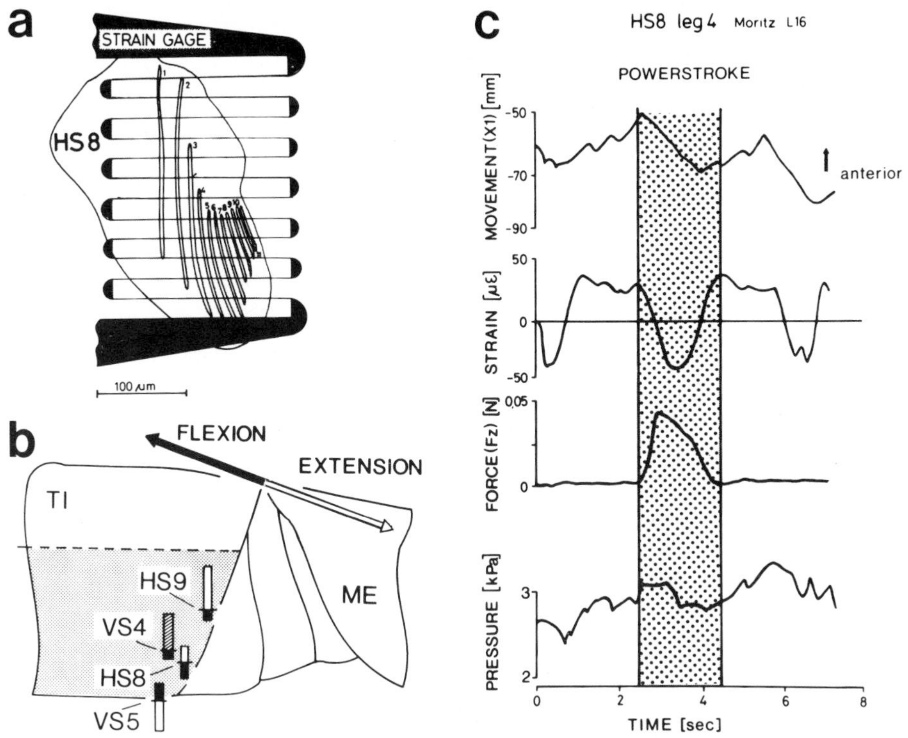

Fig. 8a–c. Strains at the site of lyriform organs. **a** miniature strain gauge glued onto the cuticle (here at site of organ HS-8 on tibia) for in vivo strain measurements. **b** during flexion induced by muscle forces negative strains (*bars below horizontal lines* indicating site of lyriform organ) implying slit compression are measured at the site of the organs laterally on the tibia (*Ti; VS-4* anterior side; *HS-8* and *HS-9* posterior side) and positive strains (slit dilatation) at the site of the ventral organ *VS-5*. During extension induced by hemolymph pressure, the ventral organ is the only one compressed (negative strain). *Me* metatarsus. **c** mechanical parameters measured simultaneously in vivo during slow locomotion of untethered spider. *From top to bottom* leg movement (leg 4), strain at lyriform organ HS-8, ground reaction force (vertical component), hemolymph pressure. A drop of $-80\,\mu\varepsilon$ coincides with the power stroke. Hemolymph pressure remains low. (Blickhan et al. 1982)

skeleton of bird spiders during slow locomotion when inertial forces are negligible (speed 1 to 10 cm s^{-1}; stepping frequency 0.3 to 1.5 Hz) (Fig. 8).

Under these circumstances the angle between tibia and metatarsus varies only between 160° and 180° in all legs. According to the angular dependence of the sensitivity factor S this implies that cuticular strain is mainly due to muscle force (largest value ca. 200 mN) and hemolymph pressure (largest value ca. 5.3 kPa; during *fast* locomotion and jumps up to ca. 50 kPa). Whereas the absolute strain values measured at a right angle to the long axis of the organ are strikingly similar (10 to 20 $\mu\varepsilon$, during fast locomotion up to 120 $\mu\varepsilon$) at the various lyriform organs, their *time courses* differ markedly. The organs are stimulated during different phases on the stepping cycle and the main reason for having several lyriform organs at one joint is likely to be the ability to distinguish between the phases of the stepping cycle and between muscular force and hemolymph pressure (Fig. 8 b). The general similarity of the strain amplitudes is interpreted as an indication of uniform strain distribution and uniform strength of the skeleton under the conditions of slow locomotion.

For the second leg this means in more detail: organ HS-8 is compressed during the powerstroke and therefore appears to be well suited to monitor the moments generated by the flexor muscles. At organ HS-8 negative strain rises already at the onset of the swing phase and changes into positive strain midway during the stance phase. At organ VS-4, compressional strain did not occur in any phase, and it is concluded that this organ is not stimulated at all during slow locomotion. Organ VS-5 is exceptional not only with regard to its location ventrally on the distal tibia. It is the only organ not dilated by an increase in the hemolymph pressure but compressed (during the return stroke) and appears well placed to monitor hemolymph pressure.

A comparison among different legs shows great similarity in the time courses of the strains but remarkable differences in strain amplitudes. These are much higher in leg 4 than in the other legs. Thus at organ HS-8, strain values of $-80\ \mu\varepsilon$ are found which is more than four times the value at the corresponding site on leg 2. This is explained by the smaller angle taken by leg 4 with respect to the ground which increases the vertical component (40 mN) of the ground reaction force. Hemolymph pressure also reaches higher values in leg 4 than in leg 2. Unlike the situation in leg 2 it rises from about 2.2 kPa to up to 5.3 kPa at the onset of walking. According to the mechanical sensitivities S, the share attributable to the hemolymph pressure in the strain measured at the site of organ HS-8 amounts to ca. 25%.

8 Behavioral Significance

The behavioral significance of slit sensilla is closely linked to locomotion which is also known in insect campaniform sensilla (Zill and Moran 1981 a, b; Zill et al. 1981). This is no surprise, since the majority of them are found on the legs, many close to joints or muscle-attachment sites. Also, the majority of slit sensilla so far studied electrophysiologically seem well suited for a proprioceptive function. Their frequency range is large and includes low frequencies down to fractions of 1 Hz (Bohnenberger 1981; Barth and Geethabali 1982). Some additional properties of lyriform organ HS-8 support this argument: here high increment sensitivity seems to be more important than a large range of absolute stimulus amplitudes typical of far-reaching exteroreceptors such as eyes, nose, and ears (Fig. 5 b). This seems appropriate for proprioceptors, which are often parts of feedback loops monitoring *small* deviations from the desired state.

Our knowledge of the role of slit sensilla in behavior is fragmentary. Surprisingly, in spiders normal, undisturbed locomotion is not at all or only slightly affected by the ablation of lyriform organs or even more drastic sensory deficits. This may be interpreted either as a prevalence of the central nervous system in the control of this behavior or an impressive redundancy of the proprioceptive sensory input (see Seyfarth, Chap. XII, this Vol.). On the other hand, the ablation of only a few or even one lyriform organ on four legs does clearly interfere with ideothetic orientation in *Cupiennius* (Barth and Seyfarth 1971; Seyfarth and Barth 1972; Seyfarth et al. 1982; see also Görner and Claas, Chap. XIV; Seyfarth, Chap. XII, and Mittelstaedt, Chap. XV, this Vol.). Lyriform organs are the first and so far the only identified receptors shown to be involved in this remarkable orientation behavior, which does not rely on landmarks but on memorized information from previous walking sequences. Collection of the memorized data and possibly also precise course control depend at least partly on proprioceptive input from lyriform organs.

Another behavior depending on the intactness of certain lyriform organs is leg reflexes (Seyfarth 1978a, b; see also Seyfarth, Chap. XII, this Vol.). It is not the resistance reflexes, which are widespread among arthropods, which are affected by ablations. Instead the destruction of lyriform organs results in a failure to elicit certain synergic reflexes which activate muscles augmenting the imposed movement and thus relieving the strain.

As amply demonstrated by the vibration-sensitive metatarsal lyriform organ and the pretarsal slits, slit sensilla may also have exteroceptive function in behavior (see Barth, Chap. XI, this Vol.).

9 Concluding Remarks

Arachnid slit sensilla nicely demonstrate the complex mechanical phenomena occurring in the cuticular armor of the arthropods and their use for the control of various behaviors. Despite the advances now made in an understanding of the principles underlying "strain" detection in the spider exoskeleton there are several fields of research which have almost not been touched. With the new knowledge and techniques available, the ground now seems prepared for their investigation.

One main line of future research should concern the central nervous system. Valuable insights into the general organization of the central nervous system of arachnids and in particular the spiders have been gained from neuroanatomical work (see Babu, Chap. I, this Vol.; Babu and Barth 1984). Almost nothing, however, is known at the single neuron level about the specific path taken by the information from the many slit sensilla and about the integration, convergence and interaction of their activities with those of other sensory modalities. The beginning made with respect to the spider vibration sense is promising (see Barth, Chap. XI, this Vol.). Both the fine neuroanatomy based on single-cell stains and the electrophysiological recording from identified interneurons may eventually give us an idea of what the concerted activity of hundreds or thousands of slits distributed over the skeleton "sounds" like to the brain of a spider or an other arachnid.

Our knowledge about the involvement of slit sensilla in behavior is just a beginning. Studies on idiothetic orientation and many ablation experiments without resulting in drastic behavioral deficits have amply shown that their role may be quite subtle and often hard to detect. A larger variety of behaviors such as feeding, silk production, and web construction and the potential involvement of slit sensilla should be considered.

The biomechanical studies which have so far concentrated on one representative joint region should be extended to other areas of the exoskeleton, including the relatively soft cuticle of the spider opisthosoma and other arachnids. Starts, stops, and jumps presumably causing strains much higher than those occurring during slow locomotion await further analysis. Finally, for all those interested in the mechanical design of a cuticular exoskeleton, the slit sensilla will prove to be an excellent guide: their distribution and arrangement, shaped during evolution by predominantly mechanical selective pressures, are likely to be a mirror image of the mechanical situation they have to cope with in a particular area.

Acknowledgments. The generous financial support of the author's laboratory by the Deutsche Forschungsgemeinschaft and the loyal collaboration of my students and associates over the years are gratefully acknowledged. I thank Mrs. H. Hahn for expert help with the preparation of the figures, Mrs. U. Ginsberg for typing the manuscript, and Dr. E.-A. Seyfarth for critical comments.

References

Babu KS, Barth FG (1984) Neuroanatomy of the central nervous system of the wandering spider, *Cupiennius salei* (Arachnida, Araneida). Zoomorphology 104:344–359

Barth FG (1967) Ein einzelnes Spaltsinnesorgan auf dem Spinnentarsus: seine Erregung in Abhängigkeit von den Parametern des Luftschallreizes. Z Vergl Physiol 55:409–449

Barth FG (1969) Die Feinstruktur des Spinnenintegumentes. I. Die Cuticula des Laufbeines adulter häutungsferner Tiere (*Cupiennius salei* Keys.). Z Zellforsch 97:139–159

Barth FG (1970) Die Feinstruktur des Spinnenintegumentes. II. Die räumliche Anordnung der Mikrofasern in der lamellierten Cuticula und ihre Beziehung zur Gestalt der Porenkanäle (*Cupiennius salei* Keys., adult, häutungsfern, Tarsus). Z Zellforsch 104:89–106

Barth FG (1971) Der sensorische Apparat der Spaltsinnesorgane (*Cupiennius salei* Keys. Araneae). Z Zellforsch 112:212–246

Barth FG (1972a) Die Physiologie der Spaltsinnesorgane. I. Modellversuche zur Rolle des cuticularen Spaltes beim Reiztransport. J Comp Physiol 78:315–336

Barth FG (1972b) Die Physiologie der Spaltsinnesorgane. II. Funktionelle Morphologie eines Mechanorezeptors. J Comp Physiol 81:159–186

Barth FG (1973) Laminated composite material in biology. Microfiber reinforcement of an arthropod cuticle. Z Zellforsch 144:409–433

Barth FG (1976) Sensory information from strains in the exoskeleton. In: Hepburn HR (ed) The insect integument. Elsevier, Amsterdam Oxford New York, pp 445–473

Barth FG (1978) Slit sense organs: "Strain gauges" in the arachnid exoskeleton. Symp Zool Soc Lond 1977 42:439–448

Barth FG (1980) Campaniform sensilla: another vibration receptor in the crab leg. Naturwissenschaften 67:201

Barth FG (1981) Strain detection in the arthropod exoskeleton. In: Laverack MS, Cosens D (eds) Sense organs, chap 8. Blacky, Glasgow, pp 112–141

Barth FG, Blickhan R (1984) Mechanoreception. In: Bereiter-Hahn J, Matoltsy AG, Richards KS (eds) Biology of the integument, vol 1. Invertebrates. Springer, Berlin Heidelberg New York Tokyo, pp 544–582

Barth FG, Bohnenberger J (1978) Lyriform slit sense organ: Threshold and stimulus amplitude ranges in a multi-unit mechanoreceptor. J Comp Physiol 125:37–43

Barth FG, Geethabali (1982) Spider vibration receptors. Threshold curves of individual slits in the metatarsal lyriform organ. J Comp Physiol 148:175–186

Barth FG, Libera W (1970) Ein Atlas der Spaltsinnesorgane von *Cupiennius salei* Keys. Chelicerata (Araneae). Z Morphol Tiere 68:343–369

Barth FG, Pickelmann H-P (1975) Lyriform slit sense organs. Modelling an arthropod mechanoreceptor. J Comp Physiol 103:39–54
Barth FG, Seyfarth E-A (1971) Slit sense organs and kinesthetic orientation. Z Vergl Physiol 74:326–328
Barth FG, Stagl J (1976) The slit sense organs of arachnids. A comparative study of their topography on the walking legs. Zoomorphology 86:1–23
Barth FG, Wadepuhl M (1975) Slit sense organs on the scorpion leg (*Androctonus australis*, L. Buthidae). J Morphol 145: (2) 209–227
Barth FG, Ficker E, Federle H-U (1984) Model studies of the mechanical significance of grouping in compound spider slit sensilla. Zoomorphology 104:204–215
Bertkau PH (1878) Versuch einer natürlichen Anordnung der Spinnen nebst Bemerkungen an einzelnen Gattungen. Arch Naturgesch 44:351–410
Biederman-Thorson M, Thorson J (1971) Dynamics of excitation and inhibition in the light adapted *Limulus* eye in situ. J Gen Physiol 58:1–19
Bleckmann H, Barth FG (1984) Sensory ecology of a semi-aquatic spider (*Dolomedes triton*) II. The release of predatory behavior by water surface waves. Behav Ecol Soc 14:303–312
Blickhan R (1983) Dehnungen im Außenskelett von Spinnen. Dissertation, Univ Frankfurt
Blickhan R, Barth FG (1979) Dehnungen und Spannungen im Außenskelett von Arthropoden. GESA-Symp, Exp Spannungsanal, Braunschweig, 21 p
Blickhan R, Barth FG (1985) Strains in the exoskeleton of spiders. J Comp Physiol (in press)
Blickhan R, Barth FG, Ficker E (1982) Biomechanics in a sensory system. Strain detection in the exoskeleton of arthropods. VIIth Int Conf Exp Stress Anal, Haifa, pp 223–233
Blickhan R, Weber W, Barth FG (1984) Strain at the site of biological strain detectors in the exoskeleton of spiders. Int Conf Exp Stress Anal, Montreal (in press)
Bohnenberger J (1978) On the transfer characteristics of a lyriform slit sense organ. Symp Zool Soc Lond 42:449–455
Bohnenberger J (1979) Das Übertragungsverhalten eines zusammengesetzten Spaltsinnesorgans auf dem Spinnenbein. Dissertation, Univ Frankfurt
Bohnenberger J (1981) Matched transfer characteristics of single units in a compound slit sense organ. J Comp Physiol 142:391–402
Chapman DM, Mosinger JL, Duckrow RB (1979) The role of distributed viscoelastic coupling in sensory adaptation in an insect mechanoreceptor. J Comp Physiol 131:1–12
Frank U (1957) Untersuchungen zur funktionellen Anatomie der lokomotorischen Extremitäten von *Zygiella x-notata*, einer Radnetzspinne. Zool Jahrb Abt Anat Ontog Tiere 76:423–460
Jonscher AK (1977) The universal dielectric response. Nature (London) 267:673–679
Kaissling K-E, Thorson J (1980) Insect olfactory sensilla: structural, chemical and electrical aspects of the functional organization. In: Hall LM, Hildebrand JG, Satelle DB (eds) Receptors for neurotransmitters, hormones and pheromones in insects. Elsevier, Amsterdam Oxford New York, pp 261–282
Kaston BJ (1935) The slit sense organs of spiders. J Morphol 58:189–209
Küppers J (1974) Measurements on the ionic milieu of the receptor terminal in mechanoreceptive sensilla of insects. Rheinisch-Westfael Akad Wiss 53:387–394
Lachmuth U, Grasshoff M, Barth FG (1985) Taxonomische Revision der Gattung *Cupiennius* SIMON 1891 (Arachnida: Araneae). Senckenbergiana biol 65:329–372
Mann JW, Chapman KM (1975) Component mechanism of sensitivity and adaptation in an insect mechanoreceptor. Brain Res 97:331–336
McIndoo NE (1911) The lyriform organs and tactile hairs of araneids. Proc Acad Nat Sci Philadelphia 63:375–418
Millot J, Vachon M (1949) Ordre des scorpions. In: Grassé T (ed) Traité de zoologie, vol VI. Masson, Paris, pp 386–436
Parry DA (1957) Spider leg-muscles and the autotomy mechanism. Q J Microsc Sci 98: (3) 331–340
Pringle JWS (1955) The function of the lyriform organs of arachnids. J Exp Biol 32:270–278
Rick R, Barth FG, Pawel A (1976) X-ray microanalysis of receptor lymph in a cuticular arthropod sensillum. J Comp Physiol 110:89–95

Seyfarth E-A (1978a) Lyriform slit sense organs and muscle reflexes in the spider leg. J Comp Physiol 125:45–57

Seyfarth E-A (1978b) Mechanoreceptors and proprioreceptive reflexes: lyriform organs in the spider leg. Symp Zool Soc Lond 42:457–467

Seyfarth E-A, Barth FG (1972) Compound slit sense organs on the spider leg: mechanoreceptors involved in kinesthetic orientation. J Comp Physiol 78:176–191

Seyfarth E-A, Bohnenberger J, Thorson J (1982) Electrical and mechanical stimulation of a spider slit sensillum: outward current excites. J Comp Physiol 147:423–432

Seyfarth E-A, Pflüger H-J (1984) Proprioceptor distribution and control of a muscle reflex in the tibia of spider legs. J Neurobiol 15:365–374

Seyfarth E-A, Eckweiler W, Hammer K (1985) A survey of sense organs and sensory nerves in the legs of spiders. Zoomorphology (in press)

Snodgrass RE (1965) A textbook of arthropod anatomy. Hafner, New York London

Speck J, Barth FG (1982) Vibration sensitivity of pretarsal slit sensilla in the spider leg. J Comp Physiol 148:187–194

Thorson J, Biederman-Thorson M (1974) Distributed relaxation processes in sensory adaptation. Science 183:161–172

Thurm U (1974) Basics of the generation of receptor potentials in epidermal mechanoreceptors in insects. In Schwarzkopff J (ed) Mechanoreception. Rheinisch-Westfael Acad Wiss 53:355–385

Thurm U (1982a) Grundzüge der Transduktionsmechanismen in Sinneszellen. In: Hoppe W, Lohmann W, Markl H, Ziegler H (eds) Biophysik. Springer, Berlin Heidelberg New York, pp 681–691

Thurm U (1982b) Mechano-elektrische Transduktion. In: Hoppe W, Lohmann W, Markl H, Ziegler H (eds) Biophysik. Springer, Berlin Heidelberg New York, pp 691–696

Thurm U, Küppers J (1980) Epithelial physiology of insect sensilla. In: Locke M, Smith D (eds) Insect biology in the future. Academic Press, London New York, pp 735–763

Thurm U, Wessel G (1979) Metabolism-dependent transepithelial potential differences at epidermal receptors of arthropods. J Comp Physiol 134:119–130

Vogel H (1923) Über die Spaltsinnesorgane der Radnetzspinnen. Jena Z Med Naturwiss 59:171–208

Wainwright SA, Biggs WD, Currey JD, Gosline JM (1976) Mechanical design in organisms. Unwin, London

Yamada H (1970) In: Evans FG (ed) Strength of biological materials. Williams & Wilkins, Baltimore

Zill SN, Moran DT (1981a) The exoskeleton and insect proprioception. I. Responses of tibial campaniform sensilla to external and muscle-generated forces in the american cockroach, *Periplaneta americana.* J Exp Biol 91:1–24

Zill SN, Moran DT (1981b) The exoskeleton and insect proprioception. III. Activity of tibial campaniform sensilla during walking in the american cockroach, *Periplaneta americana.* J Exp Biol 94:57–75

Zill SN, Moran DT, Varela FG (1981) The exoskeleton and insect proprioception. II. Reflex effects of tibial campaniform sensilla in the american cockroach, *Periplaneta americana.* J Exp Biol 63:1–13

X Sensory Nerves and Peripheral Synapses

R. F. Foelix

CONTENTS

1 Introduction

The peripheral nervous system of arachnids is usually represented as fine extensions of the condensed central nervous system (CNS) located in the prosoma (Gerhardt and Kaestner 1938). These fiber bundles emanating from the CNS constitute the major nerves which supply the appendages and the opisthosoma. Only the leg nerves have been studied in some detail. They consist mostly of *afferent* fibers coming from leg receptors, and of *efferent* fibers innervating the leg muscles. Some of the efferent fibers belong in another category; they contain small dark granules and are considered to be neurosecretory.

A few years ago numerous synaptic connections were discovered at the receptor cells and within the small sensory nerves of arachnids (Foelix 1975). As a consequence, the seemingly simple picture of their peripheral nervous system becomes now much more complex. The functional significance of these peripheral interactions is at present hardly understood.

2 Sensory Nerves: Distribution and Fine Structure

2.1 Distribution, Number, and Diameter of Fibers

The sensory cells of the various leg receptors give off axons which lie first within the hypodermis. They bundle together with axons from neighboring sen-

Université de Fribourg, Institut d'Anatomie, 1, Rue Gockel, CH-1700 Fribourg, Switzerland

silla and form small sensory nerves. These leave the hypodermis to join larger sensory nerves in the hemolymph space of the leg (Fig. 1, inset). In the arachnid tarsus we find two sensory nerves which run on either side of the leg artery. While proceeding proximally, these two nerves merge (in spiders at the level of the metatarsus) to form one large sensory nerve. This nerve, C (Rathmayer 1965), consists of thousands of small nerve fibers, whereas the motor nerve, B, contains only about 100 large axons. The actual number of sensory fibers in nerve C depends on the density of leg receptors and also on the size of the animal. In the small spider *Zygiella* we counted over 7000 sensory axons in the first leg, about 4400, 4200 and 3900 in legs 2, 3, 4, and 2200 in the palp (Foelix et al. 1980). In comparison, the large house spider *Tegenaria* had 5000 to 6000 sensory fibers in one palp (♀: 6400, ♂: 5000; Hülser, personal communication). In the antenna-like first legs of whip spiders we found over 23,000 axons comprising the two tarsal nerves (Beck et al. 1977), so that the total sensory input of one first leg can be estimated to come from about 30,000 fibers. These numbers are quite comparable to those from certain insect antennae, where 50,000 have been reported for the locust (Boeckh et al. 1976) and 65,000 for the bee (Esslen and Kaissling 1976).

The high numbers of fibers in these sensory nerves can only be achieved if the axons are of rather small diameter. For the spider *Zygiella* we found the

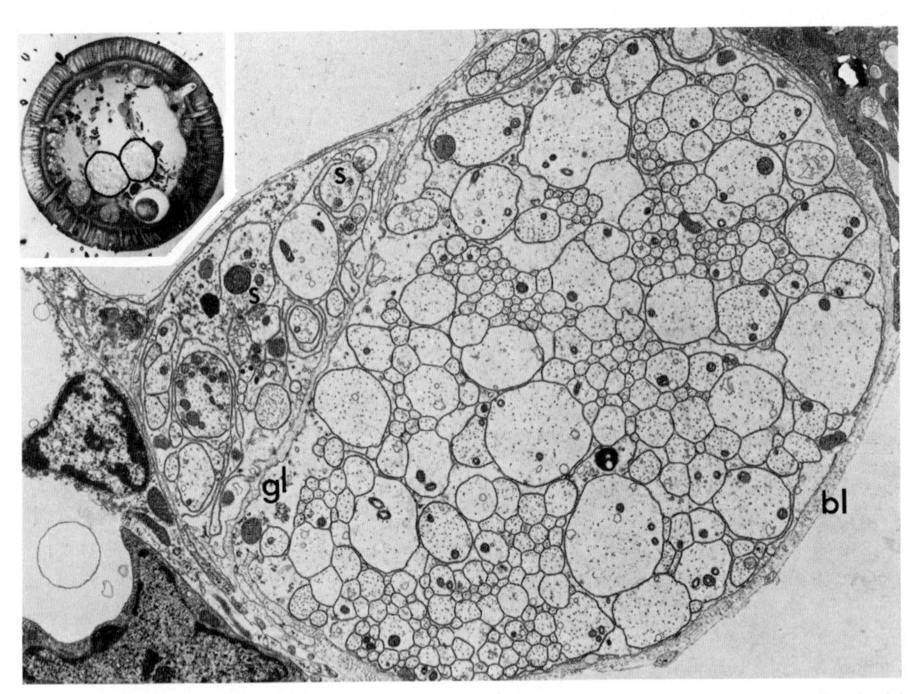

Fig. 1. *Inset* Cross-section of a spider tarsus (*Lycosa*); the two sensory nerves are outlined with ink. ×80. Cross-section of one tarsal nerve (spider *Zygiella*). About 400 sensory axons are seen, enclosed by a thin glial lamella (*gl*) and a basement lamina (*bl*). Note area on the *left*, where synaptic contacts (*s*) occur between axons. (Photo Müller-Vorholt). ×6600

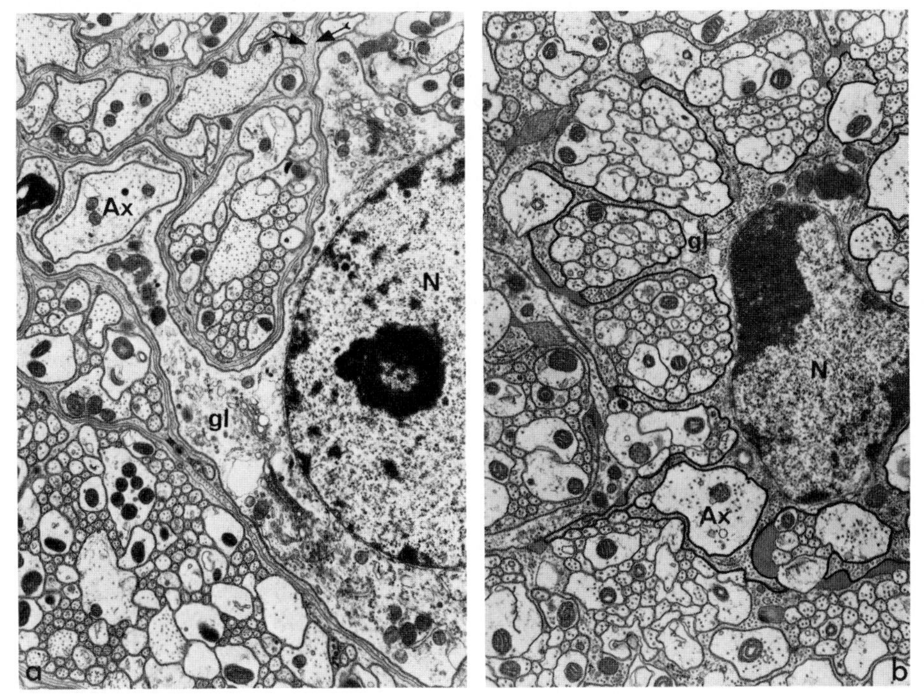

Fig. 2a, b. Comparison of sensory nerves from a whip spider tarsus (**a**) and a cockroach antenna (**b**). In both cases extensions of glial cells (*gl*) enclose bundles of axons, rarely single axons (*Ax*). The basal lamina is indicated by *arrows* in (**a**) and appears as *grey areas* in (**b**). The outline of the glial cell in (**b**) is enhanced by an ink line. **a** ×9300; **b** ×11,500

sensory axons to measure from 0.1 to 2 μm in thickness (mean: 0.36 μm), with most fibers in the range between 0.1 and 0.2 μm (Foelix et al. 1980). Similar values were noted for the whip spider's tarsal nerves, i.e., a mean diameter of 0.25 μm (Foelix and Troyer 1980). Although the fiber diameter may vary along the course of an axon, the mean values remain the same in distal and proximal regions of the leg.

2.2 Fine Structural Organization

Each sensory nerve is formed by three components: (1) sensory axons, (2) glial cells, and (3) a basal lamina (Fig. 1). Even close to their origin, when consisting of only a few fibers, the sensory nerves are already enclosed by at least one glial cell. As the number of fibers increases, glial processes extend radially into the nerve, thus giving rise to several fascicles. Each fascicle encases from a few up to several hundreds of "naked" axons (Fig. 2a). The surface of the glial cells is covered with a prominent basal lamina, which often contains distinct collagen fibrils. The basal lamina is most pronounced on the outside of the sensory nerve. At more proximal levels, a thin layer of connective tissue cells may be added, which one could call a perineurium.

The general organization of arachnid sensory nerves is thus very similar to that of insects (Fig. 2b; Lane and Treherne 1973; Smith 1968) and need not be discussed here in detail. There is, however, one feature which sets the sensory nerves of arachnids distinctly apart from those of any other arthropods, and that is the occurrence of synapses.

3 Peripheral Synapses

It is generally believed that sensory axons of arthropods proceed into the central nervous system to make their first synaptic contact with second-order neurons (Bullock and Horridge 1965). Consequently, no synapses are to be expected in the peripheral nervous system. Although this assumption may be valid for insects and crustaceans, it certainly does not apply for arachnids.

After the first discovery of peripheral synapses in the specialized giant fiber system of whip spiders (see below), it soon became apparent that synapses are a common feature in the peripheral nervous system of all arachnids (Foelix 1975). These synapses are most abundant in the periphery itself, i.e., at the level of the sensory cells and their processes. They are still frequently seen in small sensory nerves but are rare inside the main sensory nerves, where they may form occasional axo-axonal contacts. At the receptor level we find axo-dendritic, axo-somatic, axo-axonal and even axo-glial contacts (Fig. 3). Before addressing the difficult question which elements are pre- and postsynaptic, we shall take a closer look at their fine structure.

3.1 Fine Structure

The peripheral synapses of arachnids are a good model for fine structural studies of synapses in general, because they fix unusually well and certainly show a much better preservation than any central synapses. The main features of an arachnid synapse are the following (Figs. 4, 5): (1) A presynaptic membrane density in the shape of a rod (bar); (2) many synaptic vesicles which aggregate around this bar, and (3) a postsynaptic side consisting typically of two (or more?) fibers which lie opposite and symmetrical to the presynaptic bar. This type of synapse is therefore often referred to as dyad or bar synapse. It is quite common in the central nervous system of insects (Wood et al. 1979; Tolbert and Hildebrand 1981) and other arthropods (Fahrenbach 1979). The "grid" type synapse, which is typical for vertebrates (Akert and Peper 1975) but which also occurs in the CNS of insects (Wood et al. 1979), was never seen in the nervous system of arachnids, either centrally or peripherally.

3.1.1 Presynaptic Membrane Thickening

The presynaptic density is apparently the center of the synapse, because the synaptic vesicles gather in clouds above it (Fig. 4, inset).

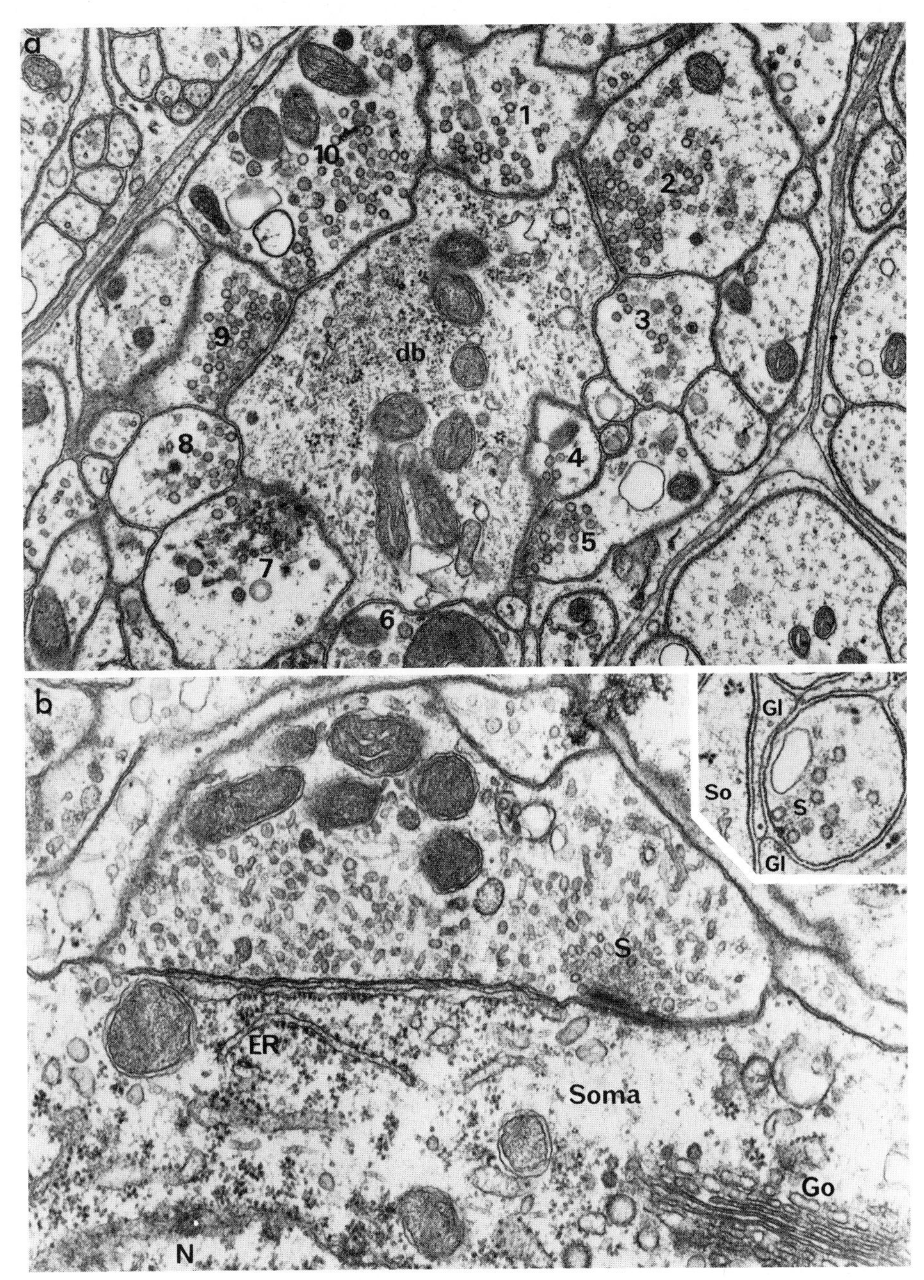

Fig. 3. a Ten axons (*1–10*) synapse on a dendritic branch (*db*) of a giant neuron in the tarsus of a whip spider (*Heterophrynus*). ×22,000. **b** Axo-somatic synapse from a tarsal hair sensillum in a spider (*Gradungula*). Note the flattened synaptic vesicles and the cisterna of the endoplasmic reticulum (*ER*), which is closely apposed to the cell membrane. *Go* Golgi apparatus; *N* nucleus, *s* synaptic contact. ×31,000. *Inset* an axo-glial synapse, where the presynaptic density faces the cleft between two glial extensions (*Gl*). *So* soma. ×38,000

This density can be up to 1 µm long, but measures only 0.05 µm in width and height. It is covered by a delicate "capping membrane" at a distance of 50–60 nm from the presynaptic membrane. Some fine fuzzy material extends from the bar and the capping membrane and seems to connect to the adjacent synaptic vesicles (Foelix and Choms 1979). Usually, only one bar is present in the presynaptic fiber, but occasionally several can be seen lying parallel to each other. In the axo-dendritic synapses of the giant fiber system in whip spiders, this seems to be the rule rather than the exception (Foelix and Troyer 1980).

3.1.2 Synaptic Vesicles

About four different kinds of vesicle can be differentiated presynaptically: (1) spherical, (2) flattened, (3) dense-core, and (4) coated vesicles. The first two types appear relatively light and represent the typical synaptic vesicles.

Several studies have related spherical vesicles to excitatory synapses and flattened (or smaller) vesicles to inhibitory synapses (Uchizono 1965, 1967; Tisdale and Nakajima 1976; Atwood and Kwan 1979). We also observed flattened (or small) vesicles mostly in axo-somatic synapses (Fig. 3); if they were spherical, then they were distinctly smaller than in axo-dendritic synapses (spider: 32 vs. 37 nm, Foelix and Choms 1979; whip spider: 51 vs. 62 nm, Foelix and Troyer 1980). In the spider CNS we noticed two kinds of synapses, one with large vesicles (37 nm, excitatory?) and one with smaller ones (32 nm, inhibitory?). Dense-core and coated vesicles appear often in the periphery of a synapse; the latter seem to bud off the presynaptic membrane (Fig. 5) and are thought to play a role in membrane retrieval (Gray and Pease 1971; Heuser and Reese 1973).

Close to the presynaptic membrane the synaptic vesicles become grouped in a single row on either side of the dense bar (Fig. 4). Apparently, the vesicles attach to specific sites of the presynaptic membrane and then fuse with it. This exocytosis can be seen as "omega" figures in electron micrographs (Fig. 4c). The bar is supposed to represent the site of transmitter mobilization and/or release (Govind and Meiss 1979).

Other typical presynaptic organelles are mitochondria, smooth endoplasmic reticulum, and microtubules, but they are usually restricted beyond the cloud of synaptic vesicles (Fig. 5). The synaptic cleft is rather unstructured and does not show any "cleft line" as in vertebrate synapses. Similarly inconspicuous is the postsynaptic side, where membrane thickening is weak or lacking altogether. Sometimes, however, cisternae of the endoplasmic reticulum are closely apposed to the postsynaptic membrane (Fig. 3b).

3.2 Peripheral Integration

As we have already seen, the typical arachnid synapse is represented by the dyad configuration, where one presynaptic fiber contacts two (or more) postsynaptic fibers. This kind of divergence is commonly found in the CNS of arthropods (Schürmann 1971, 1974; Fahrenbach 1979). Most of the peripheral axo-axonal synapses in arachnids are simple dyads (one pre- on two postsynaptic fibers), but sometimes the situation is more complex. For instance, the postsynaptic fiber may also contain synaptic vesicles and form synapses with another fiber ("serial" synapse; Fig. 4a) or even with the presynaptic fiber ("re-

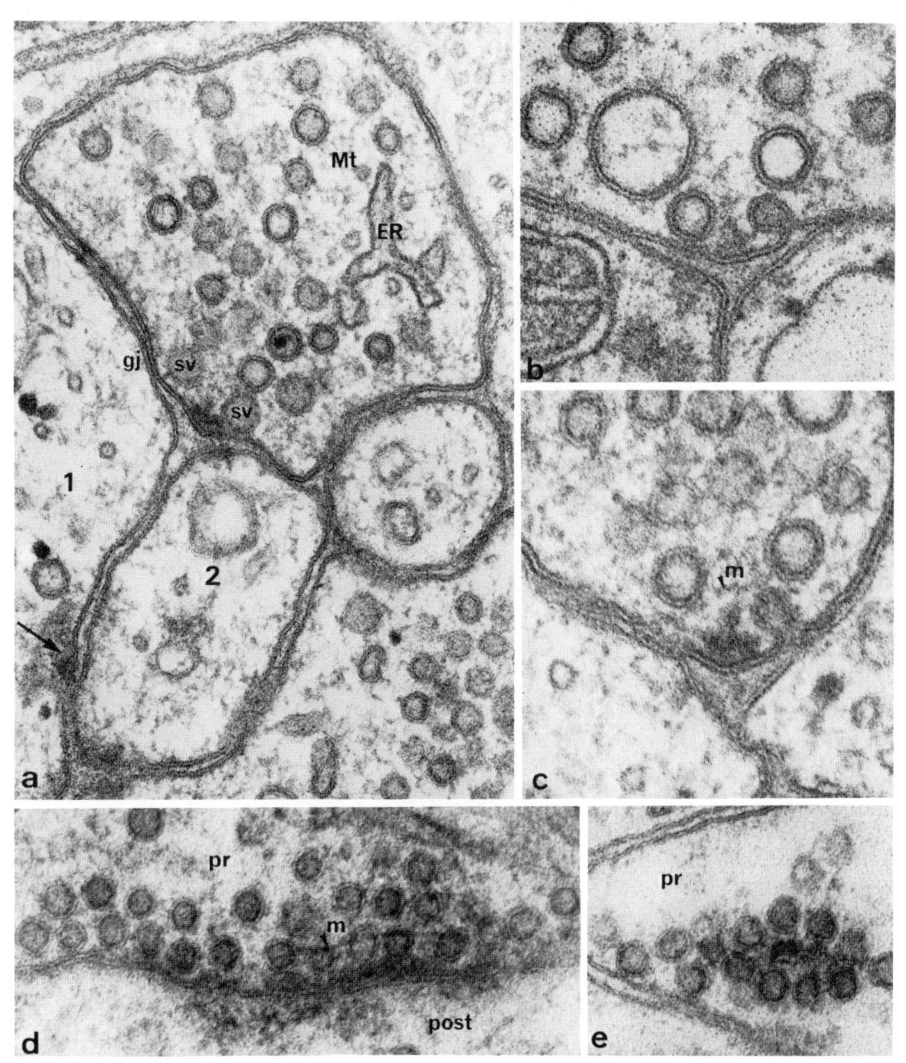

Fig. 4a–e. Fine structure of peripheral synapses in the pecten of a scorpion (*Androctonus*). **a** Typical dyad synapse with one presynaptic fiber opposing two postsynaptic fibers (*1, 2*). Synaptic vesicles gather on either side of the presynaptic density. Note that fiber 2 is also postsynaptic to fiber 1 (*arrow*). *ER* endoplasmic reticulum; *Mt* microtubule, *gj* gap junction. ×82,500. **b, c** Attachment and fusion of synaptic vesicles to the presynaptic membrane. *m* capping membrane. ×140,000. **d** Longitudinal section of bar synapse with synaptic vesicles arranged in rows on the presynaptic side (*pr*). ×90,000. **e** Tangential section of bar synapse showing part of the presynaptic density in the center. ×97,000

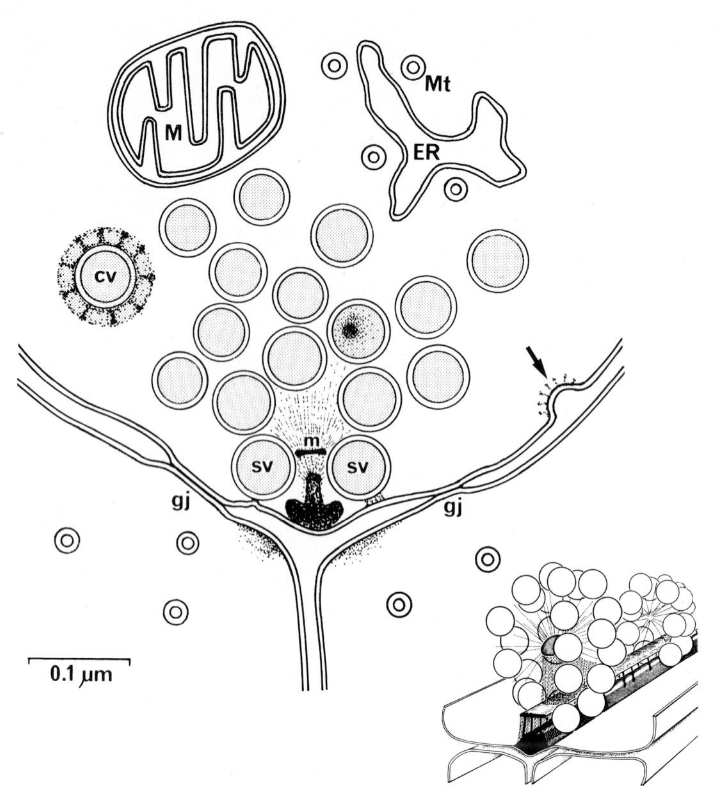

Fig. 5. Diagram of the dyad synapse. The presynaptic bar forms the center of the synapse and is covered by a thin capping membrane (*m*). Synaptic vesicles (*sv*) gather around the bar and attach to the presynaptic membrane. Further presynaptic organelles are mitochondria (*M*), smooth endoplasmic reticulum (*ER*) with associated microtubules (*Mt*) and coated vesicles (*cv*) which seem to originate from the presynaptic membrane (*arrow*). Gap junctions (*gj*) occur often on both sides of the synaptic center. *Inset* three-dimensional representation of a dyad synapse in spiders. (Foelix and Choms 1979)

ciprocal" synapse). This is the case in the plexus underlying the pecten sensilla in scorpions, which shows a high density of synaptic interactions of sensory axons.

Synaptic interactions at the receptor level have been recently demonstrated in the peripheral nervous system of *Limulus*, one of the most ancient arthropods. There, the chemosensitive sensilla establish synaptic contacts through axon collaterals (Fig. 6) and it is believed that peripheral summation occurs at these sites (Hayes and Barber 1982). Quite probably, many of the axo-axonal synapses in arachnids can be interpreted in the same way.

Axo-dendritic synapses on peripheral sensory cells are more difficult to explain. In a spider joint receptor we proposed that the presynaptic input comes from efferent fibers which might exert an inhibitory function (Foelix and Choms 1979). This conclusion was based on the small size of the synaptic vesicles, and on the observation that these alleged efferent fibers increased in num-

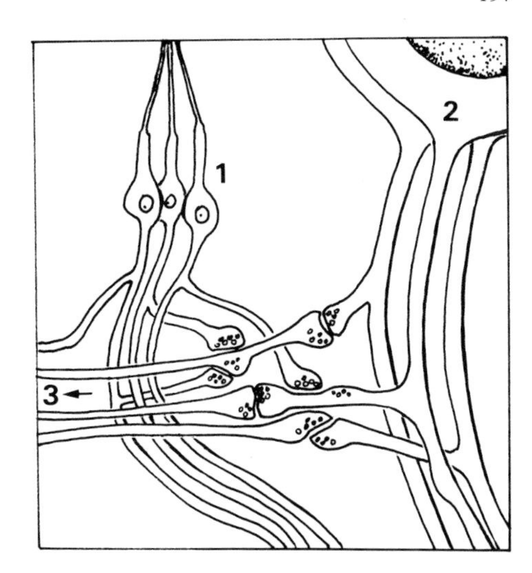

Fig. 6. Neural plexus formed by the interaction of three chemoreceptive sensilla (*1, 2, 3*) in *Limulus.* Note that synapses are formed by axon collaterals. (After Hayes and Barber 1982)

ber toward the periphery, presumably by branching. We also assumed an inhibitory function for the axo-somatic synapses, possibly controlling receptor activity. Axo-somatic synapses were seen on sensory cells of many other receptors as well, such as hair sensilla, slit sensilla and even on retinula cells of photoreceptors (scorpions: Fleissner and Fleissner 1978; spiders: Foelix and Choms 1979; horseshoe crab: Fahrenbach 1981).

It is evident that any functional interpretation of peripheral synapses in arachnids is at present largely speculative. There is only one example where the synaptic connectivity can be explained reasonably well, and that is in the giant fiber system of whip spiders (Amblypygi) and whip scorpions (Uropygi).

3.3 Giant Fibers and Associated Synapses

Whip spiders have modified first legs: they are extremely elongated (up to 25 cm) and are not used for walking. Instead, they have a purely sensory function, scanning the environment in a way similar to that of insects with their antennae. Thousands of mechano- and chemosensory hairs cover the tarsi of these legs. Their axons bundle together to form two large tarsal nerves which contain more than 20,000 fibers. Most of these axons are very thin (0.1 – 0.2 μm), but a few are exceptionally large (up to 20 μm). These giant fibers belong to interneurons whose large somata lie within the tarsus. Each giant neuron sends out 1 – 2 branching dendrites distally and these form numerous synapses (Figs. 3a, 6). The presynaptic input comes apparently from large hair sensilla (bristles), which respond to mechanical deflection (Foelix and Troyer 1980; Igelmund 1983, personal communication). The role of the giant fibers would be to transmit nerve impulses rapidly to the central nervous system (Fig. 7). Mechanical stimulation of the tip of the first leg triggers indeed a rapid withdrawal

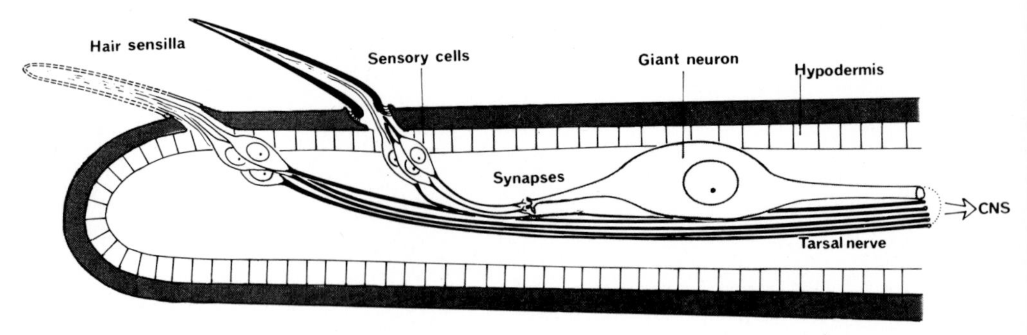

Fig. 7. Synaptic connectivity in the giant fiber system of the whip spider tarsus. Sensory axons from mechanosensitive bristles make en passant synapses with dendritic branches of a giant neuron. The sensory input is probably integrated by the giant neuron and then rapidly sent to the CNS via the giant axon. (After Foelix and Troyer 1980)

reaction from the animal. Using high speed film analysis we measured 80–120 ms as reaction times in a leg of 10 cm length. This corresponds to a conduction velocity of at least 1 m s^{-1}, which is certainly much faster than one would expect in the thin sensory axons. Recent electrophysiological recordings from these giant fibers gave values of 5–6 m s^{-1}. Even more important were the following observations: (1) Deflection of one large bristle elicits spikes in the giant fiber, and (2) input from many bristles converges on one giant axon (Igelmund 1984, personal communication). In fact, two giant neurons with overlapping sensory fields could be detected in these experiments: one in the distal tarsus which receives input from about 500 bristles, and the other, covering the more proximal part, which responds to about 1000 bristles.

This new information calls for a correction of our original wiring diagram (Fig. 7), where we omitted any synapses on the giant axon. Although we had seen such axo-axonal synapses occasionally in our electron micrographs, we had not realized how numerous they actually must be on an axon that is 10–20 cm long.

When we compare the giant fiber system of whip spiders to that of insects, crustaceans, annelids or molluscs, we notice one important difference: in all other invertebrates investigated so far, the giant neurons are always localized in ganglia (i.e., the CNS), whereas in whip spiders they lie far out in the periphery. Perhaps this evolved as a necessary consequence of having such extremely long sensory appendages.

4 Concluding Remarks

The sensory nerves of arachnids look at first glance very much like those of other arthropods, i.e., they consist of axon bundles which are enclosed by glial cells and a basal lamina. However, a closer examination reveals many synaptic contacts, especially in the very peripheral sensory nerves. At first, one is rather puzzled about the widespread occurrence of these peripheral synapses – after all, they are hardly known from insects or crustaceans (for one exception see King and Wyman 1980).

Also, we have to admit that the significance of most of these synapses cannot be explained satisfactorily. Their presence indicates, however, that a great deal of nervous integration must already take place in the periphery. This may be an ancient character, linking the arachnids to other, more primitive arthropods (such as *Limulus*) which still have diffuse peripheral nerve plexus rather than strictly centralized ganglia.

References

Akert K, Peper K (1975) Ultrastructure of chemical synapses: a comparison between presynaptic membrane complexes of the motor end-plate and the synaptic junction in the central nervous system. In: Santini M (ed) Proc Golgi Centennial Symp. Raven Press, New York, pp 521–527

Atwood HL, Kwan I (1979) Reciprocal axo-axonal synapses between excitatory and inhibitory neurons in crustaceans. Brain Res 174:324–328

Beck L, Foelix R, Gödecke E, Kaiser R (1977) Morphologie, Larvalentwicklung und Haarsensillen der Geißelspinne *Heterophrynus longicornis* Butler (Arach., Amblypygi). Zoomorphology 88:259–276

Boeckh J, Ernst KD, Sass H, Waldow U (1976) Zur nervösen Organisation antennaler Sinneseingänge bei Insekten unter besonderer Berücksichtigung der Riechbahn. Verh Dtsch Zool Ges 1976:123–139

Bullock TH, Horridge GA (1965) Structure and function in the nervous systems of invertebrates, vol II. Freeman, San Francisco London

Esslen J, Kaissling KE (1976) Zahl und Verteilung antennaler Sensillen bei der Honigbiene (*Apis mellifera* L.). Zoomorphology 83:227–251

Fahrenbach WH (1979) The brain of the horseshoe crab (*Limulus polyphenus*) III. Cellular and synaptic organization of the corpora pedunculata. Tissue Cell 11:163–200

Fahrenbach WH (1981) The morphology of the horseshoe crab (*Limulus polyphemus*) visual system VII. Innervation of photoreceptor neurons by neurosecretory efferents. Cell Tissue Res 216:655–659

Fleissner G, Fleissner G (1978) The optic nerve mediates the circadian pigment migration in the median eyes of the scorpion. Comp Biochem Physiol 61A:69–71

Foelix RF (1975) Occurrence of synapses in peripheral sensory nerves of arachnids. Nature (London) 254:146–148

Foelix RF, Choms A (1979) Fine structure of a spider joint receptor and associated synapses. Eur J Cell Biol 19:149–159

Foelix RF, Troyer D (1980) Giant neurons and associated synapses in the peripheral nervous system of whip spiders. J Neurocytol 9:517–535

Foelix RF, Müller-Vorholt G, Jung H (1980) Organisation of sensory leg nerves in the spider *Zygiella x-notata* (Clerck) (Araneae, Araneidae). Bull Br Arachnol Soc 5:20–28

Gerhardt U, Kaestner A (1938) 8. Ordnung der Arachnida: Araneae = Echte Spinnen = Webspinnen. In: Kükenthal W, Krumbach T (eds) Handbuch der Zoologie. De Gruyter, Berlin, p 426

Govind CK, Meiss DE (1979) Quantitative comparison of low- and high-output neuromuscular synapses from a motoneuron of the lobster (*Homarus americanus*). Cell Tissue Res 198:455–463

Gray EG, Pease HL (1971) On understanding the organization of the retinal receptor synapses. Brain Res 35:1–15

Hayes WF, Barber SB (1982) Peripheral synapses in *Limulus* chemoreceptors. Comp Biochem Physiol 72A:287–293

Heuser JE, Reese TS (1973) Evidence for recycling of synaptic vesicle membrane during transmitter release at the frog neuromuscular junction. J Cell Biol 57:315–344

King DG, Wyman RJ (1980) Anatomy of the giant fibre pathway in *Drosophila.* I. Three thoracic components of the pathway. J Neurocytol 9:753–770

Lane NJ, Treherne JE (1973) The ultrastructural organization of peripheral nerves in two insect species (*Periplaneta americana* and *Schistocerca gregaria*). Tissue Cell 5:703–714

Rathmayer W (1965) Die Innervation der Beinmuskeln einer Spinne *Eurypelma hentzi* Chamb. (Orthognatha, Aviculariidae). Verh Dtsch Zool Ges 1965:505–511

Schürmann FW (1971) Synaptic contacts of association fibres in the brain of the bee. Brain Res 26:169–176

Schürmann EW (1974) Bemerkungen zur Funktion der Corpora pedunculata im Gehirn der Insekten aus morphologischer Sicht. Exp Brain Res 19:406–432

Smith DS (1968) Insect cells. Oliver and Boyd, Edinburgh

Tisdale AD, Nakajima Y (1976) Fine structure of synaptic vesicles in two types of nerve terminals in crayfish stretch receptor organ: Influence of fixation methods. J Comp Neurol 165:369–386

Tolbert LP, Hildebrand JG (1981) Organization and synaptic ultrastructure of glomeruli in the antennal lobes of the moth *Manduca sexta:* a study using thin sections and freeze fracture. Proc R Soc London Ser B 213:279–301

Uchizono K (1965) Characteristics of excitatory and inhibitory synapses in the central nervous system of the cat. Nature (London) 207:642–643

Uchizono K (1967) Inhibitory synapses on the stretch receptor neuron of the crayfish. Nature (London) 214:833–834

Wood MR, Pfenninger KH, Cohen MJ (1977) Two types of presynaptic configurations in insect central synapses: An ultrastructural analysis. Brain Res 130:25–45

C Senses and Behavior

XI Neuroethology of the Spider Vibration Sense

FRIEDRICH G. BARTH

CONTENTS

1 Introduction

Spiders are not the favorite animals of most people. Despite much irrational antipathy, however, generally one finds admiration for the beauty shown by the

Zoologisches Institut der J. W. Goethe-Universität, Gruppe Sinnesphysiologie, Siesmayerstraße 70, D-6000 Frankfurt am Main 1, Federal Republic of Germany

regular cartwheel geometry of the orb web and for the swiftness and precision of spiders that are lured and guided to prey by the slightest vibrations.

This chapter deals with the vibration sense of spiders. Prey capture, courtship and escape from predators all greatly depend on this sense. This is not only true for orb-weavers and for other spiders building less conspicuous webs, but also for the many "wandering" or "hunting" spiders which receive vibratory signals through the ground, plants, and even the water surface and do not build webs at all (review Barth 1982). Spiders do not only perceive vibratory signals but also emit them as, for instance, in their highly developed courtship rituals (on acoustic communication and reproductive isolation see Uetz and Stratton 1982).

How can we explain the behavior that is guided by vibratory signals in terms of the activities of the nervous system including its sensory input stage? Courtship behavior of *Cupiennius salei* (Rovner and Barth 1981) may serve as an example to illustrate the practical background of this question.

Cupiennius is a nocturnal Central American hunting spider courting on bromeliads, agaves, and banana plants (Barth and Seyfarth 1979). Its courtship follows a stereotyped pattern of events. These bring together mates which may sit at different ends of the same plant, more than 1 m distant from each other and without visual contact. The male starts emitting vibratory signals after he has come across an area where the female has previously left an aphrodisiac pheromone. The male's signals travel through the plant and reach the female, which in turn produces her own vibratory response. Communication by reciprocal signaling goes on. The vibrations emitted by the stationary female guide the male on his way to his mate.

Among the main events to be scrutinized and the questions to be asked in this and other cases of behavior guided by vibratory signals are the following: What are the prominent physical features of the signals? How are they transmitted and modified by the various media involved? What are the physiological characteristics of the vibration receptors and of the interneurons handling vibratory input? How does the receiver recognize a particular, biologically relevant signal and distinguish it from background noise? What mechanisms are involved in the localization of the signal source and the orientation toward it?

Evidently the selective pressures at work during the evolution of spiders have led to efficient solutions of the general problems involved in the detection, identification, and localization of vibratory signals. In the following, answers will be sought to the above questions, starting with an evaluation of the vibrations relevant in the given context.

2 Signals and Noise

There are many reports describing the vibrations caused by bouncing and jerking spiders and prey entangled in the web or passing by on solid substrates; all these evidently attract the spider's attention (review Barth 1982). Only recently, however, adequate measuring techniques and critical physical evaluation have been applied to the vibratory stimuli we are interested in.

According to present knowledge, the frequency contents of prey signals and background noise that potentially interferes with the relevant messages clearly differ. Very small bandwidth and low frequencies seem to be typical of background noise, while prey signals were found to be broad-banded and to extend to much higher frequencies. Courtship signals, at least those produced by the male spiders, are, in addition, clearly patterned in the time domain (Krafft 1978; Schüch und Barth 1985).

These general conclusions are illustrated in Figs. 1 and 3b, c and discussed in the following for signals carried by plants, the water surface, and the orb web.

2.1 Plants

The male signals received by a female *Cupiennius* during *courtship* on plants and referred to in the introduction show a very regular temporal structure (Fig. 1a). The "syllable", its basic element, comes in trains of up to 50 and at intervals of ca. 350 ms. Its frequency content peaks at about 75 Hz (due to leg and abdominal oscillations) and 115 Hz (due to palpal movements). The female signal, by contrast is a single "syllable" and more irregular than the male signal as reflected by a broader frequency spectrum with most of the signal concentrated from 20 to 50 Hz. It is precisely placed into a narrow time frame after the end of the male train. Analysis of complete courtships showed that the male signaling contains several cues that are potentially used by the female for such precise timing and species recognition (Rovner and Barth 1981; Schüch and Barth 1985).

Prey signals such as those generated by a cockroach running on the pseudostem of a banana plant and effectively eliciting prey capture by *Cupiennius* differ from male courtship signals. They are more irregular in the time domain and contain both a broader band of frequencies and higher frequencies with peaks between 400 and 700 Hz. The vibrations caused by *background noise*, i.e., mainly by the wind, and measured on banana plants and bromeliads in the natural habitat of *Cupiennius* contain much lower frequencies than the courtship and prey signals. There is a peak at about 1.4 Hz and often hardly any vibration above about 10 Hz (Barth et al., in preparation).

2.2 Water Surface

Water-borne vibrations are known to be an important source of information for the semi-aquatic spider *Dolomedes triton,* which also produces them itself while courting (Roland and Rovner 1983). This spider runs toward prey when alarmed by the surface waves generated by an insect that is trapped on the water (Bleckmann and Barth 1984). Both prey- and wind-generated noise signals are known from the natural biotop of *Dolomedes* (Bleckmann and Rovner 1984; Bleckmann and Barth 1984, see also Lang 1980). Again there is a clear difference (Fig. 1b). The wind-generated vibrations are characterized by much lower frequencies (maximum rel. displacement between ca. 1.4 Hz and 8.2 Hz) than prey signals (struggling fly, fish touching water surface: ca. 5 Hz to 150 Hz).

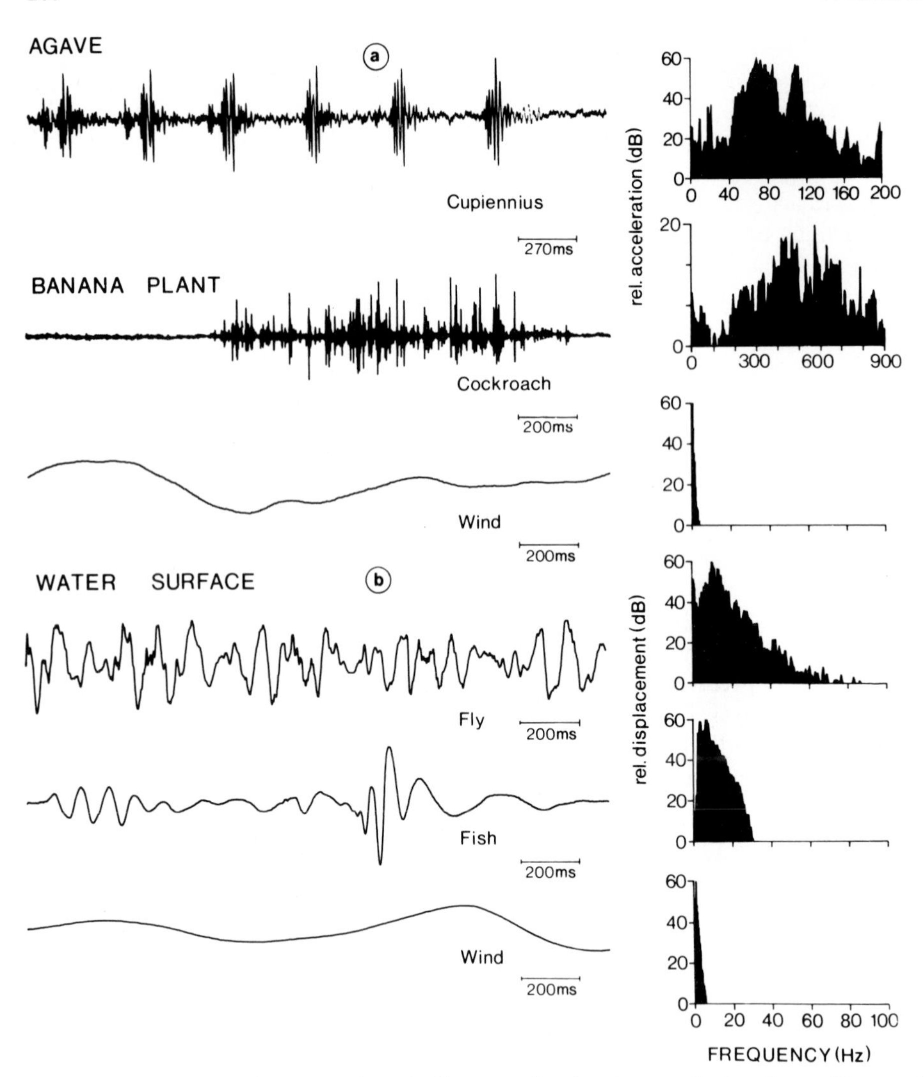

Fig. 1 a, b. Representative examples of vibratory signals relevant for hunting spiders and reaching these through plants (**a**) (agave, banana plant) or the water-surface (**b**). Original recordings on *left*, corresponding frequency spectra on *right*. High signal-to-noise ratios were obtained by recording acceleration in case of low amplitude signals containing high frequencies (first two recordings from above) and displacement in case of signals containing very low frequencies (all other recordings). Note different calibration of frequency axis in spectra (calibration of lowermost spectrum also applies to those without numerals below x-axis). Signals *from above to below:* courtship of male *Cupiennius salei* on *Agave americana* showing six syllables; cockroach (*Periplaneta americana*) running along pseudostem (height ca. 3 m, diameter ca. 20 cm) of banana plant (probably *Musa sapientum*); wind-generated vibrations of banana pseudostem (same as above); fly struggling on water-surface; fish touching water surface from below; wind-generated water surface waves (Dow lake, Ohio). Calibration of y-axis differs in different recordings. (The cooperation of H. Bleckmann, W. Schüch and E.-A. Seyfarth in gaining these data is much appreciated.)

2.3 Orb Webs

Studies with the laser-Doppler vibrometer and other advanced techniques have now also provided reliable data on vibrations generated in and transmitted by elements of the orb web of *Nuctenea sclopetaria* (Masters and Markl 1981). This araneid spider dashes from the hub of its web toward entangled prey, guided by prey-produced vibrations which have most of their energy at frequencies below 50 Hz, unless generated by buzzing insects, when components above 100 Hz become important as well. Vibrations due to wind, however, have most of their energy below 10 Hz (Masters 1984b) (Fig. 3b, c).

3 Wave Types and Signal Transmission

Vibratory signals are modified by the conducting medium on their way from the sender to the receiver. They are changed in amplitude (generally, but not always, they are attenuated) and in frequency composition and time course. The underlying physics is complex and differs with the various media and the size and shape of the transmitting structure. Here we will concentrate on boundary vibrations, for which experimental data relevant to spider substrate vibration sensitivity are available. The transfer characteristics of the substrates (plants, water surface, orb web) introduced above will be discussed in the same sequence.

Contact vibration (receiver directly touched by sender, with rhythmically varying pressure) and *near field medium vibration* (receiver stimulated by the alternating flow of air or water produced by rhythmic movements of sender), are other types of vibratory stimulation (Markl 1973, 1983). For details on sand-conducted vibrations such as those relevant to scorpions, see Brownell (1977). For a full treatment of boundary waves, textbooks of physics (Cremer et al. 1973; Skudrzyk 1971; Sommerfeld 1970) and a recent review by Markl (1983) should be consulted.

3.1 Plants

Bending waves. There are several types of boundary wave. They differ with respect to type of motion, propagation speed, and attenuation. According to the available data, *bending waves* are of particular importance not only in the plants used by insects such as "small cicadas" (Cicadina) and cydnid bugs (Michelsen et al. 1982) and leafhoppers (Keuper and Kühne 1983), but also in those, the spider *Cupiennius* is living on (Barth et al., in preparation).

In bending waves, particle motion is in a plane perpendicular both to the direction of wave propagation and to the surface of the plant which is rhythmically bent as a whole. Apart from the direction of motion, the feature generally used to identify this wave type is the propagation speed. It is in general very low as compared to that of other wave types occurring in plants (see below). Also, wave propagation is dispersive, i.e., it depends on frequency.

Phase velocity C_{ph} has to be distinguished from group velocity C_g, which refers to the carrier wave envelope of a wave "group". In rods with a diameter $d < \lambda/6$ (λ wavelength)

$$C_g = 2\,C_{ph} = 2\sqrt[4]{\frac{EJ}{m'}} \cdot \sqrt{\omega} \tag{1}$$

(E Young's modulus, J axial moment of inertia, m′ mass per unit length, $\omega = 2\pi f$, f frequency).

Since C_{ph} (and C_g) is proportional to the fourth root of both E, J and 1/m′ it varies little with large changes in the mechanical properties of the bending structure. J is proportional to the fourth power of beam radius r and m′ to the second power of r. Therefore C_{ph} is proportional to $\sqrt{r}$ and to $\sqrt{f}$.

Michelsen et al. (1982) studied bending waves in a variety of slim plant stems differing widely in mechanical properties such as soft bean plants and stiff reed and maple. Using laser-Doppler vibrometry they found propagation velocities between 36 and 95 m s^{-1} at 200 Hz and between 120 and 220 m s^{-1} at 2 kHz. Measurements on plants that are used by bushcrickets during stridulation support the concept of bending waves (Keuper and Kühne 1983).

In our measurements on banana plants[1] (which is one of the substrates used by *Cupiennius* when receiving or emitting vibratory signals) we found values below 50 m s^{-1} at 100 Hz and 500 Hz for both the leaves and the pseudostem (Barth and Bohnenberger, unpublished).

Although the available studies suggest that bending waves in plants are primarily used for the transmission of vibrations by arthropods, the biological relevance of other types of wave cannot generally be excluded.

Other types of wave potentially relevant in the given context are the following. (1) *Surface waves* like Rayleigh waves: Rayleigh waves, like bending waves, have a motion component perpendicular to the surface. Their propagation velocity, however, is much higher approaching that of transverse waves, where $C_t = \sqrt{G/p}$; ($G = E/2 \cdot (1 + \mu)$; E Young's modulus; p density; μ Poisson ratio). Rayleigh waves only occur in structures that do not prevent the wave in one or more dimensions as is the case in thin plates or slim rods. (2) *Quasi-longitudinal waves:* particle motion mainly in same direction as wave propagation, with an additional local transverse change in the diameter of the structure according to its Poisson ratio. Such waves occur in structures that are small with respect to wavelength in one or two directions such as in many plant leaves. (3) *Transverse waves:* particle motion at a right angle to wave propagation and in the plane of the structure's surface. This wave type occurs in structures able to transmit elastic shearing forces like in plates which are large with respect to the wavelength. (4) *Torsional waves:* particle motion as in transverse waves, with no component perpendicular to the surface provided the structure is rotational symmetric. This wave type is found in structures that are long as compared to their diameter.

Attenuation. The next question to be asked is that of *frequency filtering* and *attenuation* of a signal on its way through a plant. An elegant way to measure this is to apply a noise band containing all the frequencies of interest and to compare its frequency spectrum at the point of origin (or close to it) with that at various points along the plant. With *Cupiennius* in mind, we have undertaken such a study with both a banana plant[1] (*Musa sapientum*, Fig. 2) and an agave (*Agave americana*): Without considering various complex physical aspects that are reflected by certain irregularities of the attenuation curves, the main result implies that *Cupiennius* uses a substrate very well suited for its vibratory com-

---

1 I am indebted to A Michelsen, who made these measurement possible in his laboratory in Odense, Denmark.

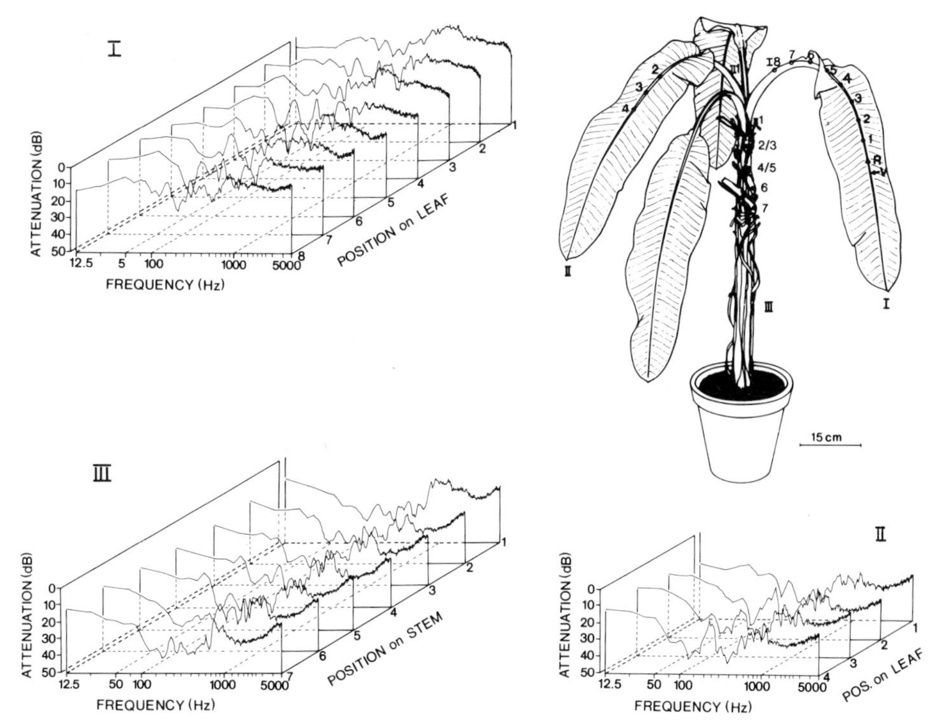

Fig. 2. Attenuation of vibratory signal traveling through banana plant (*Musa sapientum*) from its site of origin at leaf *I (V)* to leaf *II* and the pseudostem *III*. Graphs are based on spectral analysis of noise signals and laser-Doppler vibrometry and give attenuation values at different sites (*1* to *8*) on leaves and pseudostem. *OdB* (no attenuation) corresponds to value measured at reference point *R* close to *V* on leaf *I*. (Barth and Bohnenberger, unpublished)

munication. Attenuation at 75 Hz (dominant frequency of male signal) averages only 0.3 dB cm^{-1}. Even when traveling from one leaf to another or from a leaf to the pseudostem of the banana plant, typical average values are only 0.3 to 0.4 dB cm^{-1}. The values are even smaller at the dominant frequency of the female signal (for a comparison with other media see Barth 1982). This is certainly one of the reasons why reciprocal signaling during courtship occurs over distances of at least 1.5 m. Attenuation increases with frequency in a non-monotonous way. The irregularities are mainly between about 150 Hz and 1 kHz and probably due to resonances and reflections from the end of the plant. Even at 5 kHz, i.e., the highest frequency used, attenuation is surprisingly low with up to ca. 0.35 dB cm^{-1} if averaged for a signal that has traveled about 1 m from one leaf to another or down the pseudostem.

3.2 Water Surface

In water-surface waves, particle movement is along ellipses or circles (depending on water depth) in a plane perpendicular to the surface at rest. Transmission is characterized by dispersion: propagation velocity changes with frequency (wavelength). Natural signals which will always contain several or many frequency components will change their shape along their path, with the high and very low frequency components traveling increasingly ahead of frequency components between ca. 5 Hz and 20 Hz. Group velocity of such signals is minimal at about 6 Hz (17.5 cm s^{-1}) and increases toward both lower and higher frequencies (e.g., 55 cm s^{-1} at 100 Hz). Attenuation is much stronger than that of bending waves in plants and again highly dependent on the frequency. Examples for concentric surface waves are 1.67 dB cm^{-1} at 10 Hz and 8.57 dB cm^{-1} at 140 Hz (distance from source ca. 3 cm). Consequently, a water-surface wave signal will also change its spectral energy contents while traveling, and having progressed for a few cm will practically contain no more frequency components above 200 Hz. More details on the physics of water surface waves are found in Sommerfeld (1970) (see also Lang 1977, 1980; Bleckmann and Schwartz 1982; Markl 1983).

3.3 Orb Web

Vibration transmission in the orb web of *Nuctenea sclopeteria* was recently studied with the laser-Doppler vibrometer, a noncontact procedure which is of particular value when dealing with a superlight structure weighing only 1 to 2 mg such as the spider web (Masters and Markl 1981).

The three types of vibration distinguished here are *transverse* (motion perpendicular to long axis of thread and plane of web), *lateral* (motion perpendicular to long axis of thread and in the plane of the web), and *longitudinal* vibrations (motion along long axis of thread and in the plane of the web). A fourth type of vibration, *torsional waves*, which may occur in elongated structures (motion perpendicular to surface provided rotational symmetry of the structure's cross-section), could not yet be measured, and there are no data available on their biological significance.

In the *empty* web, the radii which are the main signal line in an orb web transmit *longitudinal vibrations* from the catching area to the hub over about 7 cm with a total attenuation of only 1 to 2 dB in a remarkably wide range of frequencies (1 Hz to 3 kHz). This corresponds to ca. 0.2 dB cm^{-1}, a value similar to that found for the bending waves in plants. There is even amplification by up to 10 dB above about 3 kHz. Transverse and lateral vibrations, on the other hand, are attenuated by at least 10 to 30 dB and even 50 dB (7.1 dB cm^{-1}) above about 40 Hz, where transmission curves become irregular, partly due to resonances (Masters and Markl 1981). Apart from their effective transmission, longitudinal vibrations along the radii also provide information about the radius the spider has to choose to get to the signal source: there is a particularly large difference in amplitude between the vibrated radius and its neighboring radii.

In a more relevant situation, the web is not empty but loaded by the spider and its prey or mate. Under these circumstances (weight of spider 200 mg; data refer to measurement of tarsal motion) attenuation of longitudinal vibrations

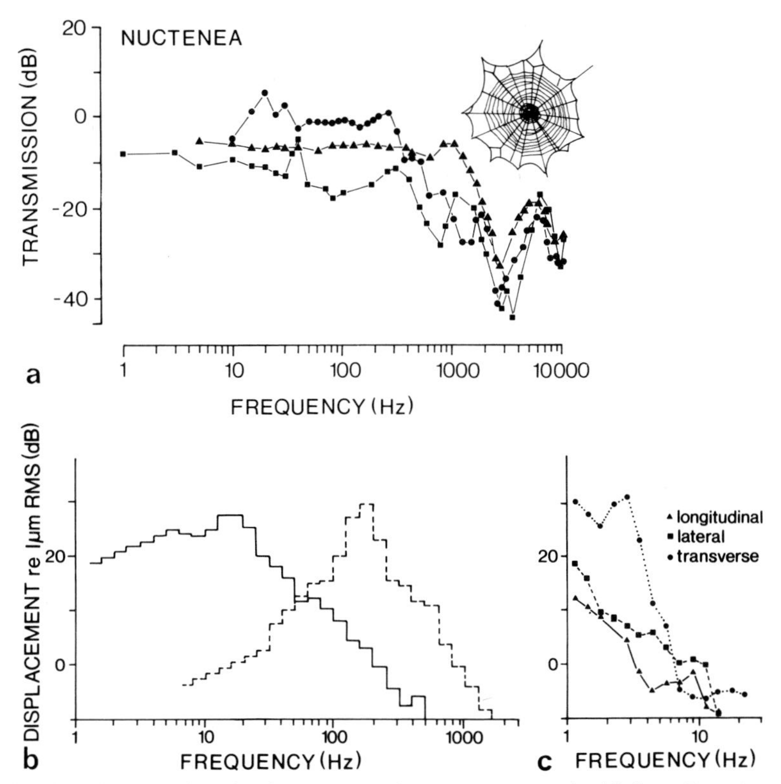

Fig. 3a–c. Vibrations in the orb web of *Nuctenea sclopetaria* measured with laser-Doppler vibrometry. **a** Transmission of longitudinal vibrations through a radius from the middle of the catching area to the tarsus of the spider at the hub (distance ca. 7 cm); curves measured for three individual animals. **b** Vibrations generated by a fly (*Calliphora erythrocephala*, 38 mg) in an otherwise empty web and measured at the hub at a distance of ca. 9 cm. *Solid line* indicates that fly was struggling but not buzzing; *dashed line* indicates buzzing fly. **c** Frequency spectrum for the three types of web vibration in wind-induced signals. Measurements made at edge of hub near one of the spider's forelegs. Given are the means (one web, eight samples averaged for each type of vibration) for natural wind gusts. (**a** Masters 1984a; **b** and **c** Masters 1984b)

(Fig. 3a) increases to about 2 dB to 18 dB (at a distance of 7 cm this amounts to ca. 2.5 dB cm^{-1}) up to ca. 300 Hz, depending on the radius and the web chosen. The transmission curve drops steeply beyond about 300 Hz at a rate of ca. 20 dB/decade, with an attenuation peak as large as ca. 40 dB between 2 and 3 kHz (Masters 1984a). Comparable measurements for transverse and lateral vibrations are not available.

4 Vibration Sensitivity

4.1 Receptors

Types. The most sensitive receptor known in spiders for vibrations transmitted through solid substrates (as opposed to air) is the *metatarsal lyriform organ* (Fig. 4). This is a compound slit sense organ consisting of up to about 21 slits

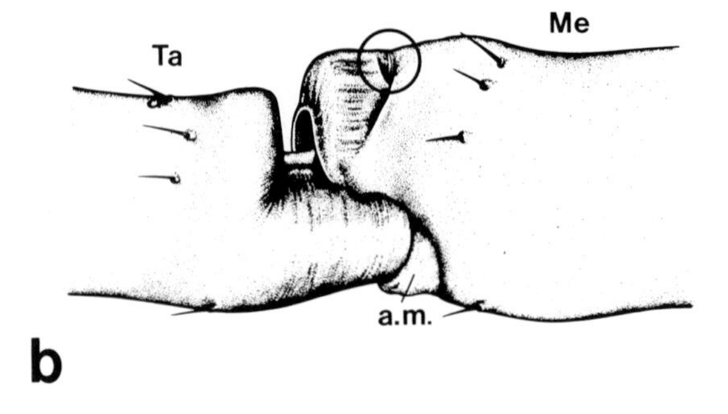

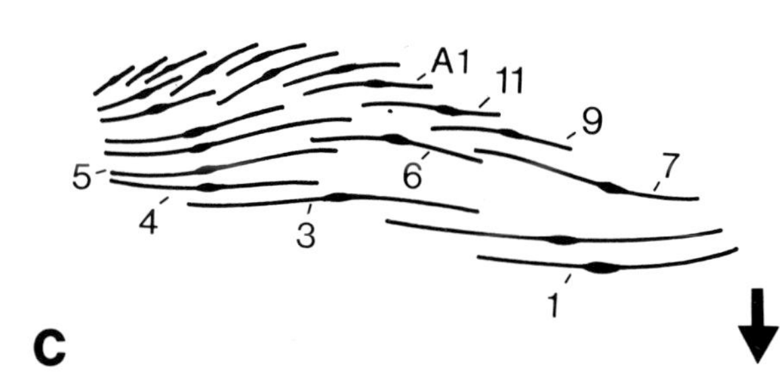

Fig. 4a–c. Metatarsal lyriform organ (*Cupiennius salei*). **a** position of organ on legs (*circle*). (Courtesy E.-A. Seyfarth); **b** details of joint between tarsus (*Ta*) and metatarsus (*Me*) in lateral view; note metatarsal organ bridging a furrow on the dorsal aspect of distal metatarsus; *a.m.* articular membrane; **c** dorsal view of slits in metatarsal organ. *Arrow* points toward tarsus; note orientation at right angle to leg long axis. (Barth and Geethabali 1982)

whose arrangement follows the same basic pattern in a variety of species from different families, irrespective of differences in slit number (van de Roemer 1980) (e.g., 21 in *Cupiennius salei*, 11 in *Salticus scenicus*). It lies dorsally on the distal metatarsus and is oriented perpendicular to the leg. The slits are compressed, that is, adequately stimulated by upward displacement of the tarsus beyond the point when it just touches the distal end of the metatarsus. In addition, dorso-lateral and even lateral displacement of the tarsus stimulates some of the slits in the organ (Barth 1972; Barth and Geethabali 1982; see also Barth, Chap. IX, this Vol.).

The other receptor with well-established sensitivity to substrate vibrations is the *pretarsal slit* (Speck and Barth 1982). This is a single slit (length in *Cupiennius* ca. 45 μm) on each ventrolateral side of the pretarsus behind the claws, and in the immediate neighborhood of the ventral articular membrane.

The participation of both the metatarsal organs and the pretarsal slits in the localization of a stimulus source (as, for instance, during prey capture) has been demonstrated experimentally (Barth 1982; Hergenröder and Barth 1983a; Klärner and Barth 1982; Bleckmann and Barth 1984). Behavioral evidence indicates the existence of still other vibration receptors, however: prey capture jumps can still be elicited after ablation of the above receptors, at least by strong vibrations (elevation of threshold; Barth 1982; Hergenröder and Barth 1983a). Among the candidate receptors serving in this additional vibrosensitive function are sensory cells coupled to the claws (*Zygiella*, Foelix 1970; *Cupiennius*, Seyfarth et al., in preparation), serrated bristles on the ventral side of the tarsus tip specialized for grasping thread (*Zygiella*, Foelix 1970), and other slit sensilla more proximally on the leg.

Sensitivity. In *Cupiennius* electrophysiologically determined threshold curves are available for about half of the 21 slits of the metatarsal organ for frequencies between 0.1 Hz and 1 to 3 kHz (Barth and Geethabali 1982). All slits behave like *high-pass filters* (Fig. 5a). Up to 10−40 Hz stimulation frequency their threshold sensitivity is low; threshold values expressed as displacement of the tarsus tip are from 10^{-3} cm to 10^{-2} cm. Thresholds decrease steeply – up to ca. 40 dB/decade – beyond 10 to 40 Hz and reach 10^{-6} to 10^{-7} cm at 1 kHz. Whereas the threshold curve roughly follows constant displacement at low frequencies, it approximately follows constant acceleration beyond about 40 Hz.

A characteristic feature of all slits studied is the lack of a pronounced tuning to limited sections of the frequency range tested, which includes the most important frequencies contained in behaviorally relevant signals (see Sect. 2). This finding also applies to the cases where the tarsus and even the whole animal (Barth and Hahn, unpublished) where not firmly coupled to the vibrator, but were free to move as in natural stimulus conditions. Tuning of different slits to different frequency bands can therefore not be used for the *discrimination of frequencies* picked up by the metatarsal lyriform organ. Frequency discrimination – as it was shown to exist in behavioral experiments (Hergenröder and Barth 1983a) – however, could be based on sensitivity differences of the various slits at particular stimulus frequencies. These differences amount to more than 20 dB. They result in frequency-dependent activity patterns of the metatarsal organ as a whole.

At least some of the slits of the metatarsal lyriform organ are sensitive to lateral as well as dorso-ventral displacements of the tarsus, with threshold curves very similar both in shape and absolute values. Using band-limited noise stimuli (bandwidth 1/3 octave, Q = 0.35) instead of sinusoidal ones (vibration of the tarsus) lowers the threshold values by up to ca. 10 dB (Barth and Hahn, unpublished).

Recordings from individual slits are only available for the metatarsal organ of the hunting spider *Cupiennius salei* (Barth and Geethabali 1982) and for the semi-aquatic spider *Dolomedes triton* (Bleckmann and Barth 1984). In *Zygiella*, an orb-weaver, summed recordings from the leg nerve were made (Liesenfeld 1961). The typical high-pass filter properties are basically the same in all cases, irrespective of the substrate the spider is living on. This finding is surprising in view of the differences in the respective joint mechanics. Whereas the metatarsus-tarsus joint in *Cupiennius* and *Dolomedes* allows considerable excursion of the tarsus both laterally and dorsoventrally, it is much stiffer in *Zygiella* and other orb-weavers like *Nephila* (van de Roemer 1980).

According to studies in *Cupiennius* (Speck and Barth 1982), the pretarsal slits are not tuned to limited frequency ranges either. Thresholds again are high

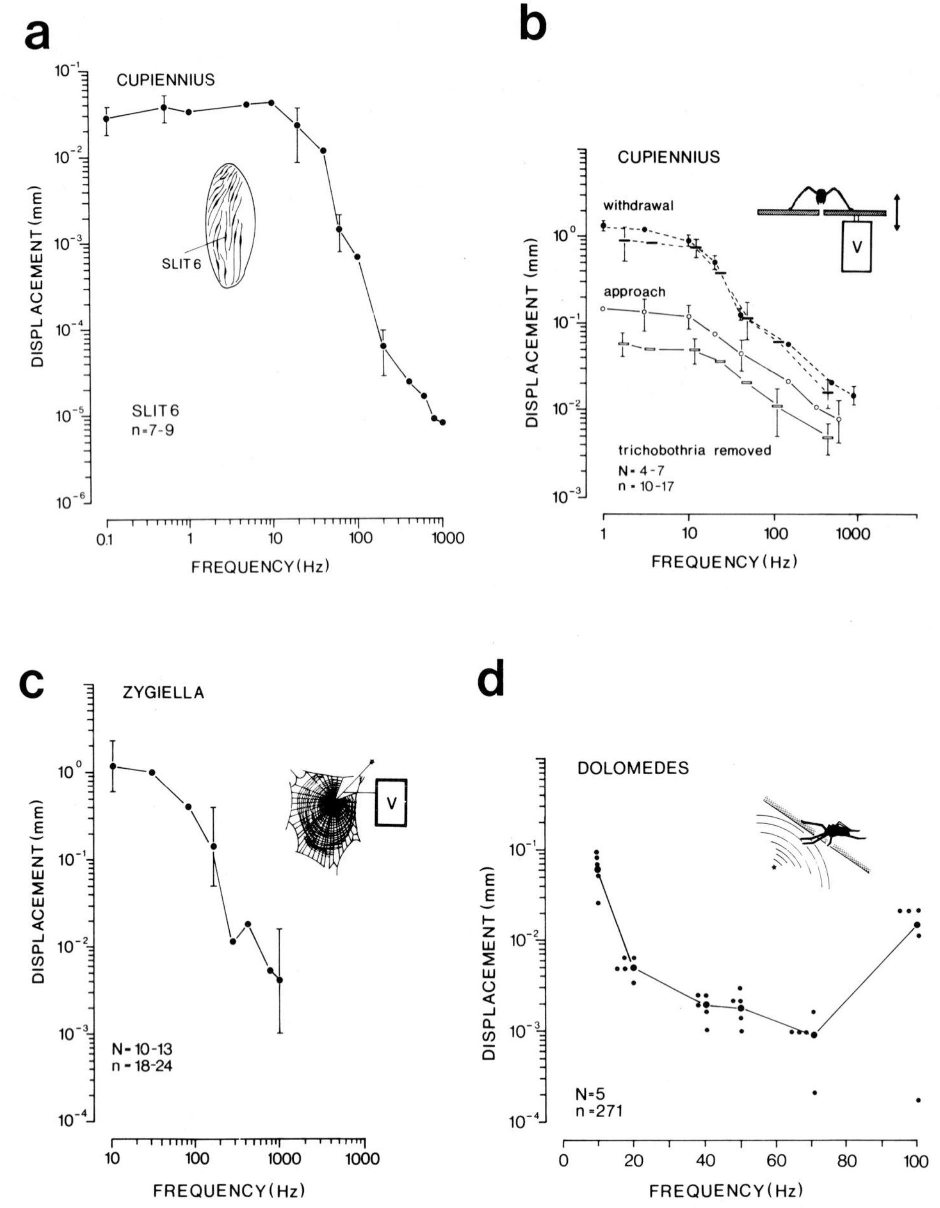

Fig. 5a–d. Vibration sensitivity of receptors (**a**) and behaviors (**b–d**) determined for spiders receiving signals through different media. **a** threshold curve of slit *6* (cf. Fig. 4c) in the metatarsal organ of *Cupiennius salei.* (Barth and Geethabali 1982); **b** behavioral threshold curves for the withdrawal and approach reaction of *C. salei* measured with sinusoidal stimuli (○, ●) and narrow noise bands (▭, ▬) respectively. (Hergenröder and Barth 1983a); **c** behavioral threshold curve for prey capture reaction of orb weaver *Zygiella x-notata.* (Klärner and Barth 1982); **d** threshold curve for prey-capture reaction of semi-aquatic spider *Dolomedes triton* elicited by water-surface waves; values represent means of the lowest five values found at each frequency among a total of 271 values. (Bleckmann and Barth 1984) In **a** to **c** standard deviation given by *vertical bars. N* number of animals; *n* number of trials

(pretarsal displacement 10^{-2} cm to 10^{-3} cm) up to ca. 40 Hz. At higher frequencies they decrease at a rate of ca. 18 dB/decade, which is less than in the case of the metatarsal lyriform organ (ca. 40 dB/decade). Accordingly, the threshold values reached at 1 kHz are only 10^{-4} cm to 10^{-5} cm. Threshold varies with different preset tension in the articular membrane. This suggests a mechanism for active sensitivity adjustment by the spider (lowering of sensitivity by lifting pretarsus).

Both in the metatarsal lyriform organ and the pretarsal slits functional morphology and electrophysiological data suggest an additional *proprioceptive role.* The pretarsal slits are also stimulated by an active downward movement of the pretarsus, resulting from muscular activity or increased hemolymph pressure, and the consequent inward buckling of the ventral articular membrane (Speck and Barth 1982). Like the pretarsal slits those of the metatarsal organ are slowly adapting to a maintained stimulus. They therefore appear well suited to monitor low frequency tarsal movement occurring during locomotion (Barth and Geethabali 1982; see also Barth, Chap. IX, this Vol.).

4.2 Interneurons

Together with the palpal and opisthosomal ganglia, the leg ganglia form the subesophageal mass of the spider central nervous system (see Babu, Chap. I, this Vol. and Babu and Barth 1984). According to elctrophysiological work there are several types of substrate-vibration-sensitive interneurons among the neurons located ventrally in the ganglia and close to the median axis of symmetry of the suboesophageal mass (Fig. 6) (Speck and Barth, in preparation). These interneurons differ with respect to their tuning (threshold) curves and their connectivity to the input from other ipsi- or contralateral legs. All threshold curves, however, differ from those found for both the behavioral and receptor response (Fig. 5) in that they are minimum curves; these interneurons are in fact tuned to restricted frequency ranges.

A frequently found "*intraganglionic*" cell type is only activated by stimulation of the ipsilateral leg (tarsus) corresponding to the respective ganglion. Three subgroups have their best frequencies in the low (ca. 80 – 100 Hz), median (ca. 200 Hz), or high frequency (ca. 900 Hz) range and their lowest threshold values at ca. 0.13 μm of tarsal displacement (peak to peak). Typically the frequency range that elicits a response is broader in case of the first and second leg ganglion than in the third and fourth. In the latter, low stimulus frequencies elicit the smallest response (e.g., neurons with best frequency at 800 Hz; frequency range in 1st ganglion 40 Hz to 1 kHz, in 3rd ganglion 100 Hz to 1 kHz, 4th ganglion 200 Hz to 1 kHz; maximum stimulus amplitude used 40 dB above threshold).

Another type of interneuron is "interganglionic" or "*plurisegmental*", that is, it is activated by stimulation of any one of the ipsilateral legs. These cells have a broad tuning curve with best frequencies between 80 Hz and 200 Hz and absolute threshold sensitivity (but not frequency range) often (but not always) gradually decreasing with stimulation of legs 1 to 4 (e.g., at 80 Hz: stimulation

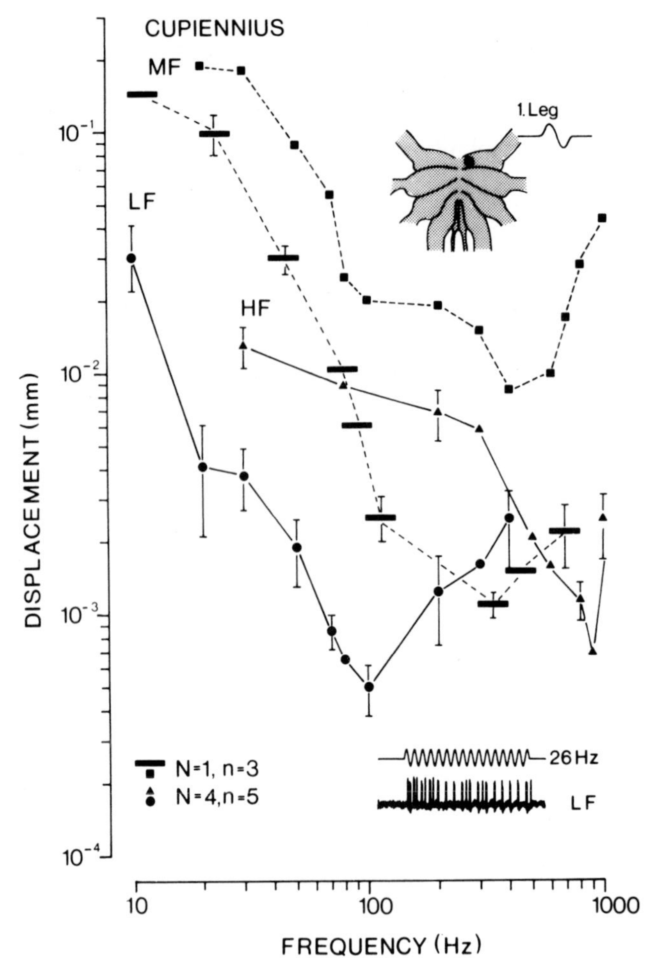

Fig. 6. Vibration sensitivity of interneurons in the subesophageal nerve mass (leg ganglia; *filled circle in inset* marks recording site) of *Cupiennius salei*. *LF*, *MF* and *HF* interneurons most sensitive in the low, middle and high frequency range; *curve with horizontal bars* was measured using narrow noise bands as stimuli instead of sinusoidal vibrations (same *MF* neuron). *Lower inset* recording of spike activity of *LF* neuron. *Vertical bars* give standard deviation; *N* number of animals; *n* number of trials. (Speck and Barth, in preparation)

of leg 1, 0.4 μm; stimulation of leg 4, 1.6 μm). Other "segmental" neurons respond to stimulation of both the ipsi- and contralateral leg of the same segment and other "plurisegmental" ones to the sensory input from any of the 8 legs.

4.3 Behavior

Behavioral threshold curves can also tell us something about the integration of sensory input by the central nervous system. They were measured in spiders receiving vibratory signals through plant tissue, a web, and the water surface.

1. *Cupiennius* responds to vibrations when sitting on a platform (Fig. 5b) divided into two mechanically independent halves, one being immobile and the other coupled to a vibrator (Hergenröder und Barth 1983a). The spider, with some of its legs resting on the mobile part of the platform, either moves toward

the stimulus source ("prey capture") or turns away from it (avoidance or escape) as soon as the platform is vibrated. Its decision depends on the properties of the stimulus (see Sect. 7). Similar to the receptors, the behavioral tuning shows low sensitivity up to ca. 10 Hz and steadily decreasing threshold values beyond that frequency. In the case of the approach reaction, absolute values are ca. 100 μm in the insensitive range and about 10 μm at the highest frequency tested (550 Hz). This latter value is higher than the receptor thresholds by about three powers of ten. This does not necessarily imply a generally smaller sensitivity in the behavior, however, since the reaction studied may not be the most sensitive indicator.

2. *Zygiella x-notata*, an orb-weaver, is a particularly convenient experimental animal. Most of the time it sits in its retreat in the web periphery, with one or both of its front legs on the signal thread through which it receives vibrations generated anywhere in the web. Upon experimental vibration of the signal thread it dashes toward the hub. The behavioral threshold curve again looks like that of a high-pass filter and again the threshold values are high compared with the receptor thresholds (Klärner and Barth 1982; Liesenfeld 1961) (Fig. 5c). In case of transverse sinusoidal vibrations of the signal thread (perpendicular to the plane of the web; stimulus applied about 6 cm away from the spider) the threshold values are about 1 mm at 10 Hz, and 3 to 4 μm at 1 kHz. With longitudinal vibration, these values decrease by 3 to 4 dB. In contrast to *Zygiella*, the threshold curve for the responses of *Nephila clavipes* to vibrations of a radius is a minimum curve with lowest values between 280 and 420 Hz (Klärner and Barth 1982).

3. The semi-aquatic spider *Dolomedes triton* is most sensitive to single frequency, water-surface waves between 40 Hz (2.1 μm) and 70 Hz (1.0 μm), with higher thresholds at both lower and higher frequencies (ca. 60 μm at 10 Hz; ca. 14 μm at 100 Hz) (Fig. 5d) (Bleckmann and Barth 1984).

Considering the frequencies contained in natural signals and the sensitivity of the spiders to various substrate vibration frequencies, a few examples can be quoted which look like special adaptations in the frequency domain. One is the low sensitivity of the substrate vibration receptors to low frequencies; this appears to be a filter that keeps background noise away from the information-processing system (Barth and Geethabali 1982). Another example is that, among the spiders tested, the metatarsal organ of *Cupiennius* is the most sensitive at frequencies between 40 Hz and 100 Hz, which is the range of its courtship signals (Barth and Geethabali 1982). Also in *Cupiennius*, one type of the vibration-sensitive interneurons was found with a best frequency between 80 and 100 Hz. Potentially it is particularly important for the processing of the male courtship signal (Speck and Barth, in preparation). Finally, in *Dolomedes* the behavioral threshold curve has a minimum between 40 and 70 Hz (Bleckmann and Barth 1984). This may be a special adaptation to the stimulus propagation properties of the water surface, where high frequencies are unlikely to occur under natural conditions (Lang 1980). The argument is supported by the difference to findings in *Zygiella* and *Cupiennius*, which both show a high-pass characteristic in their behavior that is similar to that of the receptors (Klärner and Barth 1982; Hergenröder and Barth 1983a).

5 Orientation

Following adequate stimulation, spiders turn toward substrate-borne vibrations which indicate a potential prey. The turning movement is very precise with respect to the turning angle in both orb-weavers and spiders hunting on the water surface (Klärner and Barth 1982; Bleckmann and Barth 1984). At least for web spiders, this seems to reflect the unlikeliness of a change in position of the prey within the time needed for an approach even from many body lengths away. *Cupiennius* employs a different strategy. It usually waits motionless in ambush on its plant until prey has come within a few centimeters of its body periphery. It then quickly turns just far enough to grasp its victim with its front legs and pulls it toward its chelicerae (Melchers 1967). Speed and short duration of the turn seem to be of utmost importance since the prey, warned by its own sensitive vibration receptors and free to move, might otherwise escape.

If *Zygiella x-notata* receives a prey stimulus from the web through the signal thread, it runs down to the hub, turns toward the stimulus, and then follows the radii toward the prey. The mean "error angle" $\Delta\alpha$ (difference between stimulus angle α_s and turning angle α_t) is only 3.6° (± 7.7 SD), irrespective of the stimulus angle. *Nephila clavipes*, which usually sits in the hub while at rest, orients with similar precision (7° ± 8.2° SD), on average also turning slightly short of α_s, in particular at large values of α_s (α_s, 0° to 50°: $\Delta\alpha = 3.9° \pm 5.6$ SD; α_s 50° to 100°: $\Delta\alpha = 10.1° \pm 9.1$ SD) (Klärner and Barth 1982). *Dolomedes triton* turns toward a source of water waves with similar precision (9.3° ± 1.3 SD), also with a surprising independence of stimulus direction (Bleckmann and Barth 1984). This is not so with *Cupiennius salei*, where the error angle increases significantly the more the stimulus comes from the rear (stimulation under first leg $\Delta\alpha = 12°$, under fourth leg $\Delta\alpha = 80°$) (Hergenröder and Barth 1983b).

Behavioral evidence suggests that the "fishing" spider *Dolomedes triton*, when starting its run, has extracted information from water-surface waves not only on goal direction but also on goal distance (Bleckmann and Barth 1984).

Little is known on how the spiders accomplish such orientation behavior. Parameters contained in vibrational stimuli which can provide the information needed are discussed in the following.

5.1 Cues Contained in Stimuli

Angular Orientation. In the web the task appears easy. Once the spider detects the radius vibrating most, it simply has to run along it to get to the entangled, and thus stationary, prey. The most relevant question then seems to be which types of vibration caused by the prey are most appropriate to tell the spider the proper radius. *Longitudinal waves* are the most "directional" ones: the difference amplitude between the vibrations of a radius coupled to the vibration source and its neighbors is much larger in longitudinal waves than in the lateral and transverse vibration components (Masters and Markl 1981). Just like the low attenuation values (see above), this points to the particular relevance of longitudinal waves in orb webs.

The water surface, as well as plants, seem to present a more difficult task. They lack the "track" automatically leading to the goal once it has been chosen correctly.

1. *Amplitude gradients* between the stimuli arriving at the various legs due to different traveling distances can be used by *Cupiennius salei* for its angular orientation (Hergenröder and Barth 1983b; see below). According to the attenuation values measured for banana plants (Fig. 2), which is a favorite dwelling site of *Cupiennius* (Barth and Seyfarth 1979), and a diagonal leg span of ca. 10 cm, gradients to be expected under natural conditions amount to about 4 dB for the frequencies prominently contained in the courtship signals. On the water surface, and taking *Dolomedes triton* and *D. fimbriatus* as the examples studied (Bleckmann and Barth 1984), the amplitude differences at the different legs (distance 4 to 7 cm) amount to about 15 dB at 40 Hz, which is the frequency these spiders most readily respond to.

2. *Time differences* as small as 0.2 ms between the arrival of vibratory signals at different receptors (between events of the same phase, respectively), which depend on propagation velocity and distance between the receivers (i.e., the leg tips), are an important indicator of target angle for the desert scorpion, *Paruroctonus mesaensis,* on and in sand (here surface-Rayleigh-waves seem to be of particular importance; Brownell and Farley 1979b). They also elicit oriented behavior in *Cupiennius* which – with its legs vibrated at different times (interval used 4 ms) – always turns toward the leg first stimulated (Hergenröder and Barth 1983b). Time differences are even more likely to indicate the target angle in *Dolomedes,* where they are as high as 200 ms or more in the spider's most sensitive frequency range, due to the low propagation velocity of water-surface waves (Bleckmann and Barth 1984).

Taking a propagation speed of 50 m s^{-1} of the waves relevant in a banana plant, the time difference between leg tips 10 cm apart in *Cupiennius* amounts to 2 ms. Propagation velocities of water-surface waves are only 23 to 40 cm s^{-1} at 10 to 140 Hz (Sommerfeld 1970). At 40 Hz, which is in the most sensitive frequency range of *Dolomedes,* the difference in arrival time at legs 4 to 7 cm apart amounts to ca. 200 ms. By carefully controlled phase-shifted stimulation, Wiese (1974) demonstrated the backswimmer's (*Notonecta*) ability to resolve *phase differences* between the water-surface waves reaching its legs simultaneously. No similar experiments are yet available for spiders.

Clearly, many more experiments are needed before it can be decided which parameters are actually used by spiders for angular orientation. Measurement of the gradients occurring in the relevant substrates between the spider legs in situ and of their variation due to heterogeneities of the media and changes in the position of the spider are necessary. This also applies to the mechanisms underlying the estimation of source distance.

Source Distance. Absolute amplitude of the vibratory movement alone is no reliable indicator of source distance: it depends on distance as well as on the energy put into the signal by the sender, and in many situations it will be subject to resonances, standing waves, and reflections hard to predict for the spider.

On the water-surface, time differences may in theory be used to determine target distance, if interpreted as a measure of the curvature of the water wave front which decreases with the distance it has traveled from its source. Surface-feeding fish are known to have this ability (Hoin-Radkovski et al. 1984). For *Dolomedes* there is no direct proof yet. Due to decreasing time differences, the accuracy of such a mechanism should decrease with both increasing source distance and increasing frequency (increase of propagation velocity).

The dispersive nature of the propagation of polyfrequency wave signals implies still another possibility of making use of the physics of the stimuli for an estimate of goal distance. Since high frequencies travel faster than low ones (above 13 Hz in water-surface waves) and are in addition more attenuated, the shape of complex (i.e., nonsinusoidal) signals changes with distance. The surface-feeding fishes, *Aplocheilus lineatus* and *Pantodon buchholzi* can indeed resolve and process such differences in frequency modulation and, probably, also in amplitude modulation (Bleckmann and Schwartz 1982; Bleckmann et al. 1984; Hoin-Radkovski et al. 1984).

In the desert scorpion different solutions were proposed to the problem of distance determination. At target distances of up to 15 cm, the amplitude gradient across the sensory field of the eight legs could well be the cue. This amplitude gradient, however, sensitively decreases with source distance. Larger distances could in theory be determined from the difference in time of arrival of surface (Rayleigh) waves and compressional waves at the two classes of receptors which are supposed to selectively respond to them, i.e., tarsal slit sensilla and tarsal hair sensilla (Brownell 1977; Brownell and Farley 1979a, b).

5.2 Neuronal Networks

In addition to the analysis of the relevant signal parameters a quite different approach to the mechanisms underlying orientation behavior is making use of behavioral experiments designed to find out about the rules governing the central nervous integration of the sensory inputs from the eight legs. Such experiments were carried out with *Cupiennius* (Hergenröder and Barth 1983b).

In field conditions, *Cupiennius* usually sits and waits until its prey comes very close to its legs and within the reach of a quick jump. When standing with one or more legs on an experimental platform that begins to vibrate, it turns toward the platform if the stimulus resembles prey-generated vibrations. Its turning angle depends on the particular combination of legs that are stimulated. For simplicity, stimulation of the legs was simultaneous and with the same intensity. The comparison of turning angles in conditions where different legs were vibrated individually and in combination with other ipsi- or contralateral legs, directly indicates the contribution of individual legs to the "turning command". The turning angle can be quantitatively predicted by the application of an inhibitory network based on the following rules (Fig. 7).

a) Single Leg Stimulation. The mean-error angle (difference between stimulus and turning angle) increases from the front legs to the hind legs. This implies a stronger weight of the sensory input from the front legs in establishing the turning angle in cases of stimulation of leg combinations. The respective weighting factors F, which are ratios between turning (β) and stimulus (α) angle, are 1.2 for the first leg[2] and 0.6, 0.4 and 0.4 for the following legs.

---

2 If one of the first legs is stimulated alone, F is reduced to 0.6, which implies that the spider even in this case does not turn beyond the target.

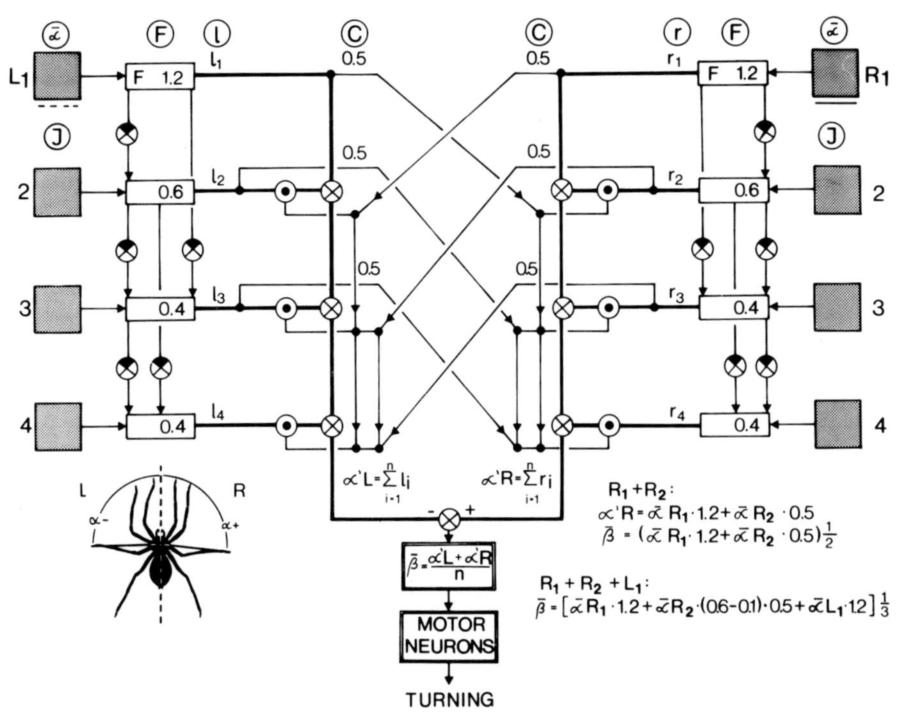

Fig. 7. Inhibitory network derived from behavioral experiments with *Cupiennius salei* and describing possible central interactions of the sensory inputs from the eight legs in determining the turning angle toward a source of vibration. Equations on *lower right* exemplify the calculation of the turning angle β for a situation in which right legs R_1 and R_2 alone and right legs R_1 and R_2 together with left leg L_1 are stimulated. *F* factor "weighting" the input ($\bar{\alpha}$, stimulus angle, i.e., angle formed by long axis of the animal and long axis of the stimulated leg) from different legs, *J* additive constant quantifying ipsilateral inhibition; *C* multiplicative constant quantifying contralateral inhibition; ⊗ addition, with reversal of sign at black quadrant; ⊙ multiplication; *r* and *l* input values modified by *F* and *J*. For more details see text. (Hergenröder and Barth 1983b)

b) Stimulation of Leg Combinations. The stimulation of anterior legs reduces F of all posterior legs on the same side by J = 0.1, which is an additive ipsilateral inhibition. Stimulation of legs on both lateral sides of the body reveals a reciprocal, multiplicative contralateral inhibition, again acting exclusively from front to back and reducing F by the factor C = 0.5.

c) Calculation of Turning Angle. Stimulus angles are assigned positive and negative signs for the right and left legs, respectively; they are weighted according to the above rules and added for the right and left legs. The sum of both values is divided by the number of stimulated legs, and thus indicates the turning angle (see Fig. 7).

Additional properties of this inhibitory network are revealed by a stimulating situation in which the spider sits on two movable plates and different legs receive one of two signals differing either in amplitude or onset time. Such conditions reflect the natural signal propagation, which requires time and involves attenuation. In both cases the spider behaves as if only the legs receiving the earlier or larger vibration were stimulated. It seems likely, then, that *Cupiennius* makes use of time and amplitude differences under natural conditions. The minimal gradients necessary were not determined yet, but a time difference of 4 ms and an amplitude ratio of 3 were found to be sufficient to see an effect.

The block diagram developed for *Cupiennius* (Fig. 7) partly both agrees and differs with similar, though only qualitative connectivity models conceived for the scorpion *Paruroctonus* (Brownell and Farley 1979b) and the backswimmer *Notonecta* (Murphey 1973). All three models require an inhibitory network with both ipsi- and contralateral inhibition. In the backswimmer a gradation of the inhibitory effects similar to those in *Cupiennius* was found (Murphy 1973). A peculiarity of *Cupiennius* is the differential weighting of the sensory inputs from the various legs for determining turning angle. Another feature only found in the *Cupiennius* model, that is, the exclusively backward direction of the inhibitory effects, is not likely to be a distinctive feature under most natural conditions in which the various legs are stimulated with time and amplitude differences. Evidence for this conjecture is derived from the above observation that animals turn towards legs stimulated earlier or stronger – just as if their sensory input inhibited the inputs from both more anterior and posterior legs (Hergenröder and Barth 1983b).

5.3 Active Localization

Localization of objects hanging motionless in the web is common among orb weavers (Liesenfeld 1956). These spiders actively vibrate the web and thus detect objects as light as 0.05 mg (*Zygiella*). The mechanisms involved in this capacity are not yet fully understood. From experiments with threads under increased tension, we concluded that the information used by the spider is not derived from thread tension itself, but from properties of the vibratory echo (Klärner and Barth 1982).

6 Identification

Who is it that vibrates? For a preying spider it is important to identify prey signals among irrelevant background noise. For the male *Cupiennius* courting on a banana plant and emitting vibratory signals himself, it may be life-saving to be identified as a male by the female and not confused with prey and killed. Some answers to the neuroethological questions involved are just beginning to accumulate.

6.1 Prey, Enemy, or Mate?

Cupiennius, when resting on a movable platform with one or more legs and stimulated with different vibrations, does indeed decide the above questions: It either approaches the source of substrate vibrations (predatory behavior) or withdraws from it (escape behavior). In other words, it does identify. Various signal-parameters were found important for the release of one or the other behavior (Hergenröder and Barth 1983a)[3].

For sinusoidal stimulation, thresholds for the withdrawal reaction are higher by up to about 20 dB than for the approach reaction. Noisy signals, even if limited to a small frequency band (bandwidth 1/3 octave, $Q = 0.35$), reduce the threshold values for the approach reaction by up to 10 dB, whereas those for the escape reaction remain almost unaffected (Fig. 5b). The number of withdrawal reactions increases steadily with increasing amplitude of sinusoidal stimuli. Approaches, on the other hand, are most numerous at particular amplitudes, each characteristic for the frequency chosen. Using narrow-band noise signals the two reaction types again differ. Whereas the number of approaches markedly increases with bandwidth, that of withdrawals is only little affected by it. Shifting the frequencies from low to high values shows that withdrawals are unlikely to occur at low frequencies. The proportion of the two behaviors can be predicted from the relation of the stimulus amplitude to their respective threshold curves (Hergenröder and Barth 1983a).

Summarizing these and similar data one would predict a *prey signal* to be attractive if it is of low amplitude, nonsinusoidal, made up by a wide range of frequencies and irregular in the time domain. It is readily seen that the prey signals presented in Fig. 1 do indeed differ with regard to these properties from the respective background noise. The decrease of the threshold values with increasing frequency – measured both in the behavior (for the approach reaction) and in the substrate-vibration receptors – indicates that the high frequency components (i.e., higher than those providing most of the energy content of the power spectrum) of prey signals are involved in their identification and may indeed be of particular importance. These arguments also apply to semi-aquatic spiders like *Dolomedes*, where the power spectrum of vibrations produced by a fly trapped on the water-surface differs markedly from that of background noise by being much broader and extending to higher frequency components (Lang 1980; Bleckmann and Barth 1984).

Similar arguments may be applied to *courtship signals*. Why does the female *Cupiennius* not misinterpret the male signal as that of a prey, although its main frequency component (ca. 80 Hz) is within the range of prey signal frequencies? As the male approaches, the increasing amplitude of its signal decreases the likelihood of aggressive female behavior, since it is increasingly likely to reach and then surpass the threshold value for the withdrawal reaction. In addition, a signal containing only a narrow band of frequencies with a very "orderly" temporal pattern such as the male courtship vibrations differs greatly from a typical prey signal (Fig. 1a) (Schüch and Barth 1985).

Recent behavioral studies using synthetic vibrations (Schüch and Barth, in preparation) have indeed shown that various temporal parameters contained in the male signal may vary only within narrow limits and still effectively elicit the

3 In the experiments described in the following, trichobothria were removed.

female's vibratory response (Rovner and Barth 1981). Thus, for example, a duration of the syllable (see Fig. 1a) of 70 ms (duration of silent pause between syllables kept constant) is most effective (natural value ca. 100 ms), whereas almost no female responses at all are recorded at values below ca. 40 ms and above ca. 250 ms. Similarly, a pause between the syllables (duration 70 ms) lasting for less than ca. 50 ms or more than ca. 400 ms renders the male signal ineffective. There is also a clear "optimum" for the ratio pause duration/syllable duration and more complex temporal relationships such as those described by Schüch and Barth (1985) are currently being examined.

The existence of a narrow frequency filter in the female courted by a male clearly emerges from the same type of experiment, where the number of her responses is related to the frequency chosen to make up the male syllables. The female *Cupiennius* most often responds if the frequency is close to the natural value, i.e., at ca. 100 Hz. She does not respond at all if it is either below ca. 50 Hz or above ca. 210 Hz, even though the other signal parameters are kept at their natural values (Schüch and Barth, in preparation).

6.2 Background Noise and Kleptoparasites

One of the main consequences of the low threshold sensitivity of the substrate vibration receptors to frequencies below about 30 Hz may be to filter out background noise, which is characterized by very low frequencies (Figs. 1 and 5a). The same conclusion comes from the behavioral threshold curves, both for the withdrawal and approach response (Fig. 5b, c).

Some animals have obviously learned to make use of this and in this way to prevent the spider's prey capture or aggressiveness. We have repeatedly watched animals like grasshoppers, frogs and salamanders in the field when coming very close to *Cupiennius* or even passing by it as if undetected. The reason appears to be their slow and regular gait, which elicits plant vibrations of only very low frequencies (Barth et al., in preparation).

Some insects have even found a safe way into spider webs as commensals or kleptoparasites. They also move in a distinctive slow and "careful" way, in particular as long as the web owner is motionless (review: Barth 1982). *Argyrodes*, a spider kleptoparasite in the web of *Nephila*, is even known to guard against a sudden release of tension when it "cuts out" the stolen prey from the web. The end of the cut radius is held by the first legs which are then slowly stretched (Vollrath 1979a, b). Low frequencies and the lack of fast transients also appear to be a safe route for young commensal spiderlings in their mother's web (Tretzel 1961).

6.3 Frequency and Amplitude Discrimination

For *Cupiennius* there is behavioral evidence both for frequency and for amplitude discrimination. It derives from experiments in which the spider's reaction was tested with stimuli of constant amplitude and varying frequency and vice versa (Hergenröder and Barth 1983a). Vibration frequency, however, was shown to be an unlikely basis for prey species discrimination in the spider web, because of much overlap and variability of the various signals. Instead, maximal signal amplitude at impact seems to be a more reliable parameter (Suter 1978).

7 Multimodal Interactions

Spider behavior in its natural context is probably never determined by substrate vibrations alone. Inputs from other sensory channels were shown to be relevant even in the cases where substrate vibrations are a sufficient condition.

7.1 Trichobothria

The foremost attention has to be given to the trichobothria, which respond to the slightest movement of the air, including the vibrations in the near-field of a vibration source (see Reißland and Görner, Chap. VIII, this Vol.; Barth and Blickhan 1984). Trichobothria affect predatory behavior in different ways with spiders living on different substrates. Whereas stimulation by air-borne prey stimuli elicits predatory behavior in spiders, like *Cupiennius,* living on plants (Hergenröder and Barth 1983 a) or on densely woven sheet webs as in *Tegenaria* and *Agelena* (Görner and Andrews 1969), this is not so in orb weavers like *Zygiella* and *Nephila* (Klärner and Barth 1982). In orb weavers invariably a thread-borne signal is required to indicate prey entangled in the web and to elicit prey-capture behavior; this seems reasonable since a jump from the web into the air may not be an effective way of catching prey. Air-borne stimuli are ineffective. They rather elicit startle responses, and it may be that the spider associates them with its own predators such as wasps and birds (Klärner and Barth 1982). In semi-aquatic spiders like *Dolomedes,* prey capture can be elicited by air-borne stimulation alone, provided the buzzing fly is no more than 10 cm away (Bleckmann and Barth 1984). Here – as well as in the case of plant-living spiders – it appears a good strategy to wait until the potential prey is within reach of a quick maneuver. Not entangled and immobilized by the sticky web and warned by its own sensitive receptors, the prey's chance for a successful escape from the predator is quite realistic – as can indeed be observed in the field.

A comparison of behaviors toward substrate-borne vibrations of intact spiders and spiders lacking trichobothria shows surprising effects. Since trichobothria are deflected by the air-borne sound generated by the substrate motion, their ablation affects the behavior of *Cupiennius* on the vibrating platform (see Fig. 5 b). The thresholds for the withdrawal reaction and the reaction times increase significantly. Those for the approach reaction remain unchanged, however, which indicates a special relationship between the presence of trichobothria and the withdrawal response (Hergenröder and Barth 1983 a).

Surprisingly, the percentage of responses to substrate vibration is dramatically *in*creased by removal of the trichobothria. This observation and the result of experiments in which air-borne stimulation with a buzzing fly was used in addition to the stimuli produced by substrate vibration has led to conclusions about the central-nervous interaction of both sensory systems (Hergenröder and Barth 1983a, b). When both systems receive either a "prey-like" stimulus or a stimulus eliciting withdrawal, their effects add to, but when trichobothria receive stimuli "unlike prey", they inhibit the approach reaction that would otherwise be triggered by substrate vibration alone.

In *Dolomedes* the orientation toward a source of water-surface waves becomes "worse" after ablation of all trichobothria with respect to angular orientation and running speed, but improves with respect to the spider's estimate of distance. Again then, the trichobothria are involved in the response, even if only surface waves stimuli are applied (Bleckmann and Barth 1984).

7.2 Vision

The importance of the interaction of several sensory systems under natural conditions is underlined by the observation that in *Cupiennius*, as well as in *Dolomedes*, blinding *in*creases the readiness to respond to vibration stimuli. On the other hand the movement of an object or a shadow not indicating prey *de*creases the readiness of the spider to respond to substrate vibration as do airborne nonprey stimuli received through the trichobothria. This shows that vision is another sensory channel involved in the shaping even of behavior obviously dominated by vibratory input.

8 Concluding Remarks

Spiders have proved to be a good choice for neuroethological studies of the vibration sense. Principles underlying the signal and information flux are gradually emerging. Knowing both the reach of the signals and the sensitivity of the vibration receptors we can now give a good estimate of the active space of the vibratory communication system. We seem to have found central nervous filters matched to various biologically significant vibration frequencies. Analysis of the releasing mechanisms with synthetic signals and knowledge of the natural vibratory Umwelt of the spiders underline the importance of both spectral and temporal patterns in the identification process. There is also a first crude idea about "hardware" properties of the central nervous system which may underly oriented behavior.

Nevertheless, we are still far from a full understanding of what we would like to know the shaky world of spiders and the ways in which evolution has adapted them to make such obviously rich use of vibratory signals.

Thus the *sensory pathway* dealing with substrate vibrations has been followed electrophysiologically only to the first stage of integration in the subesophageal nerve mass. All further stages of processing, as well as the genesis and control of the motor programs, are still a terra incognita at the neuronal level. The inhibitory network developed from behavioral studies to explain the turning behavior of *Cupiennius* toward sources of vibration provides hypotheses on some of the principles involved, which will be helpful for future work on the central nervous system. Also, recent studies into the neuroanatomy of the spider central nervous system (Babu and Barth 1984) will be of great assistance, in particular if they are extended by single neuron stains following electrophysiological recordings.

The complex interaction of the *various sensory systems*, including that for air-borne vibrations with that for substrate vibrations, deserves much more attention than so far received, both at the level of the stimuli and of multimodal interneurons and central nervous sensory convergence.

Another field of research to be extended is the *physics* and biological significance of the various wave types generated and traveling in the different media used by different spiders, including not only the soil, leaves, and tree trunks, but also a multitude of webs differing in shape and mechanics. Solving the problem of signal localization calls for precise measurements of amplitude and time differences actually found under natural stimulus conditions. The available data, although only a modest beginning, point to the particular value of field work and information which ties up behavior with ecology.

Acknowledgments. The author's research and that of his students and associates was generously supported by grants of the Deutsche Forschungsgemeinschaft (SFB45/A4). Prof. Dr. A. Michelsen kindly shared his laboratory facilities in Odense to measure the data presented in Fig. 2. The help of Mrs. H. Hahn with the preparation of the figures, the secretarial work of Mrs. U. Ginsberg, and the critical comments of Drs. E.-A. Seyfarth and H. Bleckmann on the manuscript are gratefully acknowledged.

References

Babu KS, Barth FG (1984) Neuroanatomy of the central nervous system of the wandering spider, *Cupiennius salei* (Arachnida, Araneida). Zoomorphology 104:344–359

Barth FG (1972) Die Physiologie der Spaltsinnesorgane. II. Funktionelle Morphologie eines Mechanoreceptors. J Comp Physiol 81:159–186

Barth FG (1982) Spiders and vibratory signals: Sensory reception and behavioral significance. In: Witt PN, Rovner JS (eds) Spider communication: mechanisms and ecological significance. Princeton Univ Press, Princeton, NJ, pp 67–122

Barth FG, Blickhan R (1984) Mechanoreception. In: Bereiter-Hahn J, Matoltsy AG, Richards KS (eds) Biology of the integument, vol I. Springer, Berlin Heidelberg New York, pp 554–582

Barth FG, Geethabali (1982) Spider vibration receptors: Threshold curves of individual slits in the metatarsal lyriform organ. J Comp Physiol 148:175–185

Barth FG, Seyfarth E-A (1979) *Cupiennius salei* Keys. (Araneae) in the highlands of Central Guatemala. J Arachnol 7:255–263

Bleckmann H, Barth FG (1984) Sensory ecology of a semiaquatic spider (*Dolomedes triton*). II. The release of predatory behavior by water surface waves. Behav Ecol Sociobiol 14:303–312

Bleckmann H, Müller U, Hoin-Radkovski I (1984) Determination of source distance by surface-feeding fishes *Aplocheilus lineatus* (Cyprinodontidae) and *Pantodon buchholzi* (Pantodontidae). In: Varju D, Schnitzler U (eds) Orientation and localization in engineering and biology. Springer, Berlin Heidelberg New York, pp 66–68

Bleckmann H, Rovner JS (1984) Sensory ecology of a semiaquatic spider (*Dolomedes triton*). I. Roles of vegetation and wind-generated waves in site selection. Behav Ecol Sociobiol 14:297–301

Bleckmann H, Schwartz E (1982) The functional significance of frequency modulation within a wave train for prey localization in the surface-feeding fish *Aplocheilus lineatus* (Cyprinodontidae). J Comp Physiol 145:331–339

Brownell PH (1977) Compressional and surface waves in sand used by desert scorpions to locate prey. Science 197:4303–4304

Brownell P, Farley RD (1979a) Detection of vibrations in sand by tarsal sense organs of the nocturnal scorpion, *Paruroctonus mesaensis.* J Comp Physiol 131:23–30

Brownell P, Farley RD (1979b) Orientation to vibrations in sand by the nocturnal scorpion *Paruroctonus mesaensis:* mechanism of target localization. J Comp Physiol 131:31–38

Cremer L, Heckl M, Ungar EE (1973) Structure-borne sound. Structural vibrations and sound radiation at audio frequencies. Springer, Berlin Heidelberg New York

Foelix RF (1970) Chemosensitive hairs in spiders. J Morphol 132:313–334

Görner P, Andrews P (1969) Trichobothrien, ein Ferntastsinnesorgan bei Webespinnen. Z Vergl Physiol 64:301–317

Hergenröder R, Barth FG (1983a) The release of attack and escape behavior by vibratory stimuli in a wandering spider (*Cupiennius salei* Keys.) J Comp Physiol 152:347–358

Hergenröder R, Barth FG (1983b) Vibratory signals and spider behavior: How do the sensory inputs from the eight legs interact in orientation? J Comp Physiol 152:361–371

Hoin-Radkovsky I, Bleckmann H, Schwartz E (1984) Determination of source distance in the surface-feeding fish *Pantodon buchholzi* Pantodontidae. Anim Behav 32:840–851

Keuper A, Kühne R (1983) The acoustic behavior of the bushcricket *Tettigonia cantans* II. Transmission of airborne-sound and vibration signals in the biotope. Behav Proc 8:125–145

Klärner D, Barth FG (1982) Vibratory signals and prey capture in orb-weaving spiders (*Zygiella x-notata, Nephila clavipes;* Araneidae) J Comp Physiol 148:445–455

Krafft B (1978) The recording of vibratory signals performed by spiders during courtship. Sym Zool Lond 42:59–67

Lang HH (1977) Mechanismen der Beuteerkennung und der intraspezifischen Kommunikation bei der räuberischen Wasserwanze *Notonecta glauca* L und ihre Rolle bei der Aufrechterhaltung der Populationsstruktur. Dissertation, Univ Konstanz, Konstanz

Lang HH (1980) Surface wave discrimination between prey and nonprey by the backswimmer *Notonecta glauca* L (Hemiptera, Heteroptera). Behav Ecol Sociobiol 6:233–246

Liesenfeld FJ (1956) Untersuchungen am Netz und über den Erschütterungssinn von *Zygiella x-notata* (Cl) (Araneidae) Z Vergl Physiol 38:563–593

Liesenfeld FJ (1961) Über Leistungen und Sitz des Erschütterungssinnes von Netzspinnen. Biol Zentralbl 80:465–475

Markl H (1973) Leistungen des Vibrationssinnes bei wirbellosen Tieren. Fortschr Zool 21: 100–120

Markl H (1983) Vibrational communication. In: Huber F, Markl H (eds) Neuroethology and behavioral physiology. Springer, Berlin Heidelberg New York Tokyo, pp 332–353

Masters MW (1984a) Vibrations in the orb web of *Nuctenea sclopetaria* (Araneidae). I. Transmission through the web. Behav Ecol Sociobiol 15:207–215

Masters MW (1984b) Vibrations in the orb web of *Nuctenea sclopetaria* (Araneidae) II. Prey and wind signals and the spider's response threshold. Behav Ecol Sociobiol 15:217–223

Masters MW, Markl H (1981) Vibration signal transmission in spider orb webs. Science 213:363–365

Melchers M (1967) Der Beutefang von *Cupiennius salei* Keys. Z Morphol Oekol Tiere 58:321–346

Michelsen A, Fink F, Gogala M, Traue D (1982) Plants as transmission channels for insect vibrational songs. Behav Ecol Sociobiol 11:269–281

Murphey RK (1973) Mutual inhibition and the organization of a nonvisual orientation in *Notonecta.* J Comp Physiol 84:31–69

Roemer van de A (1980) Eine vergleichende morphologische Untersuchung an dem für die Vibrationswahrnehmung wichtigen Distalbereich des Spinnenbeines. Diplomarbeit, Univ Frankfurt, Frankfurt am Main

Roland Ch, Rovner JS (1983) Chemical and vibratory communication in the aquatic pisaurid spider *Dolomedes triton* (Araneae: Pisauridae). J Arachnol 11:77–85

Rovner JS, Barth FG (1981) Vibratory communication through living plants by a tropical wandering spider. Science 214:464–466
Schüch W, Barth FG (1985) Temporal patterns in the vibratory courtship of a wandering spider (*Cupiennius salei* Keys.). Behav Ecol Sociobiol (in press)
Skudrzyk E (1971) The foundations of acoustics. Springer, Wien
Sommerfeld A (1970) Vorlesungen über theoretische Physik, Bd 2, Mechanik der deformierbaren Medien. Akad Verlagsges, Leipzig
Speck J, Barth FG (1982) Vibration sensitivity of pretarsal slit sensilla in the spider leg. J Comp Physiol 148:187–194
Suter RB (1978) *Cyclosa turbinata* (Araneae, Araneidae): Prey discrimination via web-borne vibrations. Behav Ecol Sociobiol 3:283–296
Tretzel E (1961 a) Biologie, Ökologie und Brutpflege von *Coelotes terrestris* (Wider) (Araneae: Agelenidae). II. Brutpflege. Z Morphol Oekol Tiere 50:375–524
Uetz GW, Stratton GE (1982) Acoustic communication and reproductive isolation in spiders. In: Witt PN, Rovner JS (eds) Spider communication: Mechanisms and ecological significance. Princeton Univ Press, Princeton, NJ, pp 123–158
Vollrath F (1979 a) Behavior of the kleptoparasitic spider *Argyrodes elevatus* (Araneae, Theridiidae). Anim Behav 27:515–521
Vollrath F (1979 b) Vibrations: their signal function for a spider kleptoparasite. Science 205:1149–1151
Wiese K (1974) The mechanoreceptive system of prey localization in *Notonecta*. J Comp Physiol 92:317–325

XII Spider Proprioception: Receptors, Reflexes, and Control of Locomotion

Ernst-August Seyfarth

CONTENTS

1 Introduction

Historically, the question of how animals control their own movements has been a major concern of behavioral physiology. Much of the early and recent work has focused upon the recurring debate about central versus peripheral control of repetitive behavior such as locomotion. While many rhythmic behaviors such as walking, flying, or breathing can be maintained in the absence of patterned sensory input, there is now apparent consent among neurobiologists that peripheral feedback by way of sensory receptors (such as proprioceptors) is essential for fine control of certain behavior patterns and for reacting to the environment as the situation may demand (Delcomyn 1980; Selverston 1980).

Arthropods and their rhythmic behaviors have played a central role in such research (see, e.g., reviews by Hoyle 1976; Bowerman 1977; Wendler 1978; and in Herreid and Fourtner 1982). In comparison with insects and crustaceans,

Zoologisches Institut der J.-W.-Goethe-Universität, Gruppe Sinnesphysiologie, Siesmayerstraße 70, D-6000 Frankfurt am Main 1, Federal Republic of Germany

however, little has been done with the arachnids. Notable exceptions ies of scorpions (see Root, Chap. XVII, this Vol.) and work such as Wı. (1967) ablation experiments with tarantulas.

Spiders show some unique features in their motor behavior (e.g., web-building), in their complement of sensory receptors (such as lyriform slit sense organs), and in their mode of limb movement (interaction of leg-muscle contraction with hydraulic mechanisms) that set them apart from the other arthropods. These and other peculiarities are worth studying in their own right as well as in comparison with other arthropod groups.

This chapter is not an exhaustive review. The emphasis will be on recent morphological and electrophysiological studies of leg receptors and sensory leg nerves in spiders. This will be followed by a brief discussion of certain proprioceptive reflexes. I will then summarize the literature on spider locomotion and outline some of our recent results on tethered walking and sensory ablation experiments. Finally, problems and prospects of current and future research regarding central and reflex control are highlighted.

2 Gross Morphology of Spider Legs, Muscles, and Nerves

Spider legs consist of seven segments with seven joints (Fig. 1 a). Movement at each joint is generally in one preferred plane – either flexion/extension or promotion/remotion. The two most proximal articulations allow considerable mobility in both planes. There are two "knees" which predominate in flexion/extension movements, one between femur and patella, and another between tibia and metatarsus. The main promotion/remotion in the distal part of the leg occurs in the patella/tibia joint. Extension at the two knees and the metatarsus/tarsus joint is by hydraulic force; all other segments have one or several sets of muscular agonists and antagonists.

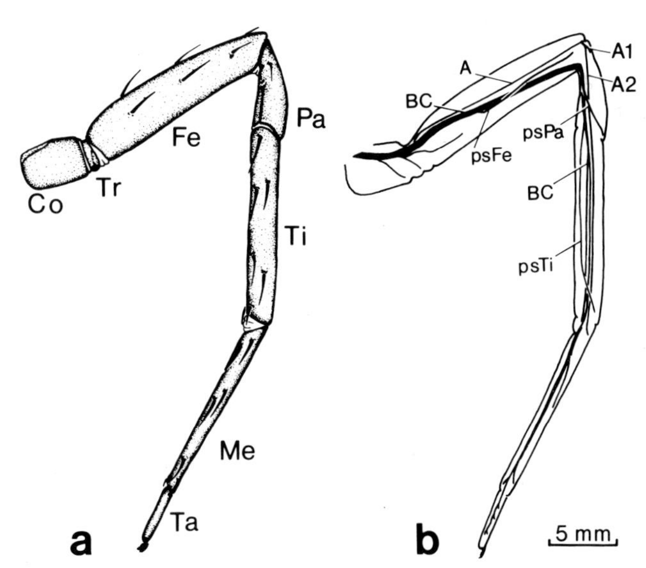

Fig. 1 a, b. Walking leg of the Central American wandering spider, *Cupiennius salei* Keys. **a** Posterior aspect of right foreleg; note large spines at Ti/Me and Me/Ta joints reaching ventrally across articulations. *Co* coxa; *Tr* trochanter; *Fe* femur; *Pa* patella; *Ti* tibia; *Me* metatarsus; *Ta* tarsus. **b** Arrangement of main leg nerves and of major sensory nerves (posterior view). The small leg nerve (*A*) branches into two parts near Fe/Pa joint (*A1, A2*); main leg nerves B and C run through leg as one bundle (*BC*). *psFe, psPa, psTi* posterior sensory nerve-branches in *Fe, Pa,* and *Ti.* (After Eckweiler 1983)

Fig. 2a–c. Typical examples of the three known proprioceptor types in spider legs. **a** Lyriform slit sense organ. *Left* stereoscan micrograph of organ HS-9 located on posterior tibia of *Cupiennius; right* camera-lucida drawing from cobalt-filled preparation; each of the 9 slits is innervated by 2 sensory neurons (in this example, the second neurons of the short slits 6 to 9 had remained unstained and were therefore not drawn). (After Seyfarth and Pflüger 1984). **b** Hairs and large spines bridging the tibia/metatarsus joint (*Ti/Me*). *Left* stereoscan micrograph, anterior view of slightly flexed joint region; *right* each tactile hair is innervated by 3 bipolar sensory neurons. *A* axon; *S* soma; *D* dendrite; *T* mobile hair shaft. (Eckweiler 1983). **c** Internal joint receptors. *Left* schematized view of posterior metatarsus/tarsus joint (*Me/Ta*); the branching dendrites from a cluster of receptor cells (*R*) end beneath the articular membrane; *S* single slit-sense organ; *D* large ventral spine (Eckweiler 1983). *Right* longitudinal section of internal joint receptor R10 in femur/patella joint (*Fe/Pa*). Ten sensory cells form ganglion (*G*) and send branching dendrites (*D*) into hypodermis (*Hy*) beneath joint membrane (*Jm*). The axons of R10 join the small leg nerve (*A*), which branches into 2 parts distally (*1* and *2*). *En* endocuticle; *Ex* exocuticle. (After Foelix and Choms 1979).

The number of leg muscles involved in active movements is much greater in spiders than in other arthropod groups. Insects typically have about 10 leg muscles and crustaceans and scorpions about 16 per walking leg, whereas spiders have at least 30 (Ruhland and Rathmayer 1978). This comparison does not even take into account that extension of several joints in spiders is by hydraulic and not by direct muscular force. Furthermore, each of the 30 spider leg muscles is generally innervated by more motor neurons than in other arthropods. Crustaceans have about 35 motor neurons for the muscles of each leg (Page 1982); in spiders, with their highly polyneural innervation of each muscle, 35 neurons (or more) may supply just the seven femur muscles – that is, the muscles in only a single leg segment. The functional significance of these anatomical features is not at all obvious. But could it be that spiders move their legs with more subtlety than other arthropods and that peripheral control is correspondingly refined?

Each of the four paired leg ganglia in the subesophageal nerve complex (see Babu, Chap. I, this Vol.) sends out three main leg nerves, A, B, and C (nomenclature of Rathmayer 1966). The small leg nerve (A) carries both sensory and motor fibers and ends in the metatarsus. Nerve B, consisting only of motor axons, and nerve C, the largest, run through the leg together (Fig. 1b). Nerve C is purely sensory and contains thousands of ascending axons from the various leg receptors (see also Foelix, Chap. X, this Vol.).

3 Proprioceptor Types

Three types of mechanoreceptors have been described near the joints of spider legs; they are stimulated during joint motion and hence can act as proprioceptors (Seyfarth and Pflüger 1984). These include (1) numerous slit sense organs in the leg exoskeleton, primarily single slits and compound ("lyriform") organs located in the immediate joint vicinity, (2) hundreds of "trichoid" sensilla (sensory hairs and several large cuticular spines) bridging articulations, and (3) several clusters of multiterminal sensory neurons, whose dendrites end between hy-

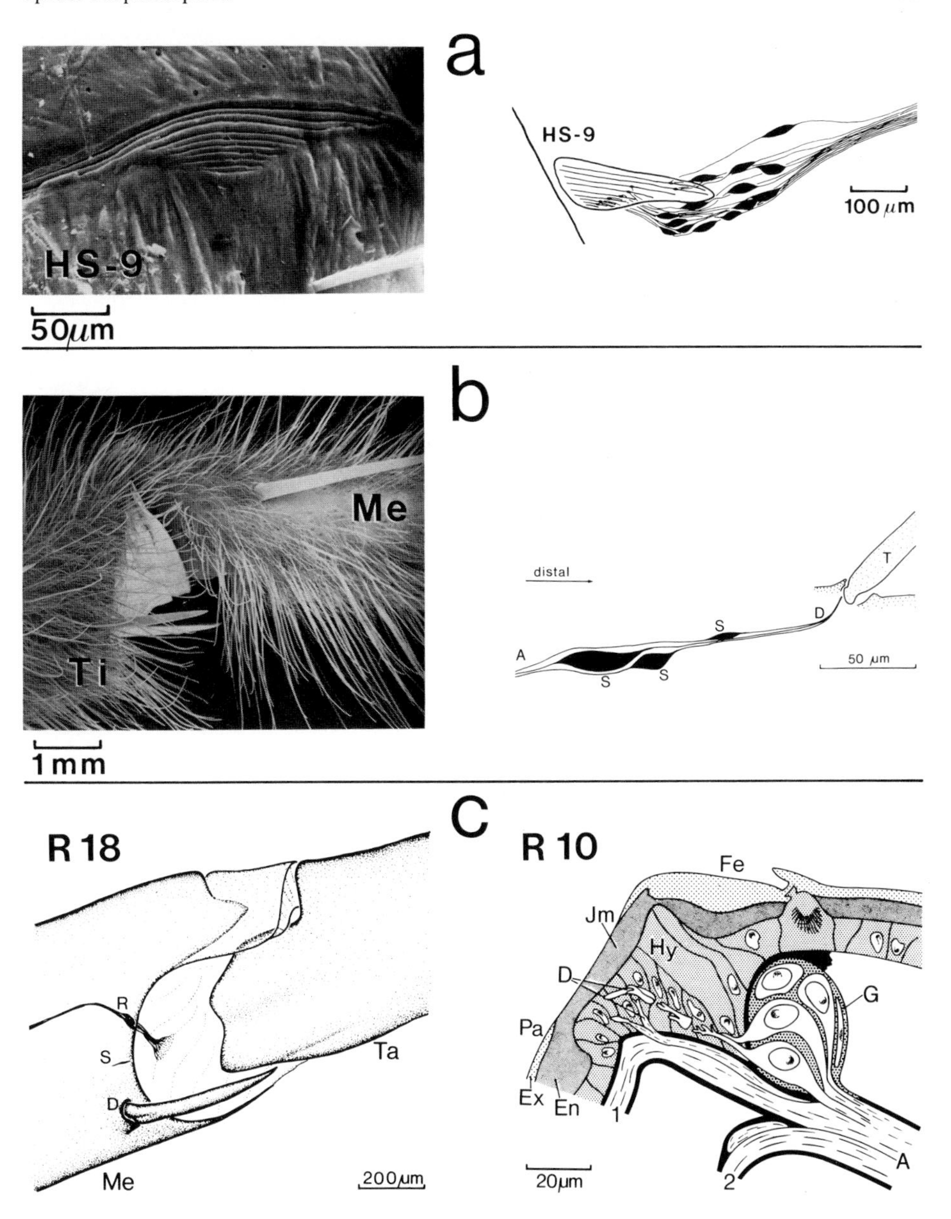

podermal cells beneath the articular membranes ("internal joint receptors"). Figure 2 shows typical examples of each proprioceptor type and details of their cellular arrangement.

All leg joints are supplied with at least one, and usually several sets of each, proprioceptor type. In addition, the tarsal claws and part of the scopulae, i.e., the hair tufts underneath the claws of many wandering spiders, are innervated by sensory neurons (Foelix 1970, and Chap. VII, this Vol.). Table 1 summarizes

Table 1. Important features of proprioceptor types found in the spider leg

Type	Number of organs per leg	Type of sensillum	Adequate stimulus	Response types	Possible proprioceptive function	Reference
Slit sense organs	Ten compound organs plus numerous single slits	Two bipolar sensory cells per slit (2–29 slits per compound organ)	Load-induced cuticle strains	Fast-adapting and slowly adapting units	Monitor strains in exoskeleton; perceive muscle forces	Reviews: Barth (1976; 1981) Barth and Blickhan (1984) Barth, Chap. IX; this Vol.
Joint-hairs and spines	Several hundred; often interdigitating	Trichoid; 3 bipolar sensory cells per hair	Displacement of hair shaft	Fast-adapting units	Signal joint angles; perceive ground contact	Foelix and Chu-Wang (1973) Harris and Mill (1977) Foelix, Chap. VII, this Vol.
Internal joint receptors	At least 18 groups	Clusters of multiterminal sensory cells (3–13 per group)	Displacement and strain of joint membrane	Fast-adapting and slowly adapting units	Monitor direction and rate of joint movement; signal joint position	Parry (1960) Rathmayer (1967) Rathmayer and Koopmann (1970) Mill and Harris (1977) Foelix and Choms (1979) Seyfarth and Pflügler (1984)

studies of the topography and morphology, fine structure, and sensory physiology of each receptor type. Slit sense organs and trichoid sensilla are treated in other chapters of this book (Barth, Chap. IX, and Foelix, Chap. VII, this Vol.). Foelix and Choms (1979) provide a detailed ultrastructural study of an internal joint receptor in the femur/patella joint of *Zygiella* and *Cupiennius*, and their results may apply to other internal joint receptors. The following points not contained in the survey of Table 1 should be mentioned:

1. The number and location of proprioceptive organs are remarkably similar in various araneid species – hunting spiders and orb weavers (see also Foelix 1982).

2. In all three receptor types the number of electrophysiologically detectable units is (at least) one less than that determined morphologically (Harris and Mill 1977; see also Foelix, Chap. VII, this Vol.).

3. We have no indication that spiders (or other arachnids) possess complex sensilla with scolopales such as the chordotonal organs typical of crustaceans and insects.

There is, however, some circumstantial evidence that another internal type of proprioceptor might exist in the spider leg. Morphological data suggest that the patella/tibia joint has a muscle receptor organ (MRO). Parry (1960) found a small muscle in *Tegenaria* legs and Ruhland and Rathmayer (1978) reported a pair of small muscles in *Dugesiella* legs with fiber diameters in the range of 18–30 μm – that is, much thinner than in normal leg musculature. The pair in *Dugesiella* legs originates near a nerve cell on the dorsal wall of the patella. The posterior member of the muscle pair terminates in connective tissue of the tibia, while the anterior fibers continue to form a thin tendon, which in turn inserts on the tendon of the claw depressor muscle in the metatarsus. It is conceivable that these putative muscle receptor organs monitor movement of the most distal three leg segments and of the claws – perhaps in an arrangement similar to the "tendon receptor organ" that spans two proximal segments in the legs of *Limulus* (Hayes and Barber 1967).

4. The most proximal joint in spiders between the prosoma and coxae provides great passive mobility and is operated by a great number of muscles (Palmgren 1981). Yet little is known about the proprioceptors at this joint. In *Cupiennius*, we have recently found groups of slightly curved, innervated hairs on the anterior and posterior sides of the coxae; these hairs are deflected when the bulging pleural membrane "rolls over" them during lateral leg movements (Fig. 3). The similarity of these distinctly grouped hairs to the trochanteral hairplates of insects (Pringle 1938; Markl 1962; Pflüger et al. 1981) is striking. In addition, single slit sense organs in the sternum may be stimulated during displacement of the coxae (Barth and Libera 1970). Our preliminary cobalt stains show that there are also internal joint receptors at this joint (Seyfarth et al. 1985).

5. We have recently used cobalt backfilling of sensory nerves in the legs of *Cupiennius* to clarify the innervation pattern of the various proprioceptors in each leg segment (Eckweiler 1983; Seyfarth and Pflüger 1984; Seyfarth et al. 1985). As a rule, axons from the sensory neurons in the joint region form ascending nerves (often bilateral pairs) that enter the main leg nerve shortly before reaching the next proximal joint. This geometry is especially apparent in the tibia and femur (Fig. 1b). The situation is not as uniform in the shorter leg segments such as patella and trochanter. In the latter only axons from the lyri-

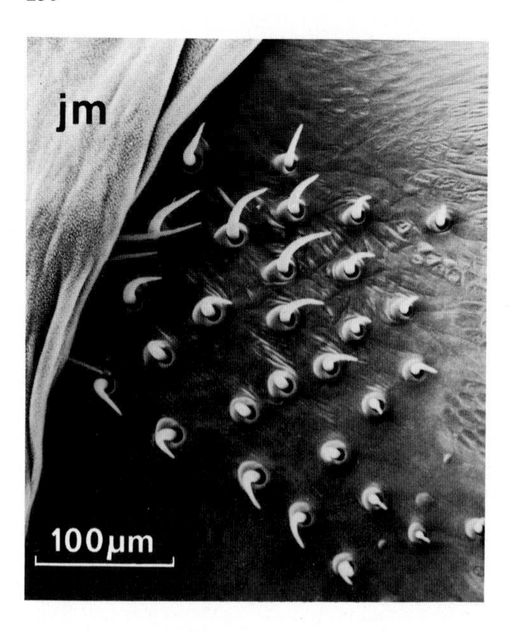

Fig. 3. Hair plate on the anterior side of the leg coxa (stereoscan micrograph; *Cupiennius salei*). When the coxa moves forward, pleural joint membrane (*jm*) folds and bends the hairs down by "rolling" over them. Unlike the other tactile hairs near joints, the trichoid sensilla of hair plates are short and slightly curved. (cf. Fig. 2b)

form organs and hair sensilla run together, while the fibers from the internal joint receptors ascend in separate small sensory nerves.

4 Proprioceptive Muscle Reflexes

Having identified various proprioceptors that are stimulated during limb movement, the next step is to look for muscle reflexes correlated with the activity of a particular type of proprioceptor; one can then check – via ablation experiments and controls – whether these sense organs actually elicit the reflex. We have simultaneously recorded the sensory and motor activity in legs of restrained but otherwise intact spiders (*Cupiennius* and *Aphonopelma*), and so far we have found two types of reflex, each of which can be traced to a particular type of proprioceptor: (1) "resistance reflexes" that are mediated by internal joint receptors, and (2) "synergic reflexes" associated with lyriform slit sense organs.

Resistance Reflexes. These oppose an imposed joint movement. They activate muscles intrinsic to the joint being stimulated and can serve as negative feedback control of limb position. Such intrinsic resistance reflexes were found in the muscles of all leg segments so far examined (patella, tibia, and metatarsus; Fig. 4 shows the arrangement of muscles in these segments). Reflex activity persists after removal of all *external* proprioceptors (i.e., hairs and lyriform organs) at the joint being stimulated (Seyfarth 1978a, b) but ceases after the sensory nerves which carry the axons of internal joint receptors are cut (Seyfarth and Pflüger 1984). Figure 5 shows the results of successive steps in an ablation ex-

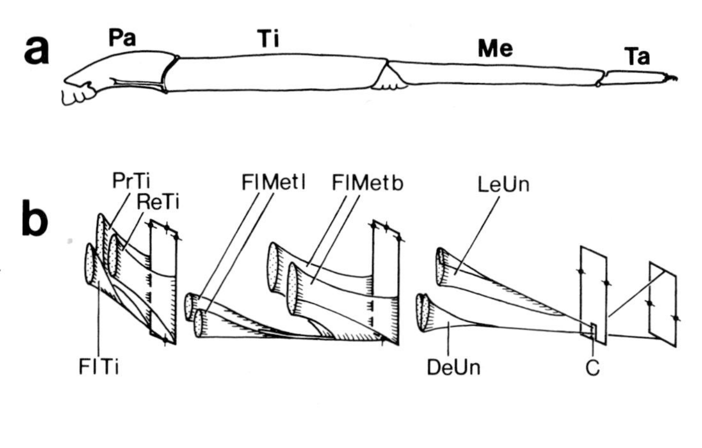

Fig. 4a, b. Arrangement of muscles in distal leg segments of *Cupiennius.* **a** Posterior view of leg; abbreviations as in Fig. 1 b. **b** Schematized diagram of the muscles and their distal insertions. Joints are drawn as *rectangles* and position of pivot axes is indicated by *dot-and-line symbols. PrTi/ReTi* promotor/remotor tibiae; *FlTi* flexor tibiae; *FlMetl/FlMetb* flexor metatarsi longus/bilobatus; *LeUn/DeUn* levator/depressor ungium; *C* "collar" serving as slide-guide for claw tendons. (After Eckweiler 1983)

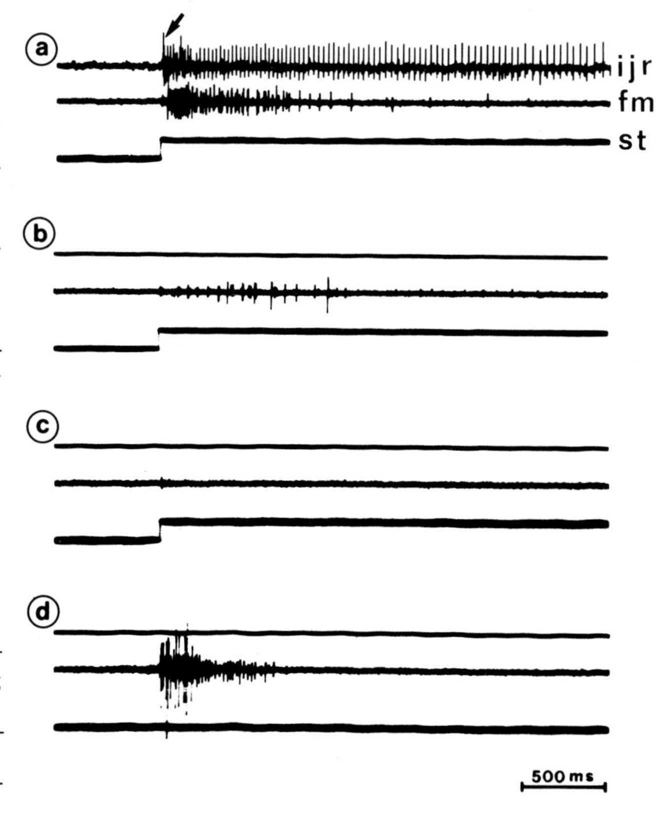

Fig. 5a–d. Resistance reflex in flexor metatarsi muscle; ablation experiment demonstrating the involvement of internal joint receptors (*ijr*). **a** Simultaneous recordings from anterior sensory nerve branch in tibia (*upper trace, ijr*), showing internal joint receptor activity *after* the ablation of all external proprioceptors (hairs and lyriform organs), and from flexor metatarsi muscle (*fm*) responding reflexly to an applied, stepwise elevation (10°) of metatarsus from horizontal starting position (trace *st*). Two units, one slowly adapting and one fast-adapting (*arrow*), are resolved in trace *ijr.* **b** After the anterior nerve was cut, sensory discharges from anterior *ijr* disappeared and reflex response was weaker. **c** Subsequent cutting of the posterior sensory nerve in tibia (carrying the axons from the posterior *ijr*) caused reflex failure. **d** flexor muscle response to air puff directed at spider body demonstrates that the muscle remained undamaged and can still contract despite the earlier sensory surgery. (After Seyfarth and Pflüger 1984)

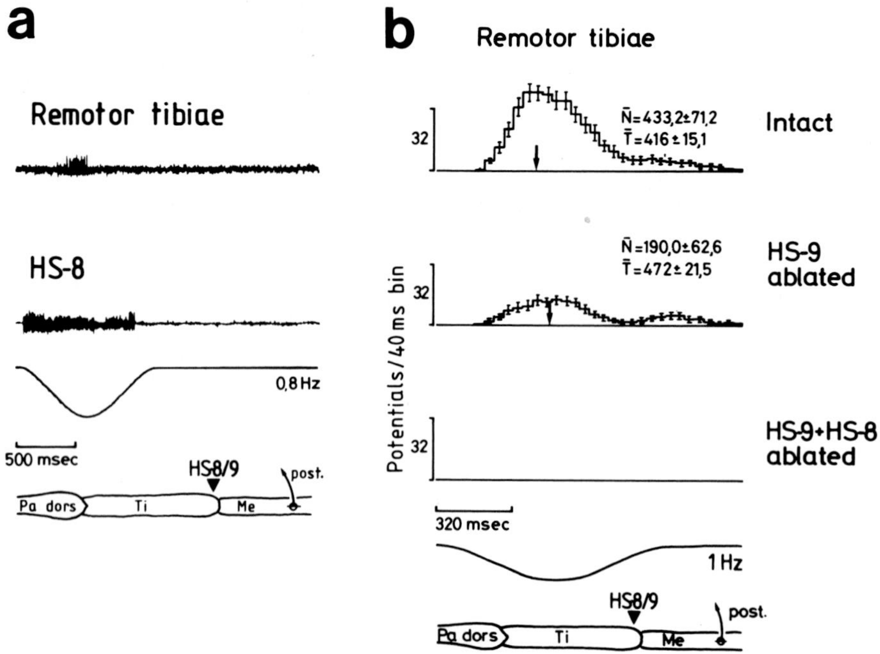

Fig. 6 a, b. Synergic reflex in remotor tibiae muscle. **a** Simultaneous extracellular recordings from the muscle (*top trace*) and from lyriform organ HS-8 (*second trace*) during imposed, sinusoidal backward movement (*third trace*) of metatarsus. The muscle response is delayed considerably with respect to that of the lyriform organ. *Bottom diagram* shows a schematized view of leg parts and the stimulus direction; organ HS-8 is located on posterior tibia close to organ HS-9 (see also Fig. 2). **b** Reflex failure after selective ablation of HS-8. *Top* intact lyriform organs; peristimulus-time-histogram for muscle responses to 10 sinusoidal mechanical stimuli. *Middle* destruction of HS-9 reduces reflex strength. *Bottom* additional ablation of HS-8 causes complete reflex failure. Other tests show that destruction of HS-8 alone causes reflex failure. *N* mean number of muscle potentials ± SE; *T* mean position of activity peak on time axis (± SE) after stimulus onset (also indicated by *arrows*). (After Seyfarth 1978 a, b)

periment demonstrating the involvement of internal joint receptors with a resistance reflex in the long tibio-metatarsal flexor muscle.

Synergic Reflexes. These act in synergy with an imposed leg movement. We found them not in muscles controlling the stimulated joint itself but rather in extrinsic muscles of the same leg. For example, *lateral* deflections of the metatarsus against the fixed tibia – which compresses and thus stimulates lyriform organs on each side of the tibia separately – causes activity of patellar muscles. Hence, in their capacity as promotor and remotor tibiae (see Fig. 4), these muscles operate so as to augment the imposed leg movement; they also act to relieve strain due to the lateral bending and so may be protective. Simultaneous sensory and muscle recordings and the successive stages of an ablation experiment demonstrate the failure of synergic reflex activity after destruction of a particular lyriform organ (Fig. 6).

Regarding hair sensilla, touching the long hairs on tarsus, metatarsus, and tibia leads to "withdrawal activity" in several muscles, moving the leg away from the stimulus (Seyfarth and Pflüger 1984). So far, we have not seen that selective stimulation of proprioceptors in one leg elicits reflexes in other ipsilateral or contralateral legs. Also, our present extracellular muscle recordings (electromyograms) resolve only excitatory motor activity and do not reveal possible inhibition of muscles. The innervation pattern of spider muscles is polyneural (Fourtner and Sherman 1973; Sherman, Chap. XVI, this Vol.) and both excitatory and inhibitory motor neurons exist in spiders as in other arthropods (Brenner 1972; Ruhland 1976). In the case of resistance reflexes, for instance, one would expect (by analogy with the situation in other arthropods) that activity of a muscle is reflexly inhibited when its antagonist is excited.

5 Spider Locomotion

5.1 General Considerations

Early observers (e.g., Dixon 1892; Kaestner 1924) noticed that spiders typically move their legs in two functional groups, resulting in the "alternating tetrapods" gait that has often been described for arachnids: the set of legs R4, L3, R2, L1 tends to alternate with L4, R3, L2, R1 (L and R are the body sides; the legs are numbered from front to rear). Orb weavers such as *Araneus* adhere to this alternating gait even during web building when they use their hind legs both for walking and for placing thread (Jacobi-Kleemann 1953). The joints and segments of the four leg pairs are anatomically very similar in most spiders; consequently the ranges of joint movement do not vary greatly from front to hind legs (Ehlers 1939; Frank 1957).

During forward walking, each leg is lifted and brought forward relative to the body (protraction or "return stroke"), then put on the ground and moved backward so that leg pairs 1 and 2 pull and leg pairs 3 and 4 push the animal forward (retraction or "power stroke"). Hence the main propulsive force is produced by flexor muscles in the forelegs and by extensors (including hydraulic mechanisms) in the hindlegs. The "stride length" of the third leg, which is often somewhat shorter than the others, is effectively increased by a greater degree of rotation about its long axis at the proximal joints (Kaestner 1924; Ehlers 1939) or by doubling the stepping rate (crab spiders; Ferdinand 1981).

5.2 Specifics of Stepping

Spiders generally do not "march" for periods as long as do many insects, so that statistical analyses of gaits are difficult. In studies of spider locomotion it has been noted that the relative timing of each leg's motion with respect to the others can vary considerably from that in the strictly "alternating tetrapods" gait and that spiders readily adapt to the loss of a leg (Kaestner 1924; Ehlers 1939; Jacobi-Kleemann 1953; Wilson 1967). Often such variable coordination between appendages has been considered a sign of prominent peripheral – that is, reflex – control of walking. Wilson (1966, 1967) used sequences of "footfall" to characterize the stepping pattern of various arthropods. His descriptive model of metachronal sequences in the stepping order of ipsilateral legs has proved useful for discussion of walking patterns and the variability of leg coordination; Delcomyn (1982) discusses critically Wilson's and other models.

In the tarantula, Wilson (1967) found several stepping orders, some of which did not fit his model and none of which were systematically correlated with stepping rate, i.e., with walking speed. The prevalent stepping order (in more than ⅔ of all data) was "4, 2, 3, 1" (Table 4 in Wilson 1967). With the exception of *Pardosa tristis* (Moffett and Doell 1980), subsequent studies have shown the same pattern to be dominant in other species (*Agelena labyrinthica:* Fröhlich 1978; various crab spiders: Ferdinand 1981). Scorpions (Bowerman 1975) use the same "4, 2, 3, 1" sequence almost exclusively, and Seyfarth and Bohnenberger (1980) have confirmed Wilson's generalization in compensated walking of tarantulas on an active spherical treadmill.

5.3 The Relative Roles of Central Control and Sensory Feedback

Various means of interfering with the normal function or sequence of leg use provide certain hints about the relative roles of central control and peripheral feedback. Several results in spiders are here compared with their analogs in similar experiments on other arthropods:

1. *Amputations of whole limbs* (often through induced autotomy) cause drastic changes in gait. For example in the tarantula, Wilson (1967) removed one foreleg and the contralateral hind leg. The operation resulted in a walking pattern in which adjacent ipsilateral legs "often" continued to alternate, but intrasegmental, contralateral legs (L2 and R2, L3 and R3) moved in synchrony, thus violating the "rule" that legs of the same body segment alternate. (Note, however, that as Wilson made explicit and as casual observation shows, this rule is often violated in normal eight-legged walking.) Such results indicate that sensory input (or its lack after ablations) causes changes of the behavior which are difficult to explain by assuming rigid, preprogrammed central commands. Insect studies have suggested, however, that such adaptive changes cannot be maintained at very high stepping rates, when central control seems to predominate (Delcomyn 1982).

2. *Tying up the middle legs* such as the third pair results in partial decoupling of the adjacent second and fourth segments and in increased tendency of the tarantula not to walk at all (Wilson 1967). Comparable findings in insects have been interpreted in the sense that the normal intersegmental transfer of neuronal information is blocked if a pattern generator receives the long-lasting, unusual sensory inputs caused by leg immobilization (Delcomyn 1982).

3. *Ablations of proprioceptors* in a leg – for example, destroying all lyriform organs at a joint or cutting major sensory nerves in the tibia and femur – has very little effect on the basic, rhythmic forward-backward motion of the leg

treated; movement of the other, intact legs remains unaffected (Parry 1960; Seyfarth and Barth 1972; Seyfarth and Bohnenberger 1980; see also the new results described below). Parallel findings in other arthropods include Bässler and Graham's (1978) report that cutting the tendon of the chordotonal organ in a leg of the stick insect does not alter the manner of leg movement significantly. Similarly, leg use is often spared following myochordotonal ablation in crustaceans (reviewed by Evoy and Ayers 1982). Such results of course emphasize the potential power of the CNS with respect to preprogrammed "pattern generators", independent of feedback. Extensions of these kinds of experiments to more drastic ablations in spiders, in our own current work, are described below.

5.4 Consequences of Proprioceptor Ablations in Tethered Walking of Spiders: Recent Results

Both to facilitate electrical recording from nerves and muscles during walking, and to simplify observation of leg use following ablations, we (collaboration with J. Bohnenberger and J. Thorson) have developed methods for tethered walking of large spiders (*Cupiennius* and tarantulas). The method and several results are described briefly here.

Tethering Methods. Figure 7 shows the tethered spider walking on a 25-cm-diameter, air-suspended spherical shell. Our method follows the earlier techniques of Carrel (1972) and Dahmen (1980) for smaller insects. The spiders require some vertical freedom, provided by the pivoted, spring-loaded beam. Moreover, they struggle against the constraint unless the attachment to the beam includes a short (1-mm) length of thread, allowing some rotational mobility beyond that allowed by sphere rotation.

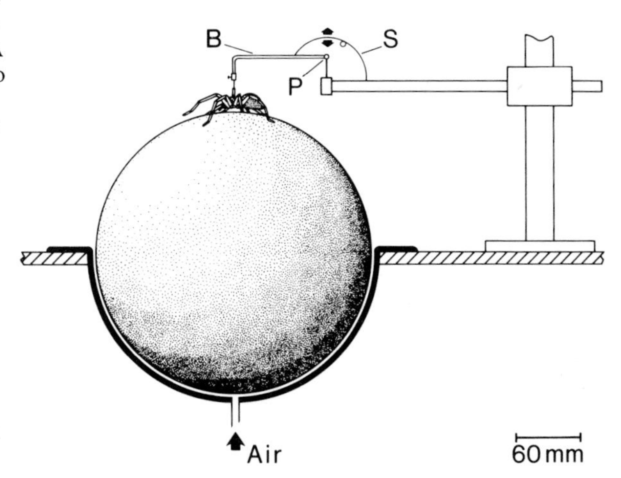

Fig. 7. Arrangement for tethered walking of wandering spiders, on a 25-cm-diameter, air-suspended styrofoam sphere. A small cardboard plate waxed to the prosoma bears a short length of thread and a rod that can be clamped (by a set screw) to a thin beam (*B*), which is pivoted (*P*) and steadied by a weak spring (*S*). The spring is adjusted so that the beam exerts no vertical force on the spider in its usual walking posture. The vertical struts below the pivot bear strain gauges for force measurement. The eyes of the spiders are covered with black paint

Free and Tethered Walking. Wandering spiders such as *Cupiennius* tend to run in short bouts of a few steps each, in nature (Seyfarth, unpublished observations in Central America), in arenas (Seyfarth and Barth 1972; Fig. 8A and B) and on the spherical treadmill. Indeed, the muscles are not equipped biochemically for sustained running (Linzen and Gallowitz 1975).

In voluntary and slow walking, gait and speed are extremely variable. More reproducible, fast bouts result from stimulation by an air puff; a fast leap (leg pairs 3 and 4 extending in synchrony) merges into several steps of approximately alternating-tetrapods gait.

Several of these bouts are shown in Fig. 8 for *Cupiennius* and the tarantula, in the form of body/substrate speed-versus-time diagrams. Note that speed fluctuates considerably even in these fast bouts – and that the changes are rarely synchronized with the stepping cycles (arrows, Fig. 8) – both in free walking on a fixed substrate (Fig. 8A and B) and in tethered walking on the various spheres (Fig. 8C–F).

Speed change is a good estimate for the energetics of walking and the "smoothness" of stepping in any gait pattern , and it offers interesting comparison with other animals. As Graham (1983) has pointed out, stick insects progress with horrendous inefficiency, often decelerating to a stop after each step. Crickets (Weber et al. 1981) modulate forward speed by 20–40% within bouts (rather like the spiders' performance in Fig. 8), whereas in man and many quadrupeds forward speed varies by only a few percent in walking and running (Cavagna et al. 1977).

The evidently natural tendency of these spiders to vary speed also allows us to demonstrate that, in tethering, choice of the correct mass ("inertial matching") of the air-suspended spherical shell in Fig. 7 is worthwhile: when a subset of the legs exerts a net forward force in free walking, the mass of the body is accelerated; with a tethered animal it is the substrate that is accelerated. It can be shown (Dahmen 1980; Weber et al. 1981) that equal forces will produce equal body/substrate accelerations (in both the free and tethered cases) if the mass of the air-suspended spherical shell is 3/2 the mass of the animal. The 9.5-g tarantula of Fig. 8 (C–F) is approximately matched inertially by a 15-g spherical shell, whereas it is unable, in these running bouts, to perform its natural (A and D) speed overshoot and fluctuation on the too-heavy 30- and 70-g spheres (E and F); accelerational reafference (and hence reflex operation) is abnormal on the latter.

Total Deafferentiation of a Joint. The resistance (Fig. 5) and synergic (Fig. 6) reflexes triggered by internal joint and lyriform receptors at the tibia/metatarsus joint are useful candidates for ablation experiments, because of their accessibility. Moreover, this joint (the "second" knee) is usually flexed prominently and is easily monitored in tethered walking.

Both tibial sensory nerves were cut in single legs of several *Cupiennius* (either leg 1 or leg 3). All hairs near the joint were broken off, for some of these may be innervated via the main leg nerve. Three *Cupiennius* females used a foreleg (and seven a third leg) following this operation, in a manner we could not distinguish from normal use. Some "skipping" of steps is seen shortly after the operation, but nearly full use, with tibia/metatarsus flexion, can occur

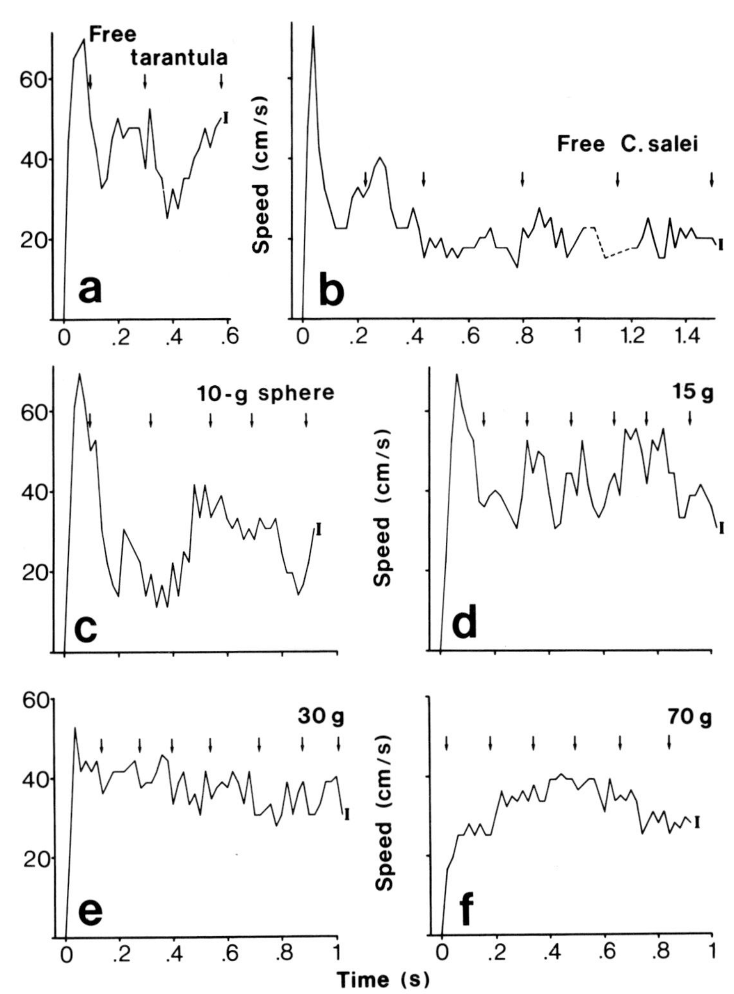

Fig. 8a–f. Speed fluctuations during air-puff-induced running bouts. The *ordinate* gives body-substrate displacement per 20-ms TV-frame interval, and *arrows* demark instants of successive lifts (beginning of protraction) of one leg, in the course of the step cycles. **a, b** Free spiders running on a fixed horizontal cardboard surface. **c–f** Tethered bouts by a 9.5-g tarantula on too-light (10 g), matched (15 g) and too-heavy (30 g, 70 g) styrofoam spheres, in the apparatus of Fig. 7. *Vertical error bars* estimated measurement error (± 0.5 mm/20 ms) on TV-screen. *Dashed lines* span missing measurements due to signal interference. All recordings were made at room temperature (20° – 24 °C)

within an hour. Not all animals show such sparing of leg use. If the operation is not done with great care, even sham operations can result in the leg being held in the air at least for many days.

Tarsus removal. The many tarsal and claw proprioceptors and the sense organs at the metatarsus/tarsus joint, are likely to be stimulated at each footfall (Barth and Geethabali 1982). To ask whether involvement of these inputs can be demonstrated in coordinated stepping, we removed (after anesthetization by chilling) a tarsus plus 1 mm of the distal metatarsus in one leg per animal, and sealed the wound with wax. The results show a variable pattern of slow recovery of leg use for legs 1 and 2, contrasted with the more rapid and complete recovery of use (from 3 hours to 2 days) for legs 3 and 4 (4 *Cupiennius* each). For the more "exploratory" legs 1 and 2, two of eight animals retained clear deficit (holding leg in air during slow walking) for 2 months, while in the others coordination improved progressively over periods of from a few to 18 days.

Hence rather drastic ablations – more severe than Parry's (1960) cutting of the small leg nerve in the femur of *Tegenaria* – can fail to demonstrate dependence on peripheral input. The experiments also underline the difficulties of interpreting and controlling for deficit when it occurs, and the need to observe behavior over days or weeks following such operations.

6 Conclusions: Current Developments and Prospects

Physiological research on spider proprioception and on the neuronal mechanisms involved in the programming and control of movement has only recently begun. So far, there have been no extensive studies systematically combining investigations of the sensory, motor and behavioral aspects of movement control. Nevertheless, applications of newly developed techniques and of well-established morphological and electrophysiological approaches offer some encouraging prospects.

6.1 Afferent Pathways

Given our present knowledge of the distribution and physiology of the various proprioceptor types in the spider leg, it is obviously of interest to ask just where the afferent fibers from particular sense organs project in the central nervous system. In view of the probably multisegmental processing of afferent information and in consideration of the known central organization in other arthropods, will such projections be confined to one side, or to one segment? Selective degeneration procedures and methods for marking sensory axons with cobalt and fluorescent dyes are now being applied to these questions (Babu, Chap. I, this Vol.).

But even the situation in the periphery – that is, at the level of the sensory neurons – may still turn out to be more interesting and complicated than anticipated. There is ultrastructural evidence in the case of the internal joint receptors (Foelix and Choms 1979) of synaptic contacts between presumably efferent fibers and the somata and dendrites of the sensory cells. Apparently hair sensilla and slit sense organs receive similar synaptic input (Foelix, personal communication; see also Foelix, Chap. X, this Vol.). It is therefore conceivable that parameters of proprioceptor function such as sensitivity and gain are under efferent control. Moreover, the preliminary reports mentioned above on the existence of one or several muscle receptor organs in spider legs await fine-structural and electrophysiological confirmation.

6.2 Coordination of Central and Reflex Control

How are muscular contraction and hydraulic extension coordinated to produce movements of the spider's individual leg segments? Methods are now available for measuring the hemolymph pressure in the legs of walking tarantulas with chronically implanted transducers (Blickhan and Barth 1985) and simultaneously recording the muscle activity in the prosoma and the legs. However, reflex connections from particular sense organs to muscles producing and modulating the hydraulic pressure have not yet been described.

A detailed analysis of motor control – that is, the interaction between putative central pattern generators and afferent signals – will require measurements of synaptic events at the level of single interneurons and motoneurons. It is therefore encouraging that extracellular and intracellular recordings from central neurons in the subesophageal complex of spiders are now being pursued, despite the difficult accessibility of the CNS and problems associated with the high hemolymph pressure (Speck, in preparation; Barth, Chap. XI, this Vol.).

Equally urgent is a search for "simple" behavior patterns that lend themselves to such analyses at the level of identifiable central neurons. If at all, intracellular recordings from the CNS will be feasible only in situations in which animals can be tethered and are relatively still. One possibly useful observation (Seyfarth and Thorson, in preparation) is that alert, standing *Cupiennius* can activate leg muscles via resistance reflexes in response to small transient displacements of the substrate. This situation may resemble that encountered by an animal when it stands on a plant leaf moved by the wind.

6.3 Ablation Experiments and Behavioral Deficit

The sensory ablations described above in wandering spiders show that considerable peripheral input can be removed without noticeable long-term change of leg and joint use. This despite the fact that spiders are often considered to depend especially upon reflex control, as compared with insects (Sect 5.3, above). Ablation experiments – the classical approach to clarification of the role particular sense organs play in a behavior – involve fundamental difficulties. If post-operative deficit is seen, then even the best-designed sham operations may not be decisive, and more careful ablations may spare function, as the history of vertebrate deafferentiation demonstrates (see the thoughtful essays by Delcomyn 1980 and DeLong 1971).

If, on the other hand, no deficit is seen following ablation, this is the clearest result, and may indicate the prevalence of central control or a redundancy of proprioceptive input that is not disrupted by the operation. However, in such absence of deficit, the assay for behavioral malfunction may simply not be sensitive enough to detect it. Ablation of lyriform organs in spiders provides a telling example. Destruction of all lyriform sensilla at a joint, or asymmetrical subsets of them, does not result in discernible abnormality of walking or turning tendency (*Cupiennius:* Seyfarth and Barth 1982), not even in films of compensated walking (tarantula: Seyfarth and Bohnenberger 1980). However, a different picture appears when one examines (in *Cupiennius*) the spiders' idiothetic orientation performance – the ability to return to retreats or to lost prey based on "memorized" information about their own previous movements (see Mittelstaedt, Chap. XV, and Görner and Claas, Chap. XIV, this Vol.). Here, ablation of lyriform organs results in a significantly poorer performance than shown by sham-operated and normal animals (Seyfarth and Barth 1972; Seyfarth et al. 1982).

Acknowledgments. Several points made in this article stem from discussions with John Thorson. I thank him and Friedrich G. Barth for their critical advice and encouragement. Johannes Bohnenberger contributed greatly to the design of our tethered walking system. I also appreciate the helpful suggestions of Ann Biederman-Thorson and Rainer F. Foelix. Klaus Hammer and Wolfgang Eckweiler kindly helped prepare the figures for publication. Parts of our research were supported by grants from the Deutsche Forschungsgemeinschaft (SFB 45/A3).

References

Barth FG (1976) Sensory information from strains in the exoskeleton. In: Hepburn HR (ed) The insect integument. Elsevier, Amsterdam Oxford New York, pp 445–473

Barth FG (1981) Strain detection in the arthropod exoskeleton. In: Laverack MS, Cosens D (eds) Sense organs. Blackie, Glasgow London, pp 112–141

Barth FG, Blickhan R (1984) Mechanoreceptors. In: Bereiter-Hahn J, Matoltsy AG, Richards KS (eds) Biology of the integument, vol I. Invertebrates. Springer, Berlin Heidelberg New York, pp 554–582

Barth FG, Geethabali (1982) Spider vibration receptors: Threshold curves of individual slits in the metatarsal lyriform organ. J Comp Physiol 148:175–185

Barth FG, Libera W (1970) Ein Atlas der Spaltsinnesorgane von *Cupiennius salei* Keys. Chelicerata (Araneae). Z Morphol Tiere 68:343–369

Bässler U, Graham D (1978) Zur Kontrolle der Beinbewegung bei einem laufenden Insekt. In: Hauske G, Butenandt E (eds) Kybernetik '77. Oldenbourg, München, pp 54–65

Blickhan R, Barth FG (1985) Strains in the exoskeleton of spiders: J Comp Physiol (submitted)

Bowerman RF (1975) The control of walking in the scorpion. I. Leg movements during normal walking. J Comp Physiol 100:183–196

Bowerman RF (1977) The control of arthropod walking. Comp Biochem Physiol 56A:231–247

Brenner HR (1972) Evidence for peripheral inhibition in an arachnid muscle. J Comp Physiol 80:227–231

Carrell JS (1972) An improved treading device for tethered insects. Science 175:1279

Cavagna GA, Heglund NC, Taylor CR (1977) Walking, running and galloping: Mechanical similarities between different animals. In: Pedley TJ (ed) Scale effects in animal locomotion. Academic Press, London New York, pp 111–125

Dahmen HJ (1980) A simple apparatus to investigate the orientation of walking insects. Experientia 36:685–686

Delcomyn F (1980) Neural basis of rhythmic behavior in animals. Science 210:492–498

Delcomyn F (1982) Insect locomotion on land. In: Herreid CF, Fourtner CR (eds) Locomotion and energetics in arthropods. Plenum Press, New York London, pp 103–125

DeLong M (1971) Central patterning of movement. In: Central control of movement. Neurosci Res Prog Bull 9:10–30

Dixon HH (1892) On the walking of arthropoda. Nature (London) 47:56–58

Eckweiler W (1983) Topographie von Propriorezeptoren, Muskeln und Nerven im Patella-Tibia- und Metatarsus-Tarsus-Gelenk des Spinnenbeins. Diplomarbeit, Fachbereich Biologie, J W Goethe-Universität Frankfurt am Main

Ehlers M (1939) Untersuchungen über Formen aktiver Lokomotion bei Spinnen. Zool Jahrb Abt Syst 72:373–499

Evoy WH, Ayers J (1982) Locomotion and control of limb movements. In: Sandeman DC, Atwood HL (eds) The biology of crustacea, vol IV. Neural integration and behavior. Academic Press, London New York, pp 61–105

Ferdinand W (1981) Die Lokomotion der Krabbenspinnen (Araneae, Thomisidae) und das Wilsonsche Modell der metachronen Koordination. Zool Jahrb Physiol 85:46–65

Foelix RF (1970) Structure and function of tarsal sensilla in the spider *Araneus diadematus*. J Exp Zool 175:99–124

Foelix RF (1982) Biology of spiders. Harvard Univ Press, Cambridge London

Foelix RF, Choms A (1979) Fine structure of a spider joint receptor and associated synapses. Eur J Cell Biol 19:149–159

Foelix RF, Chu-Wang I (1973) The morphology of spider sensilla. I. Mechanoreceptors. Tissue Cell 5:451–460

Fourtner CR, Sherman RG (1973) Chelicerate skeletal neuromuscular systems. Am Zool 13:271–289

Frank H (1957) Untersuchungen zur funktionellen Anatomie der lokomotorischen Extremitäten von *Zygiella x-notata*, einer Radnetzspinne. Zool Jahrb Anat 76:423–460

Fröhlich A (1978) Der Lauf der Trichterspinne *Agelena labyrinthica* Cl. Verh Dtsch Zool Ges 1978:244

Graham D (1983) Insects are both impeded and propelled by their legs during walking. J Exp Biol 104:129–137

Harris DJ, Mill PJ (1977) Observations on the leg receptors of *Ciniflo* (Araneida, Dictynidae). I. External mechanoreceptors. J Comp Physiol 119:37–54

Hayes WF, Barber SB (1967) Proprioceptor distribution and properties in *Limulus* walking legs. J Exp Zool 165:195–210

Herreid CF, Fourtner CR (eds) (1982) Locomotion and energetics in arthropods. Plenum Press, New York London

Hoyle G (1976) Arthropod walking. In: Herman RM, Grillner S, Stein PSG, Stuart DG (eds) Neural control of locomotion. Plenum Press, New York London, pp 137–179

Jacobi-Kleemann M (1953) Über die Lokomotion der Kreuzspinne *Aranea diadema* beim Netzbau (nach Filmanalysen). Z Vergl Physiol 34:606–654

Kaestner A (1924) Beiträge zur Kenntnis der Lokomotion der Arachniden. I. Araneae. Arch Naturgesch 90A:1–19

Linzen B, Gallowitz P (1975) Enzyme activity patterns in muscles of the lycosid spider, *Cupiennius salei*. J Comp Physiol 96:101–109

Mill PJ, Harris DJ (1977) Observations on the leg receptors of *Ciniflo* (Araneida, Dictynidae). III. Proprioceptors. J Comp Physiol 119:63–72

Moffett S, Doell GS (1980) Alteration of locomotor behavior in wolf spiders carrying normal and weighted egg cocoons. J Exp Zool 213:219–226

Markl H (1962) Borstenfelder an den Gelenken als Schweresinnesorgane bei Ameisen und anderen Hymenopteren. Z Vergl Physiol 45:475–569

Page CH (1982) Control of posture. In: Sandeman DC, Atwood HL (eds) The biology of crustacea, vol IV. Neural integration and behavior. Academic Press, London New York, pp 33–59

Palmgren P (1981) The mechanism of the extrinsic muscles of spiders. Ann Zool Fenn 18:203–207

Parry DA (1960) The small leg-nerve of spiders and a probable mechanoreceptor. Q J Microsc Sci 101:1–8

Pflüger H-J, Bräunig P, Hustert R (1981) Distribution and specific central projections of mechanoreceptors in the thorax and proximal leg joints of locusts. II. The external mechanoreceptors: hair plates and tactile hairs. Cell Tissue Res 216:79–96

Pringle JWS (1938) Proprioception in insects. III. The function of the hair sensilla at the joints. J Exp Biol 15:467–473

Rathmayer W (1966) Die Innervation der Beinmuskeln einer Spinne, *Eurypelma hentzi* Chamb. (Orthognata, Aviculariidae). Verh Dtsch Zool Ges 1965:505–511

Rathmayer W (1967) Elektrophysiologische Untersuchungen an Propriorezeptoren im Bein einer Vogelspinne (*Eurypelma hentzi* Chamb.). Z Vergl Physiol 54:438–454

Rathmayer W, Koopmann J (1970) Die Verteilung der Propriorezeptoren im Spinnenbein. Untersuchungen an der Vogelspinne *Dugesiella hentzi* Chamb. Z Morphol Tiere 66:212–223

Ruhland M (1976) Untersuchungen zur neuromuskulären Organisation eines Muskels aus Laufbeinregeneraten einer Vogelspinne (*Dugesiella hentzi* Ch.). Verh Dtsch Zool Ges 1976:238

Ruhland M, Rathmayer W (1978) Die Beinmuskulatur und ihre Innervation bei der Vogelspinne *Dugesiella hentzi* (Ch.) (Araneae, Aviculariidae). Zoomorphologie 89:33–46

Selverston AI (1980) Are central pattern generators understandable? (including open peer commentaries). Behav Brain Sci 3:535–571

Seyfarth E-A (1978a) Lyriform slit sense organs and muscle reflexes in the spider leg. J Comp Physiol 125:45–57

Seyfarth E-A (1978b) Mechanoreceptors and proprioceptive reflexes: lyriform organs in the spider leg. Symp Zool Soc Lond 42:457–467

Seyfarth E-A, Barth FG (1972) Compound slit sense organs on the spider leg: mechanoreceptors involved in kinesthetic orientation. J Comp Physiol 78:176–191

Seyfarth E-A, Bohnenberger J (1980) Compensated walking of tarantula spiders and the effect of lyriform slit sense organ ablation. Proc Int Congr Arachnol 8:249–255

Seyfarth E-A, Pflüger H-J (1984) Proprioceptor distribution and control of a muscle reflex in the tibia of spider legs. J Neurobiol 15:365–374

Seyfarth E-A, Eckweiler W, Hammer K (1985) Proprioceptors and sensory nerves in the legs of a spider, *Cupiennius salei* (Arachnida, Araneida). Zoomorphology (in press)

Seyfarth E-A, Hergenröder R, Ebbes H, Barth FG (1982) Idiothetic orientation of a wandering spider: compensation of detours and estimates of goal distance. Behav Ecol Sociobiol 11:139–148

Weber T, Thorson J, Huber F (1981) Auditory behavior of the cricket. I. Dynamics of compensated walking and discrimination paradigms on the Kramer treadmill. J Comp Physiol 141:215–232

Wendler G (1978) Lokomotion: das Ergebnis zentral-peripherer Interaktion. Verh Dtsch Zool Ges 1978:80–96

Wilson DM (1966) Insect walking. Annu Rev Entomol 11:103–122

Wilson DM (1967) Stepping patterns in tarantula spiders. J Exp Biol 47:133–151

XIII Target Discrimination in Jumping Spiders (Araneae: Salticidae)

LYN FORSTER

CONTENTS

1 Introduction

Of all the families of spiders, the Salticidae or jumping spiders undoubtedly possess the finest eyesight (Homann 1928, 1971; Land 1969a, b, 1972, 1974 and Chap. IV, this Vol.). This is reflected in their ability to detect and localise moving objects in their environment (Land 1971; Duelli 1978), to chase, creep up on and jump at prey (Drees 1952; Gardner 1964, 1966; Forster 1977, 1979a, 1982a), to recognise conspecifics (Crane 1949; Jackson 1982; Forster 1982b)

Otago Museum, Great King Street, Dunedin, New Zealand

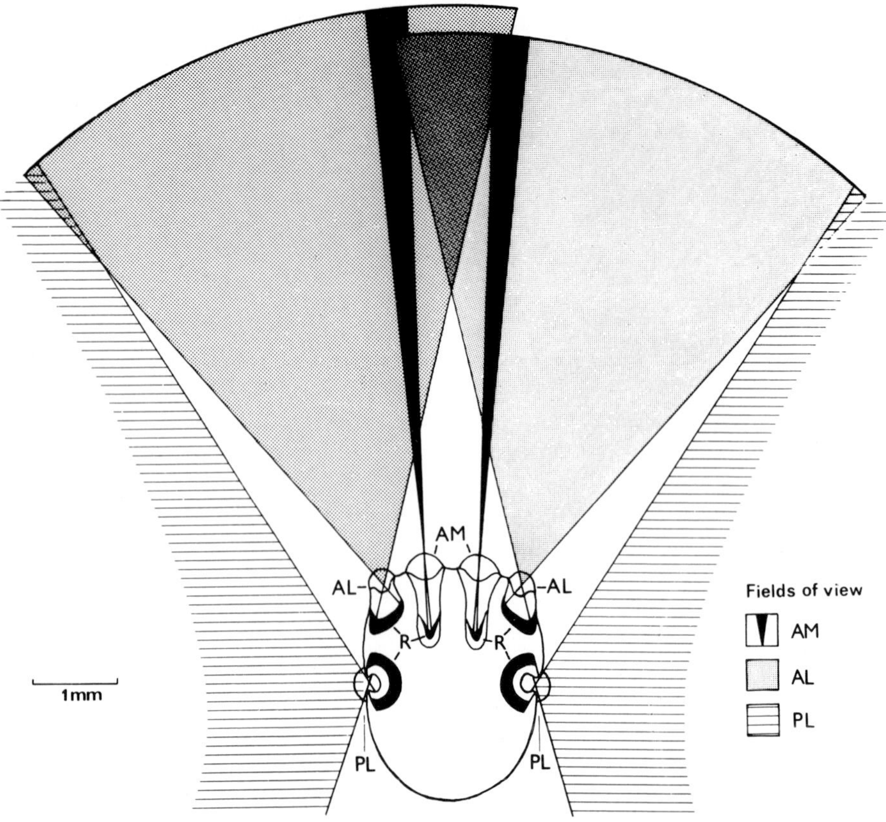

Fig. 1. Horizontal section through the head of a jumping spider showing the position, relative size, and fields of view of six of the eight eyes (diagramatic). The anterior-median (*AM*) or principal eyes have narrow fields of view but they are extended horizontally and vertically by conjugate eye movements (see Land, Chap. IV, this Vol.). The anterior-lateral (*AL*) fields overlap in front by about 25° while the posterior-lateral (*PL*) eyes survey an area of about 130° on either side of the spider. Note the elongated capsules of the principal eyes and the small pits to the rear where the layers of the retinae (*R*) lie. (Forster 1979a)

and to distinguish between prey and prospective mates (Drees 1952; Land 1974; Forster 1979a, 1982b). Six of their eight eyes (Fig. 1) mediate these tasks. The lateral eyes are primarily long-range movement detectors (Land 1971) while the peripheral regions of the anterior pair mediate and control chasing behaviour (Forster 1979a); the principal – anterior median – eyes are used in near-field approach tactics and target discrimination (Forster 1982b). For details of the structure and optics of the eyes see Land, Chapter IV, and Blest, Chapter V, this Volume.

Several features of the principal eyes – the long focal length of the lenses, the four-layered retinae and their respective chromatic sensitivities, the central foveal regions of greater receptor density in the two deepest layers, and the motility of the eye capsules – combine to mediate a range of visual behaviours often likened to those of some amphibians, birds and mammals (Crane 1949; Land 1974; Forster 1982a, b). To be successful in such behaviours, however, a jumping spider must be able to discriminate between objects in its environment. What are the visual properties which signify prey or an enemy, for instance? By what features are mates identified, and how are these features processed on the retina?

2 Visually Mediated Behaviour

2.1 Non-Discriminatory Reactions

Although the foremost tasks of a jumping spider – to escape, to catch prey, and to mate – appear to be fundamentally different, common elements limit the behavioural options involved. If, for instance, there is a movement in its surroundings, the spider swivels towards it, provided that at least two closely spaced receptors of the lateral eyes have been stimulated (Land 1971; Duelli 1978). It is immaterial whether the turn is elicited by the motion of an enemy, prey, or potential mate.

In principle, after orienting, the spider should identify the target before the subsequent behaviour is initiated, but in practice, a great deal depends on the mobility of the target. If the target departs rapidly, the spider will chase it, often persisting, in the case of *Trite planiceps*, even when the quarry is 30 cm away. Moreover, the target does not need to be identified with any certainty, except that it should be not more than twice the size of the spider itself (Forster 1979b). During the chase, which is mediated by the anterior-lateral eyes (Forster 1979a), the principal eyes track the target.

Tracking, one of four eye movements described by Land (1969b), enables the central regions of the two motile principal retinae to remain fixated on the target.

If, however, the intended target runs sharply towards the jumping spider, the jumping spider may turn and run, or scuttle backwards, a reaction probably requiring only that a large part of the retina is suddenly stimulated. In either chasing or retreating, the intended target could be prey, a conspecific and perhaps a mate, or another animal altogether, but the target acts of running away from or dashing towards provoke responses from jumping spiders not consistent with any acute discriminatory processes.

2.2 Discriminatory Tactics

Given a suitably sized target which does not loom up, or depart rapidly, a jumping spider advances either in a predatory mode (if the target has appropriate prey features) or in a courtship or agonistic mode (if the target has appropriate conspecific features). Thus discrimination hinges on the visual properties which define these target classes and the jumping spider's perception of them.

Fig. 2. A male salticid raises his forelegs in reaction to his image in a mirror. This raised foreleg stance is common to a number of salticid species while mirror image performances, which are almost identical to inter-male interactions (see Forster 1982b), testify to the visual nature of these responses

Of the eight eyes found in jumping spiders, only two, the principal – anterior median – pair (Fig. 1), are responsible for discriminating between stationary or almost stationary targets (Forster 1979a, 1982a, b). Land (1969b) showed that the retinae of these eyes scan back and forth and rotate across the image of an immobile object, thus facilitating identification and preventing fixation blindness.

Since prospective mates, other adult conspecifics, as well as prey are likely to be stationary or quasi-stationary during approach, jumping spiders must be able to distinguish between them. Distinctive behaviours exhibited with remarkable consistency towards these classes of target demonstrate this. (1) In response to prey, the spider advances, stalks, crouches and jumps (Forster 1977); (2) in response to a female, the male spider approaches with species-specific postures and movements (Crane 1949); (3) in inter-male encounters (Fig. 2) equally characteristic acts are performed (Forster 1982b) but if the principal eyes are occluded, the spider can do none of these things (Homann 1928; Crane 1949; Drees 1952; Forster 1979a).

3 Spatial Determinants of Perception

Three spider-target distances have a pronounced effect on behaviour. These are the distances at which (1) orientation, (2) target discrimination, and (3) jump (if prey) or Leg-frontal (if mate) (see Forster 1982b) occur.

3.1 Size and Developmental Contingencies

Although, to a large extent, reactive distances vary according to the size of the spider, *Euophrys parvula* males, for example, recognise mates from almost the same distance as the very much larger *Trite planiceps* (Forster 1979a and unpublished). Nor is there always a linear relationship between size and reactive distances at different ages. During development, jump-at-prey distances in *T. auricoma* increase by 20% relative to body length (Forster 1977) while recognition of prey, which reliably begins at 20 mm in the young spider, is extended to some 20 cm in the adult (Forster 1979b) thus demonstrating that the target-recognition distance has undergone a tenfold increase, while body length has barely quadrupled.

Since discrimination distances are related to the resolution of the eye, contributing optical parameters in the principal eye apparently develop allometrically, thus providing enhanced resolution and increased distance perception for the mature spider. This would be especially advantageous to the adult during sexual encounters, not only allowing for the greater probability of correct discriminations but also, by enlarging the spatial and temporal dimensions of approach, for facilitating the multiple functions of courtship (Forster 1982b).

3.2 Peripheral Detection of Targets

Orientation responses depend less on specific spider-target distances than on the angular subtense between adjacent receptors in the lateral eyes, as well as the mobility of the target. Land (1969b) showed that in *Metaphidippus aeneolus*, for example, receptor spacing is ca. 1° in the posterior-lateral eyes and that an object must move through a minimum distance of 1° at a velocity between $1° - 100°\ s^{-1}$ to be effective. Bearing in mind the diversity of prey (various larvae, ants, moths, collembola, aphids, leafhoppers, webworms, a large range of flies as well as other spiders) that jumping spiders are known to catch, it is clear that the level of activity plays a part in the distance at which peripheral detection occurs.

3.3 Discriminatory Function of the Principal Eyes

Once the spider has oriented, the distance separating it from the stimulus has a significant bearing on the spider's capacity to determine the nature of it. *Trite planiceps*, for example, can consistently distinguish between prey and a potential mate at 20 cm, but recognition of conspecifics by *Portia schultzii*, a kleptoparasitic salticid which invades other spiders' webs (Murphy and Murphy 1983), is limited to ca. 10 cm, nor do they respond to prey items at distances greater that this (Forster 1982a; Forster and Murphy, in review).

Another species, *Portia fimbriata*, is reported to discriminate between prey and potential mates at distances of up to 27 cm (Jackson and Blest 1982b), the improved resolving power of the principal eye accruing from a diffraction-limited corneal lens, and a pit lens which effectively increases the focal length of the corneal lens by one and a half times (to 1701 μm), as well as closely spaced receptors in the central area of layer 1 of the retina corresponding to a visual angle of 2.4 arc min (Williams and McIntyre 1980). This means that *P. fimbriata* has an acuity considerably better than Land (1969a) found for *M. aeneolus* (11′) and *P. johnsoni* (9′) although it is not known whether these species also possess optically useful pit lens. By implication, however, because of the much-reduced discriminatory distance of *P. schultzii,* the resolution of its principal eyes must be significantly poorer.

At 3 or 4 cm from stationary prey, the jumping spider crouches and jumps; at much the same distance, males cease their precursory overtures and direct their lowered forelegs towards the female. What causes these changes in behaviour at such discrete and predictable distances?

3.4 Depth Perception

One explanation for these spatially accurate behavioural changes is that they correspond with the projection of an appropriate image on to pre-programmed receptors in each of the anterior-lateral eyes within their regions of binocular overlap (see Fig. 1), a mechanism of depth perception referred to as the intersection theory and discussed in detail by Burkhardt et al. (1973). This proposition is supported by Blest et al. (1981), who further argue against the possibility of the retinal "staircase" of the principal eye (see their Fig. 8) functioning as a range-finder, since the depth of field at all points on the staircase is too large. Very probably the anterior-lateral eyes do act in such a capacity to keep a target within range while it is being chased (see Sect. 2.1) and perhaps contribute to spatial information at specific distances. But Forster (1979a) also showed that, when the anterior-lateral eyes of *T. planiceps* were occluded, these spiders still stalked their prey, or courted their mates from about 20 cm, hence the intersection proposal is not the full story. Evidently the principal eyes themselves possess a mechanism for depth perception despite their lack of binocularity, convergence, and accommodation (Homann 1928; Land 1969a, b).

There are numerous reports of salticids which creep up on and jump at dead or immobile objects (Heil 1936; Drees 1952; Dill 1975), either normally as in *P. schultzii* (Forster 1982a), or occasionally as in *E. parvula* (Forster and Forster 1973). Without doubt, it is the side-to-side motility of the principal retinae that allows stationary or dead flies (in the case of *E. parvula*) or quiescent spiders (in the case of *P. schultzii*) to be stalked and pounced upon, particularly at short range, since this makes near objects appear to move faster than more distant objects, a phenomenon known as motion parallax (see Wehner 1981). The effect generates both "movement stimulation" and "depth cues", more especially when horizontal disparities are large, i.e. when the target is close.

It seems likely, therefore, by virtue of the binocularity of the anterior-lateral eyes, that jumping spiders make use of the intersection technique for longer-range distance judgments but, by virtue of the motility of the principal retinae, that they rely mainly on motion parallax for short-range depth perception. In practice, intersection range-finding may only be useful when the spider is chasing a target moving at high angular velocities (see Sect. 5.3.2), whereas once the target is stationary or quasi-stationary, perhaps the spider only needs to know "what it is" and not "how far away it is" *until* it is close enough for motion parallax to come into effect. The fact that *T. planiceps*, for example, can make discriminations up to 20 cm, may merely reflect the limits of its spatial resolution and not its capacity to assess target distance at that stage of the proceedings.

3.5 Ambient Illumination

In both *P. schultzii* and *T. planiceps*, spatial discrimination progressively diminished to 30–40% of the original distances when ambient illuminances were reduced from ca. 1300 lx to less than 50 lx (Forster 1982a and unpublished). Tests showed (Forster, in preparation) that even when restricted to light intensities of ca. 10 lx and wavelengths of $\geqq$ 580 nm, all 15 *T. planiceps* test spiders stalked and caught prey visually, albeit very slowly and with prolonged pauses (mean pursuit time $5.3' \pm 6.9'$), staring at the target and occasionally rotating the prosoma from side to side. Of interest, too, is that under these conditions, peripheral swivels rarely occurred (and those that did were inaccurate), spiders mostly orienting to prey in the frontal fields of view with a series of "tantalus-like turns" (as described by Land 1974) presumably mediated by the principal eyes. At wavelengths of $\geqq$ 600 nm (and 10 lx), 93% of test spiders caught prey (mean pursuit time $8.2' \pm 8.6'$), but there were no peripheral swivels and no chases, stalks began at < 6 cm and a number of spiders repeatedly jumped "short", subsequently seizing their prey with an additional jump or lunge. In complete darkness, however, 88% of spiders seized prey, probably by means of vibratory cues (Forster 1982c).

These results suggest that only the principal eyes can function in light of longer wavelengths and/or at reduced retinal illuminances and that depth perception is apparently affected. Whether this means that one or more of the retinal layers is unable to function under these circumstances, or whether this is evidence for some receptors having a spectral sensitivity of ca. 600 nm are matters of much interest. However, Blest et al. (1981, Chap. V, this Vol.) demonstrated that receptors in layer 2 and the peripheral regions of the homogeneous receptor population of layer 1 in *Plexippus validus* contain only green receptors peaking at ca. 520 nm. If this is also true of the *T. planiceps* eye, perhaps the "inferior" pursuit and jump performances merely reflect the minimal sensitivities of these receptors under the prescribed experimental conditions, although one might have expected that the lateral eyes, with receptors peaking at 535 nm (Hardie and Duelli 1978), would also remain reactive (see also Yamashita, Chap. VI, this Vol.). More tests along such lines are required to interpret these findings in greater detail.

Since the principal eyes have optical qualities which bestow some surprising capabilities upon these spiders, it is worthwhile looking briefly at the determinants of such visual skills.

4 Optical Components of the Principal Eye

4.1 Resolution of the Eye

The lens system is responsible for image formation, but the quality of this image is established by the diffraction limit as dictated by the size of the aperture as well as aberrations of the optical system. Resolution of the eye is determined by the quality of the image produced and the fineness of the receptor grain upon which this image is projected (for a full treatment of this topic see Land 1981, and Chap. IV, this Vol., and Wehner 1981).

4.2 Refractive-Index Distribution of the Lens

Multiple Beam Interference (MBI) methods developed by Vickridge (1982) with reference to the work of Glauert (1951), Hunter and Nabarro (1952), and Tolansky (1973), indicate that the principal corneal lenses of *T. planiceps* (Fig. 3a) have a refractive-index distribution ranging from 1.385 just inside the

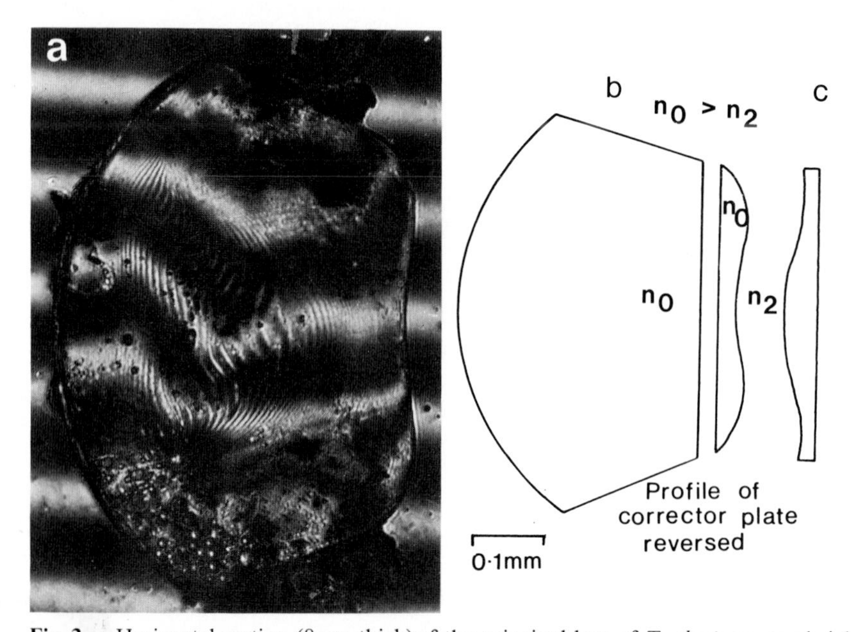

Fig. 3. a Horizontal section (8 μm thick) of the principal lens of *T. planiceps*, sandwiched between two reflective plates, showing the fringe modulations in the core of the lens. These modulations were shown to result from a refractive index gradient in the lens. (By courtesy of I. Vickridge.) **b** Diagram of horizontal section through principal eye showing rear lens portion with curvature separated from the main lens. n_0 average refractive index from Multiple Beam Interference; n_2 refractive index of vitreous (or spider ringer). (Courtesy of I. Vickridge.) **c** Schmidt corrector plate with curvature which resembles that of rear curvature of lens shown in **b** except that the curvature is reversed. Note that the figurings in the Schmidt plate are greatly exaggerated. (Born and Wolf, 1980)

cornea to a minimum value of 1.36 towards the centre, i.e. there is a lower refractive index at the core than at the periphery in contrast to the lenses of some other salticids as shown by Williams and McIntyre (1980). The MBI method also showed, on average, a smaller refractive index (1.37) than that calculated by the hanging drop (1.39), the method of measuring the focal length of a lens first used by Homann (1928, see also Land, Chap. IV, this Vol.).

The two surfaces of the lens, characterised by the radii of curvature (r_1, the corneal surface and r_2, the rear surface) separate three media (n_0 = refractive index of the lens, $n_1 = 1$ [air], $n_2 = 1.335$ [vitreous or spider ringer]). The lens has a thickness (d), and the general formula for the focal length (f) of the system is given by the following formula (Born and Wolf 1980):

$$\frac{1}{f_2} = \frac{n_0 - n_1}{r_1} + \frac{n_2 - n_0}{r_2} - \frac{d\,(n_0 - n_1)\,(n_2 - n_0)}{n_0\, r_1\, r_2}\,. \tag{1}$$

The profile of the rear surface of the lens shows considerable variation between species, being convex in *P. johnsoni* and *M. harfordi* (Eakin and Brandenburger 1971), somewhat flatter in *M. aeneolus* (Land 1969a) and slightly concave in *P. fimbriata* (Blest et al. 1981). Figure 3a shows that the rear surface of the *T. planiceps* lens is also slightly concave.

When this concave surface is modelled (Vickridge 1982), assuming r_2 to be 1500 μm and using Eq. (1) where $r_1 = 380$ μm, $d = 420$ μm and $f_2 = 993 \pm 2.9$ μm, the refractive index is shown to lie between 1.38 and 1.405 when the error limits of f_2 are taken into account. When a rear flat surface is assumed (r_2 = infinity), the refractive index lies between 1.37 and 1.39, values more consistent with those obtained from MBI methods than from the hanging drop. Thus the simple homogeneous model best assumes a rear flat surface. What effect then does the rear curvature have?

Vickridge (1982) noted that this profile in the *T. planiceps* lens (see Fig. 3b) is strongly reminiscent of the Schmidt corrector plate (Fig. 3c) (Born and Wolf 1980) which is used to pre-correct plane wavefronts entering an optical system and so compensating for spherical aberration. However, in the spider lens it is effectively reversed and presumably, therefore, would serve to increase spherical aberration.

Ray tracing through two models of the lens, the first assuming a homogeneous refractive index structure and flat rear surface, and the second using the refractive-index distribution and curvatures found in *T. planiceps* (Vickridge 1982), showed that image quality, as judged by the "circle of least confusion" for axial rays of no more than 0.188 radians, was much worse for the second model than for the first. This is further support for the view that the principal lens in this spider is working to increase spherical aberration.

Land (1969a) concluded that resolution of the principal eyes in *M. aeneolus* and *P. johnsoni* is probably limited by spherical aberration of the corneal lenses. In *P. validus* the limiting aperture is relatively narrow so that spherical aberration is not a problem and the eye is assumed to be diffraction-limited (Blest et al. 1981). However, when Williams and McIntyre (1980) examined the corneal lenses of *P. fimbriata* and several other salticids (not identified) in a hanging drop and by interference microscopy, they found that spherical aberration is corrected by a graded refractive-index distribution (core greater than periphery) in the lens, an optical device also adopted by some nocturnal spiders (Blest et al. 1981).

If, as Vickridge (1982) suggests, the principal lens of *T. planiceps* has been designed to promote spherical aberration, then some other factors, such as a trade-off between tolerable distortions of the image and depth of focus, for example, might have been at work. That this lens does, in fact, possess a large depth of focus is illustrated in Fig. 4a and b. Seemingly, there are marked inter-specific differences in the refractive-index distribution of the principal lens in this family, with varying consequences for image quality.

4.3 Properties of the Retina

The ability of a receptor mosaic to resolve an image depends on the distance between adjacent receptors as well as the focal length of the eye, and this is usually expressed as the visual angle. In the central region of layer 1 (Fig. 5) in the salticid eye, Land (1981) notes that the 2-μm receptors are packed quite closely, to the point where a greater density would impair resolution; in *P. johnsoni,* for example, the visual angle is 11 arc min. (Land 1974). However, Blest et al. (1981) found that the long receptors of this high resolution mosaic in both *P. validus* and *P. fimbriata* are even more tightly packed, but that they are constructed in a way which suggests that they function as light guides, a necessary condition if good resolution is to be achieved. In the latter species, moreover, resolution has been enhanced, not only by these innovations, but also by the use of a telephoto system which has effectively increased the focal length of the corneal lens some one and a half times (Williams and McIntyre 1980). Presumably retinal illuminance would have been reduced by these developments, a constraint no doubt aggravated by the apparent preference of *P. fimbriata* for shady habitats (Jackson and Blest 1982a). To some extent, this problem may have been overcome by the very slow reaction rates and locomotion in this species, a ploy which would serve to increase the photon sampling time of the receptors (Snyder 1977). Interestingly, slothfulness is found in other species of this genus which are apparently not confined to such dimly lit habitats; in *P. schultzii,* for example, its usefulness seems to relate to the spider's kleptoparasitic habits and the need for caution, rather than compensating for extended spatial thresholds and hence reduced ambient illuminance.

5 Visual Attributes of the Target

The distinctive sets of reactions displayed by jumping spiders to various objects in their environment is evidence of their ability to distinguish between them. Visual attributes of the target clearly include size, movement and shape, but just what dimensions do these attributes have and how are they perceived? Does colour play any part in interactions between these spiders?

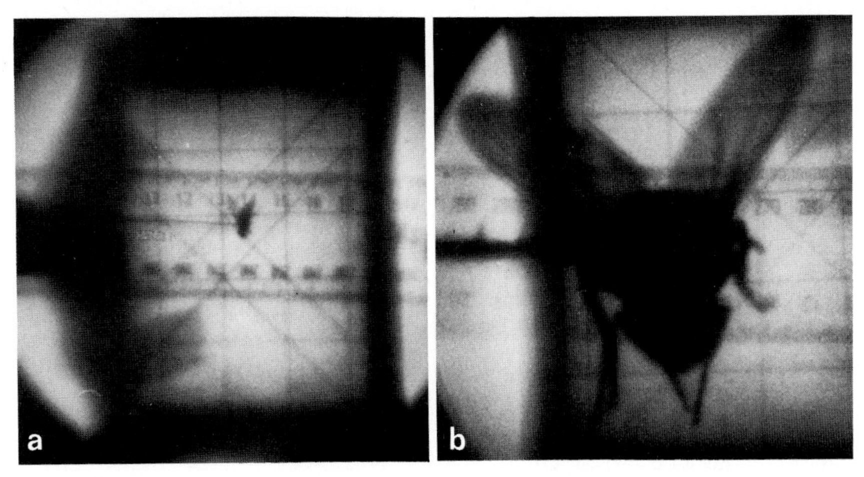

Fig. 4. **a** The image of a house fly formed by the principal lens of *T. planiceps* in a hanging drop. The fly was held against a ruler with 1 mm divisions and this was positioned at 35.6 cm in front of the lens. At this distance the lens is shown to be capable of resolving points about 0.5 mm apart, giving an angular resolution of about 5 arc min. **b** The fly has been re-positioned at 2.5 cm in front of the lens. Even at close range the whole fly is pictured, showing that the lens is surprisingly wide-angled. (Courtesy, I. Vickridge)

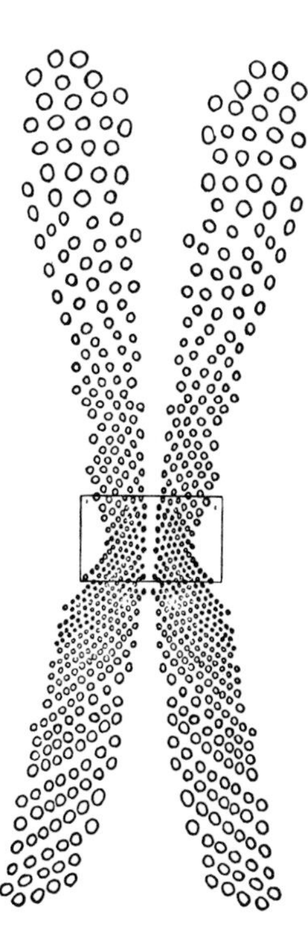

Fig. 5. The two retinae of the most proximal layer 1 of the principal eyes of *M. aeneolus* as they would appear in an ophthalmoscope. *Boxed area in the centre* encloses the area of higher receptor density referred to as the fovea. (After Land 1969 a)

5.1 Spectral Sensitivities of the Principal Eyes

The use of colour vision has been an attractive proposition ever since the Peckhams (1894) described the brightly coloured secondary sexual characters and epigamic displays of mature jumping spiders. This proposition was greatly strengthened when Land (1969a) discovered a four-layered retina in the principal eye of *P. johnsoni* and realised its potential for responding preferentially to light of different wavelengths by virtue of the spatial relationships of the layers to the longitudinal chromatic aberration effects of the corneal lens. Support for such a function in the principal eye came from De Voe (1975) and Yamashita and Tateda (1976a).

Recently Blest et al. (1981) demonstrated the absorption of wavelengths in the region of 520 nm (green) and 360 nm (ultra-violet) by different retinal layers but could find no evidence for long wavelength receptors (see also Blest, Chap. V, and Yamashita, Chap. VI, this Vol.). Nevertheless, many behavioural studies suggest that, in some species at least, long wavelengths are significant.

For example, the Peckhams (1894) tested the effectiveness of coloured markings in eliciting courtship in many species by obliterating them, Crane (1949) demonstrated the necessity of yellow patches on black and white models for releasing sexual displays in *Corythalia xanthopa*, and Kaestner (1950) found that, under appropriate conditions, *Evarcha falcata* could quite readily discriminate blue or orange stripes from grey. Consider, moreover, the green world of the *Lyssomanes* spider, and the flashing of scarlet-emblazoned palps during courtship (Forster, personal observation). Surely these attributes have visual significance?

5.2 Size of the Target

Tests with different sizes of prey, conspecifics, and two-dimensional models (see Sect. 5.4.1) established that a target lost its attractiveness and elicited retreat when it was more than one and a half times the size of the spider (Forster 1979b). Furthermore, *T. planiceps* manifested an ability to assess the absolute size of the target by always jumping at house flies and the very much smaller fruitflies from the same distance. However, Drees (1952) calculated the upper limits for prey size in *Salticus scenicus* as being equivalent to that of the spider itself but he also showed that this size limit increased as the spider's hunger intensified.

5.3 Movement of the Target

Since most jumping spiders will not react to a target, regardless of its apparent attractiveness, unless it has moved, it is not easy to separate the influence of movement from the influence of shape.

5.3.1 Stimulus Characteristics

In a wide-ranging series of experiments (Forster 1979a, b; 1982a, and unpublished), it was found that almost all test species chased (but did not stalk) and jumped at a variety of fast moving, inanimate lures of suitable size, regardless of their shape and colour. However, although motionless lures elicited stalking (but not chasing), only the "dead fly" lure induced the spider to complete the sequence by crouching and jumping. Nevertheless, if any lure not exhibiting directional movement was jiggled or twirled (quasi-stationary), spiders ran and jumped at it repetitively.

One of the exceptions mentioned above was *Portia schultzii,* a web-building species (Murphy and Murphy 1983; Forster 1982a), which did not discriminate between the inanimate and dead fly lures, stalking and lunging at all of them, provided that they were stationary, or nearly so. They seemed less affected by jiggling or twirling except that their progress towards a target was more consistent. In contrast to other salticids, moreover, they did not chase any lures, nor did they chase live, active prey.

Drees (1952), who conducted similar experiments with *Salticus scenicus,* concluded that, whereas size, movement and shape were all individually involved in triggering prey capture, response intensity (chasing ... sic) was increased by the sum of these stimulants, rather than any intrinsic properties of the prey, and that sharp vision was only necessary at distances of less than 6 cm. His conclusions were based on the belief that all post-detection behaviours are mediated by the principal eyes as implied from "blinding" tests by Homann (1928) and Crane (1949), so he endeavoured to account for his findings by ascribing all observed capabilities to the principal eyes. This, as we shall see, is not necessarily the case.

5.3.2 Angular Velocities of the Target

The solution to the problem of interpreting what often seem to be contradictory observations lies in the division of labour not just between the two anterior pairs of eyes (the principal and the anterior-lateral eyes) but also between different regions of the anterior-lateral eyes (Forster 1979a). Over much of their retinae, anterior-lateral receptors mediate turns (Land 1971) but the peripheral receptors with visual fields directed towards the frontal midline of the spider (see Fig. 1) apparently have another role (Forster 1979a). When a target is moving at more than ca. $4° s^{-1}$, it is probable that these functionally specialised regions mediate chasing behaviour, in which case discrimination of a moving target requires no more than an assessment of size. When the target is moving at less than $4° s^{-1}$, however, the principal eyes mediate behaviour; under such circumstances, directional activity is relatively insignificant while shape is a critical variable.

These findings, from studies in which the four anterior eyes of *T. planiceps* were systematically occluded (Forster 1979a), help to explain results from a number of investigations. For example, Heil (1936) found experimentally that size, shape and jerky movements were all instrumental in effecting prey capture, but he was unable to explain why *E. falcata* often stalked and jumped at dead flies, usually only from a few centimetres away. This apparent inconsistency can be accounted for if we suppose that Heil experimented with targets moving at different angular velocities, so that the observed responses were sometimes mediated by the anterior-lateral eyes and sometimes by the principal eyes.

One explanation for *P. schultzii's* inability to chase prey in the manner of most salticids (see Sect. 5.2) is that their anterior-lateral eyes do not function in the manner described here, perhaps because of a limited or non-existent overlap between these two eyes, or perhaps because the appropriate receptors have not acquired (or no longer possess) the necessary specialisation, possibilities that warrant investigation.

5.4 Shape of the Target

The problem of how animals recognise shapes and patterns and differentiate between them is of interest to neurobiologists and ethologists alike. To this end, many investigators have made use of predictable behaviour in an animal and, by presenting it with a range of two-dimensional black and white models, have attempted to measure various stimulus parameters from the reactions obtained. For example, notable studies of shape discrimination have been undertaken with octopuses (Sutherland 1960 et seq.), toads (Ewert et al. 1978), and insects such as desert locusts (Wallace 1958) and honey bees (Wehner 1967 et seq.; Cruse 1972a, b; Anderson 1977a, b, c).

Jumping spiders are excellent candidates for such methods, yet most of the information we have today has come from the work of just two people, Crane (1949) and Drees (1952). More recent investigations (Forster 1979b), in which 30 *T. planiceps* spiders were presented with "potential" prey shapes, are outlined here; it is hoped that they will lead to more extensive and systematic studies.

5.4.1 Experiments with Two-Dimensional Shapes

Thirty black, two-dimensional shapes which had an 80% stimulus contrast when attached to the white interior surface of a cylinder (27 cm diameter) were rotated at 10 mm s^{-1} and either 25 mm s^{-1} or 50 mm s^{-1} around the freely moving spider. Because angular velocities of the target vary inversely with spider-target distances, a decrease in this distance leads to an increase in angular velocity. But in the rotating cylinder, when within about 6 cm of the target, spiders generally sidled along facing the target, thus keeping the target more-or-less motionless relative to the principal eyes. However, spiders only sidled at slow target speeds; at faster speeds a greater percentage of chases and jumps were elicited by relatively unattractive targets, thus suggesting that discrimination was less accurate.

Only 12 of the 30 shapes are analysed here (see Table 1), but those not illustrated include smaller and larger versions of the discs, ellipses and squares (results of these tests are summarised) as well as a range of vertical and horizontal stripes.

Prey-catching events used to evaluate shape parameters were Capture (the spider jumped at the target) and Retreat (the spider scuttled backwards or turned and ran). From the results (Table 1) it was found that at a speed of 10 mm s^{-1}, shapes 1, 2, 4, 5 and 7 were significant ($p < 0.001$) in eliciting capture responses. Smaller discs and ellipses elicited relatively fewer responses (Fig. 6a, b); moreover, once the vertical dimension of a target became greater

Table 1. Response data for *Trite planiceps* when presented with test shapes moving at the stated speed

Shape	N	Speed mm s^{-1}	% C	% R	Speed mm s^{-1}	% C	% R
1	10	12	60	0	50	70	0
2	20	12	95	0	50	80	0
3	14	12	40	0	–	–	–
4	15	12	66	0	25	86	0
5	14	12	64	0	25	64	0
6	10	12	0	40	–	–	–
7	14	12	60	0	–	–	–
8	10	12	0	70	50	0	70
9	15	12	0	30	–	–	–
10	20	12	5	0	50	40	0
11	20	12	0	0	50	40	0
12	27	12	0	0	50	35	0

N: Number of spiders tested.
% C: Percentage captures (spiders jumped at the target).
% R: Percentage retreats (spiders retreated from the target).

than 5 mm, or its horizontal component greater than 7 mm, capture responses diminished and gradually retreat reactions were substituted. At the lower speed the solid ellipse (no. 2) was the most successful shape, but at 25 mm s^{-1} the fly shape (no. 4) evoked the greatest capture percentage (86%). However, when a hollow ellipse (no. 3) was presented, the capture percentage was markedly reduced indicating that solid figures are preferred.

Of three fly shapes (nos. 4, 5 and 6) presented, no. 4 was positioned so that, during rotation of the cylinder, its legs were masked by a grating, an effect that simulated "running". This proved most effective at a target speed of 25 mm s^{-1} when a capture percentage of 86% was recorded. A comparison of the percentage captures of this shape with the subsequent fly shape (no. 5) shows no appreciable difference at the lower speed but a significant difference ($p < 0.001$) at 25 mm s^{-1}, a finding which implicates the additional stimulation of the "running legs" in the reaction rate. Hence "flicker" played a part when targets were chased (as opposed to stalked) for at this speed sidling did not occur and since angular velocity was greater than 4° s^{-1}, "flicker" apparently had an effect on

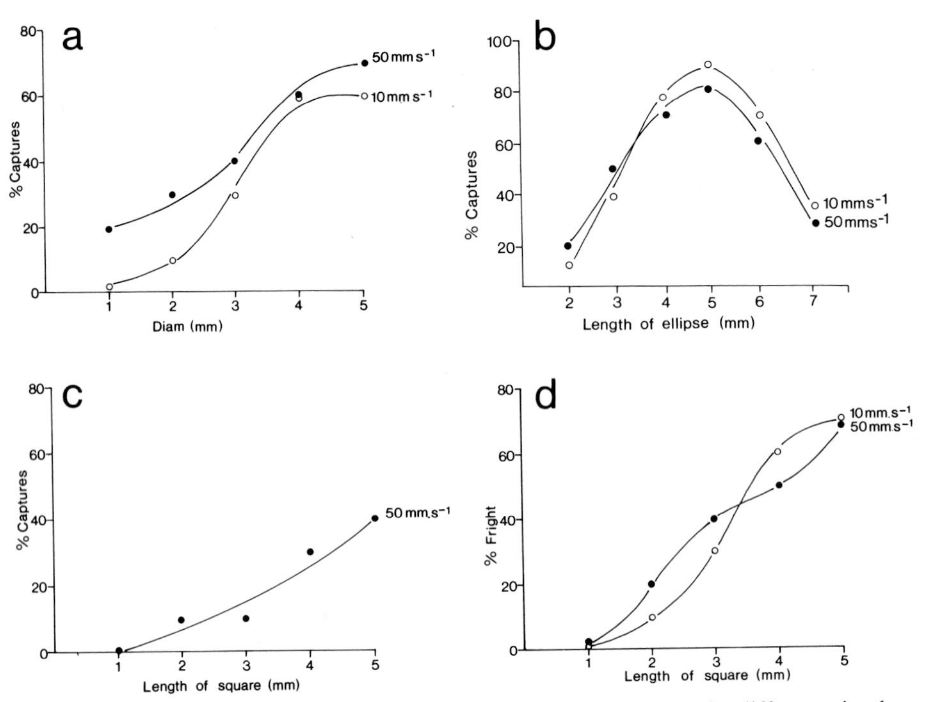

Fig. 6 a – d. Response percentages, obtained from tests with *T. planiceps* for different stimulus speeds (mm s^{-1}), plotted against the size (mm) of stimulus shapes. Captures are shown as a function of reactions to: **a** disc size; **b** ellipse size; **c** square size. Frights (Retreats) are shown as a function of reactions to **d** square size

the reactivity mediated by the anterior-lateral eyes. A rough estimate of the flicker fusion frequency placed it at about 40 flashes per s (40 f s^{-1}). If jumping spiders are indeed responding at this frequency, then their eyes must be categorised as "slow" compared to the "fast" eyes of bees, wasps and flies, whose critical flicker fusion frequencies range from 250 – 300 f s^{-1} (Autrum 1948 a, b). Of interest in this context are the experiments by Edwards (1980), who showed that *Plexippus paykulli* were more likely to attack alates or pseudoalates (ants with false wings) than ants without wings, perhaps because of an increase in "flicker" resulting from the presence of wings.

By contrast, spiders retreated in 40% of trials when presented with the subsequent fly (no. 6) which exhibited a pronounced asymmetry. Supernumery tests with skewed two-dimensional shapes support the contention that asymmetry contributes to retreat, presumably because the corresponding retinal excitation is irregular (see Sect. 7). Moreover, quite apart from its occurrence in these tests, sidling is a very common tactic in jumping spiders, which frequently sidle from lateral to frontal positions when preparing to jump at prey. One reason for this may relate to the need to obtain equivalent stimulation on the retinae of the principal eyes and also perhaps, the anterior-lateral eyes.

The diamond shape (no. 7) was an effective "prey" target, with a 66% capture rate; moreover, no retreats were elicited. Figures 6c, d show that 1 × 1 mm squares did not evoke retreat, or capture, but once a size of 2×2 mm was reached, retreats regularly occurred with the rate increasing as target size expanded. At higher speeds, however, squares elicited occasional captures. It was also found that if lines were added to the hollow ellipse (no. 9), retreats (30%) rather than captures were elicited. Evidently quite small changes in target configuration can be detected.

When the previously successful ellipse was halved (no. 10) or sub-divided (nos. 11 and 12), only higher target speeds led to captures. Since, at 10 mm s^{-1}, circular or elliptical shapes of comparable area to this half ellipse elicited an appreciable capture rate (ca. 40%) (Forster 1979b), it must be supposed that certain features of this shape, such as the angular cut-off portion, were responsible for inducing retreat. (This portion was the leading edge in these tests.) Thus, it seems likely that the "double half" (no. 11), which spiders rejected as a prey object at 10 mm s^{-1}, were seen as two shapes at this speed, rather than one. Other experiments (Forster 1979b) in which the gap in this figure was progressively narrowed and to which spiders then responded as for the ellipse (no. 2) support this view. This assumption can be applied to the results for shape no. 12 which was sub-divided into three parts. It was demonstrated, moreover, that spiders did not "capture" single stripes of similar dimensions to the central stripe in this figure but if an ellipse was divided into progressively smaller and smaller partitions, spiders eventually responded as to a complete ellipse and not as to a set of lines.

5.4.2 Collaborative Role of the Anterior-Lateral Eyes

Spiders were apparently less discriminatory at higher target velocities, as shown by the fact that only at 50 mm s^{-1} were appreciable capture rates recorded for shapes 10, 11 and 12. This may be due to the fact that, when spiders are engaged in chasing a fast-moving object, the anterior-lateral eyes play the greater part in mediating behaviour (see Sect. 5.3.2).

Experiments by Drees (1952) in which *Salticus scenicus* treated all shapes depicted in Fig. 7 as prey, suggest that they were moving at high angular velocities and hence were monitored by the anterior-lateral eyes, which means that they may not be reliable indicators of shape perception in jumping spiders.

Land (1969a) showed that the anterior-lateral eyes in *M. aeneolus* and *P. johnsoni* have poorer acuity than the principal eyes, a condition no doubt also true for *T. planiceps.* It is also probable that at the higher angular velocities cited, the principal eyes would be unable to engage in scanning movements, the pattern of eye mobility that Land (1969b) found to be crucial to the identification of objects in their fields of view. It should not be overlooked, however, that during chasing, the principal eyes are probably tracking the target, and since the central area of each retina is thereby fixated on the target (Land 1969b), this may provide an opportunity for target evaluation. Behavioural tests certainly suggest that discrimination is less acute during this phase, but this does

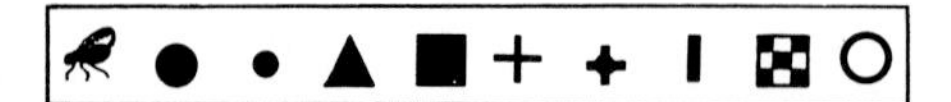

Fig. 7. These test shapes used by Drees (1952) were all successful in eliciting prey capture from *Epiblenum scenicus*

not eliminate the possibility of target evaluation altogether. Perhaps the enigmatic layer 2 (see Blest et al. 1981), of similar shape to layer 1 but with coarser receptor grain (see Land 1969a, his Fig. 5), has a collaborative role with the anterior-lateral eyes during this more active phase of hunting behaviour. Such a concomitance would provide an explanation, moreover, for the presence in the receptral processes of layer 2 of an exceptionally large number of mitochondria presupposing some greater metabolic activity in this part of the retina (Eakin and Brandenburger 1971).

Overall, these experiments with two-dimensional shapes suggest that the most favourable properties of target which are perceived by the principal eyes and to which the spider responds are: suitable size, solidity, curvaceous outlines, adequate contrast with the background, singleness, an elliptical configuration, and some degree of symmetry.

6 Parameters of Shape Perception

One of the suggestions put forward by Drees (1952) was that the amount of contour was a factor in shape perception in *Salticus scenicus*. This parameter had been used by Hertz (1929a, b) to account for the greater number of visits made by honey bees to shapes with higher contour lengths, as well as to explain why they had not been able to discriminate between simple figures such as triangles, squares and circles when they had similar outline lengths. In a series of ingenious experiments, Wehner (1967, 1969, 1971, 1972a, b) demonstrated that honey bees make use of a second parameter system, one based on the distribution of an area within a shape and dependent on the assumption that bees store a point-to-point representation of target configurations and compare them by perceptual superimposition. This shape-comparison hypothesis was modified by Cruse (1972a, b), who suggested that discrimination involved an additive interaction between these two parameters, expressed by the sum of area-distribution differences (A.D.D.) and contour-density (contour-length-to-area ratio) differences (C.D.D.), a model which was quite successful in explaining many earlier experiments.

Anderson (1977a) concluded that a bee compares shapes by remembering at least two parameters, area and contour density. He suggested that the bee measures area differences by Wehner's (1969) method of shape comparison although Wehner (1975) himself draws attention to some aspects of memory capacity in the bee as yet unresolved. Anderson (1977a) further suggested that contour density could be measured by a mechanism similar to the "flicker" detection system originally proposed by Wolf (1933). Wehner (1981), however, points out that if bees are attracted spontaneously by high spatial frequencies, whether of the whole pattern (Anderson 1977a) or within certain parts of the pattern (Anderson 1977c), there might be a need to add another term to Anderson's formula to account for this phenomenon.

The contour length, area, and density of the two-dimensional shapes presented to *T. planiceps* (see Sect. 5.4.1) were tested for their correlation with capture frequencies (see Table 1) but no significant relationships were demonstrated. The data were further treated according to the models of Cruse (1974) and Anderson (1977a) but, as might be expected, no significant relationships were revealed. Apparently, jumping spiders are not able to discriminate shapes by any of these parameters. It would be worth while, nevertheless, to assess the responses of jumping spiders to test shapes that can be altered in specified ways, as Anderson (1977a, b c) did in his experiments, a procedure that permits a more thorough analysis of the influence of the various parameters upon spatial discrimination.

One other potentially useful line of investigation emerged. "Flicker" appears to play a part in target perception in jumping spiders as well as in honey bees, judging from responses to quasi-stationary lures and the simulated movements arising from the test with "running legs" (see Sect. 5.4.2). There is a good case to suppose that jittery movement detectors (JMDs), neurons which seem to be designed to detect small jiggling movements against whole field movements and against changes in ambient illumination (see Wehner 1981), might be involved. Perhaps the "hypersensitivity" discerned by Yamashita and Tateda (1976b) in the photoreceptor cells of the principal eyes of *Menemerus confusus* following repetitive light flashes provides support for this suggestion (see also Yamashita, Chap. VI, this Vol.).

7 Retinal Model

Shape perception in jumping spiders is apparently not a function of contour length or area, or any relationship between these parameters, hence it is worth considering whether target discrimination could be based on a geometrical representation of relevant images on the retina. We know that the principal lens is capable of focusing an image of an object into the retina (see Fig. 4), but the next question concerns the extent to which the retina is capable of extracting information from this image. While we cannot as yet provide direct evidence, the retinal model proposed here offers one possibility.

This model makes use of the ophthalmoscopic representation of the foveal region of the paired first layers of the principal eyes of *M. aeneolus* (see Fig. 5) as depicted by Land (1969a). Its use here is justified by its approximate similarity to that of *T. planiceps* (Forster 1979b) and the fact that Blest et al. (1981) noted that in those species they had examined, layer 1 always possesses a regular array of receptors forming a high-resolution mosaic and appears to be the only layer suited to acute vision. Moreover, because the two foveas are effectively contiguous in image space and move in tandem, it is convenient to regard the principal eyes as a single "cyclopean" unit.

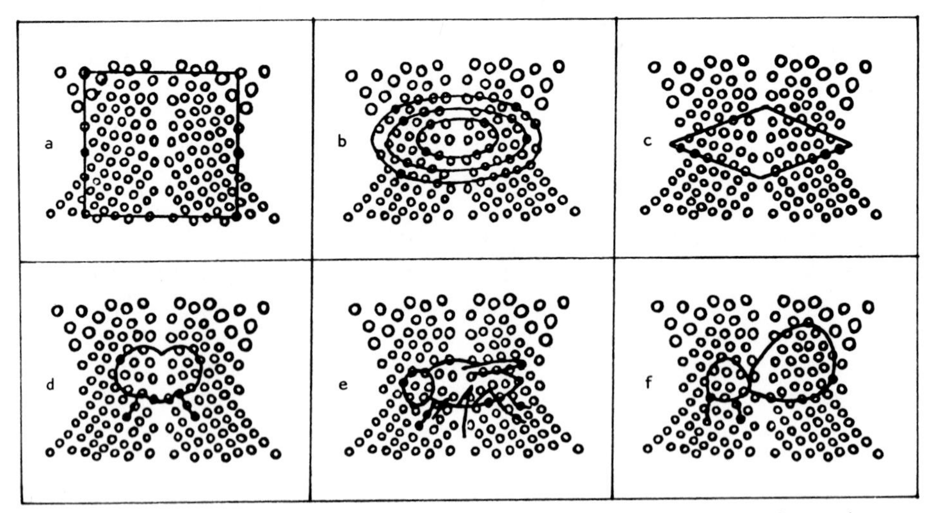

Fig. 8a–f. Six stimulus shapes (see Table 1) as they might be projected on to the "cyclopean fovea" of the principal retinae of *T. planiceps* from a distance of about 10 cm: **a** square; **b** set of ellipses; **c** diamond; **d** frontal profile of a house fly; **e** lateral profile of a house fly; **f** distorted fly shape. Note that in the tests these shapes were solid but they are shown here as outlines so that the arrangement of the receptors they overlie can be seen

Figure 8 depicts some of the stimulus shapes as they might be represented on the conjunct foveas from a distance of about 10 cm in front of the spider. Perhaps the square (Fig. 8a) simply stimulates a large number of receptors, but other tests suggest that angular objects induce retreat, although this aspect needs further study. When elliptical shapes (within the experimental size range) (Fig. 8b) are superimposed on the receptor mosaic, they fit neatly; the diamond (Fig. 8c) excites a similar set of receptors, and is frequently seen as prey. The frontal (Fig. 8d) and lateral (Fig. 8e) views of a fly occupy comparable regions but an asymmetrical shape (Fig. 8f) encompasses a greater proportion of one retina than the other. Although, in the course of scanning, images would be rhythmically shifted from one retina to the other, it is possible that each eye "remembers" the components of an image and routinely compares one view with the other. Thus, a substantial difference in visual input between the two eyes might stimulate retreat. The regular occurrence of sidling behaviour, which would expedite equivalent retinal stimulation in the principal eyes, is indirect support for this hypothesis.

What other visual consequences arise from scanning?

Blest et al. (1981) found that the distal ends of the central layer 1 receptors form a staircase (see Sect. 3.4), laterally to medially, across the horizontal axis, the purpose of which seems to be to extend the overall depth of field closer to the spider. Hence, the spider is able to receive focused images of objects at distances from about 3 cm in front of it to infinity. This means that, as the retinae sweep across the image, it will always be in focus on some part of the staircase of layer 1, within those distances. It is of some significance, too, that this near-

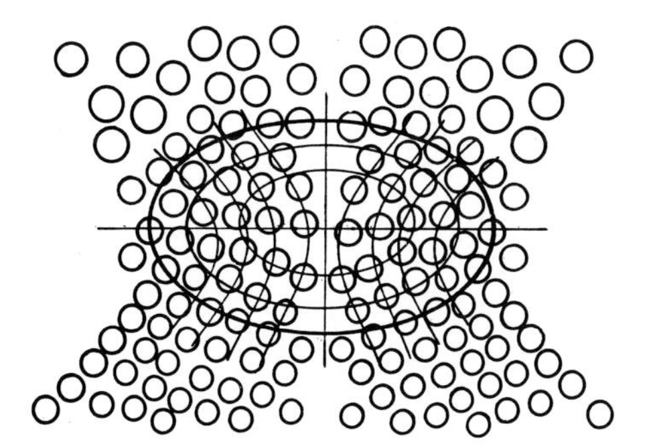

Fig. 9. The retinal model postulates that the geometrical arrangement of receptors in the "cyclopean fovea" of the principal eyes of jumping spiders resembles a set of confocal ellipses intercepted by a series of hyperbolas. When a mathematical model (after Spiegel 1959) having these characteristics is superimposed on the fovea, it is apparent that many receptors are overlain by both an elliptical edge and a hyperbola. Several sizes of prey and mate can be accommodated by this arrangement and when various leg orientations are present in the picture, retinal scanning strategies determine their relevance. Tests with stripes (Forster 1979b) suggest that only thick legs are visible, which may explain why male forelegs are often enlarged. Probably, therefore, motionless legs of flies are undetected but when legs are being groomed, their movement may be seen as "flicker"

field focal position approximates the distance at which the spatially determined jumping and leg-frontal postures occur (see Sect. 3.3).

How can the notion that "particular features of the prey have receptor correlates" explain the recognition of a prospective mate?

A system of confocal ellipses and hyperbolas modified from Spiegel (1959) suggests one solution. This model (Fig. 9) shows an interesting and appropriate goodness of fit with the disposition of receptors in the fovea of the conjunct layers 1. Many receptors fit both an ellipse and a hyperbola, i.e. a receptor lying on an intercept belongs to two systems: (1) it is one of a set of nested elliptical receptive fields, and (2) it is one of a system of hyperbolas crossing the ellipses more or less at right angles.

This arrangement of receptors presumes a simplification of pictorial input at the fovea as well as neural circuitries which reflect this retinal organisation perhaps at the first- or second-order glomeruli. It is proposed that when the elliptical circuitry alone is stimulated by an appropriate target, the spider is "instructed" to perform prey-catching behaviour, but when the second hyperbolic circuitry is activated the first is inhibited and prey-catching reactions are suppressed. Concomitantly, agonistic or courtship behaviour is initiated depending on the geometry of lines present in the picture.

The view that legs held at certain angles play a part in eliciting agonistic or courtship behaviour is supported by several investigations. For example, Drees (1952) showed experimentally that when jumping spiders were pursuing "prey" lures, the sudden elevation of wires (legs) to the lure, immediately initiated courtship display. Drees also found that the percentage of courtship responses

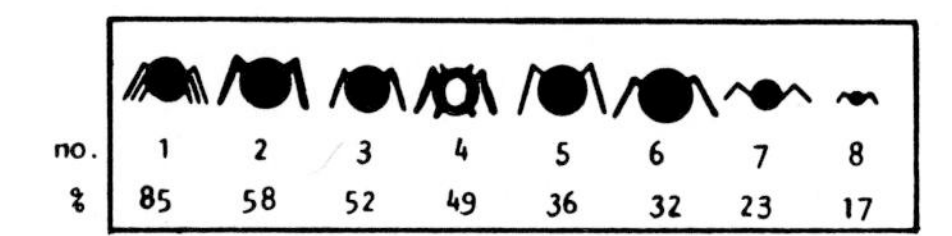

Fig. 10. Shapes which were effective in releasing courtship when presented to *Salticus scenicus* (Drees 1952). Success percentages are shown beneath each shape

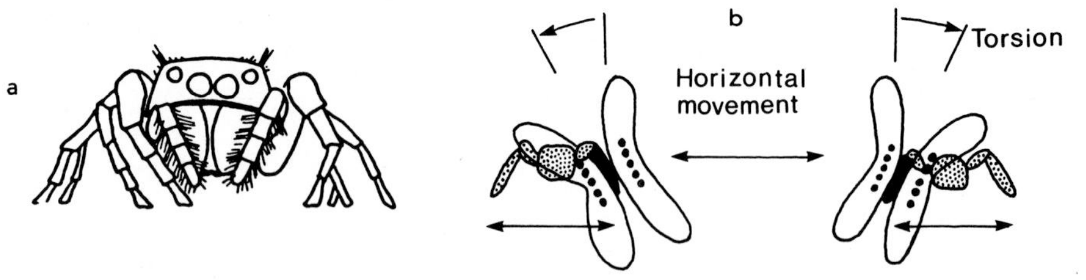

Fig. 11. a Front view of a resting female spider showing the angular relationship of the legs with the substrate. **b** Diagrams showing the manner in which line detectors in the retinae might sample the leg postures of the female spider. (Land 1972)

was greatly increased by the presence of legs on two-dimensional dummies (Fig. 10). In another study, Land (1969 b) examined leg angles in resting female salticids and found that they corresponded with the orientation of probable "line detectors" in the retina (Figs. 11 a and b).

In the central region of the jumping spider retina where the crucial features are perceived and distinguished, there are only some 100 receptors (Land 1974). Within this limitation, the shape parameters revealed by the two-dimensional tests (suitable size, solidity, symmetry, curvaceous outlines, singleness, an elliptical configuration, and adequate contrast with the background) are quite realistic since they could well represent the extent to which the retina is capable of extracting information from a "prey" image. Moreover, the receptors probably respond to off-stimuli, since white targets on black backgrounds do not elicit responses (Forster 1979 b) as Duelli (1978) also found in the anterior-lateral eyes in *Evarcha arcuata,* hence the need for adequate target contrast with the background. Mobile eye capsules not only cater for the phasic nature of photoreceptors and motion parallax but also permit a relatively small number of responding units to transmit essential spatial information.

The unique set of eyes in jumping spiders is beautifully adapted to the tasks of spatial discrimination, but it is clear that studies of shape perception in this group are decades behind comparable studies in insects, for example. It is hoped that the present review encourages neurobiologists and ethologists alike to further investigate not only "how these spiders see", but also "what they see".

8 Summary and Conclusions

The lives of jumping spiders are dominated by their visual reactions to objects in the environment. A functionally specialised set of eyes, consisting of a pair of principal eyes directed frontally, and two pairs of lateral eyes which between them cover almost the entire visual ambit, mediate these reactions. Objects which merit their attention are usually much about their own size, are dark compared to the background, and usually move. How do jumping spiders decide which objects must be avoided, or which can be pursued as prey, or courted as mates?

In the first instance, jumping spiders swivel towards unspecified objects in the environment through stimulus-dependent angles which bring these objects into the fields of view of the principal eyes. Escape responses are elicited if the intended target then looms up suddenly and consequently stimulates a large portion of the retina. Should the target recede at angular velocities greater than $4° s^{-1}$, however, most salticid species chase it, a locomotory reaction which is mediated by the anterior-lateral eyes. Targets which remain stationary, or move at angular velocities of less than $4° s^{-1}$ are evaluated by the principal eyes, and appropriate behaviour, such as prey-catching or courtship, may eventuate.

Target discrimination operates by means of a well-designed optical system and closely spaced receptors in the most proximal layers of the stratified retinae. An important property of these retinae is their conjugate mobility, manifested by four different patterns of activity executed under varying circumstances. Two of these are important to the present discussion. Tracking, in which the fovea is fixated on the target, occurs when the target is moving about, often at high angular velocities. Scanning, in which the fovea is also fixated on the target, depends on a motionless target and it is this movement which facilitates the discrimination processes. Eye movements probably prevent fixation blindness, promote depth perception by motion parallax at close range, and enable a relatively small number of receptors to transmit essential spatial information.

It is suggested that the high resolution mosaics of the two layers 1 function as a cyclopean eye, that the organisation of receptors in this "eye" resembles a set of confocal ellipses intercepted by a series of hyperbolas, and that many receptors participate in both systems. It is proposed that these two receptor systems are linked to appropriate neural sub-programmes; when the elliptical circuitry is activated by suitable prey targets, the spider is instructed to set about catching the target, but when the second hyperbolic circuitry is activated by legs held at prescribed orientations, the first system is inhibited, prey-catching behaviour is suppressed, and courtship or agonistic behaviour is initiated.

Support for this model of the retina is provided by tests in which 30 two-dimensional shapes were presented to *Trite planiceps* (Salticidae). Elliptical shapes were the most successful in eliciting prey capture; many acceptable prey items fall roughly within this elliptical configuration. Earlier studies showed that the presence of legs and elevated leg postures are instrumental in eliciting sexual displays.

The model proposed here fits the evidence at present available but there are other factors such as colour, pattern and type of movement which need to be taken into consideration in determining the basis of target discrimination. Perhaps elevated leg postures merely advertise the presence of another jumping spider, whereas colour, pattern and movement serve to announce their species identity and sexual status.

References

Anderson AM (1977a) Shape perception in the honey bee. Anim Behav 25:67–79
Anderson AM (1977b) Parameters determining the attractiveness of stripe patterns in the honey bee. Anim Behav 25:80–87
Anderson AM (1977c) The influence of pointed regions on the shape preference of honey bees. Anim Behav 25:88–94
Autrum H (1948a) Über das zeitliche Auflösungsvermögen des Insektenauges. Nachr Wiss Akad Göttingen. Math-Physik Kl Biol Physiol-Chem 8–12
Autrum H (1948b) Zur Analyse des zeitlichen Auflösungsvermögens des Insektenauges. Nachr Akad Wiss Göttingen Math-Physik Kl Biol Physiol-Chem 13–18
Blest AD, Hardie RC, McIntyre P, Williams DS (1981) The spectral sensitivities of identified receptors and the function of retinal tiering in the principal eyes of jumping spiders. J Comp Physiol 145:227–239
Born M, Wolf E (1980) Principles of optics. 6th edn. Pergamon Press, New York Paris Frankfurt
Burkhardt D, Darnhofer-Demar B, Fischer K (1973) Zum binokularen Entfernungssehen der Insekten 1. Die Struktur des Sehraums von Insekten. J Comp Physiol 87:165–188
Crane J (1949) Comparative biology of salticid spiders at Rancho Grande, Venezuela. Part IV: An analysis of display. Zoologica 34:159–214
Cruse H (1972a) Versuch einer quantitativen Beschreibung des Formsehens der Honigbiene. Kybernetik 11:185–200
Cruse H (1972b) A qualitative model for pattern discrimination in the honey bee. In: Wehner R (ed) Information processing in the visual system of Arthropods. Springer, Berlin Heidelberg New York, pp 201–206
Cruse H (1974) An application of the cross-correlation coefficient to pattern recognition of honey bees. Kybernetik 15:73–84
De Voe RD (1975) Ultraviolet and green receptors in principal eyes of jumping spiders. J Gen Physiol 66:193–208
Dill LM (1975) Predatory behaviour of the zebra spider, *Salticus scenicus* (Araneae: Salticidae). Can J Zool 53:1284–1289
Drees O (1952) Untersuchungen über die angeborenen Verhaltensweisen bei Springspinnen (Salticidae). Z Tierpsychol 9:169–309
Duelli P (1978) Movement detection in the posterolateral eyes of jumping spiders (*Evarcha arcuata*, Salticidae). J Comp Physiol 124:15–26
Eakin RM, Brandenburger JL (1971) Fine structure of the eyes of jumping spiders. J Ultrastruct Res 37:618–663
Edwards GB (1980) Experimental demonstration of the importance of wings to prey evaluation by a salticid spider. Peckhamia 2(1):6–9
Ewert JP, Borchers HW, Wietersheim AV (1978) Question of prey feature detectors in the Toad's, *Bufo bufo* (L), visual system: a correlation analysis. J Comp Physiol 126:43–47
Forster LM (1977) A qualitative analysis of hunting behaviour in jumping spiders (Araneae: Salticidae). N Z J Zool 4:51–62
Forster LM (1979a) Visual mechanisms of hunting behaviour in *Trite planiceps* – a jumping spider. NZ J Zool 6:79–93
Forster LM (1979b) Comparative aspects of the behavioural biology of New Zealand jumping spiders (Araneae: Salticidae). PhD Dissertation, Univ Otago, NZ, 402 pp
Forster LM (1982a) Vision and prey-catching strategies in jumping spiders. Am Sci 70(2):165–175
Forster LM (1982b) Visual communication in jumping spiders (Salticidae). In: Witt PN, Rovner JS (eds) Spider communication: Mechanisms and ecological significance, chap 5. Princeton Univ Press, Princeton, NJ
Forster LM (1982c) Non-visual prey capture in *Trite planiceps*, a jumping spider. (Araneae: Salticidae). J Arachnol 10:179–183
Forster RR, Forster LM (1973) New Zealand spiders. Collins, Auckland London
Gardner BT (1964) Hunger and sequential responses in the hunting behaviour of Salticid Spiders. J Comp Physiol Psychol 58:167–173

Gardner BT (1966) Hunger and characteristics of the prey in the hunting behaviour of Salticid spiders. J Comp Physiol Psychol 63 (3):475–478
Glauert AM (1951) Superposition effects in multiple beam interference fringes. Nature (London) 168:861–862
Hardie RC, Duelli P (1978) Properties of single cells in posterior lateral eyes of jumping spiders. Z Naturforsch 33c:156–158
Heil KH (1936) Beiträge zur Physiologie und Psychologie der Springspinnen. Z Vergl Physiol 23:1–25
Hertz M (1929a) Die Organisation des optischen Feldes bei der Biene I. Z Vergl Physiol 8:693–748
Hertz M (1929b) Die Organisation des optischen Feldes bei der Biene II. Z Vergl Physiol 11:107–145
Homann H (1928) Beiträge zur Physiologie der Spinnenaugen. Z Vergl Physiol 7:201–268
Homann H (1971) Die Augen der Araneae: Anatomie, Ontogenie und Bedeutung für die Systematik (Chelicerata, Arachnida). Z Morphol Tiere 69:201–272
Hunter S, Nabarro FRN (1952) The origin of Glauert's Superposition Fringes. Philos Mag 43:538–546
Jackson RR (1982) The behaviour of communication in jumping spiders (Salticidae). In: Witt PN, Rovner JS (eds) Spider communication: mechanisms and ecological significance. Princeton Univ Press, Princeton, NJ
Jackson RR, Blest AD (1982a) The biology of *Portia fimbriata*, a web-building jumping spider (Araneae: Salticidae) from Queensland: utilisation of webs and predatory versatility. J Zool (London) 196:255–293
Jackson RR, Blest AD (1982b) The distance at which a primitive jumping spider, *Portia fimbriata*, makes visual discriminations. J Exp Biol 97:441–445
Kaestner A (1950) Reaktionen der Hüpfspinnen (Salticidae) auf unbewegte farblose und farbige Gesichtsreize. Zool Beitr 1:13–50
Land MF (1969a) Structure of the principal eyes of jumping spiders (Salticidae: Dendryphantidae) in relation to visual optics. J Exp Biol 51:443–470
Land MF (1969b) Movements of the retinae of jumping spiders (Salticidae: Dendryphantidae) in response to visual stimuli. J Exp Biol 51:471–493
Land MF (1971) Orientation by jumping spiders in the absence of visual feedback. J Exp Biol 54:119–139
Land MF (1972) Mechanisms of orientation and pattern recognition by jumping spiders (Salticidae). In: Wehner R (ed) Information processing in the visual system of arthropods. Springer, Berlin Heidelberg New York, pp 231–247
Land MF (1974) A comparison of the visual behaviour of a predatory arthropod with that of a mammal. In: Wiersma CAG (ed) Invertebrate neurons and behaviour. MIT Press, Cambridge, Mass, pp 411–418
Land MF (1981) Optics and vision in invertebrates. In: Autrum H (ed) Comparative physiology and evolution of vision in invertebrates. Handbook of sensory physiology, vol VII/6B. Springer, Berlin Heidelberg New York, pp 471–592
Murphy J, Murphy F (1983) More about *Portia* (Araneae: Salticidae). Bull Br Arachn Soc 6 (1):37–45
Peckham GW, Peckham EG (1894) The sense of sight in spiders with some observations of the colour sense. Trans Wis Acad Sci Arts Lett 10:231–261
Snyder AW (1977) Acuity of compound eyes: physical limitations and design. J Comp Physiol 116:161–182
Spiegel MR (1959) Theory and problems of vector analysis. Schaum, New York
Sutherland NS (1960) Visual discrimination of shape by *Octopus:* squares and rectangles. J Comp Physiol Psychol 53:95–103
Tolansky S (1973) An introduction to interferometry. Longman, London
Vickridge IC (1982) An attempt to find the refractive index distribution of a small biological lens. B Sc (Hons) Thesis, Univ Otago, Dunedin, N Z
Wallace GK (1958) Some experiments on form perception in the nymphs of the desert locust *Schistocerca gregaria* Forskal. J Exp Biol 35:765–775
Wehner R (1967) Pattern recognition in bees. Nature (London) 215:1244–1248

Wehner R (1969) Der Mechanismus der optischen Winkelmessung bei der Biene. Zool Anz Suppl 33:586–592

Wehner R (1971) The generalisation of directional visual stimuli in the honey bee. J Insect Physiol 17:1579–1591

Wehner R (1972a) Dorsoventral asymmetry in the visual field of the bee. J Comp Physiol 77:256–277

Wehner R (1972b) Pattern modulation and pattern detection in the visual system of Hymenoptera. In: Wehner R (ed) Information processing in the visual system of arthropods. Springer, Berlin Heidelberg New York, pp 183–194

Wehner R (1975) Pattern recognition. In: Horridge G (ed) The compound eye and vision of insects. Clarendon Press Oxford, pp 75–114

Wehner R (1981) Spatial vision in arthropods. In: Vision in invertebrates. C: Invertebrate visual centres and behaviour. In: Autrum H (ed) Handbook of sensory physiology, vol VII/6C. Springer, Berlin Heidelberg New York, pp 287–616

Williams PS, McIntyre P (1980) The principal eyes of a jumping spider have a telephoto component. Nature (London) 288:578–580

Wolf E (1933) Critical frequency of flicker as a function of intensity of illumination for the eye of the bee. J Gen Physiol 17:7–19

Yamashita S, Tateda H (1976a) Spectral sensitivities of jumping spiders' eyes. J Comp Physiol 105:29–41

Yamashita S, Tateda H (1976b) Hypersensitivity in the anterior median eye of a jumping spider. J Exp Biol 65:507–516

XIV Homing Behavior and Orientation in the Funnel-Web Spider, *Agelena labyrinthica* Clerck

PETER GÖRNER and BARBARA CLAAS

CONTENTS

1 Introduction

This paper reviews what is known to date about the homing abilities of the funnel-web spider and discusses recent research in this field based on new and partly unpublished data. To provide a more comprehensive overview of the ability of spiders in general to orientate, research on other species will also be considered. A theory of the optical and idiothetic navigation of the funnel web spider will be presented in Chapter XV by H. Mittelstaedt, this Volume.

Much of our knowledge about orientation in spiders is derived from research on the funnel-web spider, *Agelena labyrinthica.* This species is widely distributed and thus easily available; moreover it is suitable for experiments.

Universität Bielefeld, Fakultät für Biologie, Postfach 8640, D-4800 Bielefeld 1, Federal Republic of Germany

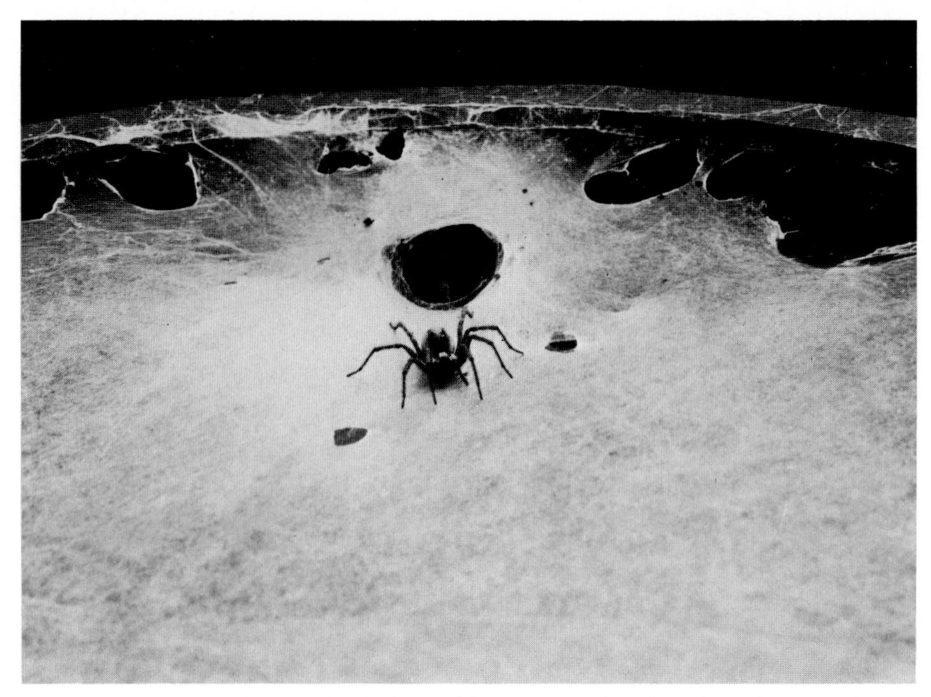

Fig. 1. *Agelena labyrinthica,* sitting in front of its retreat at the edge of a web mounted on a circular frame

Agelena spins horizontal webs which in nature are slightly sloped toward a funnel at the margin of the web. The funnel serves as a retreat where the spider spends most of its time and to which it returns after every excursion (Fig. 1). On the web the spider navigates by both external (allothetic orientation) and internal (idiothetic orientation) cues. To date it has been determined that allothetic orientation is based on optical and gravitational cues as well as mechanical cues of the substrate. As will be shown, *Agelena* stores the information about these cues in a memory. Since the spider loads its memory anew during each excursion, it is a suitable object for investigation of the properties of this memory and of the spider's ability for orientation.

2 Experimental Procedure

The data presented below derive from a *basic experimental procedure:* in a dark room two identical light sources are installed (L1 and L2), the beams of which intersect at the center of the horizontal web at an angle of 90°. L1 and L2 can be switched on alternately with a tumbler switch. At the beginning two pre-experimental runs are performed with the same light source switched on (pre-run 1 and pre-run 2, Exp. 1a, a′, Fig. 2). One light (say L1) is switched on and a fly is cast on the web. Led by the vibrations caused by the scrambling insect, the

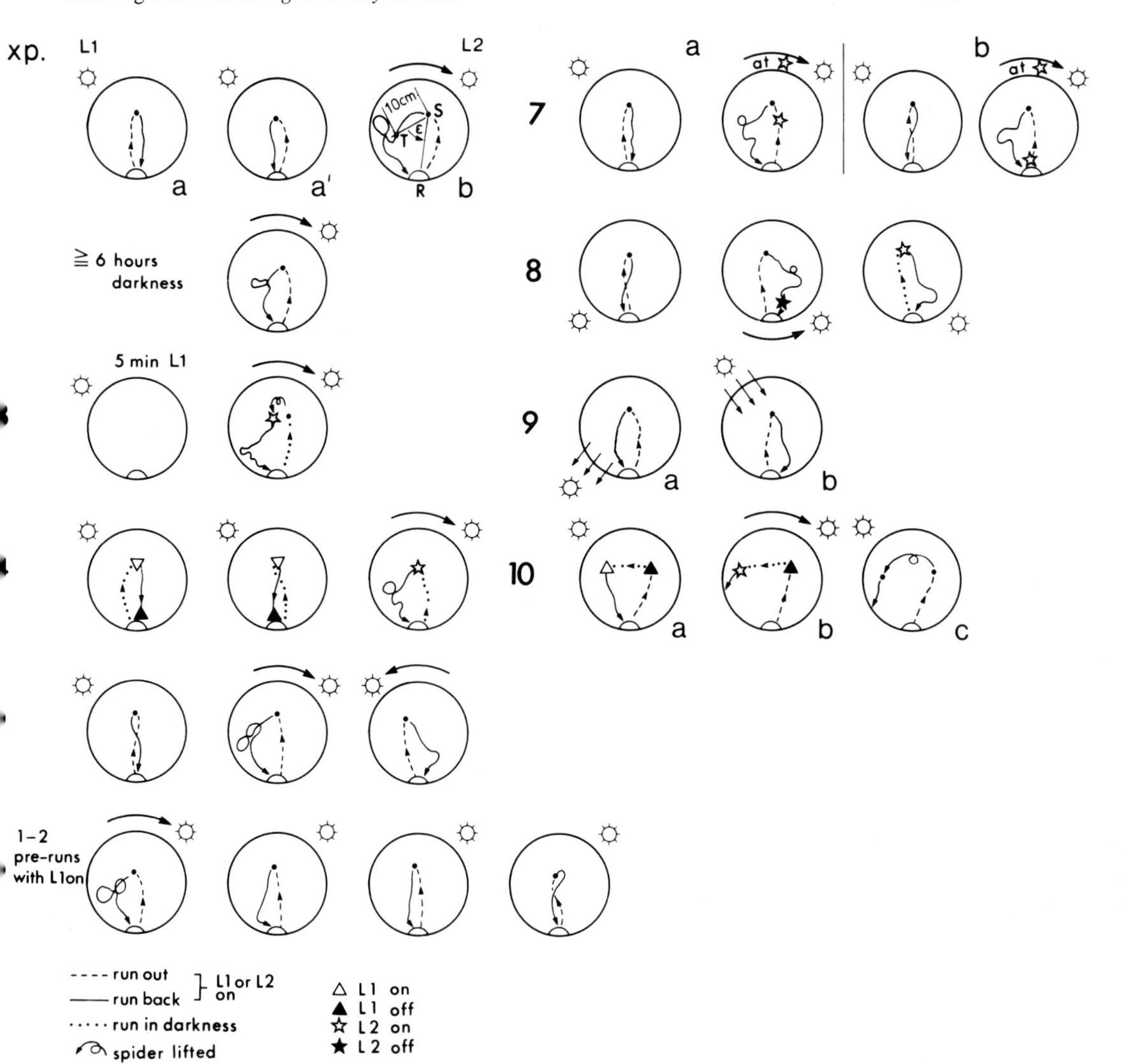

Fig. 2. Schematic illustration of the experimental procedure showing runs by *Agelena* in some of the experiments described in this chapter. The first experiment (**1**) describes the basic procedure. **a, b** Two pre-runs with fixed light source (*L1*); **c** experimental run in which the light is switched over from L1 to L2 after *Agelena* has arrived at the prey site *S*. ε starting angle between retreat *R*, starting point *S* (prey site) and point *T*, the intersection of a circle around *S* with the indirect route home. In the actual experiments (**2–10**) the position of the light source and the shape of the frame varied but are standardized here for reasons of clarity. In some of the experiments illustrated, one or two pre-runs have been omitted. Arrows in exp. 9: direction of positive (**a**) and negative (**b**) phototaxis

spider catches the fly and carries it back to its lair. In the experimental run the light source is switched from L1 to L2 just before the spider picks up its prey and starts to return (Exp. 1b, Fig. 2). Due to the change of light, *Agelena* deviates from the direct course to the retreat, running in a compromise direction between optical and other cues, but eventually arriving. The experimental run is followed by two pre-runs with L2 on and then again by an experimental run in which the light source is switched from L2 back to L1. For each run back to the lair the starting angle ε at the site of prey capture is measured (see Exp. 1b, Fig. 2). Runs obtained under different experimental conditions are normally evaluated by comparing the mean angles $\bar{\varepsilon}$ of the various deviations from the direct route back to the retreat.

Instead of comparing the starting angles $\bar{\varepsilon}$, the frequency of *run patterns* under different experimental conditions may be compared. In experimental runs on a natural web many of the runs back were curved throughout or changed abruptly. Others led straight back to the spider's lair. On the basis of these differences the runs are roughly classified in three categories: (1) a straight or slightly curved run with a deviation according to the change of direction of the light source (run pattern 1); (2) a run curved throughout or a run which at the start leads more or less toward the retreat but then changes direction abruptly (run pattern 2); (3) a fairly straight run back to the retreat (run pattern 3). On composite webs whose artificial elasticity pattern differs from those of natural webs (see Sect. 5.2), run pattern 2, with abrupt changes of direction, and run pattern 3 occurred very rarely, and the mean compromise angle increased. Since the three run patterns occur with different frequencies on different webs, and since the definition of the starting angle ε is somewhat arbitrary in curved runs, it is obvious that the mean compromise angle $\bar{\varepsilon}$ differs in experiments on different webs under otherwise similar experimental conditions.

3 Optical Navigation

Optical navigation is of major importance in the homing behavior of the funnel web spider. This is not surprising since in its natural biotope the animal is always able to make use of astronomical or other optical cues.

3.1 Orientation with Respect to a Lateral Light Source

Several authors have investigated the factors which determine the reference angle to the light source (angle between long axis of spider and line connecting its prosoma with light source) when *Agelena* returns to its retreat. Their findings suggest that there are memories which are loaded before or while the spider is moving on its web. This will be described in more detail below. (Actually there are at least three memories, as is pointed out in Chapter XV by Mittelstaedt, this Volume.)

In the experiments, the deviation from the straight route back to the lair was taken as a measure of the extent to which the spider's memory of the reference angle to the light source has been loaded. It had to be taken into account that several other factors also influence the degree of deviation. (a) Light intensity. The mean compromise angle was 66° at a light intensity of 800 lx (1180 stilb), whereas it was only 42° at 22 lx (33 stilb, Dornfeldt 1975a). (b) The angle of fixation of the light source. This influence becomes obvious if one compares runs with the light source (1) lateral and (2) right in front of or behind the spider on its way back. The compromise angle was larger in the first case (see Sect. 3.1.4). (c) The mechanical features of the web (see Sect. 5).

3.1.1 Loading the Memory

Before the spider begins an excursion on the web, it usually sits motionless in its retreat. Is the memory (of the angle of fixation of the light source)[1] loaded during this motionless phase? This problem has been considered in several investigations leading to different outcomes (Görner 1958; Moller 1970; Dornfeldt 1975b). In a series of experiments, Dornfeldt found that the memory is not loaded while the spider rests in its retreat.

In one of his experiments (Exp. 3, Fig. 2) the spider rested in its retreat in darkness. The light source (say L1) was switched on for 5 min. After L1 had been switched off, the spider was lured onto the web, lifted off (in order to prevent an idiothetic orientation, see p. 282) and set down again, after which L2 was switched on. Instead of returning directly to its retreat, the spider ran in the opposite direction from the light source L2 and missed its retreat. When Experiment 3 was repeated with L2 in a different location the outcome was the same. Dornfeldt's conclusion that the starting direction simply reflected negative phototaxis does not imply the existence of a memory.

However, using a slightly modified method Mittelstaedt (1983) recently found that the spider *Tegenaria* is indeed able to load its memory while resting in its retreat (cf. Fig. 9, Mittelstaedt, Chapter XV, this Vol.). Further experiments should be conducted to find out whether the difference between his and Dornfeldt's result is a consequence of the different spiders (*Agelena* and *Tegenaria*) or the different methods used.

The following experiments reveal that the memory of the reference angle to the light source can also be loaded by one run out to the prey, even if the distance moved is no more than 5 cm.

Spiders which had been kept in the dark for some time were lured onto the web after a light source had been switched on (Dornfeldt 1973). Before the spiders returned, the azimuth of the light source (the angle between the retreat, the center of the web and a point obtained by vertical projection of the light source on the "horizon", i.e., the edge of the web) was changed by 90° (Exp. 2, Fig. 2). The spiders deviated from the straight route back to the lair by a mean

---

1 In Chapter XV in this Volume, Mittelstaedt implies on theoretical grounds that the experiments described in this section must involve two separate memories.

angle of 62°. Approximately the same angle of deviation was obtained in experiments in which the spiders had to perform two pre-runs with the same light source before the azimuth was changed (Exp. 1, Fig. 2).

Another experiment (Exp. 4, Fig. 2) shows that the memory can also be loaded on the way back, but the mean angle of deviation after switching over the light source was considerably smaller ($\bar{\varepsilon} = 42°$) than 62° (Dornfeldt 1973). (In this run and in the one or two preceding pre-runs, the light was on only during the spider's way back to the retreat). It seems that for the spider the information gathered on its way out weighs stronger than that on its way back.

3.1.2 Reloading the Memory

The first series of experiments done up to 1970 was performed with only one pre-run. This was followed by an experimental run in which the light azimuth was changed. Thereupon the spider deviated from the straight direction to the retreat. If this experimental run was immediately followed by another (Exp. 5, Fig. 2), the spider again deviated (although to a lesser degree) from the straight route back (Görner 1958). The experiments show that the memory had been reloaded, although not completely (Moller 1970). In a series of three runs with fixed light azimuth following an experimental run (Exp. 6, Fig. 2), small deviations in the same direction as in the experimental run occurred in the first run. Obviously the spider "remembered" the azimuth of the light sources before it was switched over in the preceding run. The deviations decreased in the second run and were no longer detectable in the third. Therefore, experimental runs are now generally preceded by at least two pre-runs with a constant light source (see Exp. 1 a, a′, Fig. 2).

The reloading of the memory is not a sudden event: it takes place continuously while the spider moves on the web, as is shown in the following experiment. The light source was switched over as the spider moved from its retreat to the prey (Exp. 7 a, Fig. 2). The earlier this switch occurred, the smaller the degree of deviation from the straight direction to the retreat, i.e., the stronger the influence of the new light position. This shows that the memory was reloaded during the second part of the trip toward the prey (Görner 1958; Dornfeldt 1975 a). However, it may happen that the route from the funnel to the prey is not sufficiently long to reload the memory completely. This is apparent from experiments (Exp. 7 b, Fig. 2) in which the light source was switched over before the spider had started from its retreat – but it nevertheless deviated from the direct route to its lair (Bartels 1929; Holzapfel 1934; Görner 1958).

Since *Agelena* is also able to load its memory on the way back to the retreat (see above), the question arises whether this is also possible when the spider runs in a compromise direction after the light source has been switched over. This question is by no means trivial: the compromise angle implies that the spider "expects" to find the funnel in a different ("wrong") place. It can only correct its "error" after reaching its retreat. In the experiment (No. 8) described in Fig. 2 and in more detail in Fig. 3, it could do so only on the basis of idiothetic information (Görner, in press).

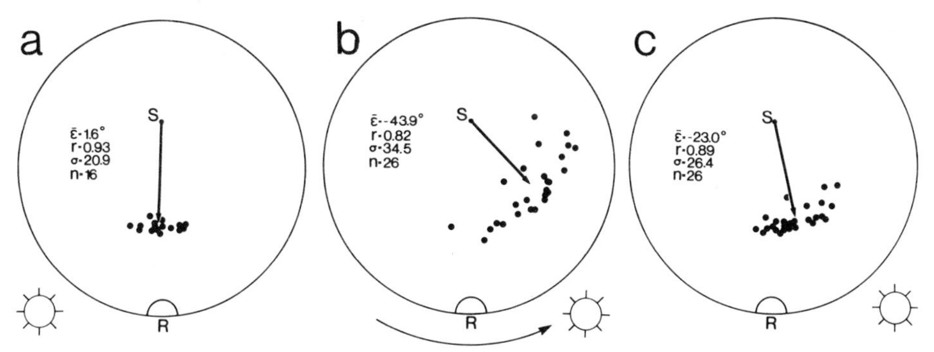

Fig. 3a–c. Reloading of *Agelena's* memory during runs in a compromise direction (Exp. 8, Fig. 2). In pre-run 2 (**a**) *Agelena* runs straight back from the starting point *S* to its retreat *R*. In the experimental run (**b**), in which the light is switched over, *Agelena* runs in a compromise direction between optical and idiothetic cues. If the light is switched off before the spider enters its retreat and switched on again (**c**) after the spider has again been lured to the starting point *S*, the mean starting angle is significantly smaller than in the preceding run. *Black dots* starting directions of eight *A. labyrinthica; straight arrows* mean vectors (r); $\bar{\varepsilon}$ direction of mean vector; σ circular standard deviation; n number of runs

3.1.3 The Functional Role of the Principal and the Secondary Eyes

Spiders have eight simple eyes: two everted principal eyes with a moveable retina and six inverted secondary eyes with a tapetum (see Land, Chap. IV, this Vol.). These morphological differences between the two eye types, as well as a comparison of their respective fields of vision, point to a difference in function. In *Agelena* the visual field of the secondary eyes covers the whole sky, including the field of the principal eyes in front of the spider extending dorsally beyond the zenith (Görner 1958; Dornfeldt 1975a). Clear differences between the principal and secondary eyes are evident in navigation by polarized light, as will be shown in Section 3.2.2. The experiments to date have not revealed whether the different types of eye have separate functions with respect to optical navigation by a light source. When Dornfeldt (1975a) covered the principal eyes with lacquer caps, *Agelena* was able to navigate by a lateral light source using the secondary eyes alone; similarly, it navigated by the principal eyes alone when the secondary eyes were blindfolded. The secondary eyes are more sensitive than the principal eyes. Even at a light intensity of 1.15 apostilb (this is 4 to 5 orders of magnitude lower than the intensity of the full moon), the spider reacted to a 90° change of the light azimuth by deviating from the straight course. With only the principal eyes intact, the light had to be 100 times more intense (111 apostilb) to cause the same deviation.

Under certain experimental conditions, the starting angle of the return run seemed to be influenced by the principal eyes. In Exp. 3 (Fig. 2), showing negative phototaxis, *Agelena* deviated from its straight way back to the retreat when the light had been switched over for a short time while the spider rested in its retreat. When the same experiment was conducted with the principal eyes covered, these negative phototactic runs no longer occurred, i.e., the spider returned straight to its retreat.

3.1.4 Phototactic Components in Optical Navigation

Like other arthropods (v. Buddenbrock 1952; Jander 1963), the funnel-web spider spontaneously orientates with respect to a light source under certain experimental conditions. Spiders set on a web and illuminated with lateral light spontaneously ran in one of several directions: 39% ran toward the light and 15% away from it, while 46% ran either to the right or left (Moller 1970). Dornfeldt (1975a) obtained a similar distribution when he lured dark-adapted spiders to the center of the web and then switched on the light. This behavior can be interpreted as positive, negative, or lateral phototaxis respectively.

In order to prevent idiothetic orientation, the spiders were lifted off the web at the prey site and set down again. This was done as follows: after catching a fly attached to a thread, the spiders were lifted from the substrate together with their prey. This procedure caused them to lose their idiothetic reference direction. Back on the web the prey was removed from the spiders' claws and another fly was presented. The spiders caught the fly and started on their return course.

The influence of positive or negative phototaxis was also revealed in the experiments on optical navigation of the funnel web spider. An influence of positive phototaxis in the run back to the lair was evident when the light was in their latero-frontal field of vision (Exp. 9, Fig. 2) whereas an influence of negative phototaxis was found when it was in their latero-caudal field of vision (Exp. 9b, Fig. 2, Moller 1970). This influence increased when the switch in the light azimuth was preceded by an interval of darkness, or when the spider ran to the prey partly or completely in the dark (Dornfeldt 1975a). A phototactic component manifest in the menotactic orientation of several arthropods led Jander (1957) to develop his compensation theory of photomenotaxis. Although his ideas are not directly applicable to optical navigation in the funnel web spider (cf. Mittelstaedt, Chap. XV, this Vol.), there are striking parallels in the way phototactic components influence the optical navigation of *Agelena* and other arthropods.

3.2 The Perception of the Pattern of Polarized Light

3.2.1 Polarized Light Navigation

Like many other arthropods, the funnel-web spider is able to analyze the plane of polarized light. This has been demonstrated under experimental conditions in the laboratory (Görner 1958, 1962). Rotation of a polarizing filter above the web caused the spiders to deviate from their route back to their retreat, turning to the right or left with a considerable scatter.

To see whether the natural polarization pattern of the blue sky is of biological significance for the spider's celestial navigation, the following experiments were performed. The spiders were taken outside, placed in an arena from which only the sky was visible, and lured onto the web. Then either the web frame was rotated by 180° or the sun was shielded and reflected from the opposite side onto the web ("Santschi's mirror-experiment"). The spiders deviated more often from their course back in the first case than in the second, showing that in their astronomical orientation the polarization pattern is more relevant than the azimuth of the sun. This is corroborated by the following experiments.

The plane and the degree of polarization differ in different regions of the blue sky, changing during the daily course of the sun. For *Agelena* the pattern of polarization around the zenith seems to be of major importance. In an outdoor experiment the spiders were lured onto the web, after which the frame was covered with a polarization filter, the direction of maximal transmission of which coincided with the e-vector of the polarization pattern in the zenith. The spiders found their way straight back to the retreat. Conversely, when the direction of maximal transmission of the filter was perpendicular to the e-vector of the polarization pattern in the zenith, the spiders often deviated from the straight path to the lair. This was also the case when local landmarks were visible and when the visible section of the sky was small.

The ability to analyze the plane of polarized light has also been demonstrated in lycosids (see Sect. 3.4) and in the orthognate spider *Aphonopelma californica* (Henton and Crawford 1966), but seems not to be common among spiders (see also Land, Chap. IV, this Vol.): in their principal eyes, salticids possess a layer of receptor cells which are well suited for the analysis of polarized light according to the arrangement of their rhabdomers (Land 1969). However, the spiders obviously do not orientate to the natural plane of polarization.

Experiments with *Phidippus pulcherrimus* were carried out using an artificial "plant" with a series of horizontally and orthogonally arranged wooden branches (Hill 1979). In the first experiment, the "plant" was illuminated from above with an artificial light source. The spider faced from below a fly attached to a thread: it ran in pursuit up the stem and reorientated to the "expected" fly (which had been removed by the observer) on one of the radial branches. It used only optical cues from the "plant" when the surround was screened with a cylinder. This was obvious from experiments in which the "plant" was turned by 90° around its long axis as soon as the spider had started to run. It ran up the stem on that radial branch which would have led it close to the prey before turning the "plant". When the surround was visible, however, the spider orientated by means of background cues in a significant fraction of runs (43%) neglecting the turning of the "plant". In another experiment the surround was screened, but instead of the lamp the blue sky above the cylinder was visible. This time, the spider orientated by means of optical cues from the "plant" only, as if the polarized light pattern had not presented an additional optical cue (although the spider sighted the prey and the blue sky from beneath).

3.2.2 The Functional Role of the Principal and Secondary Eyes

Agelena could not navigate by the pattern of polarization when its principal eyes were covered by an opaque cap (Görner 1958, 1962), although its ability to orientate by unpolarized light remained unaffected (Dornfeldt 1973). However, covering the secondary eyes did not affect the spider's ability to navigate by polarized light. It is not yet known whether the secondary eyes contribute to the perception of polarized light; however, their contribution can be of only minor importance. When the principal eyes were covered and the polarization filter was rotated by 90°, the mean direction of the runs was not affected; however, the scatter increased compared with those runs in which the plane of polarization had not been changed (Dornfeldt 1973).

The reflection pattern of the substrate may be responsible for the increase in scatter, but not for *Agelena's* ability to navigate by polarized light. (The intensity of the reflected light from the substrate changed by up to ± 20% when the plane of polarization was turned by 180°, Görner 1962.) Spiders carrying a small screen on their prosoma to prevent the perception of the polarized light from above were not influenced when the polarization filter was turned, although they perceived the reflection pattern from the substrate. On the other hand, spiders carrying a small tube on their prosoma so that they saw only the overhead polarized light orientated themselves to the plane of polarization.

Similar results have been reported in lycosids. *Arctosa* orientated well under a clear sky in the shade without visible landmarks when the secondary eyes were covered with black varnish (Magni et al. 1964). However, when the principal eyes were covered, the spiders showed only a weak tendency to run in the direction corresponding to the flight direction in their natural habitat. The experiments with lycosids again reveal that the principal eyes are crucial for analyzing polarized light. Whether the secondary eyes contribute to navigation by polarized light cannot yet be determined. Possibly the pattern of light intensity in the sky gives the spider some information about the position of the sun. Further investigations are necessary to solve this question.

3.2.3 Analysis of the Plane of Polarization

Baccetti and Bedini (1964) investigated the neuroanatomical correlates of the spider's ability to analyze the light's e-vector. In cross-section, most of the elongated sensory cells in the principal eyes of *A. variana* have a pentagonal profile with a rhabdom on every edge. Since for geometrical reasons regularly formed pentagons cannot cover an area completely, the edges of the sensory cells vary in length. On the basis of theoretical considerations, only irregular pentagonal cells respond to a certain plane of polarization with maximal excitation. The degree of excitation depends on the degree of asymmetry of the cell, but is always smaller than in cells with equally orientated rhabdomeres (which of course cannot occur in these pentagonal cells). Therefore an analysis of the e-vector by the central sensory cells seems very unlikely. Morphological data (Schröer 1974, 1975, 1976) suggest that polarized light from above is analyzed in the ventro-peripheral region of the principal eyes' retinae. These data are in accordance with the finding that under a cloudless sky the zenith is essential for navigation by polarized light in the funnel-web spider (see above). Schröer found dramatic differences between cells from different regions in the principal eyes of *Agelena gracilens.* Those located in the center and the dorsal periphery have an irregular pentagonal profile like those in the principal eyes of *Arctosa.* Unlike those of *Arctosa*, the rhabdomeres lie along only two to four of the five long edges of the sensory cells (Fig. 4a). The cells located at the ventral rim of

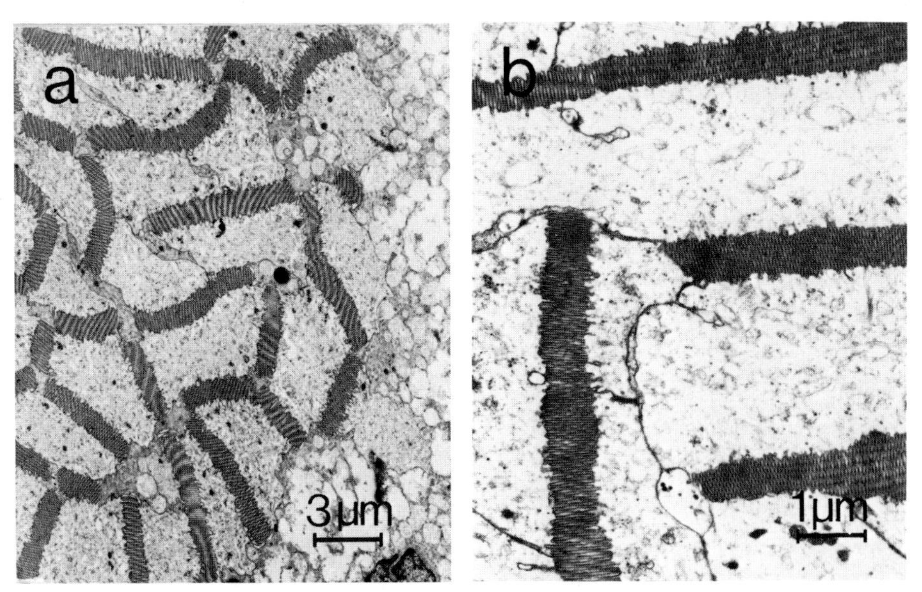

Fig. 4 a, b. Electron micrographs of the principal eye of *Agelena gracilens.* **a** Cross-section through the central area of the retina below the mass of crystal cells with an irregular arrangement of the rhabdomeres. **b** Orthogonally arranged rhabdomeres in adjacent rows of sensory cells in the ventral rim of the retina. (Courtesy of W. Schröer)

the eye, however, are more or less rectangular in profile, with rhabdomeres on only one side or on two opposite sides (Fig. 4b). Adjacent cells toward the center have their rhabdomeres at right angles to those of the periphery. This arrangement is well suited for analysis of the plane of polarization. Kirschfeld (1972) has demonstrated that simultaneous analysis of the plane of polarization requires three polarization-sensitive channels with different angles of maximal sensitivity. Since in the arrangement of the sensory cells in the ventro-peripheral region there is no indication of a three channel analyzer, *Agelena* probably performs a successive analysis. In principle, this can be done with only one channel (Kirschfeld 1972); however, as Schröer (1974) has pointed out, a system working with two perpendicularly arranged analyzers is independent of light intensity and operates on the basis of comparison. It is therefore more effective than a system using a one-channel analyzer. The principal eyes of *Agelena* are well suited for a successive analysis, since their retinae move continuously; it is not unlikely that the retinae are even rotated to scan the zenith (Schröer 1974).

Orthogonally arranged analyzers for polarized light from above directed to the zenith have recently been described for bees (Labhart 1980) and flies (Hardie 1984). Bees have about 140 anatomically specialized ommatidia on the dorsal rim of the compound eye, which consist of two populations of UV receptors having orthogonally arranged rhabdomeres and maxima of polarization sensitivities at 0° and 90°. Behavioral experiments (v. Frisch 1965; Wehner 1982) show that the dorsal rim area is important, and in many cases essential, for navigation by polarized light.

3.3 Do spiders Take into Account Changes in the Sun's Azimuth?

The azimuth of the sun and the pattern of polarization in the sky change in the course of the day. Many arthropods and vertebrates which navigate by astronomical cues are able to take these changes into account. This ability has also been shown to exist in lycosid spiders of the genus *Arctosa* (Papi 1955a, b, 1959; Papi et al. 1957; Papi and Syrjämäki 1963). Under experimental conditions where landmarks were hidden, the spiders compensated for the changing azimuth of the sun during the course of the day by using their "internal clock" so that the mean flight direction was well in accordance with the direction to the habitat.

It is not yet known whether funnel-web spiders calculate the path of the sun. This ability would be useful during "search runs" in which *Agelena* spontaneously leaves its retreat. Such runs occur up to 3 h after the spider has left its prey on the web (e.g., because of a disturbance) and fled to its lair (Bartels and Baltzer 1928; Holzapfel 1934; Görner 1958).

3.4 Optical Navigation in Other spiders

Though optical navigation is probably common among spiders, little is known about this ability in other spiders with the exception of lycosids. Wolf spiders of the genus *Arctosa* live on the edges of rivers and ponds. When thrown onto the water, they run straight back to the edge. Different populations adopt different flight directions according to the location of their habitat with respect to the water (Papi 1955a, 1955b, 1959). In sunny weather *Arctosa* navigates by astronomical cues, as was shown in an experiment using a round water-filled glass basin without visible landmarks. The spiders ran in the direction which in their natural habitat would have led them to the water's edge. Orientation by means of the polarization pattern of the sky has been demonstrated indirectly with Santschi's mirror experiment. The reflected sun caused the spiders (*A. variana*) to run in a direction which probably constituted a compromise between the two conflicting cues of sun azimuth and polarization pattern. Other experiments in a glass basin showed that, under a clear sky, lycosids were nearly as well orientated in the shade as in direct sunlight. Under an overcast sky, their flight directions were randomly distributed (Papi 1955b).

Experiments with the orb-weaving spider *Araneus diadematus* (Peters 1932) reveal that optical input is relevant for their ability to return to their hub on the vertical web. Wind scorpions (*Galeodibus olivieri* and *Galeodes barbarus*), like agelenids, are able to orientate themselves relative to a lateral light using either the two principal eyes or the four secondary ones (Linsenmair 1968). They ran spontaneously (without conditioning) at a certain angle to a lateral light source and could retain this angle for hours. The angle varied from individual to individual and for the same animal at different times of day. Intensifying the light from 0.09 to 8 stilb increased the angle; conversely, after 30 min of darkness the angle decreased. Additional experiments suggested that the principal and secondary eyes have different functions in the optical navigation of wind scorpions, but this question has not yet been answered definitively.

4 Idiothetic Navigation

4.1 Interaction between Optical and Idiothetic Cues in *A. labyrinthica*

In the experiments on optical orientation, the starting angle was interpreted as a compromise between optical and idiothetic information (e.g., Görner 1972). This interpretation requires some modification, since it did not take into account the structure of the web. Comparing return runs made on artificial and natural webs demonstrates that this factor does have an influence. On composite webs constructed from parts of natural ones (see Sect. 5.2), the mean starting angle $\bar{\varepsilon}$ is smaller than 90° (ca. 60°, see Fig. 10b) but larger than that on natural webs (ca. 45°, see Fig. 3b) (Görner and Claas 1980). Hence, we interpret ε as a compromise between optical and idiothetic information which is additionally influenced by the web structure.

Convincing proof that idiothetic navigation occurs is provided by experiments in which the spiders were lured to a prey via a detour during which the light source was switched off. At the capture site, either the former light source was switched on again or the light's azimuth was changed by 90°. In the first case (Exp. 10a, Fig. 2) the spider ran straight back to its retreat; in the second (Exp. 10b, Fig. 2) it deviated from the straight direction to the retreat as in the experimental runs (Görner 1966). Thus, *Agelena* was able to recalculate its angle toward the light source during its run in darkness as though the light had still been on. This could only have been done by making use of idiothetic information. When, instead of running the detour, the spider was lifted up (method, see p. 282) and set down where the previous detour had ended (Exp. 10c, Fig. 2) it was unable to calculate the detour. Instead, when put onto its web, it ran at an angle to the light source which would have been correct before it was lifted from the web (see also Dornfeldt 1975b and Mittelstaedt, Chap. XV, this Vol.).

4.2 Idiothetic Distance Orientation

The memories described in the preceding sections store the reference direction with respect to a light source. They can be loaded (or reloaded) in the retreat, on the way out, or on the way back. The experiments described below show that information about the actual location of the spider after path integration is also stored in a memory (see Mittelstaedt, Chap. XV, this Vol.).

There is only indirect experimental evidence for the existence of this memory. Dornfeldt (1975b) lured spiders from the retreat R to point A on the web (Exp. 11, Fig. 5), and then moved the light by 90° from L1 to L2. From A the spiders ran to point S where the prey was located (distance RA = AS). The angle RAS is approximately the same as the mean compromise angle the spiders would have run if they had started home from A. Thus, S is the location at which the memory should be emptied. Starting home from S the spiders moved about randomly for a short while suggesting that they had indeed "expected" to have arrived at their retreat.

In another series of experiments, Dornfeldt (1973) observed spiders to run roughly the same distance on their way back as that between their funnel and

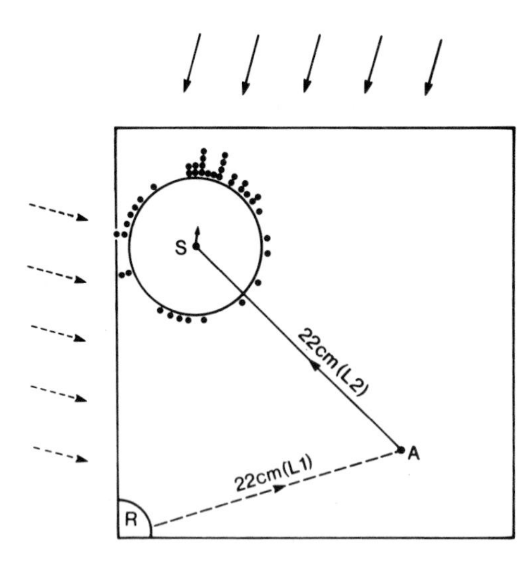

Fig. 5. Distance orientation in *A. labyrinthica*. L1 is on (*broken arrows* at *left* of diagram). The spider is lured from retreat *R* to point *A*, at which time the light source is switched over to L2 (*solid arrows* on *top* of diagram). The spider is lured to *S*, lifted up and set down again. The starting directions (*dots*) are distributed nearly at random, indicating that the animal's location memory is empty. (After Dornfeldt 1975 b)

the prey, and then sometimes suddenly change their course. He interpreted these turns as an indication that the spiders were searching for their retreat. The same "searching turns" occurred when the spider *Tegenaria* sp. orientated by means of idiothetic input only (see Mittelstaedt, Chap. XV, this Vol.).

Idiothetic distance orientation has also been found in two other families, salticids and ctenids, and seems likely to exist in lycosids as well.

Sitting on a horizontal bar, the salticid *Phidippus pulcherrimus* faced a fly at a distance of about 25 cm (Hill 1979, Fig. 6a). When the spider turned on the bar to pursue the prey, the fly – which was attached to a thread – was removed and the light switched off. When it was switched on again after a short time, the spider had run a certain distance and reorientated in the direction where it presumed the prey to be. The initial orientation angle (see Fig. 6b) measured when the spider first faced the prey differed significantly from the later reorientation angle. However, the latter was in accordance with the calculated (movement-compensated) reorientation angle which the spider should have adopted if the prey had been visible (Fig. 6c, d). *Phidippus'* ability to calculate the direction to the prey during the pursuit run by means of idiothetic information and to reorientate accurately ("route-referent orientation", to use Hill's term) is remarkable, since ths spider did not actually run the distance to the prey but rather calculated it solely from visual input.

The wandering spider *Cupiennius salei* returned to a prey from which it had been chased away, even without the help of external directional cues (Barth and Seyfarth 1971; Seyfarth and Barth 1972; Seyfarth et al. 1982). The return path was interrupted by short stops and turns. Reaching the capture site, from which the prey had been removed, the spider turned sharply, obviously "searching" for the lost prey. These "searching loops" were taken as an indication of the spider's ability to estimate distances correctly. For chasing distances of up to 20 cm, all spiders would have reached the prey, i.e., they came within 5 cm of

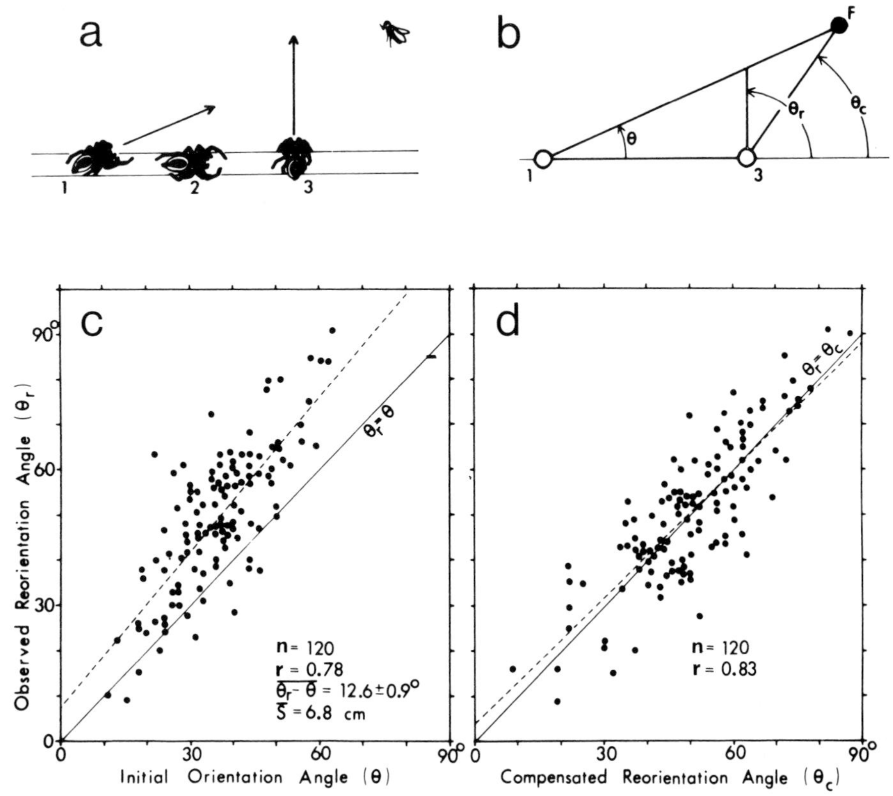

Fig. 6a–d. Idiothetic distance orientation in a female *Phidippus pulcherrimus.* **a** The spider faces a fly (*1*), runs in pursuit in darkness (*2*) and reorientates in darkness (*3*). **b** Schematic drawing of **a. c** The observed reorientation angle Θ_r (for definition see **b**) differs by about 13° from the initial orientation angle Θ. **d** Θ_r is in agreement, however, with the compensated reorientation angle Θ_c which the spider should take if it calculates the distance it has run from 1 to 3. *Broken line* linear regression of Y on X. (After Hill 1979)

the capture site. With increasing distance the scatter of the starting angles increased while the number of successful returns decreased but still exceeded 60% at distances of more than 40 cm. Like *Agelena, Cupiennius* is able to compensate for a detour. When driven away from the prey through a semi-circular corridor, it returned from the exit to the capture site by the shortest route (Fig. 7a). The sharp turns of these runs (indicating the start of a "searching loop") began at a distance only slightly larger than the distance from the start to the prey. When the lyriform slit sense organs on all leg femora were experimentally destroyed, the scatter of the return runs increased. Although the spiders started off more or less correctly, they eventually drifted off the ideal route and less than 50% of the runs reached the capture site (Fig. 7b). The searching loops of most of the spiders were performed at an approximately correct distance, however, indicating that the spider's path-integration ability has not been impaired.

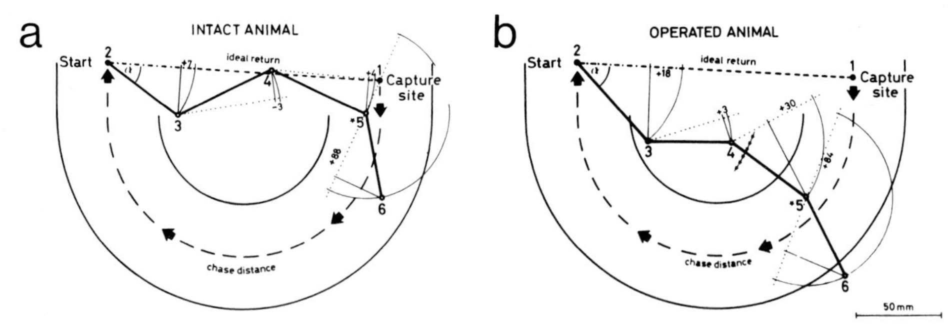

Fig. 7 a, b. Return runs of *Cupiennius salei* after having been chased from the capture site (*1*) through a semi-circular corridor (*arrows*). **a** On its return run from (*2*), the intact spider takes a short cut to the capture site. **b** The spider whose lyriform organs on all femora were ablated misses the capture site. *Large numbers* stops; *small numbers* "walking error". (The "walking error", calculated from each turning point as $e_n = v_n (1 - \cos \alpha)$, is a measure of the accuracy of the spider's ability to find the shortest way from the start to the capture site and a measure to determine the onset of the searching loops). In *5* the spider starts "searching loops" (large walking error); α starting angle. (Seyfarth et al. 1982)

The significance of the lyriform slit sense organs for idiothetic navigation is also shown by experiments with wolf spiders (Görner and Zeppenfeld 1980). Female *Pardosa amentata* were anesthetized with CO_2 and their cocoons removed. After recovery, they meandered in circles in the vicinity of the releasing point. When driven 5 cm away, their searching runs were directed toward the release point. (The experiments were done with red light to exclude optical orientation.) If the lyriform organs on the coxae of the second and third legs were covered with glue, however, the searching runs were no longer directed toward the releasing point. Covering the organs on the coxae of the first and fourth legs had very little effect. Obviously, the sensory organs on the second and third legs are of major importance for perception of leg movement. This seems plausible, since during locomotion the amplitude of horizontal movement of the second and third legs is much greater than that of the first and fourth legs (Ehlers 1939; Fröhlich 1978).

5 Orientation by Means of Directional Cues of the Web

5.1 Return Runs on Natural Webs

Baltzer (1930) and Holzapfel (1934) found that web elasticity is relevant to the spider's orientation. If a web attached to a square frame was deformed into a rhombus, *Agelena* ran along the line of highest tension instead of back to its retreat (Fig. 8). Obviously the spider is able to perceive variations in web tension. Does it also use this ability to find its way under natural conditions? Holzapfel (1934) investigated the elasticity of the web at different points. She found that elasticity decreases with increasing distance from the retreat, which is surrounded by a densely woven area. Consequently, the spider should be able to

Dotted line Agelena's route from *R* to *S*. When the spider arrived at *S*, the frame was experimentally deformed to a rhombus. *Solid line Agelena's* route from *S*. (Baltzer 1930)

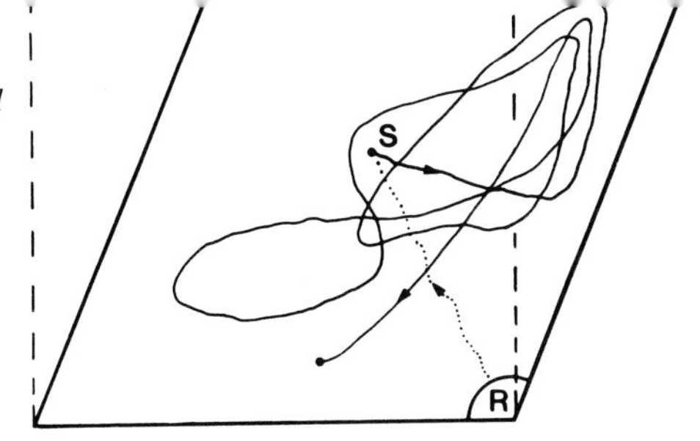

find its retreat by running against the gradient of elasticity. Indeed, *Agelena* sometimes finds its way to its retreat without having been able to store information from other orientational cues or to navigate by means of idiothetic information. For example, this is the case when a male runs straight to the funnel of a female's web (Holzapfel 1935).

There is further evidence for orientation by means of special features of the substrate. (1) On some webs spiders deviated consistently from the direct route back to the retreat (unpublished). Since all other cues remained constant, this behavior must have been a function of the substrate. (2) When the position of the light source was changed during the experimental run, the spiders normally missed the retreat. After meandering in circles, they sometimes suddenly headed straight back to their funnel. This suggests that they are able to "switch" from optic-idiothetic orientation to orientation by means of parameters of the web.

A puzzling question is why the spiders did not orientate themselves by means of the web pattern in experiments in which they were lifted from the web in darkness. Back on the web they started in any direction, i.e., no use of orientational cues was discernible (Görner 1966; Kurth 1972; Moller 1970; Dornfeldt 1975a). It may be that the spiders have to run a certain distance before they are able to "switch" to orientation according to the elasticity pattern of the web.

5.2 Return Runs on Composite Webs

Since the substrate obviously influences the course of the spiders, a composite web was constructed from parts of natural ones (Görner and Claas 1980). At the periphery the web was connected either to the natural or to an artificial funnel. The spiders accepted these artificial webs. As on natural webs, they ran in a compromise direction between optical and idiothetic cues in the experimental run. Most of the runs occurring belonged to pattern 1 (see Sect. 2), the more or less straight course with a deviation of about 60° from the directions to the retreat. Very few curved runs (pattern 2) and direct runs to the retreat (pattern 3) occurred. The same run patterns were obtained in an experiment in which the spiders started their return course from the periphery lateral to the retreat so that they traversed the web (Exp. 12 and 13, Fig. 9a, b).

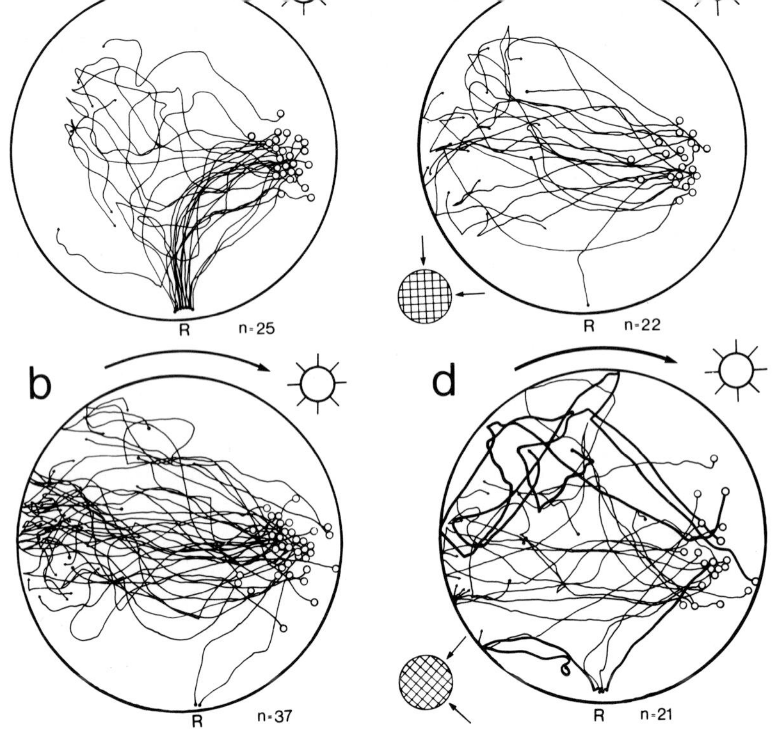

Fig. 9 a – d. Return courses of 14 *A. labyrinthica*, starting at the periphery of the web after the light has been switched over in the experimental run. On a natural web (**a**) the spiders follow the change of the light azimuth in only 11 out of 25 runs. In the other cases, the spiders either run directly back to the retreat (pattern 3) or else start in an optical-idiothetic compromise direction and then turn round toward the retreat. (This type of run can be regarded as a special case of run pattern 3.) On a web composed of pieces of natural webs (**b**), most runs traverse the web in the compromise direction. If the composite web is placed on nylon threads strung either perpendicular and parallel to a diameter of the web through the retreat (**c**) or at an angle of 45° and 135° (**d,** see *insert*), the spiders follow in some of their runs the exact direction of the threads (*heavy lines* in **d**)

We conclude from these experiments that in a conflict situation, the spider may orientate itself by the pattern of the web either at once or after a short run in a compromise direction. (In his theory on optical and idiothetic orientation of *A. labyrinthica* Mittelstaedt (1978) presented a computer simulation which showed run patterns similar to those obtained from *Agelena* in the case of conflict between optical and idiothetic input. However, in his simulation the influence of the web was not taken into account. See also Kroll 1983 and Wanger 1984.)

This hypothesis is corroborated by another series of experiments in which a composite web was placed on a net of nylon threads strung in a round frame. In the experimental run the spiders sometimes followed one thread quite exactly for a while, then changed their course to follow another perpendicular thread (Exp. 14, Fig. 9c, d).

Agelena's ability to follow the line of highest tension in an experimentally deformed web or to run against the gradient of elasticity can be interpreted as a simple taxis ("baso-taxis", Jander 1963). It is also possible that spiders are able to navigate by the elasticity pattern of the web as they do by optical and idiothetic cues, i.e., during their runs they may store and integrate information about orientational cues of the substrate. However, there are no experiments to date which prove this hypothesis. It is not yet known either how *Agelena* perceives the tension or elasticity pattern of the web.

It may measure the extent to which the web gives way under its legs and, by comparing these values, discriminate between points of higher and lesser elasticity. It may also take into account the time-, frequency- and amplitude spectra of the vibrations reflected from the edge of the web (see also Barth, Chap. XI, this Vol.).

6 Navigation by Means of Gravity

6.1 Gravity Navigation in *A. labyrinthica?*

Under natural as well as laboratory conditions, most webs made by the funnel-web spider slope toward the retreat (Bartels 1929; Holzapfel 1934). This suggests that gravity may be used as a cue for orientation. (In the experiments described above, only horizontal webs were used. Thus an orientation by means of gravity was excluded. Flattened horizontal webs were obtained by lifting the margin of natural webs onto the top edge of the frame.)

Bartels (1929) showed that *Agelena* does in fact orientate by means of gravity. He tilted the web in such a way that instead of the retreat, the frame corner adjacent to it was the lowest point within the web. Under these conditions, the spider running back from its prey headed toward the lowest point instead of the funnel. Bartels and Holzapfel were able to show that *Agelena* stores information about its direction with respect to gravity during its runs on the web. Nothing is known to date about the properties of this type of memory, however.

If gravity orientation is organized similarly to optical and idiothetic navigation, this should become obvious under similar experimental conditions.

Preliminary experiments (unpublished) do suggest the possibility that detours are calculated in gravitational orientation as well. In these experiments, a segment of the web was tilted up by either 45° or 90° (Fig. 10). The spiders were blindfolded and then lured from the retreat to the prey on the elevated part of the web either directly or via a detour. With the segment turned up by 90°, the spiders ran a short distance straight down and gradually took a direct route back to their retreat. With the segment sloped by 45°, the spiders returned by almost the shortest route. Although these experiments do not completely rule out either idiothetic orientation or orientation by means of the pattern of elasticity of the web, they suggest that *Agelena* is able to navigate by gravity.

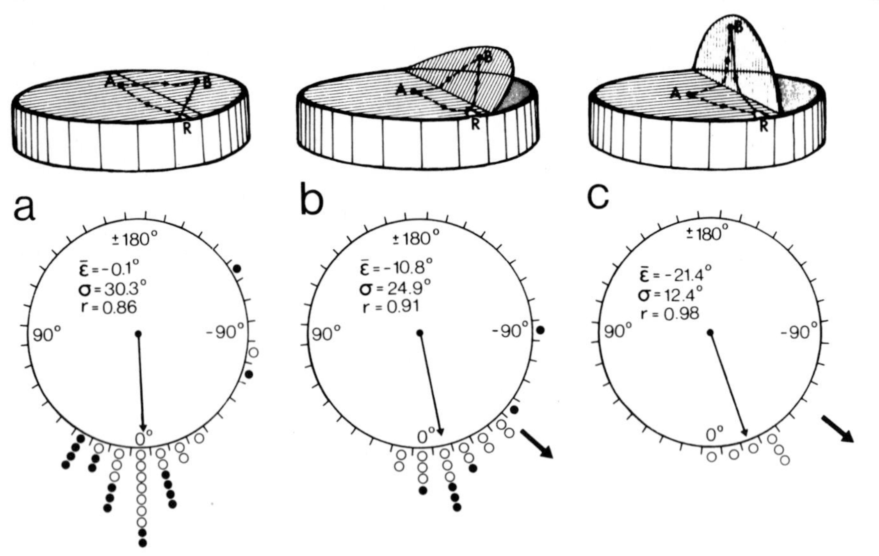

Fig. 10a–c. Single runs and starting directions of seven *A. labyrinthica* on a horizontal web (**a**), and on a segment of a web which is inclined 45° (**b**) or 90° (**c**); *filled circles* direct course to prey site *B; open circles* detour via *A* to *B; thin arrow* mean vector of starting directions; *thick arrow* direction to gravity. All but one of the spiders were blindfolded

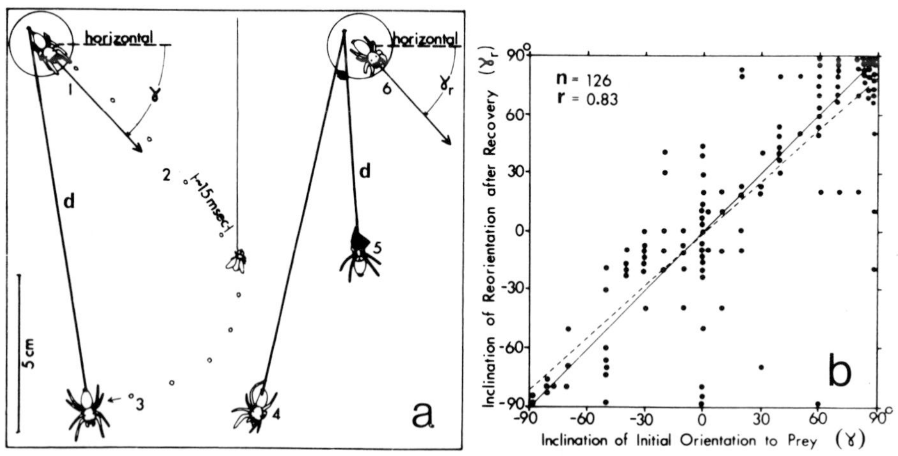

Fig. 11a, b. Gravity orientation in *Phidippus pulcherrimus.* **a** The spider sitting on a vertical disc, facing a fly (*1*), jumping at but missing the fly because the latter has been removed (*2–4*), climbing its dragline *d* (*5*); and reorientating (*6*). **b** Correlation between initial orientation angle to the prey and reorientation angle to the "expected" prey. (From Hill 1979).

6.2 Gravity Orientation in Other Spiders

From the experiments of Peters (1932) it seems likely that orb weaving spiders are able to orientate by means of gravity. There is, however, no convincing evidence of this ability (see also La Guelte 1969). Support for gravity orientation in Salticidae has been presented by Hill (1979) in an elegant experiment. He showed that *Phidippus pulcherrimus* is able to take the same position with respect to gravity on a vertical disc which it had taken before it jumped to catch a fly (for further details see Fig. 11). Because visual cues could not be used for determination of its position on the disc, the spider probably oriented by means of gravity.

7 Conclusions

Although a great deal of data about the orientational faculties of *Agelena* have been gathered and the theoretical analysis of this material has progressed a considerable distance, we are still far from having a comprehensive understanding of the mechanism that governs *Agelena's* homing behavior. Only a few of the sensory organs involved in allothetic and idiothetic orientation are known. Gravity receptors, for example, have not been found in spiders to date; receptors involved in idiothetic orientation have only been dealt with in detail in ctenids, and hardly anything is known about either data processing in the CNS of the spider or the neuroanatomical correlates of the different memories. The amazing faculties of the funnel-web spider in finding its way about its web are unlikely to be exceptional among spiders. Indeed, the results of the few investigations carried out on orientation in other spiders suggest that they may have still other, as yet undiscovered, capabilities.

Acknowledgments. Our thanks are due to Dr. H. Mittelstaedt and the colleagues of our department for many valuable discussions, to Dr. Ursula Görner and U. Will for translating the manuscript, and to Dr. Diana Forsythe for critical reading and correcting of the English draft. We are grateful to Martina Koppers and K. Weigel for technical assistance. Supported by a grant of the Minister für Wissenschaft und Forschung des Landes Nordrhein- Westfalen to P. Görner.

References

Baccetti B, Bedini C (1964) Research on the structure and physiology of the eyes of a lycosid spider. I. Microscopic and ultra-microscopic structure. Arch Ital Biol 102:97–122

Baltzer F (1930) Über die Orientierung der Trichterspinne *Agelena labyrinthica* (Cl.) nach der Spannung des Netzes. Rev Suisse Zool 37:363–369

Bartels M (1929) Sinnesphysiologische und psychologische Untersuchungen an der Trichterspinne *Agelena labyrinthica* (Cl.). Z Vergl Physiol 10:527–591

Bartels M, Baltzer F (1928) Über Orientierung und Gedächtnis der Netzspinne *Agelena labyrinthica.* Rev Suisse Zool 35:247–258

Barth FG, Seyfarth E-A (1971) Slit sense organs and kinesthetic orientation. Z Vergl Physiol 74:326–328

Buddenbrock W von (1952) Vergleichende Physiologie, vol I. Sinnesphysiologie. Birkhäuser, Basel

Dornfeldt K (1973) Die Bedeutung der Haupt- und Nebenaugen für das Heimfindevermögen der Trichterspinne *Agelena labyrinthica* (Clerck) mit Hilfe einer Lichtquelle. Dissertation, Freie Univ Berlin

Dornfeldt K (1975a) Die Bedeutung der Haupt- und Nebenaugen für die photomenotaktische Orientierung der Trichterspinne *Agelena labyrinthica* (Cl.). Z Tierpsychol 38:113–153

Dornfeldt K (1975b) Eine Elementaranalyse des Wirkungsgefüges des Heimfindevermögens der Trichterspinne *Agelena labyrinthica* (Cl.). Z Tierpsychol 38:267–293
Ehlers M (1939) Untersuchungen über Formen aktiver Lokomotion bei Spinnen. Zool Jahrb Syst 72:373–499
Frisch K von (1965) Tanzsprache und Orientierung der Bienen. Springer, Berlin Heidelberg New York
Fröhlich A (1978) Verhaltensphysiologische Untersuchungen zur Lokomotion der Spinnen *Agelena labyrinthica* Cl. und *Sitticus pubescens* F. Dissertation, Freie Univ Berlin
Görner P (1958) Die optische und kinästhetische Orientierung der Trichterspinne *Agelena labyrinthica* (Cl.). Z Vergl Physiol 51:111–153
Görner P (1962) Die Orientierung der Trichterspinne nach polarisiertem Licht. Z Vergl Physiol 45:307–314
Görner P (1966) Über die Koppelung der optischen und kinästhetischen Orientierung bei den Trichterspinnen *Agelena labyrinthica* (Cl.) und *Agelena gracilens* C. L. Koch. Z Vergl Physsiol 53:253–276
Görner P (1972) Resultant positioning between optical and kinesthetic orientation in the spider *Agelena labyrinthica* Clerck. In: Wehner R (ed) Information processing in the visual systems of arthropods. Springer, Berlin Heidelberg New York
Görner P, Claas B (1980) The influence of the web on the directional orientation in the funnel-web spider *Agelena labyrinthica* Clerck. Verh Dtsch Zool Ges 1979, 316. Fischer, Stuttgart
Görner P, Zeppenfeld Chr (1980) The runs of *Pardosa amentata* (Araneae, Lycosidae) after removing its cocoon. 8th Int Congr Arachnol, Vienna. Egermann, Vienna, pp 243–248
Hardie RC (1984) Properties of photoreceptors R7 and R8 in dorsal marginal ommatidia in the compound eyes of *Musca* and *Calliphora*. J Comp Physiol 154:157–165
Henton WW, Crawford FT (1966) The discrimination of polarized light by the *Tarantula*. J Comp Physiol 52:26–32
Hill DE (1979) Orientation by jumping spiders of the genus *Phidippus* (Araneae: Salticidae) during the pursuit of prey. Behav Ecol Sociobiol 5:301–322
Holzapfel M (1934) Die nicht-optische Orientierung der Trichterspinne *Agelena labyrinthica* (Cl.). Z Vergl Physiol 20:55–116
Holzapfel M (1935) Experimentelle Untersuchungen über das Zusammenfinden der Geschlechter bei der Trichterspinne *Agelena labyrinthica* (Cl.). Z Vergl Physiol 22:656–690
Jander R (1957) Die optische Richtungsorientierung der roten Waldameise (*Formica rufa* L.). Z Vergl Physiol 40:162–238
Jander R (1963) Insect Orientation. Annu Rev Entomol 8:95–114
Kirschfeld K (1972) Die notwendige Anzahl von Rezeptoren zur Bestimmung der Richtung des elektrischen Vektors linear polarisierten Lichtes. Z Naturforsch 27b:578–579
Kroll C (1983) Zur Theorie der Navigation durch Wegintegration: Lösungen im Ortsfrequenz- und im Winkelbereich bei gegebenem Leistungskatalog. Diplomarbeit, Tech Univ München
Kurth B (1972) Die Orientierung der Trichterspinne *Agelena labyrinthica* (Clerck) nach Parametern des Netzes. Diplomarbeit, Freie Univ Berlin
Labhart T (1980) Specialized photoreceptors at the dorsal rim of the honeybee's compound eye: Polarizational and angular sensitivity. J Comp Physiol 141:19–30
Land MF (1969) Structure of the retinae of the principal eyes of jumping spiders (Salticidae: Dendryphantinae) in relation to visual optics. J Exp Biol 51:443–470
Le Guelte L (1969) Learning in spiders. Am Zool 9:145–152
Linsenmair KE (1968) Zur Lichtorientierung der Walzenspinnen (Arachnida, Solifugae). Zool Jahrb Physiol 74:254–273
Magni F (1966) Analysis of polarized light in wolf-spiders. In: Bernhard CG (ed) The functional organization of the compound eye. Pergamon Press, Oxford London
Magni F, Papi F, Savely HE, Tongiorgi P (1964) Research on the structure and physiology of the eyes of a lycosid spider. II. – The role of different pairs of eyes in astronomical orientation. Arch Ital Biol 102:123–136
Magni F, Papi F, Savely HE, Tongiorgi P (1965) Research on the structure and physiology of the eyes of a lycosid spider. III. – Elektroretinographic responses to polarized light. Arch Ital Biol 103:136–158

Mittelstaedt H (1978) Kybernetische Analyse von Orientierungsleistungen. In: Hauske G, Butenand E (eds) Kybernetik 1977. Oldenbourg, München Wien, pp 144–195

Mittelstaedt H, Mittelstaedt M-L (1973) Mechanismen der Orientierung ohne richtende Außenreize. Fortschr Zool 21:46–58

Moller P (1970) Die systematischen Abweichungen bei der optischen Richtungsorientierung der Trichterspinne *Agelena labyrinthica*. Z Vergl Physiol 66:78–106

Papi (1955a) Astronomische Orientierung bei der Wolfsspinne *Arctosa perita* (Latr). Z Vergl Physiol 37:230–233

Papi F (1955b) Ricerche sull' orientamento astronomico di *Arctosa perita* (Latr) (Araneae Lycosidae). Publ Stn Zool Napoli 27:76–103

Papi F (1959) Sull' orientamento astronomico in specie del gen. *Arctosa* (Araneae Lycosidae). Z Vergl Physiol 41:481–489

Papi F, Serretti L, Parrini S (1957) Nuove ricerche sull' orientamento e il senso del tempo di *Arctosa perita* (Latr.) (Araneae Lycosidae). Z Vergl Physiol 39:531–561

Papi F, Syrjämäki J (1963) The sun-orientation rhythm of wolf spiders at different latitudes. Arch Ital Biol 101:59–77

Peters H (1932) Experimente über die Orientierung der Kreuzspinne *Epeira diademata* Cl. im Netz. Zool Jahrb Abt Allg Zool Physiol 51:239–288

Schröer W-D (1974) Zum Mechanismus der Analyse polarisierten Lichtes bei *Agelena gracilens* CL Koch (Araneae, Agelenidae). I. Die Morphologie der Retina der vorderen Mittelaugen (Hauptaugen). Z Morphol Tiere 79:215–231

Schröer W-D (1975) Polarised light detection in an Agelenid spider, *Agelena gracilens* (Araneae, Agelenidae). Proc 6th Int Arachnol Congr, Amsterdam, pp 191–194

Schröer W-D (1976) Polarisationsempfindlichkeit rhabdomerialer Systeme in den Hauptaugen der Trichterspinne *Agelena gracilens* (Arachnida: Araneae: Agelenidae). Entomol Germ 3:88–92

Seyfarth E-A, Barth FG (1972) Compound slit sense organs on the spider leg: Mechanoreceptors involved in kinesthetic orientation. J Comp Physiol 78:176–191

Seyfarth E-A, Hergenröder R, Ebbes H, Barth FG (1982) Idiothetic orientation of a wandering spider: Compensation of detours and estimates of goal distance. Behav Ecol Sociobiol 11:139–148

Wanger J (1984) Navigationsmotorik bei Trichterspinnen. Dissertation, Tech Univ München

Wehner R (1982) Himmelsnavigation bei Insekten. In: Bossard HH (ed) Neujahrsbl Naturforsch Ges Zürich. Naturforsch Ges, Zürich

XV Analytical Cybernetics of Spider Navigation

HORST MITTELSTAEDT

CONTENTS

1 Aims and Means

The navigation of the funnel web spider is a complex behavioral performance: it depends on the animal's motivation, which in turn changes seasonally, diurnally, and by dint of various vicissitudes including the consequences of its own success of failure; it uses allothetic sources of spatial information such as the overall light distribution, specifically the sun and the polarization pattern of the sky, the structure and the inclination of the web, as well as idiothetic sources such as stored records of the animal's own movements provided by proprioceptors or efference copies (see also Görner and Claas, Chap. XIV, this Vol.); it employs large arrays of receptors and effectors connected by a central nervous organization which might lead to perfect navigation, were it not for the ever-present influence of noise at all levels of the system.

Consequently, it is impossible to predict the navigational behavior at any moment, e.g., the exact trajectory in space and time of a spider which has, for example, captured a fly at the periphery of the web and is just starting to move, prey in claw. Nor would it ever be possible in retrospect, after the spider has completed its excursion, to retrace all the processes which actually shaped its course. How, then, do we come to understand the homing performance of the

Max-Planck-Institut für Verhaltensphysiologie, D-8131 Seewiesen, Federal Republic of Germany

animal? By extracting, from observation of the temporal behavior, exactly those time-invariant properties of the underlying system which make the navigational performance possible. This search is guided by questions such as: what information is needed, how is it gained, and how is it processed to yield the performance. In fact, ours is an engineering task in reverse: whereas the scientific engineer tries to attain a desired performance from his knowledge of the theory of the system, the biologist tries to develop the theory of the system from his knowledge of its performance.

As the history of this kind of research shows, its first steps can successfully be done using common sense to design experiments, and common language to formulate conclusions. Very soon, however, as we shall find out below, the limits of such means become apparent. In order to control one's assumptions, sharpen one's conclusions, and formulate precise hypotheses about the inner structure of the information-processing system, one needs the methodological and theoretical tools of modern information technology. Yet, since the biologist's task is the inverse of the engineer's, he needs a different method of approach. Whereas the latter is free to use any ways and means to achieve at least a sufficient, or at best an optimal performance, the former must disclose the way the performance is attained in the biological case at hand. Therefore, instead of suggesting merely sufficient or even optimal solutions, I shall try to follow a dialectical sequence of deductions and experimental cross-checks, thereby climbing, as it were, the logical tree of the theoretically possible alternative solutions to our problem.

2 Basic Alternatives

The female of the funnel-web spider (family Agelenidae) spins a flat, more or less horizontal web covering an area of 300 to 1000 cm² with a downsloping open end tube situated normally, but not necessarily, in one corner, her retreat. Alarmed and guided solely by mechanical cues caused by a prey, thence she darts out to capture, and thither she dashes back. The primary alternative initiates the first question: is homing achieved with or without information about the spatial relation between spider and retreat? Clearly, when the outbound path was straight, an about-face at the point of return would suffice (cf. Bartels 1929); yet, as shown by Görner (1958, see also Chap. XIV, this Vol.), *Agelena* can home by the shortest way after detours on the way out, in any starting position, and from any location. In these cases, then, the spider must have had information about its relation to home.

This opens the next alternative: is this information gained at the point of return ("on-site information") or on the way to it ("en-route information")? The alternative may be resolved by passive transport, under cueless conditions, to the "site of release". *Agelena* as well as *Tegenaria* is perfectly able to home in total darkness, yet, if the spider is lifted from her retreat in darkness and set down somewhere on the web (Moller 1970; Dornfeldt 1972) or if, again in the dark, she is lured out to a prey and lifted with the prey to some other place on the web, she appears to be totally disoriented. In the first case, she mostly starts scurrying around; in the second, she mostly takes a straight but randomly oriented course (Dornfeldt 1972; Kurth 1974). It would be premature, however, to exclude "on-site information" altogether. A statistical analysis of the second case in *Tegenaria* (Manert 1983) shows a small but higher than chance number of perfect returns in an otherwise circularly homogeneous distribution. Also, upon hitting the border of the web, which in this case was perfectly circular, she usually returned alongside it, yet chose in a significant ($p < 0.001$) number of cases (72%) the shorter one of the two ways (see also Görner and Claas, Chap. XIV, this Vol.). The decisive point in our case is, however, that although on-site cues should abound on a web constructed by the user herself, and be easily accessible to an animal equipped with highly developed mechanoreceptors (see Chaps. VII to IX, XI and XII, this Vol.), the spider homes rather poorly when forced to use them, yet almost perfectly when en-route information is available. The spider even homes, in darkness, from the first excursion on a web spun by another animal or assembled from various pieces on top of an artificial netting by the experimenter.

Opening our third alternative, en-route navigation can be done by "piloting" or by "path integration". In the first case the information consists of a sequence of stored local features picked up on the way out, in the second in a running computation of the present location from the past trajectory. The first allows, but also constrains, its user to retrace his outward path; the second permits return on the shortest line and even detours on the way. In large samples of return runs in *Tegenaria* one occasionally finds a few cases which seem to copy the detours of the outbound excursion; in general, however, the shape of the outbound trajectory is not correlated with the shape of the inbound trajectory. There can be no doubt, then, that, for a fast, efficient and reliable homing performance, path integration is both necessary and sufficient.

3 Basic Constituents of Path Integration

In order to compute one's present location from one's past trajectory, it seems intuitively evident that one must somehow take account of all rotatory as well as translatory movements made on the way. But from here on, sheer commonsense begins to fail us: For instance, would it be sufficient to compute the mean of the summed rotations weighted by the respective path segments? In the example of Fig. 1 this would lead to

$$\varphi_{res} = \frac{\Delta\varphi_1 \cdot \Delta s_1 + (\Delta\varphi_1 + \Delta\varphi_2)\,\Delta s_2 + (\Delta\varphi_1 + \Delta\varphi_2 + \Delta\varphi_3)\,\Delta s_3}{\Delta s_1 + \Delta s_2 + \Delta s_3} = 34.5^\circ . \qquad (1)$$

At the start for home, a rotation about the angle $\Delta\varphi_{home}$

$$\Delta\varphi_{home} = \varphi_{res} - (\Delta\varphi_1 + \Delta\varphi_2 + \Delta\varphi_3) \pm 180^\circ \qquad (2)$$

would then turn the spider very closely into the right direction. However, a few detours and loops on the outward excursion will result in disastrous deviations. Hence, integration, over the path, of the *angles* by which the spider deviates

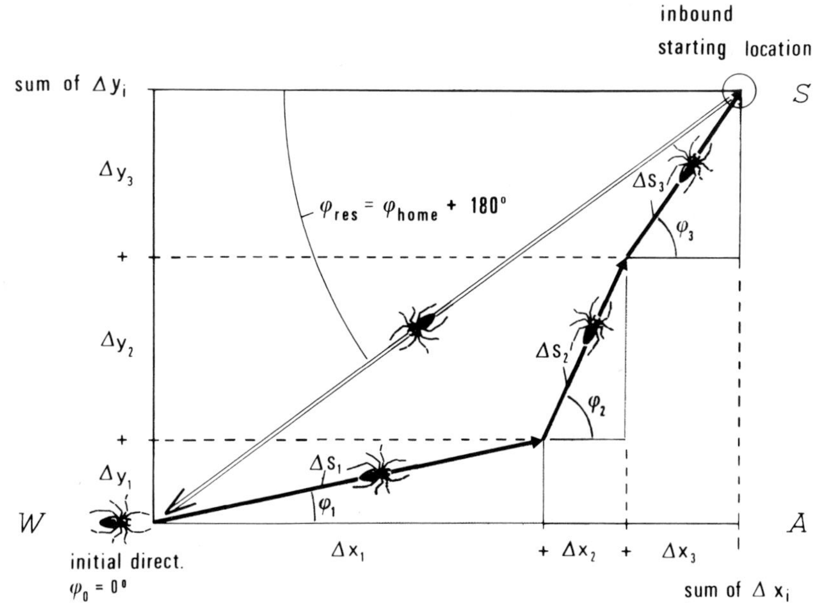

Fig. 1. Simplified example of homing to show geometrical basis of path integration. *W* initial *location* = place of the spider at the beginning of the outbound excursion; *W – A* initial *direction* = direction of the spider's long (X-)axis at that moment. φ deviation of the spider's X-axis from initial direction; Δs length of straight path segments; *S* location of the spider at the beginning of the return. If the path segments Δs_i and the orthogonal components of φ, $\sin \varphi_i$ and $\cos \varphi_i$, are known, the coordinate values of the animal's location may be computed as $\Sigma \Delta y_i = \Sigma \Delta s_i \sin \varphi_i$ and $\Sigma \Delta x_i = \Sigma \Delta s_i \cos \varphi_i$. Their quotient yields the tangent of φ_{res}, which is the inverse of the direction φ_{home} into which the spider must head for a perfect return. In the general case, the Δs_i must be sufficiently small or summation be replaced by continuous integration

from the starting direction, while tolerable for moderately curved trajectories, is insufficient to explain the animal's actual performance. As may be seen from Fig. 1, the mathematically correct solution to our problem is

$$\tan \varphi_{res} = \frac{\sum \Delta y_i}{\sum \Delta x_i} = \frac{\sum \Delta s_i \sin \varphi_i}{\sum \Delta s_i \cos \varphi_i} \tag{3}$$

that is, one needs the vector sum of the path segments or, in the general continuous case, the two integrals, over the path, of the sines and the cosines of the deviations φ_i from the starting direction φ_0

$$\tan \varphi_{res} = \frac{\int \sin \varphi_i \, ds}{\int \cos \varphi_i \, ds} . \tag{4}$$

This solution could be implemented by actually computing the two integrals of Eq. (4) separately, that is, by representing one's location in Cartesian coordinates. Possibly, it may also be realized by summation of the vectors in a joint angular representation, that is, in polar coordinates. Before we can tackle this, our fifth alternative, however, we first need to know the source of the angular information and the way it is processed.

4 The Angular Information

The angular information may either be idiothetic (Mittelstaedt and Mittelstaedt 1973), that is, derived from a stored record of the animal's own rotations, or allothetic, that is, originating from sources which are unaffected by the animal's movements. As to the condition of total darkness, this (6th) alternative is decided by luring the spider to a fly and then rotating the frame of a perfectly horizontal, circular web about the vertical axis through its center, relative to the laboratory. *Agelena* (Görner 1958) and *Tegenaria* (Mittelstaedt 1978) nevertheless return to the retreat as perfectly as without rotation. This outcome also excludes inertial origin of the idiothetic information, thus also deciding the 7th and opening the 8th alternative: is the idiothetic information based on inflow or outflow, that is, on mechanoreceptive signals which monitor the movements of the legs, or from efference copies of the commands which control them? Unfortunately, by contrast with the respective state of affairs in *Cupiennius* (Barth and Seyfarth 1971; Seyfarth and Barth 1972; Seyfarth et al. 1982), this important problem has not yet been tackled in *Agelena* or *Tegenaria*. Howbeit, there remains no doubt that the spider navigation system can compute a representation of the angular deviation from some starting reference – based on nothing but a record of its rotations relative to the substratum.

If the spider loses contact with it, however, the value of this information is in jeopardy: it could hardly be supposed to be still correct, if, say, the spider fell through a hole and clambered up again on its twistable safety thread. Indeed there is evidence that its influence on the return run is actively shunted: as mentioned above, the animal resorts to searching and on-site information if it is lifted from the web in darkness. Careful analysis shows, furthermore, that the deviation of the return direction from home is then not correlated with the angular difference between lift-up and drop-down. Conclusive evidence on this point will result from the subsequent inquiry into alternatives for angular information.

5 Interaction of Idiothetic and Visual Angular Information

If the animal is lifted from the web in sunlight, it returns straight home, more or less. If, in addition, the web is rotated by 90° when the spider is in mid-air, it runs in the reverse of the compass direction of the outbound excursion missing the home direction by – 90°. If it is not lifted during the rotation, it runs mostly into intermediates between the home and the reversed compass direction. Taken together, this proves the point made above about the inactivation of the idiothetic information after a lift-off. It also shows the influence of visual angular cues. But how do they interact with the idiothetic ones?

When many runs of this type are plotted with reference to the home direction, the pattern resulting in the third experiment (Fig. 2) resembles neither a

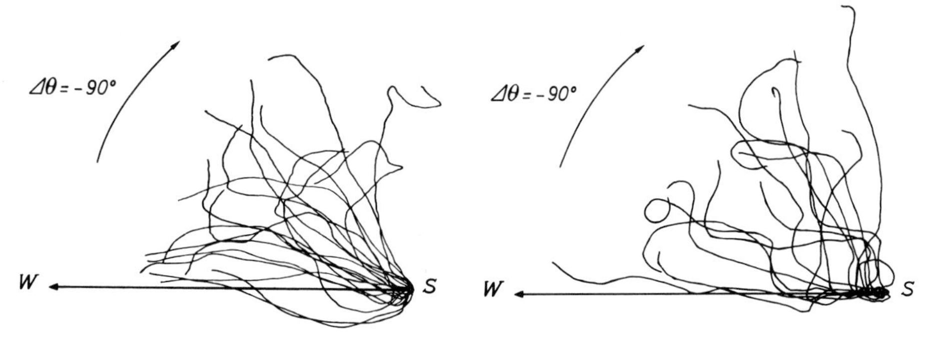

Fig. 2. *Left side* return trajectories of a female of *Agelena labyrinthica* (recorded in collaboration with P. Görner) after a shift of light direction of $\Delta\Theta = -90°$. The abscissa coincides with the normalized inbound excursions (not shown). *Right side* computer simulation of this kind of test based on the filter frequency theory. (Mittelstaedt 1978; cf. later in the text.) *S* and *W* see Fig. 1

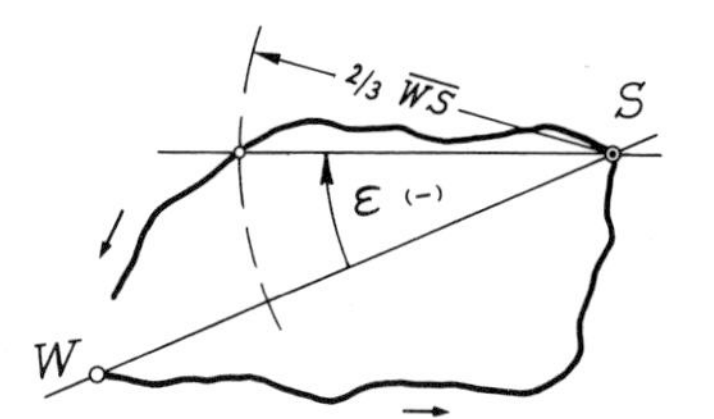

Fig. 3. Definition of homing error ε. *W* outbound, *S* inbound starting points. ε the angle subtended, at *S*, by *W* and the intersection of the return path with a circle about *S* with radius of ⅔ the distance $\overline{WS}$

superposition of the patterns of the first two, nor a shift of one of them into an intermediate direction. A crude but nevertheless rather instructive way of evaluating such a pattern is to determine the "homing error" ε as the angle, at the point of return, subtended by the retreat and the intersection of the return trajectory with a circle of a radius of two thirds of the home distance (Fig. 3), and sum the home deviations vectorially, treating them as unit vectors. In the laboratory experiments, the sun is replaced by two projectors with controllable shutters, and the spider trajectory is recorded automatically for computer evaluation. As was first shown by Moller (1970), the system behaves as if the mean homing error, $\bar{\varepsilon}$, were the resultant of two vectors, one pointing homeward and the second deviating from home by the angle $\Delta\Theta$ of the light shift, or, mathematically,

$$\mathrm{E} \sin \bar{\varepsilon} + \mathrm{L} \sin (\bar{\varepsilon} - \Delta\Theta) = 0 \tag{5}$$

and hence

$$\tan \bar{\varepsilon} = \frac{\sin \Delta\Theta}{\frac{\mathrm{E}}{\mathrm{L}} + \cos \Delta\Theta} \tag{6}$$

where E/L is the relative magnitude of the idiothetic vector E to the light vector L. Whereas the former is assumed to be constant, the latter proved to be a monotonic, probably logarithmic or power function of the light intensity (Dornfeldt 1975 a). The homing error after a lift-off obtains with $E = 0$.

6 A Cartesian Solution

If the visual system indeed provided variables which are proportional to the sine and cosine of the deviation β of the spider's long axis from the azimuth of the sun, and the idiothetic system respectively of the deviation σ of the long axis from some direction on the web, e.g., its midline, the theoretical requirements laid down in Eqs. (3) and (4) would be met in a rather clear-cut way: to begin with the light, let the sine and cosine components of β be separately integrated over the path s during the entire excursion and the result be permanently cross-multiplied with the actual components of β during the return trip, then a rotatory command r_{light} to the legs is obtained which will turn the animal into the home direction.

$$r_{light} = L \sin\beta \int L \cos\beta \, ds - L \cos\beta \int L \sin\beta \, ds. \tag{7}$$

If all operations are executed correctly during the return, the two integrals will be zero when the spider has again arrived at the retreat. The same would be true for an idiothetic homing command r_{idio}:

$$r_{idio} = E \sin\sigma \int E \cos\sigma \, ds - E \cos\sigma \int E \sin\sigma \, ds. \tag{8}$$

If both commands are added, the relations of Eq. (6) would result indeed (with E/L replaced by E^2/L^2). In an analysis of the dynamics of spider navigation, Wanger (1984) has shown that a system of this kind does not only yield the mean values of the experimentally found homing errors $\bar{\varepsilon}$, but, with plausible assumptions about the stochastics of the intervening noise, also leads to similar patterns of homing trajectories as those shown in Fig. 2.

Although the present, "Cartesian", solution to our problem appears to be sufficient, at least for an explanation of the results given above (cf. Fig. 4), it cannot be relied on before we are sure about its place on the "logical tree" of possible alternatives, specifically on the "polar-Cartesian" alternative (the fifth) left open above. I suggest postponing the decision further, however, until we have explored some important consequences of the present solution. Because they pertain also to the "polar" alternatives, this can be done without loss of generality.

7 Consequences

A system operating according to Eq. (7) provides homing independent of the initial azimuth β_0 of the sun. The same is true for Eq. (8) as to the initial value of the idiothetic azimuth σ_0, and even if both are combined additively, provided this is done *after* the formation of the motor commands r_{light} and r_{idio} of Eqs.

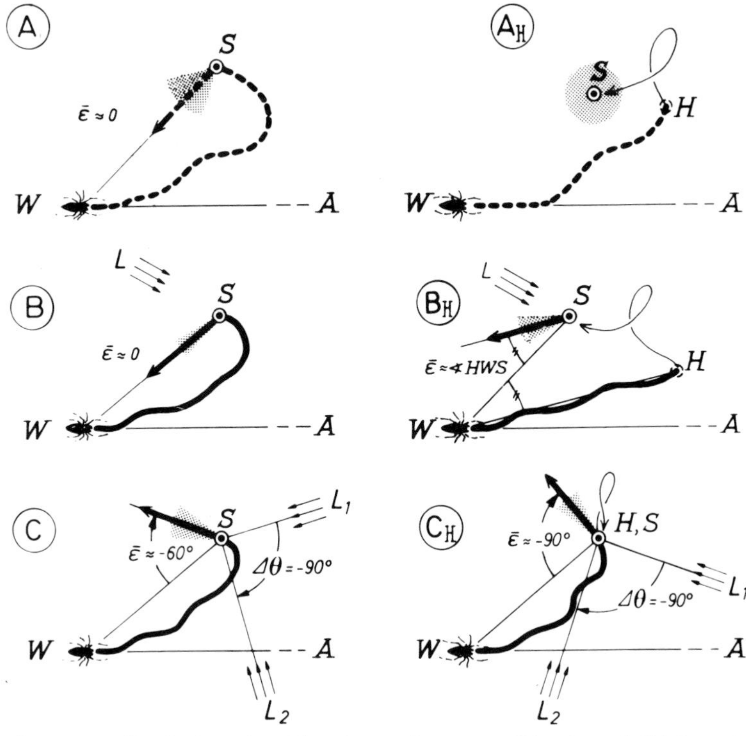

Fig. 4. Test paradigms revealing basic traits of homing performance. *W* outbound, *S* inbound starting points. *Dotted line* runs in total darkness. *Loop from H to S* lift of the spider from point *H* to *S; W–A* initial direction of the spider's long axis; ε see Fig. 3; $\Delta\Theta$ angular shift of light direction from source *L1* to *L2; grey zone at S* standard deviation of homing errors ε_i (approximate values!). For details see text

(7) and (8):

$$r = r_{light} + r_{idio} . \tag{9}$$

Since the sun's azimuth changes by more than 180° on a mid-European summer day, this would be a very valuable property of such a system. It has consequences, however, which can be readily tested, for instance by the paradigms shown in Fig. 5. In the experiment D the outbound excursion is divided into two legs, the first to the right in the light, the second across the web in the dark. Before the inbound start at S, the same light is switched on again. Equation (9) now predicts a deviation of the return path to the right, since r_{idio} alone would direct the spider homeward, whereas r_{light} alone would direct it into the reverse direction of the first leg. In fact, however, the spider heads straight home! One might suspect, of course, that an intelligently constructed system should realize, as it were, that the visual information must now be misleading if summed as usual, and hence disregard it in this case. That should indeed be possible, provided the two modalities were separately integrated and hence all four integrals be available for separate cancelation. An additional light shift just before the

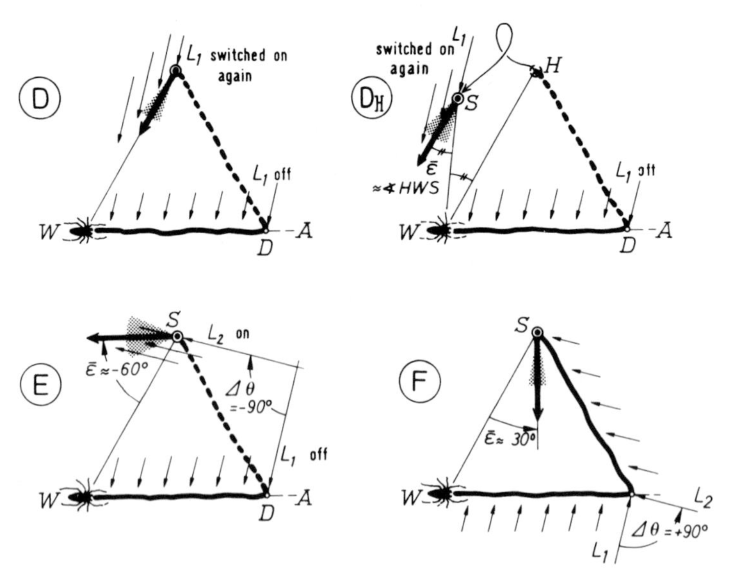

Fig. 5. Test paradigms of crucial experiments **D, E, F.** In all four, the spider is led on a two-leg path along the sides of a nearly equilateral triangle. Visual and/or idiothetic conditions are changed midway and/or at the point of return. **D** point where light was turned off; all other symbols as in Fig. 4

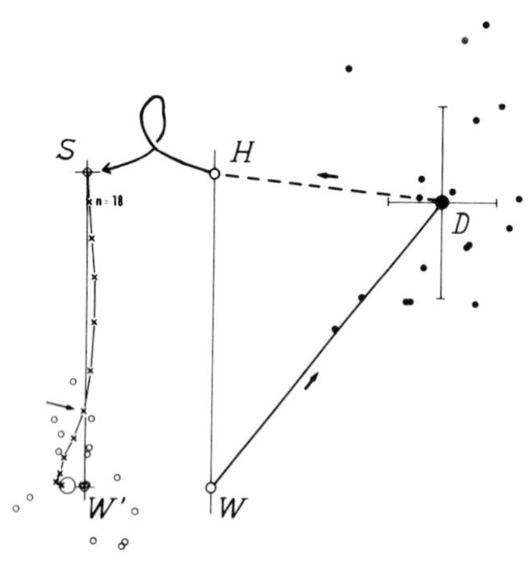

Fig. 6. Eighteen runs under test paradigm DH of a house spider (*Tegenaria*). Under light from source L1, the spider is led on a more or less winding path to *different* points D_i, from there in the dark to points H_i, lifted there from the web and dropped down at points S_i. In the right diagram, all 18 recordings are *normalized* with respect to distance and direction of $W_i - H_i$. *Small full circles* points D_i; *D* their mean with standard deviations (nota bene of the normalized recordings; in absolute terms they would be still larger!). The return paths (*left diagram*) are segmented with 1/6 of $\overline{WH}$, and the coordinated segments x---x summed vectorially with reference to the direction *H–W* through *S*. Around the sixth segment (*arrow*) runs begin to arrive at the home location (*small open circles*) causing a convergence of the mean vectors with the mean home location (*large open circle*). Neither the mean vectors nor the single runs show any noticeable correlation to the *form* of the preceding path *W–D–H*, whereas the mean homing performance is nearly perfect

inbound start (Fig. 5E), however, does result in exactly the same deviation from home as in Fig. 4C, showing that the light influence is *not* canceled. One might try to catch at a straw by assuming that the integration is *always exclusively idiothetic* and the light modality is merely used for the computation of the final return direction, additively of course in order to satisfy Eqs. (5) and (6). The experiment of Fig. 5F shows, however, that the light influence is operative all the time. Otherwise the amount of the return path deviation should be as in Fig. 4C. In fact it is significantly smaller [and, by the way, close to the deviation predicted by Eqs. (7), (8), (9)]. To top it all, in addition to the treatment of the test in Fig. 5D, the spider is lifted from the web after a detour in the dark, before the initial light is turned on again (see Fig. 5DH). The hypothesis of Eq. (9) now predicts a return in the reverse direction of the leg traversed in the light (i.e., $\bar{\varepsilon} = \sphericalangle SWD$). In fact, the animal runs in the direction $\bar{\varepsilon} = \sphericalangle SWH$ which would have led it straight home if it had been set down where it was lifted! This, as well as tests D and E, appears to strictly falsify the additive solution (Eq. 9).

Before jumping to conclusions, though, we ought to be quite sure that no loophole remains, for instance the possibility, however remote, that in the experiments (Dornfeldt 1975b) a "menotactic" light angle may have been trained-in by long sequences of return runs from the same inbound starting point S, or as a consequence of the peçuliar geometric constraints resulting from runs in equilateral or isosceles triangles. Thus we have repeated the critical experiments with the house spider *Tegenaria*, yet tried to shuffle those conditions as much as possible. An example of paradigm DH is given in Fig. 6. Although the situation was further aggravated by working under a reduced influence of the light, which (expectedly) caused an increase in variance, the average inbound direction in paradigm DH turned out to be perfectly parallel to the line W – H, thus corroborating the earlier results. Hence we seem to be in the precarious position of having to salvage or to abandon the explanatory vehicle, which so far has served us so well.

8 Conclusions

Taking stock of the situation, one discovers that the source of the predicament is a general problem of any additive bimodal convergence: Although otherwise beneficial, the bimodal summation would wreck the navigational performance, if one party is lacking or merely weakened during a detour en route (and how often may that happen under the shifting sunlight of a windy day!). Actually, however, the system behaves in tests D, DH, and E as if the amplitude of the joint input components (the *length* of the "resultant input vector") were always of standard magnitude, no matter whether both parties are operative or merely one. Yet, almost miraculously, the deviations in tests C, E, and F are nevertheless proportional to the relative amplitudes of the rivaling parties! Obviously it would even be no help if the light input were intensity-independent.

To resolve these seemingly irreconciliable contradictions, I have suggested (Mittelstaedt and Mittelstaedt 1973, 1978, 1983a) that *before integration* the components are *normalized*, that is, mathematically, divided by a common factor F:

$$F = \sqrt{(L \sin \beta + E \sin \sigma)^2 + (L \cos \beta + E \cos \sigma)^2}. \tag{10}$$

Hence, the amplitude parameters L and E in Eqs. (7) and (8) will be replaced by L/F and E/F respectively. In order to understand how this normalization may solve our problems, let us consider the special case that, at the beginning of an excursion, the light azimuth $\beta_0 = \sigma_0 = 0$, so that both may be substituted by the starting direction φ_0 of the spider. Hence the idiothetic input angle σ may be substituted by φ, and the light input angle β by $\varphi - \Delta\Theta$. Let the normalized components be summed to become

$$y_{norm} = \frac{E \sin \varphi + L \sin (\varphi - \Delta\Theta)}{F} \tag{11a}$$

and

$$x_{norm} = \frac{E \cos \varphi + L \cos (\varphi - \Delta\Theta)}{F} \tag{11b}$$

respectively, and the normalization factor F be

$$F = \sqrt{[E \sin \varphi + L \sin (\varphi - \Delta\Theta)]^2 + [E \cos \varphi + L \cos (\varphi - \Delta\Theta)]^2}. \tag{12}$$

The rotatory command r (to the left: positive) after integration and cross-multiplication then reads

$$r = y_{norm} \int x_{norm}\, ds - x_{norm} \int y_{norm}\, ds. \tag{13}$$

This solution directly yields the outcome of all tests reviewed, since

$$F = \sqrt{E^2 + L^2 + 2\,EL \cos \Delta\Theta} \tag{14}$$

and, consequently, the amplitudes of x_{norm} and y_{norm} become unity in all cases. As may easily be seen, this is even true after a light shift, so that even then, (e.g., in tests C and E) the integrals become zero after a deviating run at a distance equalling the distance WS. In these experiments the spider does then, in fact, show the typical sharp turns at this point (Dornfeldt 1972; see also Mittelstaedt 1978, Fig. 9 p. 167). In test F, however, the predicted "home distance" is much larger, usually even extending beyond the web's border. Indeed, on the return run the spider is often in danger of falling out of the web at full speed! The most striking evidence results if in test F the light shift is done in the opposite direction ($\Delta\Theta = -90°$, when the spider is otherwise led as in Fig. 5F). By the incidental (unwitting!) combination of an equilateral triangle as standard path and a value of $E/L = \cot 60°$, the predicted "home point" does coincide with the inbound starting point S: and indeed, instead of dashing off, the spider lingers or scurries around it (Fig. 7).

9 Prerequisites

Normalization is, then, a necessary condition for the explanation of the results of Fig. 5. As a consequence, however, the independence of the performance

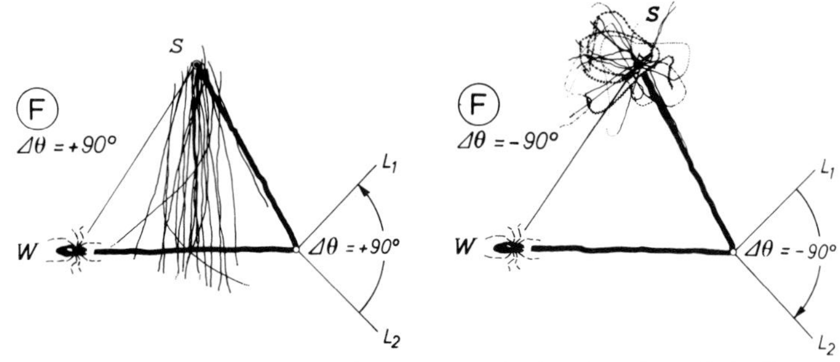

Fig. 7. Recordings of the return trajectory in paradigm F (see Fig. 5 and text) made by Dornfeldt (1975b). The spider is lured along the base of an (almost) equilateral triangle under L_1, and along the right side under L_2 ($\Delta\Theta = +90°$ or $-90°$) to the tip of the triangle, where it is left to its own devices. Note the striking difference seen when (*left*) the light shift $\Delta\Theta$ had been positive (in the same direction as the change of path) or negative (*right*). According to the theory (Mittelstaedt 1973, 1978 ,1983), the home location computed by the spider's navigation system after the light shift is at twice the height of the triangle from the tip "downward" in the first case, whereas at, or very close to, the tip of the triangle in the second – provided $E/L \approx \cot 60°$

from the initial azimuths, discussed above, is no longer guaranteed. This can be most easily seen if the azimuth of the sun is in opposition to the initial position of the spider: $\beta_0 = \sigma_0 + 180°$, and $L = E$; or consider the outcome of test D_H or B_H with $\beta_0 = \sigma_0 - 90°$ (the simplest way is to use Eqs. (11) to (14), but let the light be shifted by $\Delta\Theta = 90°$ *before* the beginning of the outbound excursion!). Consequently, it must be secured that, at the outbound start, the deviation of the visual components from their reference is equal to the deviation of the idiothetic components from theirs, in other words: that the two "vectors" point in the same direction.

With this latter condition we encounter another general problem of a bimodal convergence. Also, only under this condition may the organism reap the benefits which the additive convergence would then entail, namely an increase of the signal-to-noise ratio due to the increase of the input amplitudes. How may the problem be solved? I have suggested (Mittelstaedt 1978) that the actual values of at least one set of input components, preferably the visual ones, are stored at a given moment ("registered"), to serve as a permanent reference, until the need for a new adjustment arises; for instance when the sun deviates from its initial azimuth by an intolerably large angle. Figure 8 shows the definition of the input angles. In this article, all angles are defined, in a righthanded system, with reference to the *web* X-axis [whereas, nota bene, the earlier publications (Mittelstaedt 1978, 1983a) used an *animal*-fixed reference system]. Let the angle σ_0 be the deviation of the spider's long axis from the X-axis of the web which it had at the crucial moment of storage of the visual information, and let β_0 be the angle between the long axis and the light source azimuth at that moment. If the light input is computed from the *difference* between the angle β_0 and the current light angle β_i, and if the idiothetic input is, respectively, proportional to $\sigma_i - \sigma_0 = \varphi$, then the two inputs would be equal as long as the light azimuth remains constant: $\sigma_i - \sigma_0 = \beta_i - \beta_0 = \varphi$, if $\Delta\Theta = 0$. Since the light may shift en route from L1 to L2 by $\Delta\Theta$, however, the light input would in general be proportional to $\varphi - \Delta\Theta$. Hence the conditions of Eqs. (11) to (14) are realized at any initial azimuth of the sun and of the spider.

It must be ascertained, however, that the light reference β_0 has always been properly registered. In the experiments on *Agelena* (Görner 1958; Dornfeldt 1972), it was attempted to ensure this by means of three training runs under the light source L1 before each critical test. A series of experiments to be reported elsewhere in detail has shown that this suffices in most cases. In order to find out whether the reference has been correctly registered after a training run in the light, we turn it off before the start, lure the spider out in the dark – nota bene to a *different* location than in the preceding training run – lift it from the web, and switch the same light on just before or shortly after the animal drops onto the web again. The spider will now run in the right direction, namely homeward, if it has come down where it was lifted (yet in a *different* direction to the light than in the training run!), if and only if the reference is correctly registered (Fig. 9). By means of this check it has been possible to show that, when the spider sits "at attention" in front of its retreat, not more than 7 s of exposure to a light source, at any arbitrary light azimuth, are sufficient to set a new reference. On the other hand, it may happen that the animal stubbornly sticks to its old reference even after three consecutive training runs under a new light direction. This explains, by the way, why, if the light direction is changed *before* the outbound start and kept constant during the entire run, the spider – after lift-off and set-down at the same spot – sometimes deviates in the same direction as in Fig. 4C (yet by a smaller angle than in test C, if E/L < 1, by a larger angle, if E/L > 1!). Also, some puzzling earlier results under such conditions (Bartels and Baltzer 1928) may thus find their explanation.

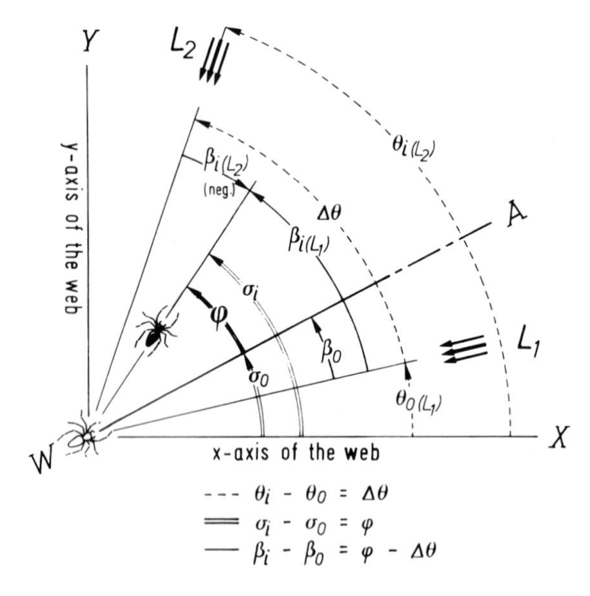

Fig. 8. Definition of input angles. *Web-fixed* right-handed reference system. *Abscissa* X-axis of web; *ordinate* Y-axis of web. Line *W–A* direction of the spider's X-axis at the moment of "registration" (reference setting; see text). $\beta_{i\,(L1)}$, $\beta_{i\,(2)}$ angles between present X-axis of spider and light direction from source L_1 or L_2 respectively. β_0 angle between spider X-axis and light direction (L_1) at the moment of "registration". σ_i present angle between spider's and web's X-axes; σ_0 angle σ at the moment of registration of β_0. $\Delta\Theta$ angular difference between light directions from L_1 to L_2, to left positive [in the diagram all angles except $\beta_{i\,(L2)}$ are positive]. $\sigma_i - \sigma_0 = \varphi$ idiothetic input angle; $\beta_i - \beta_0 = \varphi - \Delta\Theta$ light input angle

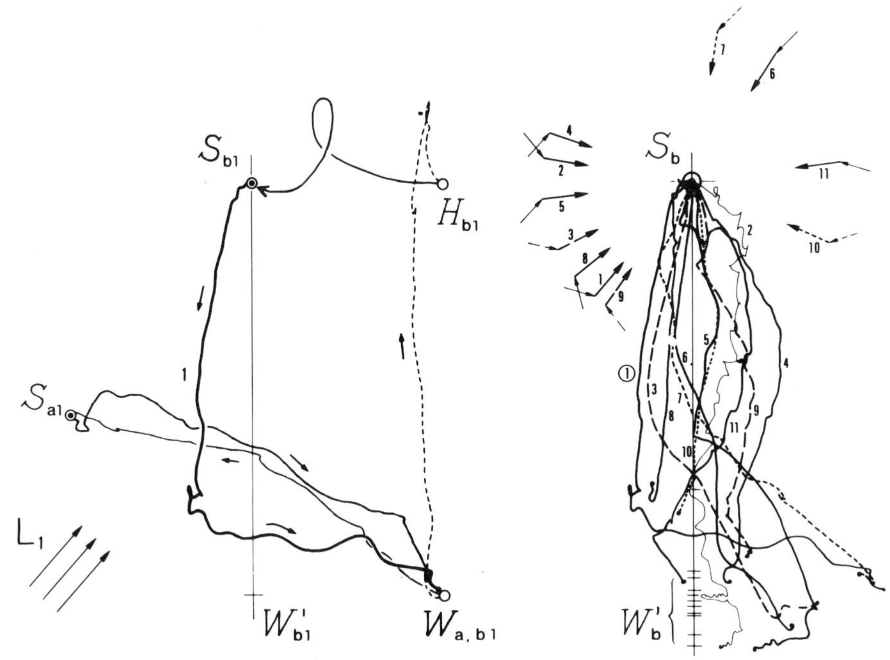

Fig. 9. Experimental check of reference setting in the house spider *Tegenaria*. After a training run *a* under a light source L_1 (inbound start from point S_{ai}), the spider in a subsequent run *b* is lured out in *darkness* to a *different* spot H_{bi}, lifted, and the light source L_1 turned on again just before or immediately after the spider drops down at point S_{bi} (see left side diagram, which shows preceding training run to S_{ai}, and subsequent test from S_{bi} of run 1). Eleven cases of successful "registration" have been selected from a large number to demonstrate the independence of the performance from the light directions (*right side* diagram; *hinged arrows*, the rear part signifying the light direction at the preceding run, the front part that at the runs shown). All inbound runs are drawn with reference to the directions $W_{bi} - H_{bi}$ of the paths in darkness, with the set-down points S_{bi} as common origin S_b, and the distance W_{bi}, H_{bi} is marked by crossbars (W'_{bi}) near the base. All runs would have led the spider homeward, if it had been set down where it was lifted (i.e., if $H_{bi} = S_{bi}$). Since this was normally not the case, "searching turns" occur at about the expected distance (i.e., around W'_{bi})

10 Space or Spatial Frequency

We are now in a position to tackle the postponed fifth alternative: "polar" versus "Cartesian" solutions, representation of directions by spatial maps or by Fourier components, space versus spatial frequency. This indeed concerns a problem of paramount interest in the neurobiology of the visual system. Visual information processing, at least in its primary stages, appears to be done in neural layers where visual angle is mapped into neuron position, that is, where space is represented spatially. In the funnel-web spider the very inputs, at any rate, the rhabdomeres of the eight eyes, map the panorama of the surrounding light distribution in such a way. Eventually, however, the information thus represented *must be transformed into scalar variables fit to control a muscle.* In our case, then, the alternative is an inclusive one: the question is just *where* the transformation occurs.

In my own hypothesis of 1977 (Mittelstaedt 1978) this is supposed to be done at the very beginning by a spatial low pass filter which provides the first 3 or 5 harmonics through a simple weighting process (see also Mittelstaedt 1982, 1983b). After the reference setting, they are summed with the respective idiothetic components, and then normalized by an internal feedback loop which also permits to extract the first harmonics, that is, just two variables proportional, at $\Delta\Theta = 0$, to $\sin\varphi$ and $\cos\varphi$ respectively. Only those two are then integrated over the path variable s, which may be procured by proprioceptors of the legs or by an efference copy[1], thus satisfying Eqs. (13) and (14). Computer simulation of this hypothesis shows a reasonably good fit with the empirical data, particularly with the pattern seen in test C (see Figs. 2 and 4): for bifurcation phenomena are due to occur, if and when the higher harmonics are out of phase at the point of summation.

Alternatively, by realizing the equivalent mathematical operations of cross- and autocorrelation, reference setting, summation, normalization, and integration may also be done on a map. Figure 10 shows, for the visual system, the activity of a circle of neurons after integration, as it might look at the end of the excursion of Fig. 1. In order to determine the home direction, this pattern should now be cross-correlated with the pattern of the *actual* light distribution, which would be a slender single peak in our example. The resulting maximum would then nearly coincide with φ_1, however, instead of with what is wanted, namely: $\varphi_{res} = \varphi_{home} + 180°$. As has been shown by Kroll (1983), the resulting constant error would be the greater the more the path directions diverge, and the more the higher spatial harmonics invade the frequency spectrum of the light (or idiothetic) distributions. The pith of the matter is that information about the spider's true relative location resides in the first harmonics only, and nothing but these can eventually be admitted in a correct computation of the home course.

Consequently the final cross-correlation cannot be done before extraction of the first harmonics, e.g., by smoothing the mapped distributions in analogy to the weighting process suggested above. And this affords a conclusive way of falsifying the correlation hypothesis: If the spider had made an outbound excursion in the beams of *two identical* light sources impinging from opposite directions ($\beta_1 = \beta_2 + 180°$), no first harmonics would be present on the map of Fig. 10; and the spider, if it had been deprived of idiothetic information all the way, would be bound to rotation or search. This is true even if the idiothetic information were only canceled at the inbound starting point. The pattern on the integrated map would then contain a first harmonic indeed. However, the latter would have no counterpart in the pattern of the *actual* distribution after a

---

1 The latter alternative is interesting because it may result in a direct dependence of the idiothetic influence on the spider's distance from the computed "home" (cf. Kroll 1983) and hence may help to explain the pattern of test C and its modifications (cf. Dornfeldt 1972, p. 72, Fig. 11 and Görner 1985, in press).

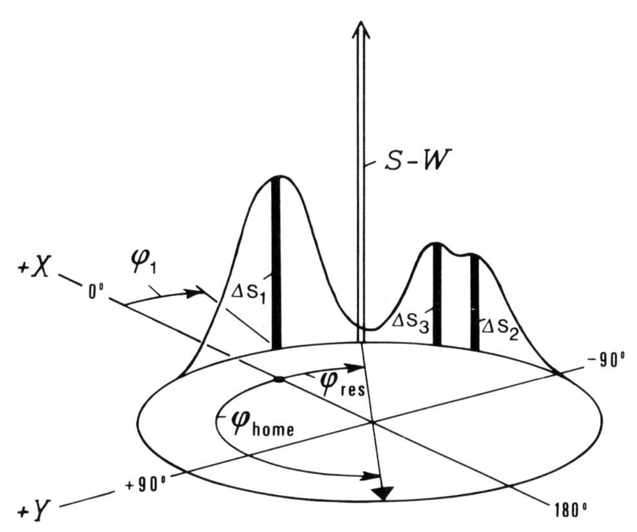

Fig. 10. Path integration in polar coordinates. At the outbound start, a remote light source, say the sun, is assumed to have been located in the direction W–A of the abscissa of Fig. 1 ($\beta_0 = \varphi_0 = 0$). The signals of a uniform circular array of light receptors have been integrated over (here: multiplied by) the path segments Δs_1 to Δs_3 of Fig. 1 and are then stored in a circular "retinotopic" representation. *X-axis* internal representation of the spider's long axis; forward (0°) positive; *Y-axis* internal representation of spider's transverse axis; to left (+90°) positive (note that in the transition from a *web fixed* to an *animal fixed* coordinate system the sign of φ changes) *Black columns* Δs_1 to Δs_3 state of the system in case of nonoverlapping light receptors and a point source. *Smooth curves* state of the system in case of a fuzzy light source, sensory overlap or a Gaussian jitter of the animal's course (no quantitative precision intended)

lift-off, which contains only even number harmonics. Hence its contribution to the result of the cross-correlation would be zero throughout (because $\int_0^{2\pi} [\sin(\varphi - \tau) \sin 2n\varphi] \, d\varphi = 0$). The frequency filter hypothesis of 1977, however, would have the animal then turn into the home direction or its reverse, thereby permitting rotation through obtuse as well as acute angles. This has, in fact, been observed in pilot experiments on *Tegenaria* (Mittelstaedt 1978). Because of the import of the decision, a repetition, and a parallel test in *Agelena*, is certainly called for. If secured, the result would indicate that the transformation from maps to components occurs *before* the integration over the path variable. Reference setting and bimodal convergence, however, may well be executed on maps; in fact, such a hybrid system appears to be the most appealing, if not the most likely.

11 Conclusively Inferred System Properties

Although many details have had to be omitted (which may be found in Mittelstaedt 1978 for the structure of the system, in Kroll 1983 for its mathematics, and in Wanger 1984 for its dynamics), at last, I have come to the present-day

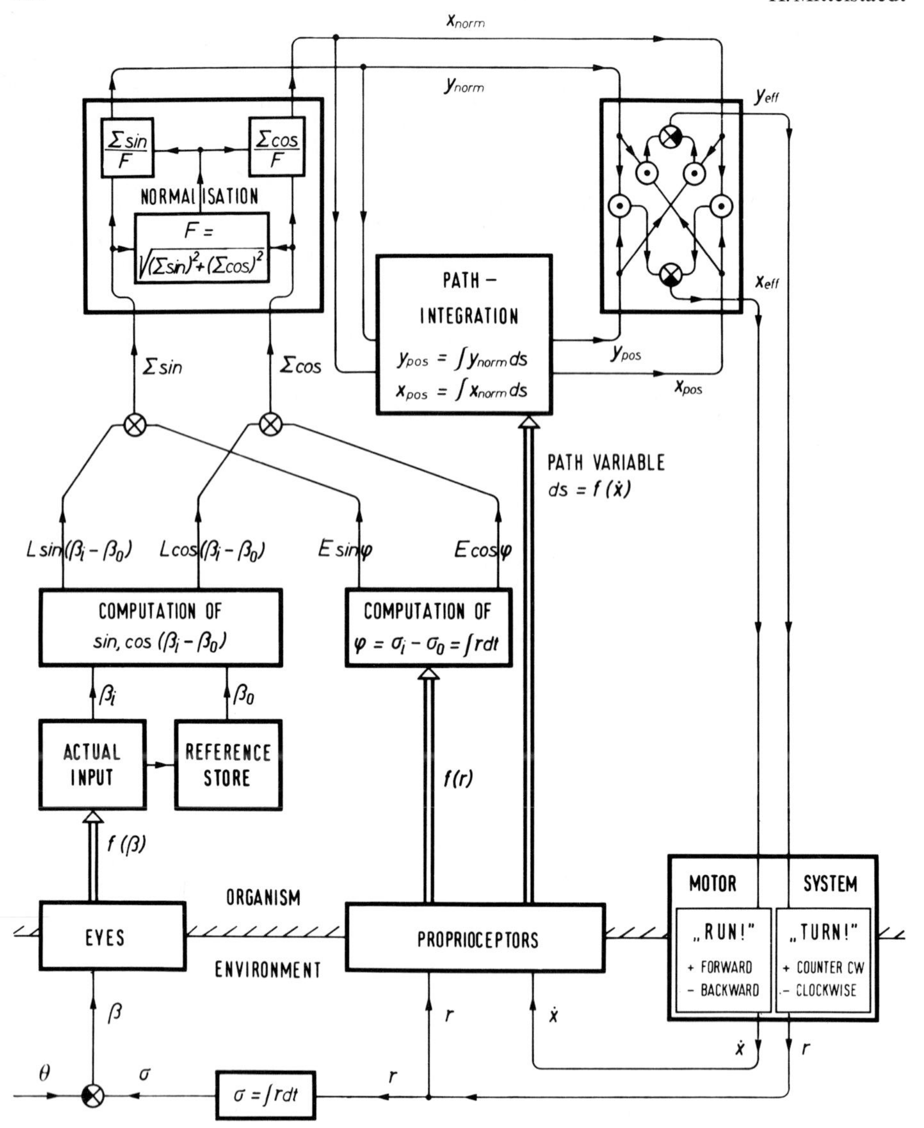

Fig. 11. Block diagram of the signal-processing operations which are shown to be *necessary* for the experimentally demonstrated navigation performances of the funnel-web spider (see last paragraph). The diagram illustrates how they could be connected in a "Cartesian" solution such that the entire control system is also *sufficient;* constraints: the spider moves on a flat horizontal web devoid of on-site cues; the higher spatial harmonics in the inputs, the quantitative temporal properties of the system, and the sideways movements of the spider are disregarded (for attempts at lifting these constraints see Mittelstaedt 1978; Kroll 1983; Manert 1983; Wanger 1984). *Boxes* signal processing operations; *lines* variables (*arrows* direction of causality); ⊗ addition (reversal of sign at black quadrant); ⊙ multiplication. x_{eff} efferent command leading to a translatory velocity $\dot{x}$ in the direction of the spiders long axis (forward positive). y_{eff} efferent command leading to a rotatory velocity r (counter-clockwise positive). For other symbols see text and Fig. 8

end of a long way – with apologies to my learned readers for taxing them with the tedium of logical exertions. Hopefully it has become clear, though, why the procedure is necessary as well as rewarding: certainly, more often than not, discoveries are made by intuition rather than method, and new lines of thought are developed from "models" which are neither sufficient nor necessary. If the quest for a physical explanation of mental and behavioral performances is to succeed, however, a methodological approach is needed which demands and affords conclusiveness and consistency. The gains are insight into the workings of a complex system which could hardly be won otherwise.

To summarize them in the present case: any physiological realization of the spider navigation system must meet the following requirements (cf. Fig. 11):

1. There have to be at least three different stores: one for the retention of the reference angle β_0, one for the running computation of the idiothetic angle φ, one for the result of the path integration, that is, the representation of the spider's computed location (y_{pos}, x_{pos}) at any moment (and, in addition, one for the coordinates of the location of a deposited prey to which the spider wants to return later – probably just another integrator to be set at zero before the animal leaves that location).
2. The idiothetic and light inputs must be normalized such as to satisfy Eq. (14).
3. The reference storage and the computation of the difference $\beta_i - \beta_0$ has to be performed before the bimodal convergence, and the normalization must precede the path integration.
4. If the normalization is done by internal feedback (as in Mittelstaedt 1978), joint integration of the idiothetic and the light inputs and joint computation of the motor commands is a necessary condition. (Yet, if the normalization is done by feedforward division, albeit through a, necessarily, joint factor (F), the two modalities could in principle be integrated separately, although that does not seem very likely. But then, all integrals must be cross-multiplied or cross-correlated with the actual (normalized) inputs of *both* modalities in order to correctly compute the motor commands.)
5. Independently of whether the computation of the relative location is done in space or spatial frequency, it must be *exclusively* based on the first harmonics. Their production must occur *before* the integration, if indeed, deprived of idiothetic information, the animal is still able to turn into the home direction ε_{home} or alternatively into ε_{home} plus the multiples of $2\,\pi/n$ (with n even) under a light distribution which contains only even number harmonics.

Note added in proof:
This decisive ability has been further corroborated in recent experiments under two opposing light sources of equal intensity.

Acknowledgments. I wish to thank Peter Görner, Clemens Kroll, Michael Manert, Werner Mohren, Michael Potegal and Sepp Wanger for stimulating discussions, and, last but not least, the many temporary and permanent staff members whose help has made this work possible.

References

Bartels M (1929) Sinnesphysiologische und psychologische Untersuchung an der Trichterspinne *Agelena labyrinthica* (Cl.) Z Vergl Physiol 10:527–593

Bartels M, Baltzer F (1928) Über Orientierung und Gedächtnis der Netzspinne *Agelena labyrinthica*. Rev Suisse Zool 35:247–258

Barth FG, Seyfarth EA (1971) Slit sense organs and kinesthetic orientation. Z Vergl Physiol 74:306–328

Dornfeldt K (1972) Die Bedeutung der Haupt- und Nebenaugen für das Heimfindevermögen der Trichterspinne *Agelena labyrinthica* mit Hilfe einer Lichtquelle. Dissertation, Freie Univ Berlin

Dornfeldt K (1975 a) Die Bedeutung der Haupt- und Nebenaugen für die photomenotaktische Orientierung der Trichterspinne *Agelena labyrinthica* (Cl.). Z Tierpsychol 38:113–153

Dornfeldt K (1975 b) Eine Elementaranalyse des Wirkungsgefüges des Heimfindevermögens der Trichterspinne *Agelena labyrinthica* (Cl.). Z Tierpsychol 38:267–293

Görner P (1958) Die optische und kinästhetische Orientierung der Trichterspinne *Agelena labyrinthica* (Cl.). Z Vergl Physiol 41:111–153

Görner P (1985) Goal orientation without directional cues in the funnel-web spider *Agelena labyrinthica* Clerck. (in press)

Kroll C (1983) Zur Theorie der Navigation durch Wegintegration: Lösungen im Ortsfrequenz- und im Winkelbereich bei gegebenem Leistungskatalog. Diplomarbeit, Tech Univ München

Kurth B (1974) Die Orientierung der Trichterspinne *Agelena labyrinthica* (Cl.) nach Parametern des Netzes. Diplomarbeit, Freie Univ Berlin

Manert M (1983) Rechnergestützte Untersuchungen über das Verhalten der Trichterspinne bei der Suche der Warte nach Hochheben bei Dunkelheit. Diplomarbeit, Tech Univ München

Mittelstaedt H (1978) Kybernetische Analyse von Orientierungsleistungen. In: Hauske G, Butenand E (eds) Kybernetik 1977. Oldenbourg, München Wien, pp 144–195

Mittelstaedt H (1982) Einführung in die Kybernetik des Verhaltens. In: Hoppe W, Lohmann W, Markl H, Ziegler H (eds) Biophysik. Springer, Berlin Heidelberg New York, pp 822–830

Mittelstaedt H (1983 a) The role of multimodal convergence in homing by path integration. Fortschr Zool 28:197–212

Mittelstaedt H (1983 b) Introduction into cybernetics of orientation behavior. In: Hoppe W, Lohmann W, Markl H, Ziegler H (eds) Biophysics. Springer, Berlin Heidelberg New York, pp 794–801

Mittelstaedt H, Mittelstaedt M-L (1973) Mechanismen der Orientierung ohne richtende Außenreize. Fortschr Zool 21:46–58

Moller P (1970) Die systematischen Abweichungen bei der optischen Richtungsorientierung der Trichterspinne *Agelena labyrinthica*. Z Vergl Physiol 66:78–106

Seyfarth E-A, Barth FG (1972) Compound slit sense organs on the spider leg: Mechanoreceptors involved in kinesthetic orientation. J Comp Physiol 78:176–191

Seyfarth E-A, Hergenröder R, Ebbes H, Barth FG (1982) Idiothetic orientation of a wandering spider – compensation of detours and estimates of goal distance. Behav Ecol 11:139–148

Wanger J (1984) Navigationsmotorik bei Trichterspinnen. Dissertation, Tech Univ München

D The Motor System

XVI Neural Control of the Heartbeat and Skeletal Muscle in Spiders and Scorpions

R. G. SHERMAN

CONTENTS

1 Introduction

Neural control of muscle systems in arachnids is poorly understood in comparison to the other large classes of arthropods, namely the crustaceans and insects. The vast majority of arachnid studies have involved spiders and scorpions, and they have been directed primarily to the muscles of the heart and leg. In this report, the existing information on the neural control of these muscles will be summarized.

2 Neural Control of the Heartbeat

There are basically two different mechanisms for the initiation of the heartbeat. Either muscle cells in the heart itself spontaneously exhibit action potentials that initiate rhythmic contractions of the myocardium or neural elements intimately associated with the myocardium generate periodic bursts of action potentials which are conducted to the myocardial cells, causing them to contract. In the former case, the heartbeat is myogenic; in the latter, the heartbeat is neurogenic. Well-known examples of myogenic hearts occur in molluscs and vertebrates. Perhaps the best-known neurogenic hearts are those of the lobster and horseshoe crab (Prosser 1973).

Department of Zoology, Miami University, Oxford, Ohio 45056, USA

2.1 Spiders

The gross morphology of the heart will not be described; the reader is referred to Foelix (1982) for spiders and to Randall (1966) for scorpions in this respect.

2.1.1 Origin of the Heartbeat

The first indication as to the origin of the heartbeat in spiders came from the finding of a cardiac ganglion on the heart of *Geolycosa missouriensis* (Sherman and Pax 1967), *Heteropoda venatoria* (Wilson 1967) and *Scodra calceata* (Legendre 1968). The ganglion is a thread-like structure that lies mid-dorsally on the heart and extends nearly its entire length (Fig. 1). Demonstration that the cardiac ganglion contains neuron somata and numerous axons came from staining longitudinal sections through the mid-line of the heart with silver (Sherman and Pax 1968). This procedure also showed that axons arising from neuron somata in the cardiac ganglion occur in the myocardium.

Subsequent examination of dozens of different spider species revealed that the presence of a cardiac ganglion on the heart is commonplace in spiders (Sherman et al. 1969). A detailed histological study of the cardiac ganglion in *Geolycosa missouriensis* and *Eurypelma marxi* showed that there are an average of 57 neurons in the cardiac ganglion of the former and 85 in the latter (Bursey and Sherman 1970).

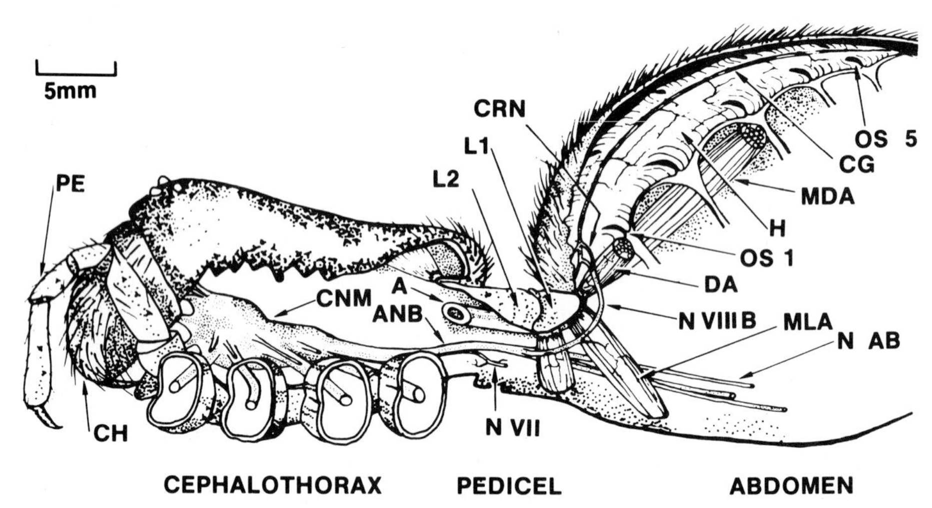

Fig. 1. Lateral view of *Eurypelma marxi* showing the relationship between the cephalothoracic (prosomal) nerve mass (*CNM*), abdominal (opisthosomal) nerve bundle (*ANB*), nerve VIII B, heart (*H*), cardiac ganglion (*CG*) and the cardioregulatory nerve (*CRN*). The opposite member of the paired nerve VIII B cannot be seen. *PE* pedipalp; *CH* chelicera; *A* anterior aorta; *L* lorum; *DA* depressor "abdominis"; *MLA* musculus longitudinalis anterior; *MDA* musculus dorsoventralis "abdominis"; *OS* ostium; *N AB* continuation of the abdominal nerve bundle. (Drawing by F. Gonzalez-Fernandez)

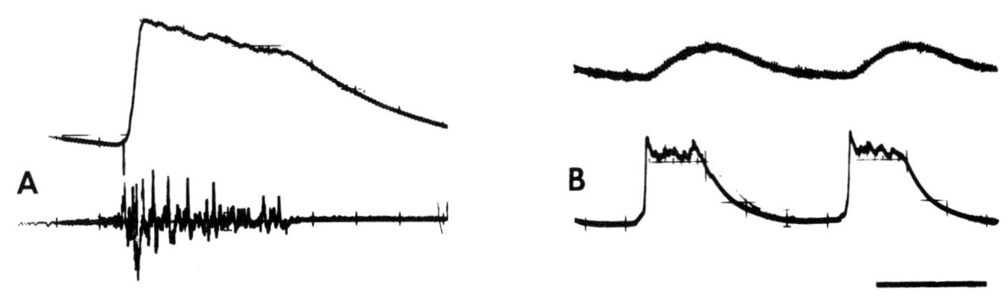

Fig. 2. Recordings that show the sequence of events in a heartbeat. **A** Simultaneous recording of cardiac ganglion electrical activity (*lower trace*) and the depolarization of a myocardial cell (*upper trace*) during a heartbeat. A burst of impulses in the ganglion begins just prior to the myocardial cell depolarization which is maintained until the burst ends. The myocardial cell depolarization is 35–45 mV in amplitude and consists of summating excitatory postsynaptic potentials. **B** Simultaneous recording of the myocardial cell membrane potential (*lower trace*) and contraction of the heart (*upper trace*). Two heartbeats are shown. The prolonged depolarization of the myocardial cell starts just before and continues throughout the contractile phase of the heartbeat. (Bursey and Sherman, 1970)

The first direct physiological information about the origin of the spider heartbeat came from making external electrical recordings of heart activity during the heartbeat. Recordings from isolated hearts immersed in artificial physiological solution and from hearts left intact in the animal showed that periodic bursts of electrical impulses are associated with each heartbeat (Fig. 2A) (Sherman and Pax 1968). This type of electrical activity is characteristic of neurogenic hearts, and is unlike the electrocardiogram of myogenic hearts. The observation that nerve impulses can be recorded from a ganglion that has been removed from the heart and immersed in physiological solution is evidence that they originate in the cardiac ganglion. Impulses cannot be recorded from the heart itself, which ceases to beat once the ganglion is removed.

2.1.2 *Function of the Cardiac Ganglion*

The burst of impulses in the cardiac ganglion initiates the heartbeat as seen in intracellular recordings of the myocardial cell membrane potential during a heartbeat. Just preceding a heartbeat, and persisting throughout the systolic phase, is a prolonged depolarization of each cell in the myocardium (Fig. 2B) (Bursey and Sherman 1970). The depolarization occurs immediately after the onset of a burst of impulses in the cardiac ganglion (Fig. 2A). It is not an action potential such as occurs in myogenic hearts. Instead, the depolarization is a series of excitatory postsynaptic potentials (EPSP's) that sum together to produce the prolonged depolarization of 0.5 s or more.

Direct demonstration that impulses from the cardiac ganglion elicit EPSP's in the myocardial cells comes from recently conducted experiments in which the cardiac ganglion was electrically stimulated between burst of impulses (Gonzalez-Fernandez and Sherman, unpublished). At the same time, the mem-

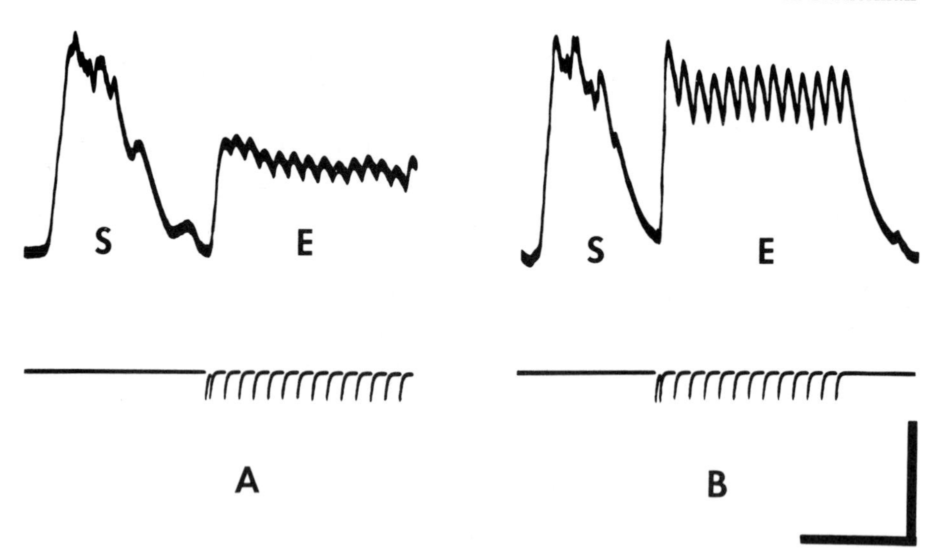

Fig. 3. Recording of the transmembrane potential of a myocardial cell (*upper trace*) and time of electrical stimulation (*downward deflections*) of the cardiac ganglion (*lower trace*). The ganglion was electrically stimulated (*E*) at a frequency of 8 Hz just after the occurrence of a normal heartbeat (*S*). An excitatory postsynaptic potential (EPSP) is elicited in the myocardial cell by each stimulus. The EPSP's sum together at this frequency to produce a maintained depolarization similar to that elicited by a burst of impulses from the cardiac ganglion. In **A** a stimulation voltage of 3 V was used, which activated one motor axon to the myocardial cell. In **B** a second motor axon was recruited along with the first by increasing the stimulation voltage to 4.2 V. This indicates that the cell receives input from two different cardiac ganglion neurons. The vertical calibration line equals 10 mV; the horizontal line equals 1 s. (Gonzalez-Fernandez and Sherman unpublished)

brane potential of a myocardial cell was recorded with microelectrodes. Individual EPSP's occur during the stimulation on a 1:1 relationship with each stimulus. By using a stimulation frequency of 8 Hz, a maintained depolarization is produced as a result of summation of the EPSP's (Fig. 3). Upon increasing the stimulus voltage, a greater depolarization, occurs in a stepwise fashion, indicating that each myocardial cell receives excitatory synaptic input from more than one neuron in the cardiac ganglion. In the example shown in Fig. 3, the myocardial cell receives input from two motor neurons, since only two discrete levels of depolarization could be elicited by stimulation of the ganglion.

Neuromuscular synapses are present on the myocardial cells (Fig. 4), and may be found at more than one location along the surface of the myocardial cell (Sherman 1973a; Ude and Richter 1974). The latter is consistent with the physiological evidence for multiple neuronal input to each myocardial cell. At least some of the neuromuscular synapses presumably are made by motor neurons leaving the cardiac ganglion. Axons have been observed to leave the ganglion and penetrate the myocardium (Fig. 5). However, one cannot rule out the possibility that other neurons besides those in the cardiac ganglion may make direct synaptic contact with myocardial cells.

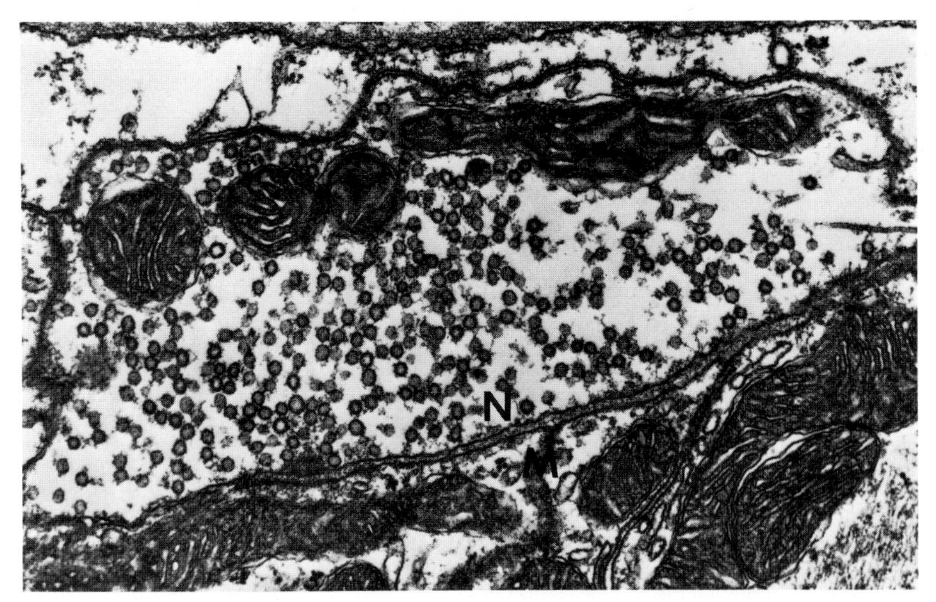

Fig. 4. Transmission electron micrograph of a neuromuscular synapse in the myocardium (*M*). Numerous spherical synaptic vesicles are prominent in the nerve terminal (*N*). ×30,000. (Sherman 1978)

2.1.3 Regulation of Cardiac Ganglion Activity

While the cardiac ganglion functions to initiate the heartbeat, to coordinate the synchronous contraction of the myocardial cells and to set the basal rate and strength of the heartbeat, it is clear that both extrinsic neurons and a variety of chemicals can influence the activity of the heart. The literature contains reports on the effects of several different kinds of chemicals that alter the rate and strength of the heartbeat (Sherman and Pax 1970a, b; Ude and Richter 1974). Some of these may be released by intrinsic and extrinsic nerves that connect to the heart, while others may serve a hormonal function.

Decades ago, Carlson (1905) reported that electrical stimulation of the spider prosomal nerve mass leads to inhibition or acceleration of the heartbeat. This suggests that neurons in the prosoma innervate the heart directly, since a ventral nerve cord is absent in spiders. The neurons would modulate the periodic bursting activity of the cardiac ganglion to alter the rate of the heartbeat.

A neural connection between the central nervous system and the heart was found by Legendre (1968) and Gonzalez-Fernandez and Sherman (1979), who traced branches of nerve VIIIb in the opisthosoma to the cardiac ganglion. Nerve VIIIb arises in the opisthosoma from the main ventral opisthosomal nerve bundle that extends posteriorly from the prosomal nerve mass (Fig. 1) (see also Babu, Chap I, and Legendre, Chap III, this Vol.).

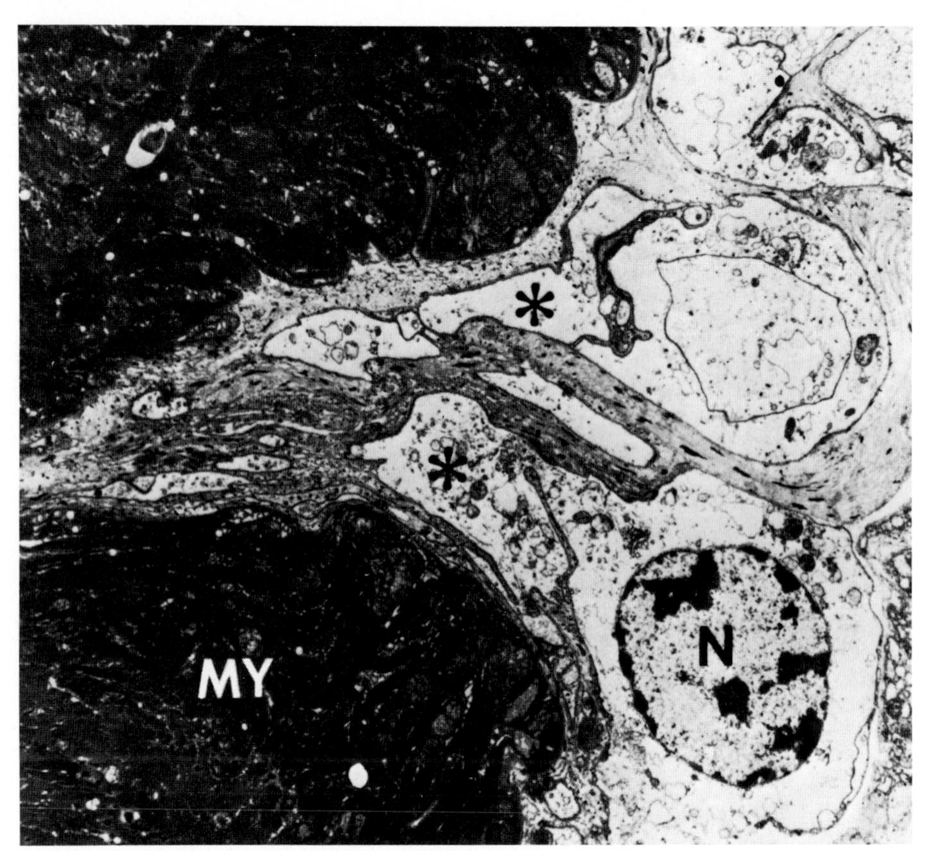

Fig. 5. Transmission electron micrograph showing that axons (*asterisks*) from neurons (*N*) in the cardiac ganglion extend into the myocardium (*MY*). ×5500. (Sherman 1973b)

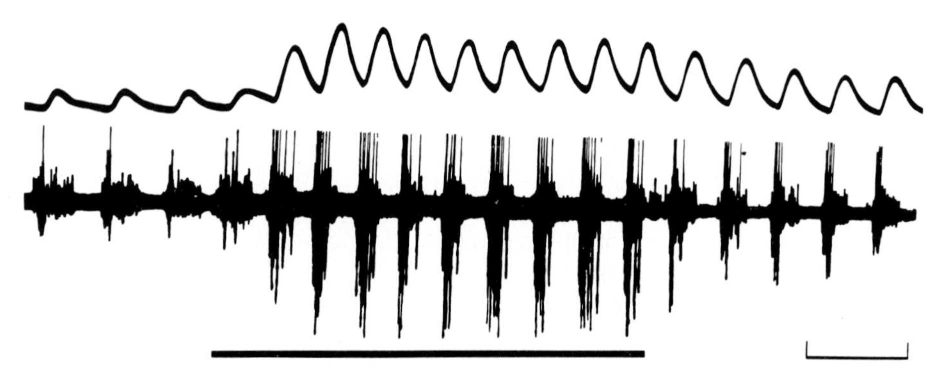

Fig. 6. Effects of an acceleratory axon on cardiac ganglion electrical activity (*lower trace*) and contractions of the heart (*upper trace*). The cardioaccelerator axon was stimulated at 50 Hz during the period shown by the line. The calibration line equals 4 s. (Gonzalez-Fernandez and Sherman 1984)

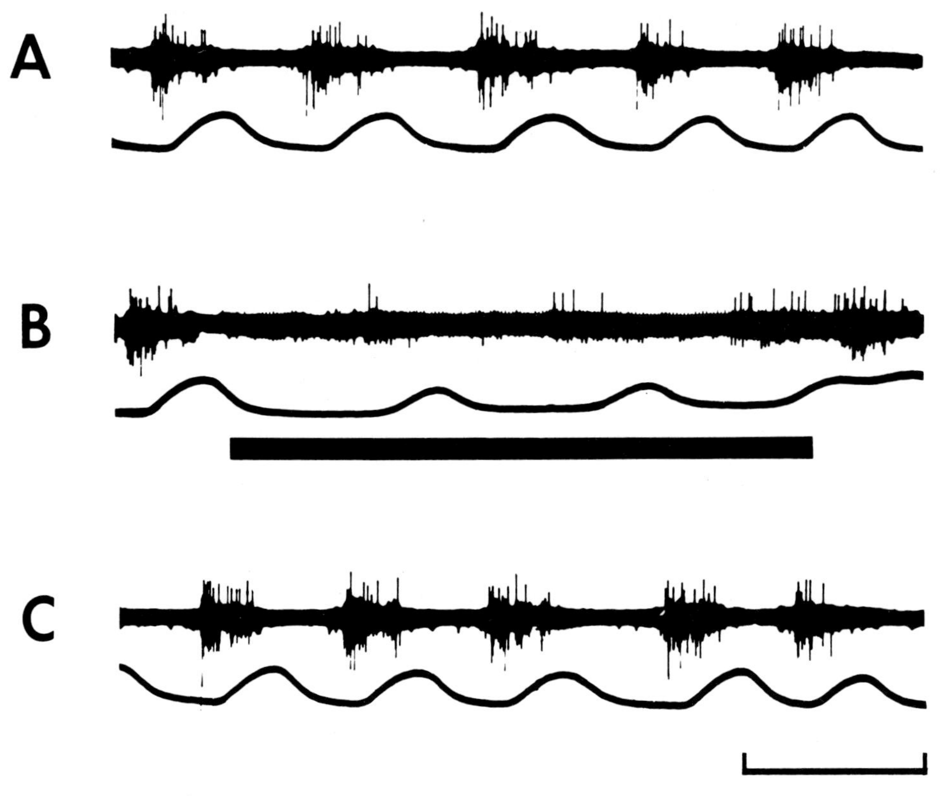

Fig. 7. The effects of an inhibitory axon on cardiac ganglion electrical activity (*upper trace*) and contractions of the heart (*lower trace*). The cardioinhibitory axon was stimulated at 20 Hz during the period shown by the line in **B.** The recording is continuous from **A** to **C.** The calibration line equals 2 s. (Gonzalez-Fernandez and Sherman 1984)

Electrical stimulation of either the opisthosomal nerve bundle or nerve VIII b produces inhibition and acceleration of the heartbeat (Gonzalez-Fernandez and Sherman 1984). Stimulation of other nerves which arise from the main opisthosomal nerve has no effect on the heart rate. Severing the branch of nerve VIII b that connects to the cardiac ganglion abolishes the effects of electrical stimulation. These observations demonstrate that two branches, one from each member of the paired nerve VIII b, function as cardioregulatory nerves.

The actions of the cardioregulatory neurons on cardiac ganglion activity are illustrated in Figs. 6 and 7. When an inhibitory axon is stimulated, it reduces the number of impulses in each ganglionic burst and increases the interval between bursts (Fig. 7). This decreases the rate and strength of the heartbeat. Activation of an acceleratory axon has the opposite effects. It increases the number of impulses in each burst and decreases the inter-burst interval (Fig. 6).

Thus far, the spider heart does not appear to differ significantly from the well-known neurogenic hearts of the lobster and horseshoe crab in terms of basic mechanisms.

2.2 Scorpions

2.2.1 Origin of the Heartbeat

Police (1902) first described the presence of a nerve on the dorsal surface of the scorpion heart. Zwicky and Hodgson (1965) demonstrated that in *Uradacus novaehollandiae* this nerve is a cardiac ganglion. They found neuron somata in the nerve and noted that axons arising from some of the somata leave the ganglion to innervate the myocardium. Naidu and Padmanabhanaidu (1975) have reported a cardiac ganglion in *Heterometrus fulvipes.*

Physiological studies have shown that a burst of electrical impulses in the ganglion precedes each contraction of the heart (Zwicky and Hodgson 1965; Naidu and Padmanabhanaidu 1975). Removal of the cardiac ganglion causes a cessation in beating. Transection of the ganglion in several places leads to asynchronous contractions of the affected areas of the myocardium.

These observations strongly suggest that the heartbeat in scorpions is neurogenic, and that the cardiac ganglion functions to initiate and to coordinate the heartbeat. Additional studies employing microelectrode recording methods and transmission electron microscopy are needed to further elucidate the role of the cardiac ganglion in the heartbeat.

2.2.2 Regulation of the Heartbeat

Zwicky (1968) has demonstrated the presence of cardioregulatory neurons in *U. novaehollandiae.* He found that agitation of the scorpion leads to cardiac acceleration; this effect is abolished by transecting the ventral nerve cord anterior to the first free ventral ganglion. By transecting the cord and segmental nerves at other locations, Zwicky concluded that the cardioacceleratory nerves reach the first free ganglion and travel to the heart via the first pair of segmental nerves. From similar transection experiments coupled with electrical stimulation of the ventral nerve cord, he concluded that the cardioinhibitory fibers exit to the heart in segmental nerves from the first two pairs of free ventral ganglia. Presumably, the cardioregulatory axons innervate the cardiac ganglion, and perhaps other structures in the heart, but this has not been demonstrated directly in scorpions.

Only a few studies have tested the actions of well-known neurotransmitter chemicals on the scorpion heart. Curiously, acetylcholine reduces the heart rate (Kanungo 1957; Zwicky 1968). This is surprising, since neurogenic hearts typically are excited by acetylcholine (Prosser 1973). Eserine alone is excitatory on the heart; however, it prevents the effects of exogenously applied acetylcholine and potentiates the inhibition produced by the electrical stimulation of the segmental nerves. Clarification of these effects awaits a careful investigation of the sites of action of these chemicals, for there are several locations in the neurogenic heart system where they may be active.

Choline esterase activity has been reported in cardiac muscle of *H. fulvipes* (Venkatachari and Naidu 1969; Naidu 1974). Eserine inhibits the esterase activity. The possible role of acetylcholine in the scorpion neurogenic heartbeat deserves further study.

3 Skeletal Neuromuscular Systems

The gross anatomy of the skeletal muscles will not be reviewed here. The reader is referred to Ruhland and Rathmayer (1978) and Bowerman and Root (1978) for details of walking leg muscle morphology in spiders and scorpions respectively (see also Seyfarth, Chap. XII, this Vol.).

3.1 Spiders

3.1.1 Muscle Ultrastructure

The ultrastructure of spider skeletal muscles has been described by Kawaguti and Kamishima (1969), Zebe and Rathmayer (1968) and Sherman and Luff (1971). The fibers are transversely striated, contain thick and thin myofilaments, show an H zone in the center of the A band, and have from 8 to 12 thin myofilaments around each thick myofilament. A sarcoplasmic reticulum (SR) and a transverse tubular system (TTS) are very abundant. Dyads are prevalent and the SR and T-tubule membranes facing each other show specialized structures associated with the dyadic junction (Franzini-Armstrong 1974).

Since the length of sarcomeres has been shown to be inversely related to contraction speed in some crustacean muscle fibers (Jahromi and Atwood 1969), sarcomere length in spiders is of particular interest. Zebe and Rathmayer (1968) examined four different leg muscles in *Dugesiella hentzi* and found sarcomeres ranging from 3.0 to 7.3 μm. Kawaguti and Kamishima (1964) reported that the femoral flexor muscle of *Heteropoda venatoria* has sarcomeres from 3.5 to 5.5 μm in length. Neither study reported if this range occurs in each muscle or if some muscles characteristically have shorter sarcomeres than others. The first known attempt to determine if spider muscles differ in sarcomere length was done by Sherman and Luff (1971). They measured sarcomeres in the two tarsal claw muscles, the levator praetarsi and the depressor praetarsi, in *Eurypelma marxi*. Sarcomeres in the former average 3.2 + 0.2 μm and in the latter, 6.2 + 0.4 μm. To date, there have not been any attempts to directly relate contraction speed with sarcomere length in arachnids.

3.1.2 Motor Innervation

In the first physiological study of the innervation of an arachnid skeletal muscle, Rathmayer (1965) found that the basitarsus flexor muscle fibers in *D. hentzi* are innervated by three excitatory neurons (Fig. 8). All three elicit EPSP's of similar amplitude and time course, but they differ in the degree of facilitation. Two of them produce fast contractions of the whole muscle, while

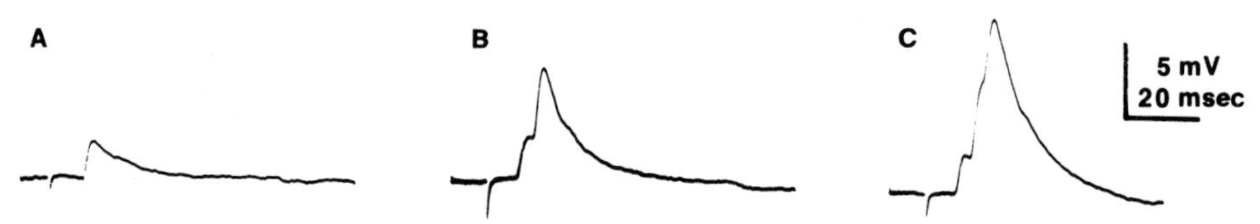

Fig. 8. Excitatory postsynaptic potentials (EPSP's) recorded from the basitarsus flexor muscle of the spider *Eurypelma (Dugesiella) hentzi*. In **A** one motor axon was stimulated; in **B** a second motor axon was recruited along with the first resulting in two EPSP's that summed together. In **C** a third motor axon was recruited, resulting in a third EPSP's adding to the first two. (Rathmayer 1965)

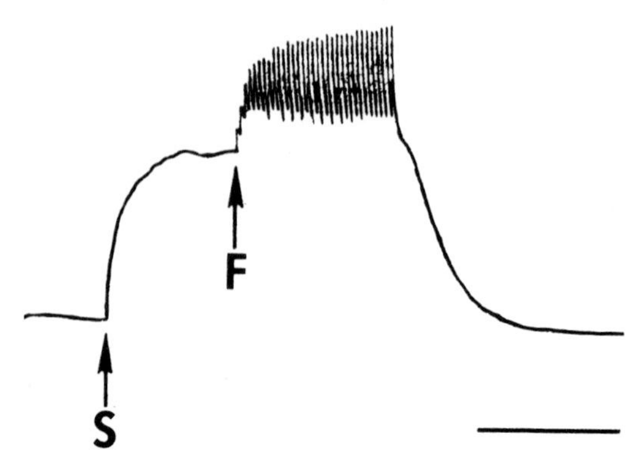

Fig. 9. Whole muscle tension recording from the tarsal claw levator muscle of *Eurypelma marxi*. During the recording, the strength of the electrical stimulation of the motor nerve was gradually increased. A slow-graded contraction (*S*) was evoked first, followed later by fast, twitch contractions (*F*) evoked at a higher stimulus strength. Stimulation frequency is 3 Hz; calibration line equals 10 s. (Sherman 1973b)

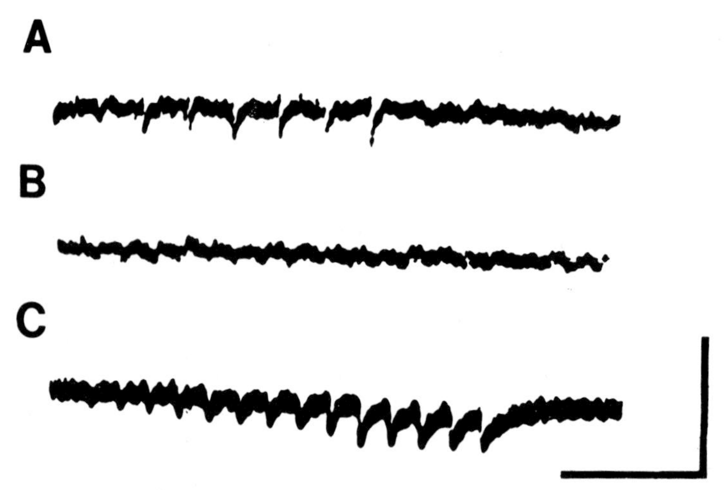

Fig. 10. Hyperpolarizing inhibitory postsynaptic potentials (IPSP's) recorded from the promotor tibiae muscle in *Dugesiella hentzi*. **A** shows IPSP's evoked by low frequency axon stimulation. **B** shows that the IPSP's are abolished by adding picrotoxin, a known GABA blocker, to the saline. In **C** the axon was stimulated in normal saline at a higher frequency than in **A.** Note the progressive increase in IPSP amplitude, indicating facilitation. Vertical calibration equals 500 μV; horizontal calibration equals 100 ms in **A** and **B,** and 200 ms in **C.** (Brenner 1972)

the third elicits a slow contraction. Both slow-graded and fast muscle contractions also were found for the claw levator muscle of *E. marxi* (Fig. 9) (Sherman 1973b).

Transmission electron microscopic studies have revealed between 1 and 12 axonal profiles associated with each region of synaptic contact on spider skeletal muscle fibers (Zebe and Rathmayer 1968; Sherman and Luff 1971). Neither of the above studies determined the total number of neurons to each muscle.

In a morphological study, Land (1969) reported that each of the six muscles that move the anterior median eye in salticids is innervated by a single motor axon. This suggests that each muscle represents a single motor unit. Melamed and Trujillo-Cenóz (1971), studying the anterior median eye muscles in lycosids, found only two muscles involved. One was innervated by three neurons and the other by one neuron.

Rathmayer (1966) demonstrated that each muscle fiber in the leg flexor tarsi major and minor in *Eurypelma (Dugiesella) hentzi* is innervated at places along its entire length and that each receives input from three excitatory motoneurons. Thus, the innervation of these muscle fibers can be characterized as both multiterminal and polyneuronal. The available information suggests that this is true in general for spider skeletal muscle fibers, except in those cases where some muscles seem to be innervated by only one neuron.

The first report of peripheral inhibition in spider skeletal muscle was published by Brenner (1972). In studying the promotor tibiae muscle in *D. hentzi,* he recorded small hyperpolarizing potentials which were evoked by electrical stimulation of the motor nerve (Fig. 10). The potentials underwent facilitation as the stimulation frequency was raised. Brenner also was able to record a decrease in the amplitude and the facilitation of EPSP's when the inhibitory axon was activated prior to the excitatory axon. The application of gamma-aminobutyric acid (GABA), a known crustacean inhibitory neuromuscular transmitter, to the preparation reduced substantially the amplitude of neurally evoked slow muscle contractions. GABA also reduced the membrane input resistance of slow muscle fibers, and this effect was prevented by picrotoxin, a known GABA blocker.

There is evidence for the occurrence of an inhibitory motoneuron to the superficial fibers in the depressor praetarsi muscle of *D. hentzi,* in addition to the three excitatory motoneurons that are present (Ruhland 1976).

3.1.3 Synaptic Ultrastructure

The ultrastructural features of neuromuscular synapses are basically the same for all of the spider muscles examined except for the number of axon terminals (Zebe and Rathmayer 1968; Melamed and Trujillo-Cenóz 1971; Sherman and Luff 1971; Sherman 1973b).

Each motor axon gives off numerous branches which extend over the muscle fiber surface perpendicular to the long axis of the muscle (Fig. 11A). Each branch forms multiple synaptic contacts. The smaller axon branches are embedded beneath the basement membrane portion of the sarcolemma in a

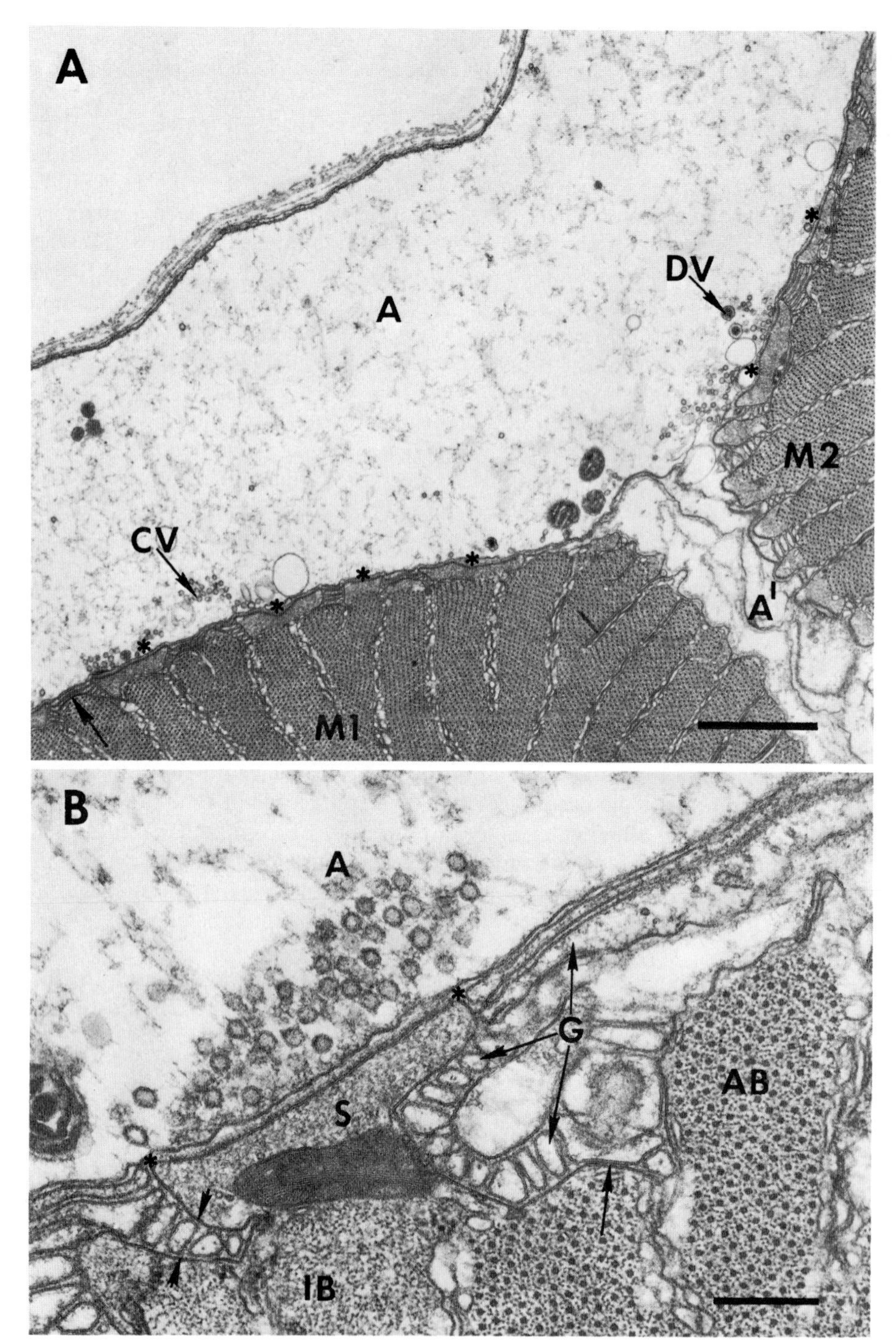
A
A
DV
M2
CV
A'
M1
B
A
G
AB
S
IB

Fig. 11. Neuromuscular synapses formed by a motor axon and two muscle fibers in the claw-levator muscle of *Eurypelma marxi.* In **A** the main axon (*A*) gives off a branch (*A'*) between the muscle fibers (*M1, M2*). Numerous synaptic contacts (*asterisks*) are formed by the axon. Glial processes (*unlabeled arrow*) occur between adjacent synapses. Both clear-cored (*CV*) and dense-cored (*DV*) vesicles are present in the axon. **B** is an enlarged view of an area of synaptic contact between a motor axon and a muscle fiber in the tarsal claw-levator muscle of *E. marxi.* The axon (*A*) is surrounded by glial elements (*G*) except in the synaptic cleft (*asterisks*). The sarcoplasm (*S*) beneath the synapse has a granular appearance and lacks myofilaments. Gap junctions formed by glial and muscle fiber membranes are indicated by the *unlabeled arrow.* I band and A band contractile regions of the muscle fiber are designated *IB* and *AB* respectively. (Sherman 1973b)

trough formed by an indentation of the muscle fiber or by a slight expansion of the basement membrane outward from the fiber surface. The axon branches are enveloped by glial cell processes except at the point of synaptic contact between axon and muscle fiber.

The synapses have the usual details, including clusters of spherical, electron-lucent vesicles of 45–60 nm diameter, axon and muscle fiber membrane densities and dense material in the synaptic cleft (Fig. 11B). Omega figures suggestive of synaptic vesicles releasing their contents were observed by Melamed and Trujillo-Genóz (1971). Less numerous vesicles, larger in diameter and electron-opaque, also occur at some synapses (Atwood et al. 1971). See also Foelix, Chap. X, this Vol., on synapses found in the peripheral nervous system.

3.1.4 Neuromuscular Transmission

Ultrastructural studies of neuromuscular synapses in spider skeletal muscles have demonstrated the occurrence of numerous synaptic vesicles, some actually preserved while apparently releasing their contents (Zebe and Rathmayer 1968; Melamed and Trujillo-Cenóz 1971; Sherman and Luff 1971; Sherman 1973b). This is morphological evidence for a quantal release process for neurotransmission in spiders.

There also is physiological evidence for the quantal release of the neurotransmitter at different leg muscles of *D. hentzi.* Brenner and Rathmayer (1973) recorded randomly occurring spontaneous miniature excitatory postsynaptic potentials in the muscle fibers (Fig. 12). By focal extracellular recording of synaptic current, they were able to study transmitter release events at single synaptic regions. The nerve evoked release of transmitter occurred in a quantal fashion and followed Poisson's theorem at the synapses studied. The facilitation of transmitter release at these synapses could be explained purely as an increase in the probability of release of transmitter packets (presumably the synaptic vesicles).

The external calcium ion concentration greatly affects the amplitude of the EPSP's in spider leg muscle. Doubling it with respect to normal results in a two-fold increase in the amplitude of the EPSP, while reducing it by 50% reduces

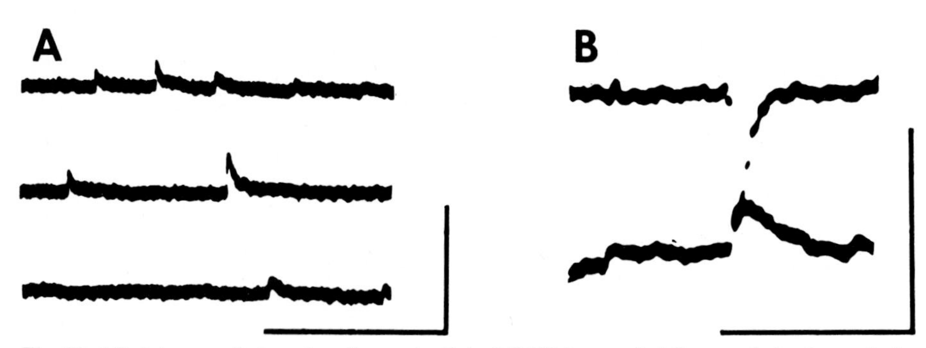

Fig. 12. Miniature excitatory junction potentials (MEJP's) recorded from a skeletal muscle in *Dugesiella hentzi.* Several MEJP's are shown in **A. B** shows a simultaneous extracellular (upper trace) and intracellular (lower trace) recording of a MEJP. *Vertical calibration* equals 0.5 mV for extracellular trace and 1.0 mV for intracellular traces; *horizontal line* equals 20 ms. (Brenner and Rathmayer 1973)

the EPSP amplitude by one-half (Rathmayer 1965). The mechanism by which calcium ions affect EPSP amplitude has not been examined in spiders. Based on such studies on other animals, calcium ions may increase the postsynaptic membrane input resistance and/or increase the probability of transmitter release.

The identity of the neurotransmitters at spider skeletal muscle synapses is unknown. There is some evidence that GABA is the transmitter chemical released by inhibitory motoneurons (Brenner 1972). It reduces postsynaptic membrane input resistance, EPSP amplitude and slow muscle contractions. Picrotoxin, a known GABA antagonist, blocked the neurally evoked inhibitory postsynaptic potentials and the effect of GABA on membrane resistance.

3.2 Scorpions

3.2.1 Muscle Ultrastructure

The initial ultrastructural description of a scorpion skeletal muscle was made by Auber-Thomay (1963), who examined the tibia extensor in *Buthus occitanus* and *Euscorpius carpathicus.* The muscle fibers possess sarcomeres about 3.0 μm in length. H zones are present, and 10 to 12 thin myofilaments surround each thick myofilament. A similar description was provided for the claw-closer muscles of *Leiurus quinquestriatus* by Gilai and Parnas (1970, 1972), who also reported the presence of an abundant SR and TTS, with dyads formed in a manner very similar to those in spiders. The area of TTS membrane was calculated to be six times that of the surface membrane.

The most extensive study of muscle ultrastructure for any arachnid was done by Root and Bowerman (1981), who examined seven leg muscles in the scorpion *Paruroctonus mesaensis.* Their results conform to those of previous studies, and in addition, indicate that the sarcomere lengths of fibers in all seven muscles did not differ substantially.

Thus far, it appears that skeletal muscle fibers in spiders and scorpions do not display the range of diversity that occurs in crustacean skeletal muscle fibers. There do exist some differences in sarcomere length, but it is unknown whether these differences are functionally significant.

3.2.2 Motor Innervation

Gilai and Parnas (1970) determined that in *L. quinquestriatus* the long claw-closer muscle receives four excitatory neurons, while the short claw closer receives six excitors. Each muscle fiber is innervated by two excitors. One evokes small EPSP's that facilitate and elicit small twitch contractions. The other axon elicits spikes which produce large contractions.

Two types of electrical response were also found in the tibia extensor and flexor muscles of *Centruroides gracilis,* EPSP's, and fast spikes (Fountain 1970). Each muscle fiber in these muscles is innervated by up to three excitatory neurons.

Root and Bowerman (1979), using electrophysiological methods, studied seven different leg muscles in *P. mesaensis* and found that each muscle was innervated by three to seven excitatory neurons. They determined that each muscle fiber receives a motoneuron that elicits fast spikes which overshoot (termed a fast motoneuron) or a fast motoneuron and a slow motoneuron that produces EPSP's that facilitate production of graded spikes (see Fig. 3a in Root, Chap. XVII, this Vol.).

In an extensive study of the motoneurons that innervate the 16 muscles in each walking leg of *P. mesaensis,* Bowerman and Burrows (1980) identified each motoneuron as to the muscle it innervated by recording soma spikes and then staining the motoneuron using the cobalt sulfide method (see also Root, Chap. XVII, this Vol.). They confirmed the finding of up to seven excitatory motoneurons to each muscle, located classes of leg muscle motoneurons in the subesophageal ganglion that correspond to each leg movement, and determined that the motoneurons have a cellular morphology that is similar to that of other arthropods.

None of the studies of scorpion skeletal muscle innervation has found any evidence for the existence of peripheral inhibition.

3.2.3 Synaptic Ultrastructure

The only published descriptions of neuromuscular synapses in scorpion skeletal muscle are those by Smith (1971) and Root and Bowerman (1981). The details for scorpions are very similar to those noted above for spiders. Especially noteworthy, however, is the presynaptic rod that overlies a row of synaptic vesicles located next to the presynaptic membrane of the axon (Smith 1971). Smith proposes that the rod may be a microtubule which might be involved in vesicle alignment at the transmitter release site (see also Foelix, Chap. X, this Vol.).

3.2.4 *Neuromuscular Transmission*

Mechanisms of transmitter release have not been directly studied for any scorpion. Indirect indications that transmitter release may occur in packets of transmitter come from the finding of synaptic vesicles at the neuromuscular synapse (Smith 1971; Root, personal communication) and the recording of miniature excitatory postsynaptic potentials in scorpion skeletal muscle fibers (Gilai and Parnas 1970).

There is a paucity of neuromuscular transmitter candidates for scorpions. Gilai and Parnas (1970) could detect no effect of GABA, picrotoxin, and glutamate on electrical or mechanical activity even after treatment of muscle fibers with pronase to eliminate diffusion barriers.

4 Conclusions

Heart muscle in both spiders and scorpions is controlled by at least two neural networks. The cardiac ganglion initiates and coordinates the heartbeat; inhibitory and excitatory cardioregulatory nerves arising from the CNS affect the rate and strength of the heartbeat.

Many, but not all, skeletal muscles in the legs of scorpions and spiders show fast and slow types of contraction, which suggests that two types of excitatory motoneuron are present. There is evidence for direct inhibitory control of some spider leg muscles, but none exists as yet for scorpion muscle.

There are numerous aspects of the neural control of arachnid muscle for which there is very little information. For example, no neurotransmitter has been identified for any arachnid neuromuscular system; and studies are needed on other arachnids besides spiders and scorpions.

References

Atwood HL, Luff AR, Morin WA, Sherman RG (1971) Dense-cored vesicles at neuromuscular synapses of arthropods and vertebrates. Experientia 27:816–817

Auber-Thomay M (1963) Remarques sur l'ultrastructure des myofibrilles chez des scorpions. J Microsc 2:233–236

Bowerman RF, Burrows M (1980) The morphology and physiology of some walking leg motor neurons in a scorpion. J Comp Physiol 140:31–42

Bowerman RF, Root TM (1978) External anatomy and muscle morphology of the walking legs of the scorpion *Hadrurus arizonensis*. Comp Biochem Physiol 59:57–63

Brenner HR (1972) Evidence for peripheral inhibition in an arachnid muscle. J Comp Physiol 80:227–231

Brenner HR, Rathmayer W (1973) The quantal nature of synaptic transmission at the neuromuscular junction of a spider. J Gen Physiol 62:224–236

Bursey CR, Sherman RG (1970) Spider cardiac physiology. I. Structure and function of the cardiac ganglion. Comp Gen Pharmacol 1:160–170

Carlson AJ (1905) Comparative physiology of the invertebrate heart. IV. The physiology of the cardiac nerves in the arthropods. Am J Physiol 15:127–135

Foelix RF (1982) Biology of spiders. Harvard Univ Press, Cambridge

Fountain RL (1970) Neuromuscular properties of scorpion walking leg muscles. M S Thesis, Univ Miami, Coral Gables, Fla

Franzini-Armstrong C (1974) Freeze fracture of skeletal muscle from the tarantula spider. J Cell Biol 61:501–513
Gilai A, Parnas I (1970) Neuromuscular physiology of the closer muscles in the pedipalp of the scorpion *Leiurus quinquestriatus*. J Exp Biol 52:325–344
Gilai A, Parnas J (1972) Electromechanical coupling in tubular muscle fibers. I. The organization of tubular muscle fibers in the scorpion, *Leiurus quinquestriatus*. J Cell Biol 52:626–638
Gonzalez-Fernandez F, Sherman RG (1979) Cardioregulatory nerves in a spider. Neurosci Abstr 5:246
Gonzalez-Fernandez F, Sherman RG (1984) Cardioregulatory nerves in the spider *Eurypelma marxi* Simon. J Exp Zool 231:27–37
Jahromi SS, Atwood HL (1969) Correlation of structure, speed of contraction, and total tension in fast and slow abdominal muscle fibers of the lobster (*Homarus americanus*). J Exp Zool 171:25–37
Kanungo MS (1957) Cardiac physiology of the scorpion *Palamnaeus bengalensis* C. Koch. Biol Bull Woods Hole 113:135–140
Kawaguti S, Kamishima Y (1969) Electron microscopy on the long-sarcomere muscle of the spider leg. Biol J Okayama Univ 15:73–86
Land MF (1969) Movements of the retinae of the jumping spiders (Salticidae: Dendryphantinae) in response to visual stimuli. J Exp Biol 51:471–493
Legendre R (1968) Sur la présence d'un nerf cardiaque chez les Araignées Orthognathes. C R Acad Sci 267:84–86
Melamed J, Trujillo-Cenóz O (1971) Innervation of the retinal muscles in wolf spiders (Araneae-Lycosidae). J Ultrastruct Res 35:359–369
Naidu VD (1974) On the point of origin of heart beat in the scorpion, *Heterometrus fulvipes* C. Koch. Curr Sci 20:643–645
Naidu VD, Padmanabhanaidu B (1975) Physiology of the scorpion heart: Part I – Experimental analysis of neurogenic nature of the heart beat in the scorpion *Heterometrus fulvipes* C. Koch. Indian J Exp Biol 13:22–26
Police G (1902) Il nervo del cuore nello Scorpione. Boll Soc Nat Napoli 15:146–147
Prosser CL (1973) Comparative animal physiology, 3rd edn. Saunders, Philadelphia
Randall WC (1966) Microanatomy of the heart and associated structures of two scorpions, *Centruroides sculpturatus* Ewing and *Uroctonus mordax* Thorell. J Morphol 119:161–180
Rathmayer W (1965) Neuromuscular transmission in a spider and the effect of calcium. Comp Biochem Physiol 14:673–687
Rathmayer W (1966) Die Innervation der Beinmuskeln einer Spinne, *Eurypelma hentzi* Chamb. (Orthognata, Avicularidae). Verh Dtsch Zool Ges Jena Zool Anz Suppl 29:505–511
Root TM, Bowerman RF (1979) Neuromuscular physiology of scorpion leg muscles. Am Zool 19:993
Root TM, Bowerman RF (1981) The scorpion walking leg motor system: muscle fine structure. Comp Biochem Physiol 69:73–78
Ruhland M (1976) Untersuchungen zur neuromuskulären Organisation eines Muskels aus Laufbeinregeneraten einer Vogelspinne (*Dugesiella hentzi* Ch.). Verh Dtsch Zool Ges 1976:238
Ruhland M, Rathmayer W (1978) Die Beinmuskulatur und ihre Innervation bei der Vogelspinne *Dugesiella hentzi* (Ch.) (Araneae, Aviculariidae). Zoomorphologie 89:33–46
Sherman RG (1973a) Ultrastructural features of cardiac muscle cells in a tarantula spider. J Morphol 140:215–242
Sherman RG (1973b) Unique arrangement of glial membranes between adjacent neuromuscular synapses in a spider muscle. J Cell Biol 59:234–238
Sherman RG (1978) Insensitivity of the spider heart to solitary wasp venom. Comp Biochem Physiol 61A:611–615
Sherman RG, Bursey CR, Fourtner CR, Pax RA (1969) Cardiac ganglia in spiders (Arachnida: Araneae). Experientia 25:438–439
Sherman RG, Luff AR (1971) Structural features of the tarsal claw muscles of the spider *Eurypelma marxi* Simon. Can J Zool 49:1549–1556

Sherman RG, Pax RA (1967) A physiological and morphological study of a spider heart. Am Zool 7:190–191

Sherman RG, Pax RA (1968) The heartbeat of the spider *Geolycosa missouriensis.* Comp Biochem Physiol 26:529–536

Sherman RG, Pax RA (1970a) Spider cardiac physiology. II. Responses of a tarantula heart to cholinergic compounds. Comp Gen Pharmacol 1:171–184

Sherman RG, Pax RA (1970b) Spider cardiac physiology. III. Responses of a tarantula heart to certain catecholamines, amino acids, and 5-hydroxytryptamine. Comp Gen Pharmacol 1:185–195

Smith DS (1971) On the significance of cross-bridges between microtubules and synaptic vesicles. Philos Trans R Soc London Ser B 261:395–405

Ude J, Richter K (1974) The submicroscopic morphology of the heart ganglion of the spider *Tegenaria atrica* (C. L. Koch) and its neuroendocrine relations to the myocard. Comp Biochem Physiol 48A:301–308

Venkatachari SAT, Naidu VD (1969) Choline esterase activity in the nervous system and the innervated organs of the scorpion, *Heterometrus fulvipes.* Experientia 25:821–822

Wilson RS (1967) The heartbeat of the spider *Heteropoda venatoria.* J Insect Physiol 13:1309–1326

Zebe E, Rathmayer W (1968) Elektronenmikroskopische Untersuchungen an Spinnenmuskeln. Z Zellforsch 92:377–387

Zwicky KT (1968) Innervation and pharmacology of the heart of *Urodacus,* a scorpion. Comp Biochem Physiol 24:799–808

Zwicky KT, Hodgson SM (1965) Occurrence of myogenic hearts in arthropods. Nature (London) 207:778–779

XVII Central and Peripheral Organization of Scorpion Locomotion

THOMAS M. ROOT

CONTENTS

1 Introduction

Studies of invertebrate locomotion have recently become successful model approaches for the analysis of behavior because invertebrates offer the advantage of being relatively simple, and locomotion, although it is a complex behavior, is stereotyped and easily comparable among different animals.

The scorpion used in our studies, *Paruroctonus mesaensis* offers several advantages for the study of locomotion. For example, it is a large, durable animal, easy to maintain in the laboratory, and the cuticle of its leg is quite transparent, allowing the viewing of leg muscles and even some nerves without dissection. Also, the scorpion central nervous system, like that of most arachnids, is highly cephalized, in contrast with the nervous systems of the crustaceans and insects whose locomotion patterns have been studied more extensively (Root 1985). Of particular interest, is that the scorpion's leg nerves enter the large subesophageal ganglion, which contains sensory and motor centers for many behaviors. In contrast, the leg nerves of crustaceans and insects typically enter separate ganglia in the ventral nerve cord. Therefore, the generation and coordination of walking in the scorpion may rely upon different neuronal mechanisms.

The descriptive work on scorpion locomotion was begun by Bowerman (1975a, b). Since that time there have been published studies of various aspects of the behavior, leg proprioceptors, the anatomy of the leg motor system, and of some motor centers in the brain. Current work includes studies of leg muscle

---

Biology Department, Middlebury College, Middlebury, Vermont 05753, USA

physiology, the role of leg receptors in shaping motor output (see also Seyfarth, Chap. XII, this Vol.), and recordings from the brain. After a brief review of the walking behavior, I will describe some of these sensory, motor, and central nervous system mechanisms that we believe are controlling locomotion in the scorpion.

2 Walking Behavior

To document the walking behavior, scorpions are filmed at various angles and data are recorded for such features as when each of the legs step, how far laterally the legs are held relative to the body, how far each leg is raised as it is brought forward, and the joint angles between any two leg segments. Utilizing cinematographic techniques, Bowerman (1975a) found, for example, that the scorpion uses a strict pattern of leg movements (Fig. 1A). In 97% of the cases, the gait was the stepping sequence of legs 4, 2, 3, 1 respectively, with a tendency for the fourth and second legs to step together. Since the fourth and second leg pair on one side steps simultaneously with the third and first leg pair on the opposite side, the result is that two tetrapods of legs alternate with one another during forward walking. Actually, each of the legs in a tetrapod does not step at exactly the same time, but rather, there is a slight latency between the step of each leg that allows a smooth, forward movement.

Each of the eight legs goes through a similar cycle of movement, which is composed of a step phase, when the leg is on the ground, and a swing phase, when the leg is off the ground. However, each leg accomplishes these phases differently (Fig. 1B). For example, the fourth and third legs are raised very slightly off the ground as they are brought forward and put back down. However, the second leg, and to a greater extent the first leg, are raised much higher as they are brought forward, perhaps to sense obstructions in its pathway.

Another difference was noted in the angles of certain joints during walking (Root and Bowerman 1978). For example the femur-patella joints in the third and fourth legs extend during their stance phase, but this same joint in the first leg flexes during its stance phase. In the second leg, this joint undergoes a complicated movement consisting of a sequence of flexion, extension, followed by another flexion and extension. Finally, although the joints go through rather complex series of movements, the leg tip remains parallel to the body as the animal moves forward (Fig. 1C). Therefore the motor program for each leg must be a complicated series of muscle movements unique to each leg.

The strongly patterned gait observed in the scorpion is the result of strictly timed latencies in the movements of the legs. Standardized latency data, expressed as the latency to step between two legs divided by the time for a complete cycle of leg movement, is termed a phase value. The phase value ranges from 0.0 to 1.0, and a value of 1.0 indicates that two legs are stepping exactly together, while a value of 0.5 indicates that two legs are stepping exactly in alternation.

When the phase values for all eight legs are studied, some interesting relationships describing leg coordination are observed. For example, if the phase

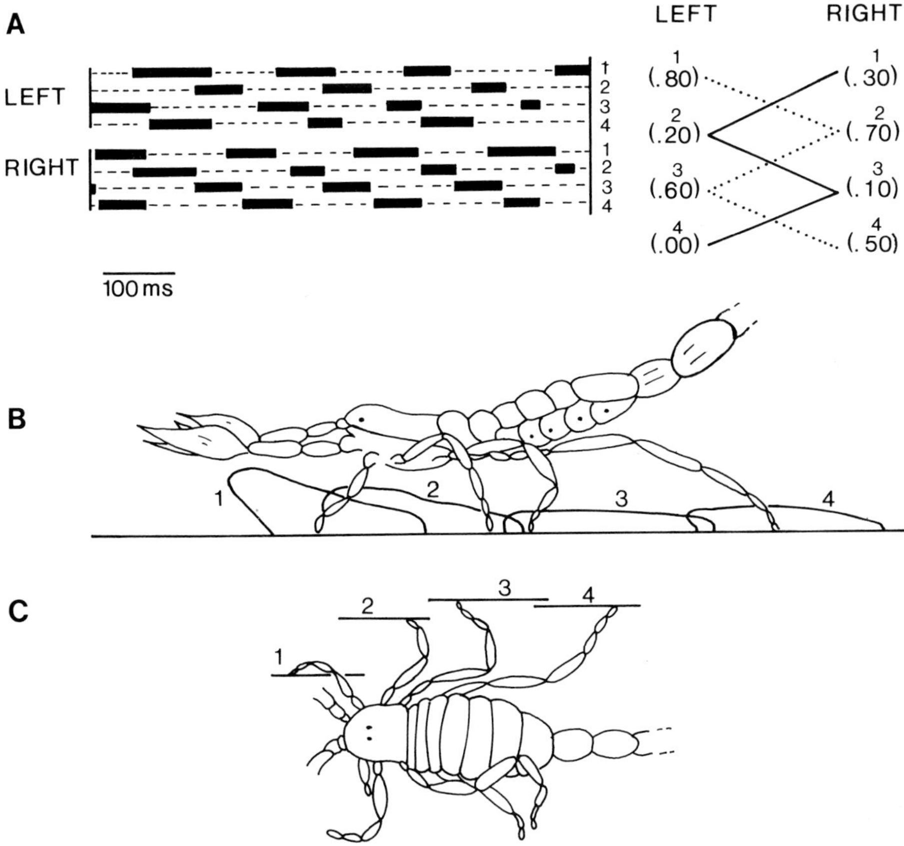

Fig. 1 A–C. Walking behavior in the scorpion. **A** Stepping patterns illustrate changes in gait as an animal prepares to stop. *White spaces* indicate when each leg is on the ground and *black spaces* indicate when each leg is off the ground. Prior to stopping, the fourth and second legs on one side and the third and first legs on the opposite side step approximately together (Bowerman 1981 a). The phase relationships between all eight legs are shown to the *right* of the stepping sequence. The phase values are expressed relative to the left fourth leg, which is assigned a starting values of 0.0. Adjacent contralateral legs have differences in phase values of approximately 0.5, and adjacent ipsilateral legs have differences in phase values of approximately 0.6. As a result of the phase differences, two sets of diagonally coupled legs step together (Bowerman 1975 a). **B** Viewed from the side, each of the legs is lifted off the substrate differently as they are brought forward (Root and Bowerman 1978). **C** Viewed from above, the leg tips remain parallel to body axis when they are on the substrate. (Root and Bowerman 1978)

values for all legs are plotted relative to the left fourth leg, which is assigned a value of 0.0 (Fig. 1 A), then it becomes clear that adjacent contralateral legs have phase values approximating 0.5 (i.e., they step in strict alternation), and adjacent ipsilateral legs have phase value differences approximating 0.6 (i.e., they step in a time slightly greater than alternation). Bowerman (1975 a) observed that these phase relationships are very constant over a wide range of walking speeds, indicating tight coupling between the neural control centers for each leg, and suggesting that this coupling is responsible for the almost invariance of the 4, 2, 3, 1 gait.

Ablation experiments, in which one or two of the legs were removed, caused slight, predictable modifications in the leg coordination and gait of the animal (Bowerman 1975 b). These experiments suggested that although the central nervous system generates the basic movements of the legs, peripheral sensory feedback is also important in maintaining normal locomotion. It was found, for example, that following the removal of one leg, a scorpion was able to reorganize its normal pattern of leg coordination to achieve a new stepping pattern for stable forward movement. Specifically, removing one leg caused a shift in the phase between ipsilateral legs, and those contralateral legs anterior to the site of leg removal. The result was a new gait that insured that two legs on each side of the animal were always on the ground together.

Other changes from the normal gait of the scorpion were found when the movements of the legs during starting, stopping, turning, and backward walking were studied (Bowerman 1981 a). During the first few cycles of leg movement before a stop or after a start, the normal 4, 2, 3, 1 gait could be altered dramatically. Similarly, a different pattern of leg movements was used for backward walking, and three different classes of turn (spin turn, arc turn, and pivot turn), all utilizing different gaits, were also observed.

Therefore, although during normal forward locomotion the scorpion utilizes a rather stereotyped, coordinated pattern of stepping which suggests strong neural coupling, each of the legs can also be utilized differently as the circumstances of the environment dictate, and it is likely that sensory systems within each of the legs are utilized to aid coordination.

3 Sensory Mechanisms

It is clear from the findings of the behavioral studies that some sensory influences are involved in the ability of the scorpion to alter its gait. In an attempt to understand how these sensory systems control leg movements, we have identified some of the leg receptors that we believe may be involved (see also Seyfarth, Chap. XII, this Vol.).

One group about which we currently know very little are the slit sensilla. We have confirmed the location of many of the slit sensilla in the legs of *Paruroctonus mesaensis* that Barth and Wadepuhl (1976) described for the scorpion *Androctonus australis*. We have begun to study some of the grouped slits near the leg joints, and although we believe that they may provide timing information about when a leg is on the ground, and also initiate reinforcement reflexes to muscles that are active during the stance phase, we do not yet have good physiological data.

Bowerman (1981 b) has recorded from a group of unidentified receptors in the tarsus which may signal when the leg touches the substrate (Fig. 2A). He located them in the last tarsal segment and by tracing the sensory nerve branches with cobalt sulfide, he was able to find a series of bipolar sensory cells

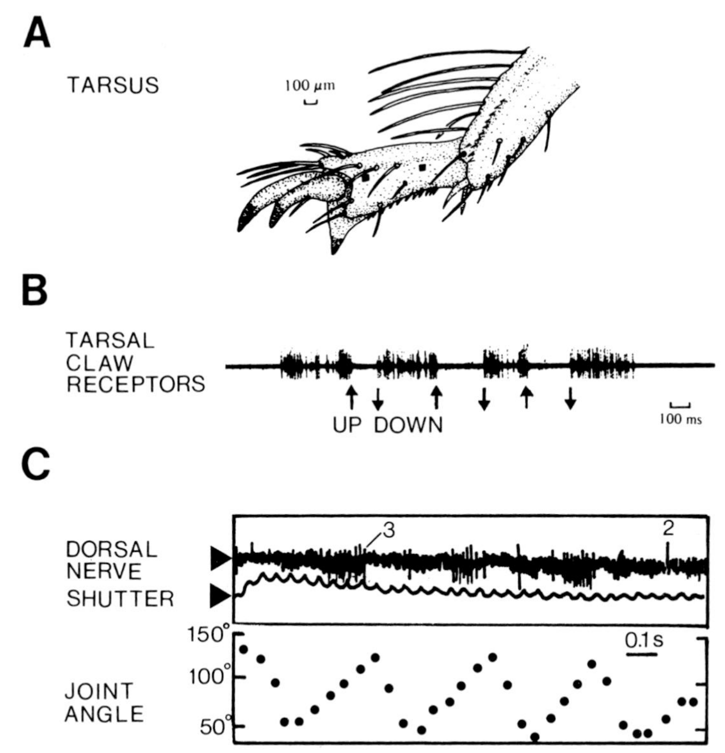

Fig. 2A–C. Sensory receptors in the pretarsus of scorpion leg. **A** *Solid boxes* indicate sites for recording electrodes (Bowerman 1981b). **B** Recordings from the tarsal claw receptors during walking show that these receptors are only active when the leg is on the ground (*downward arrows*). *Upward arrows* indicate when a leg is off the ground (Bowerman 1981b). **C** Recordings from the femur-patella joint receptors (*upper trace*), movements of the shutter from a movie camera (*middle trace*), and determinations of the joint angle from the movie film (*lower record*). Two units are identified (*2*) and (*3*) among the few sensory units in the record. These units are active primarily at different times during extension of this joint. (Weltzin 1981)

that are involved. Whether these belong to hair sensilla or sensory slits is still an open question. Recently, Foelix and Schabronath (1983) described the structure of the tarsal sensilla. In addition to finding some receptors that are apparently chemoreceptive, they also described mechanoreceptive hairs and bristles, and some slit sensilla.

By implanting fine wire electrodes near the nerves in the tarsus, Bowerman (1981b) was able to record electrical activity from the tarsal nerve during walking (Fig. 2B). He found that these receptors were active in bursts only when the leg was on the ground, and therefore were an accurate monitor of leg position during the stance phase. This suggests that the tarsal claw receptors could be assisting the switching between the groups of leg muscles responsible for the stance and the swing phase.

Weltzin (1981) described the location of other groups of receptors at the leg joints, and studied in particular those at the femur-patella joint. He back-filled the dorsal leg nerve branch with cobalt sulfide, and identified several bipolar cells that innervate the hypodermis at this joint. With suction electrode recordings from the dorsal leg nerve of restrained animals, he characterized single units whose activity was correlated with different phases of the flexion and extension of this joint. Also, by implanting wire electrodes near the dorsal leg nerve in walking animals, he was able to simultaneously record the activity of these receptors while filming the joint's movement. The response of two identified cells are shown in Fig. 2C. Cell 2 fired throughout the extension of the joint, and its firing frequency increased with faster rates of extension. Cell 3 also fired during extension, but did not initiate firing until extension was approximately 30% completed, and its firing frequency also increased at faster extension rates. Bowerman (1976) recorded from sensory nerve branches at several leg joints in isolated legs and found a similar pattern of sensory activity from unidentified units. In response to imposed movements, he found that each joint had several sensory cells that could define the position, direction and rate of movement of the joint based on their firing rates. In conjunction with Weltzin's study, this work suggests that the leg joints have many receptors that provide specific and sometimes redundant information about the location of the leg joints, and the direction and velocity of their movement.

Therefore, we believe that the slit sense organs on the leg surface may sense loads on the legs, the touch receptors in the tarsus may respond to contact of the leg with the ground, and receptors in the membranes at the leg joints may monitor the position and movements of the joints. We are continuing to study the structure and function of several of these sensory systems. We are also analyzing the phasic influence of several different receptors upon the motor program in walking animals, by simultaneously recording from leg muscles and sensory nerve branches before and after perturbation of these receptors.

4 Motor Mechanisms

The movements of each leg are controlled by 15 muscles. Seven of these muscles, which are located in the trochanter, femur, and patella segments, are of particular interest, since they control the major elevation and depression movements and the major flexion and extension movements of the legs. When examined with light and electron microscopy, it was found that the muscles were very similar in structure (Root and Bowerman 1981). Each muscle had tubular muscle fibers with only minor differences in their fiber diameters, in their sarcomere lengths, and in the structure of their transverse tubule systems and sarcoplasmic reticulum.

In addition, all of the muscles that we examined were very similar to one another. For example, the mean muscle fiber diameters of these seven muscles were very similar, ranging from only 18 to 21 μm. Similarly, the mean sarcomere lengths were also very similar, ranging only from 2.5 to 4 μm. The number of thin or actin filaments surrounding each thick or myosin filament was very consistent, as were the details of the T-tubule system and the sarcoplasmic reticulum. Therefore, it appears that the muscles are structurally quite uniform.

There is no obvious structural difference in the fibers such as is seen in the slow and fast muscle fibers of insects and crustaceans.

Recent histochemical work, however, suggests that the fibers within a muscle may have different capacities to produce and sustain tension (Root and Gatwood 1981). When stained with the enzyme myosin ATPase, for example, several fibers within the muscle stain very darkly for the enzyme, whereas others stain very lightly, suggesting differences in the ability of these fibers to provide rapid energy for contraction. Similarly, when stained for two oxidative enzymes (alpha glycerophosphate dehydrogenase and succinic dehydrogenase) there were also differences, some muscle fibers staining very lightly, some staining very darkly, and others with intermediate levels of stain. Therefore, the fibers within a muscle may be capable of producing different levels of tension quickly, and of sustaining contraction.

We have also noted differences in the innervation pattern of different muscle fibers on the basis of electrophysiological recordings (Root and Bowerman 1979). By stimulating a motor nerve and recording intracellular potentials from muscle fibers, we characterized their innervation on the basis of their synaptic potentials. We found two types of innervation, similar to those observed by Gilai and Parnas (1970) for the pedipalp closer muscles of the scorpion *Leiurus quinquestriatus*. Some muscle fibers were innervated only by a single "fast" motoneuron, while other muscle fibers were innervated by both a single "fast" and a single "slow" motoneuron (Fig. 3A). The fast motoneuron elicited a brief, all or none, over-shooting spike potential. The slow motoneuron elicited smaller and longer postsynaptic potentials which showed considerable facilitation. In most of the muscles examined, there was a mixture of these two types of innervation, but there was no evidence for inhibitory motoneurons. We are currently examining the question of peripheral inhibitory motoneurons in more detail, since they are a common feature of crustacean and insect neuromuscular systems.

Bowerman (1981b) studied the basic elevation and depression motor program of the leg by recording from the trochanter-femur elevator and trochanter-femur depressor muscles with fine wire electrodes as the animal was free to walk (Fig. 3B). As expected, these two muscles alternate rather strictly with one another. The depressor muscles show bursts of activity when the leg is on the ground, concomitant with the tarsal claw receptors, and the elevator muscles show bursts of activity when the leg is off the ground. Changes in walking speed were accomplished primarily by changes in the duration of bursts in the depressor muscle, while the duration of bursts in the elevator muscle remained relatively constant, except at the fastest walking speeds, when the activity of both muscles was reduced by similar amounts.

Interesting differences were also noted at the switching points between activities in these two muscles (Fig. 3C, D). The depressor muscles are always turned off prior to the elevator muscles being turned on, but the converse is not true. Although the elevator muscles may be turned off prior to activity in the depressor muscles, in some instances these two muscles are co-activated, and there are also instances in which the depressor muscles are turned on prior to cessation of activity in the elevator muscles. Therefore, although this basic

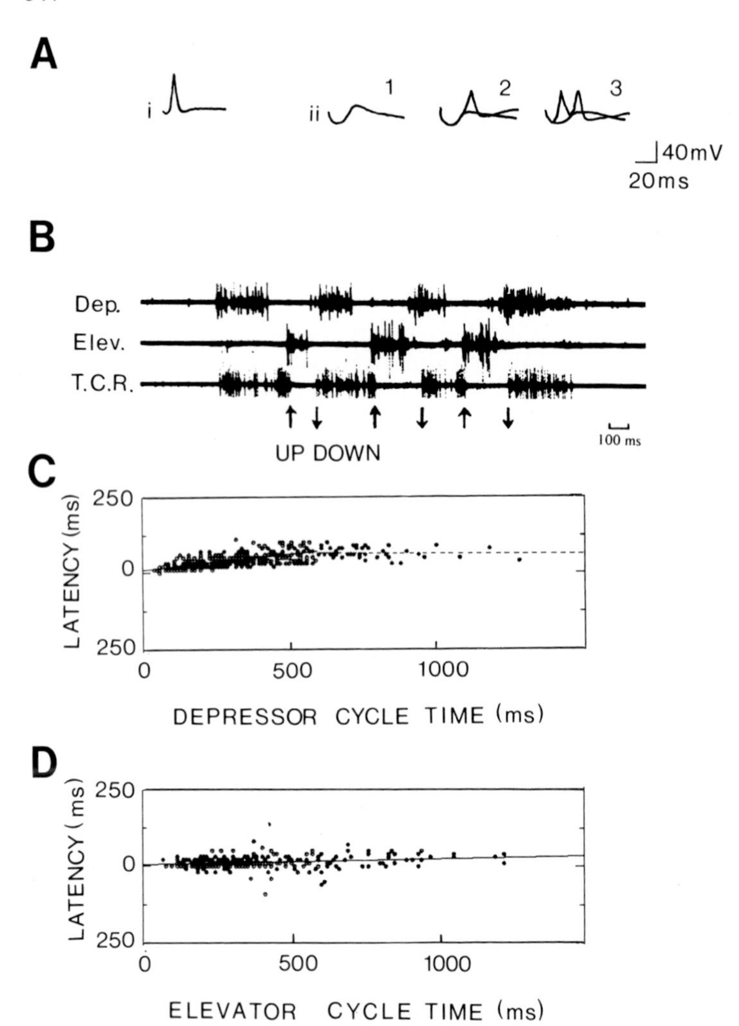

Fig. 3 A – D. The leg motor system. **A** Intracellular potentials recorded from single muscle fibers illustrating two innervation patterns (*i*). In the first pattern, only one response, a fast spike-like potential was elicited with increasing intensities of stimulation (*ii*). In the other pattern, at low stimulating intensities an excitatory postsynaptic potential was elicited (*1*), which became a graded spike with repetitive stimuli (*2*). With increasing stimulus intensities a second motoneuron was recruited that caused a fast spike-like potential (*3*) (Root and Bowerman 1979). **B** Simultaneous electrical recordings from the trochanter-femur depressor muscle (*upper trace*), trochanter-femur elevator muscle (*middle trace*), and tarsal claw receptors (*lower trace*). The two muscles alternate in their activity, and the depressor muscles are active when the tarsal claw receptors are active (i.e., when the leg is on the substrate) (Bowerman 1981 b). **C** Plot of the latency between the cessation of activity in the depressor muscle and the initiation of activity in the elevator muscle, as a function of depressor cycle time (i.e., walking speed). The latency values are always positive, indicating that these two muscles are never coactive (Bowerman 1981 b). **D** Plot of the latency between the cessation of activity in the elevator muscle and the initiation of activity in the depressor muscle, as a function of cycle time. There are numerous occasions when these latency values are zero (i.e., the muscles are co-activated), and numerous occasions when the latency values are negative (i.e., the depressor muscle is activated prior to the end of activity in the elevator muscle). (Bowerman 1981 b)

motor program consists of rather strict alternation between the elevator and the depressor muscles, there appear to be different neural mechanisms for achieving faster walking speeds, and there also appear to be different mechanisms for ceasing activity at the end of the step phase than at the end of the swing phase.

5 Central Nervous Mechanisms

Although there is much information about some of the sensory and motor mechanisms involved in locomotion, only recently have there been attempts to probe into the central nervous system in scorpion locomotion. All eight leg nerves contain both motor and sensory fibers, and the nerves enter into the large subesophageal ganglion of the brain region. Using cobalt sulfide, and in some cases horseradish peroxidase, we have determined the location of the motoneuron cell bodies in the subesophageal ganglion. Two groups of motoneurons are located for each leg ventrally in the subesophageal ganglion, near the entrance of each leg nerve. Currently, it is not known how these groups of motoneurons are interconnected.

Bowerman and Burrows (1980) were able to inject cobalt sulfide into single motoneuron cell bodies and study their cyto-architecture (Fig. 4A). The structure of these cells is comparable to those in other arthropod nervous systems. The cell body gives rise to a neurite that is only 2 to 9 μm in diameter and free of dendritic processes for the first 150 to 200 μm of its length. The dendritic branches arise from the next 1000 μm segment, branch profusely, and are somewhat planar. The long axon passes from this region uninterrupted into the leg nerve.

Bowerman and Burrows (1980) were also able to record from single motoneurons by impaling them with glass microelectrodes in the brain while the animal was free to move its legs. They studied the connection between individual motoneurons and leg muscles by stimulating the cell, noting the muscular response, and also often recording simultaneously from the muscle. Examples of two cell responses are shown in Fig. 4B. The spontaneous synaptic activity of one cell (upper trace) reveals considerable depolarizing and hyperpolarizing synaptic potentials. In another motoneuron (lower traces) spontaneous spike potentials were correlated with potentials recorded from a muscle.

Future work will examine the connections between these and other motoneurons and between the motoneurons and interneurons in the brain. The close proximity within the brain of all of the motoneurons for a single leg, and of the motor centers for the eight different legs makes possible neural control mechanisms that rely upon direct communication between motoneurons. By studying the functional connections between motoneurons we can answer this question.

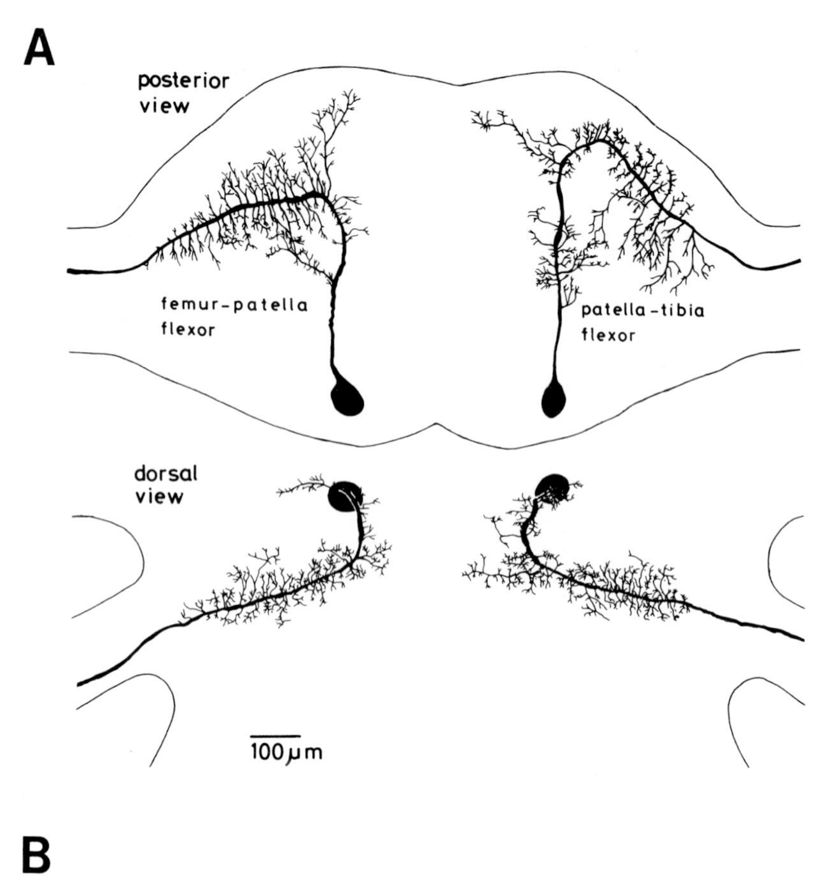

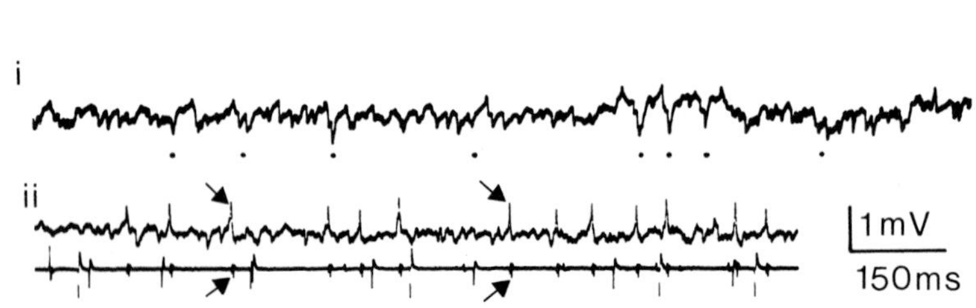

Fig. 4A, B. Motoneurons in the central nervous system. **A** Camera lucida drawings of two motoneurons in the subesophageal ganglion which were stained with cobalt sulfide. **B** Intracellular recordings from the somata of motoneurons in the subesophageal ganglion. **i** Recording showing spontaneous synaptic activity (indicated by *dots*). **ii** Simultaneous recordings from a motoneuron soma (*upper trace*) and from the muscle it innervates (*lower trace*). The activity of one spike in the motoneuron is phase-locked to the activity of a spike from the muscle record (*arrows*). (Bowerman and Burrows 1980)

6 Conclusions

To date, we have completed detailed studies of leg movements during walking, major descriptive work on the sensory and motor systems of the legs, and experimental studies of the role of some receptor systems, leg muscles and motoneurons in locomotion. Our current work is to simultaneously study ongoing muscle and receptor activity during walking to understand the phasic role of proprioceptive feedback in shaping the motor program. We also will be studying how locomotion is initiated and maintained by control centers in the brain. Because of the advances being made on several different fronts, it is anticipated that the control of scorpion locomotion may be understood on many different levels.

References

Barth FG, Wadepuhl M (1976) Slit sense organs on the scorpion leg (*Androctonus australis* L., Buthidae). J Morphol 145 (2):209–228
Bowerman RF (1975a) The control of walking in the scorpion. I. Leg movements during normal walking. J Comp Physiol 100:183–196
Bowerman RF (1975b) The control of walking in the scorpion. II. Coordination modification as a consequence of appendage ablation. J Comp Physiol 100:197–209
Bowerman RF (1976) Electrophysiological survey of joint receptors in walking legs of scorpion, *Paruroctonus mesaensis.* J Comp Physiol 105 (3):353–363
Bowerman RF (1981a) Arachnid locomotion. In: Herried CF II, Fourtner CR (eds) Locomotion and energetics in arthropods. Plenum Press, New York London, pp 73–102
Bowerman RF (1981b) An electromyographic analysis of the elevator/depressor motor programme in the freely-walking scorpion, *Paruroctonus mesaensis.* J Exp Biol 91:165–177
Bowerman RF, Burrows M (1980) The morphology and physiology of some walking leg motor neurons in a scorpion. J Comp Physiol A 140:31–42
Foelix RF, Schabronath J (1983) The fine structure of scorpion sensory organs. I. Tarsal sensilla. Bull Br Arachnol Soc 6 (2):53–67
Gilai A, Parnas I (1970) Neuromuscular physiology of the closer muscles in the pedipalp of the scorpion *Lieurus quinquestriatus.* J Exp Biol 52:325–344
Root TM (1985) (in press) Scorpion neurobiology. In: Poli GA (ed) Biology of the scorpionida. Stanford Univ Press, Stanford
Root TM, Bowerman RF (1978) Intra-appendage movements during walking in the scorpion *Hadrurus arizonensis.* Comp Biochem Physiol 59A:49–56
Root TM, Bowerman RF (1979) Neuromuscular physiology of scorpion leg muscles. Am Zool 19 (3):993
Root TM, Bowerman RF (1981) The scorpion walking leg motor system: muscle fine structure. Comp Biochem Physiol 69A:73–78
Root TM, Gatwood SC (1981) Histochemistry of scorpion leg muscles. Am Zool 21 (4):941
Weltzin R (1981) Dorsal leg nerve joint receptors in restrained and freely-walking scorpions. MS Thesis, Univ Wyoming, Laramie

E Neurobiology of a Biological Clock

XVIII Neurobiology of a Circadian Clock in the Visual System of Scorpions

GÜNTHER FLEISSNER and GERTA FLEISSNER

CONTENTS

1 Introduction

What does a circadian clock look like in terms of its anatomy and physiology? This is a challenging question for neurobiologists, on which we have been focusing our research for the past several years.

Basic features of the circadian clock, such as free-running under constant conditions, synchronization by Zeitgeber, and other properties have been intensively studied for about the last 30 years. Although several general reviews have been published (Aschoff 1981; Block and Page 1978; Brady 1974; Bünning 1973; Cloudsley-Thompson 1978; Enright 1980; Pittendrigh 1974; Rusak and Zucker 1979; Saunders 1982; Winfree 1980), the interior of the circadian clock has largely remained a mystery. Only little is known about the physiological mechanisms involved, about the site and nature of the oscillators, and about the interactions between them. Even the true function of the Zeitgeber is not fully understood. Amongst the invertebrate studies, those on the opisthobranch snail *Aplysia* seem to be the closest to the single neuron level of the clock mechanisms (Jacklet and Rolerson 1982; Jacklet et al. 1982). Within the arthropods, the scorpion is the first where this search for the clock has been brought down to the cellular level within its CNS (Fleissner and Heinrichs 1982). This chapter

Zoologisches Institut der J.-W.-Goethe-Universität, Arbeitskreis Neurobiologie Circadianer Rhythmen, Siesmayerstraße 70, D-6000 Frankfurt am Main 1, Federal Republic of Germany summarizes our knowledge on the circadian clock controlling the sensitivity of the eyes in the fat-tailed scorpion *Androctonus australis* (Buthidae, Scorpiones). *Androctonus* is found in desert areas and oases of North Africa (Vachon 1952, 1953). It is extremely well-adapted to arid environments and therefore an ideal candidate for long-term experiments. Unfortunately it is "one of the two most deadly species in the world" (Cloudsley-Thompson 1956, for details about its venom see Goyffon et al. 1982 and Zlotkin et al. 1978).

The first documentation of an endogenous sensitivity change in an arthropod eye goes back to Kiesel who described rhythmic changes of eye glow in the moth *Plusia gamma* as early as in 1894. Later this phenomenon was also investigated in crayfish (reviewed by Welsh 1938), and in 1940 Jahn and Crescitelli were the first to report an endogenous ERG rhythm in the compound eyes of beetles. It was, however, about 30 years before this subject was investigated again, when a circadian rhythm of retinomotoric processes was demonstrated anatomically in the eyes of *Tenebrio molitor* (Wada and Schneider 1968). Long-term recordings over weeks and months of the ERG rhythms of crayfish (Arechiga and Wiersma 1969) and scorpions (Fleissner 1971) were then the start of neurobiological work which consequently led to the search for the circadian clock itself (crayfish: for review Arechiga et al. 1973; Arechiga 1977; Larimer and Smith 1980; *Limulus:* for review Barlow 1983; scorpions: for review see this chapter; insects: Bennett 1983; Köhler and Fleissner 1978; Fleissner 1982; Tomioka and Chiba 1982).

2 Circadian Rhythms in the Visual System of the Scorpion

If any eye is kept long enough in darkness, its sensitivity usually increases to a maximum (for rev. Autrum 1981). For a physiologist, such a dark-adapted eye seems to be in a well-defined state, however, in the *median eyes* of the scorpion there is an additional endogenous factor modifying the sensitivity (Fig. 1a). This is reflected in both single-cell recordings and in mass responses of the retina.

Electroretinogram (ERG) recordings offer many advantages for long-term observations. They not only allow precise measurements of the sensitivity of an eye, but are also harmless to the animals and may be tolerated for many months without any sign of damage. It has therefore been our chosen method in the search for answers to most of our questions.

Fig. 1a–d. Circadian rhythm of sensitivity in the dark-adapted median and lateral eyes of the scorpion, *Androctonus australis.* Unless indicated otherwise, all experiments are in continuous darkness (DD). In long-term experiments (**c, d**) the darkness was only interrupted by short test light flashes of 30 ms duration every half hour. **a** Electroretinograms (ERG) of a median eye: Circadian day state (*left*) and circadian night state (*right*), same light stimulus, DC recording. △, ○ refer to the intensity-response curves in **b. b** Intensity-response curves of ERG-on-effects during circadian day (△) and night (○), demonstrating a sensitivity change of 4 log units in the median eye. *Arrows in insets* mark the registration time within the circadian period. **c** Time course of ERG-amplitudes simultaneously recorded in a median and a lateral eye during several circadian periods. Note the flat but clearly circadian oscillation in the lateral eye showing a less than twofold change in sensitivity, whereas the median eye sensitivity varies by a factor of > 1000 in a square wave-like time course. **d** Simultaneous ERG recordings from both median eyes of the same scorpion. Only one eye (○) is illuminated for 2 h (*white bar*). Note the synchrony of the ERG rhythms in both eyes, which is not disturbed by unilateral exposure to light. (**a** and **b** after Fleissner 1974; **c** Fleissner 1977d; **d** Fleissner 1977a)

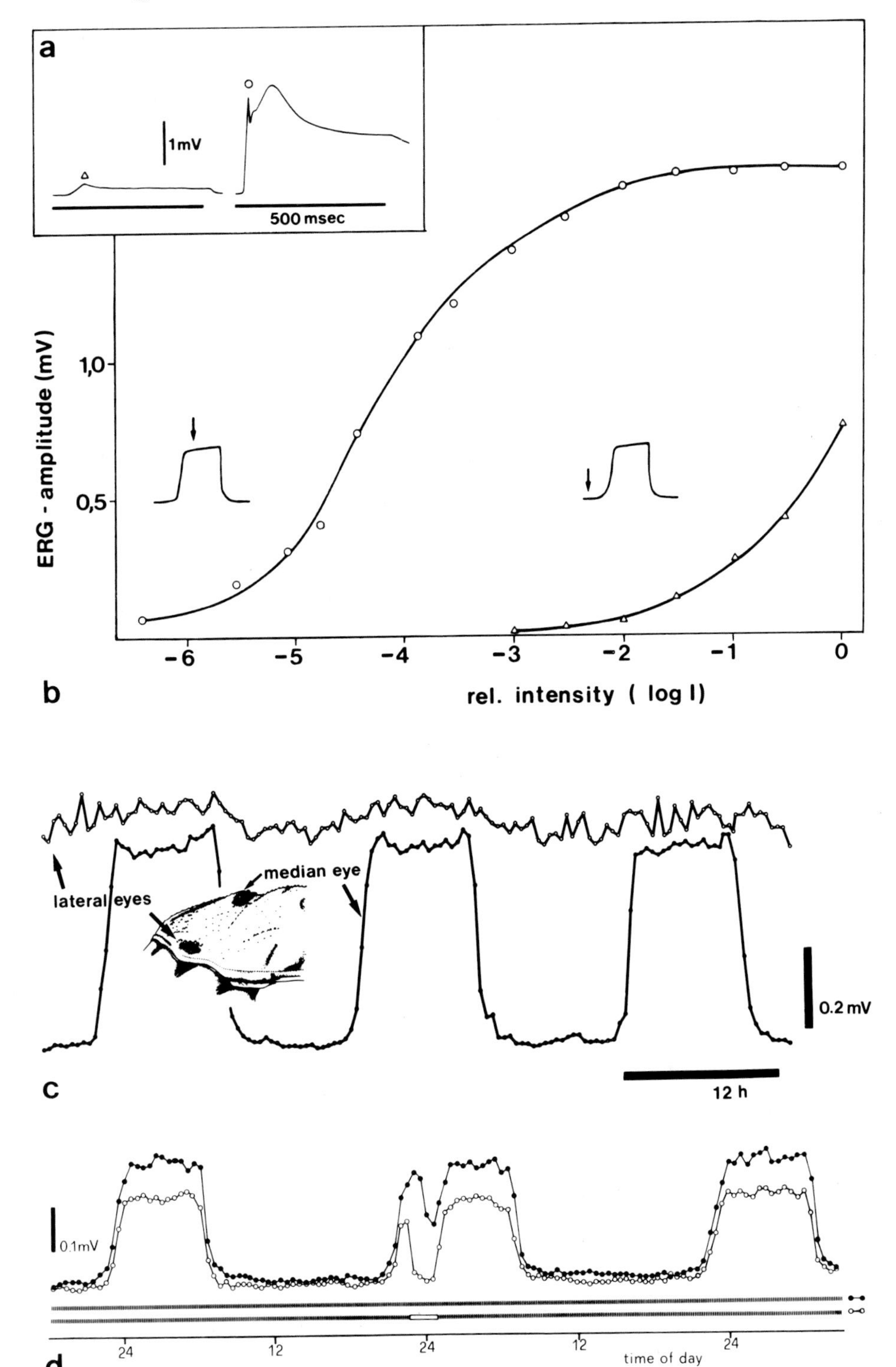
a
1mV
500 msec
ERG - amplitude (mV)
1,0
0,5
-6
-5
-4
-3
-2
-1
0
b
rel. intensity (log I)
lateral eyes
median eye
0.2 mV
12 h
c
0.1mV
24
12
24
12
24
time of day
d

Under constant conditions the time course of this endogenously controled sensitivity change reveals an extremely clear-cut circadian rhythm (Figs. 1, 2). The state of low sensitivity (the circadian day state) alternates with a state of about 1000- to 10,000-fold higher sensitivity (the circadian night state) (Fig. 1 b, Fleissner 1971, 1972, 1974). Both states are about 11 h long with transition phases between them lasting 2 to 3 h. In the steep middle part of the transition phase the sensitivity change can reach up to two log units per hour (Fig. 1 c, Fleissner 1975, 1977 d). The resulting nearly square wave time pattern enables us to measure the circadian period length with an accuracy of about 1 min (better than 1 in 1000).

A strong criterion for a true circadian rhythm is a phase shift of at least one circadian period length against the 24-h scale under constant conditions (Lohmann 1964). Using this definition the scorpion was the first arthropod in whose eyes a true circadian sensitivity rhythm in this strict sense could be demonstrated (Fleissner 1971). Its sensitivity rhythm shows all the characteristics of a circadian oscillation (e.g., free-running, synchronization by Zeitgeber cycles (Fig. 1 d, 2): Fleissner 1971, 1974, 1975, 1977 a, b; temperature-compensation of τ. Michel and Fleissner 1983).

The circadian ERG rhythms in the eyes of *Androctonus* are remarkably rigid and tightly coupled to one another, which suggests that only one clock may be responsible for them. The rhythm appears to remain stable throughout the life-time of an animal (recording period with an individual scorpion more than 8 months). Under constant conditions of continuous darkness (DD) the maximum value of sensitivity is attained within the first circadian period. The rhythm is always exactly the same in both median eyes, and in the intact animal we could never see any phase-angle differences between them. If only one eye is exposed to light for some hours, the contralateral eye shows exactly the same phase-shift (Fig. 1 d). Both eyes thus remain oscillating in precise synchrony (Fleissner 1972, 1977 a). This rigidity of the circadian ERG rhythm contrasts with that of other arthropods, e.g., some species of beetles which have loose circadian systems. Their ERG rhythm can completely disappear and reappear spontaneously, and it can even exhibit different period lengths in the left and right eye of the same animal (Köhler and Fleissner 1978).

In contrast to the median eyes, the *lateral eyes* only alter their endogenously controlled sensitivity by a factor of about 2 or even less (Fleissner 1975). Although the endogenous adaptation is comparatively small, a circadian rhythm is also clearly recognizable (Fig. 1 c). In every one of the ten lateral eyes it is in complete synchrony with the rhythm in the median eyes.

Since the innate period length of a circadian clock generally differs from 24 h, synchronization by an external *Zeitgeber* is necessary. Without this the rhythm would free-run against the environmental cycles and important advantages of the circadian clock would be lost. The ERG rhythm of *Androctonus* can be synchronized with light intensities as low as about 0.1 mlx (Fig. 2) (Fleissner 1977 b). Both the median and the lateral eyes are able to mediate the Zeitgeber information (Fleissner 1977 a, b). They represent an interesting combination of different features: the median eyes exhibit the distinct circadian sensitivity rhythm and are designed for good image processing (Jander 1965; Linsenmair 1968). In contrast, the lateral eyes are obviously sacrificing image

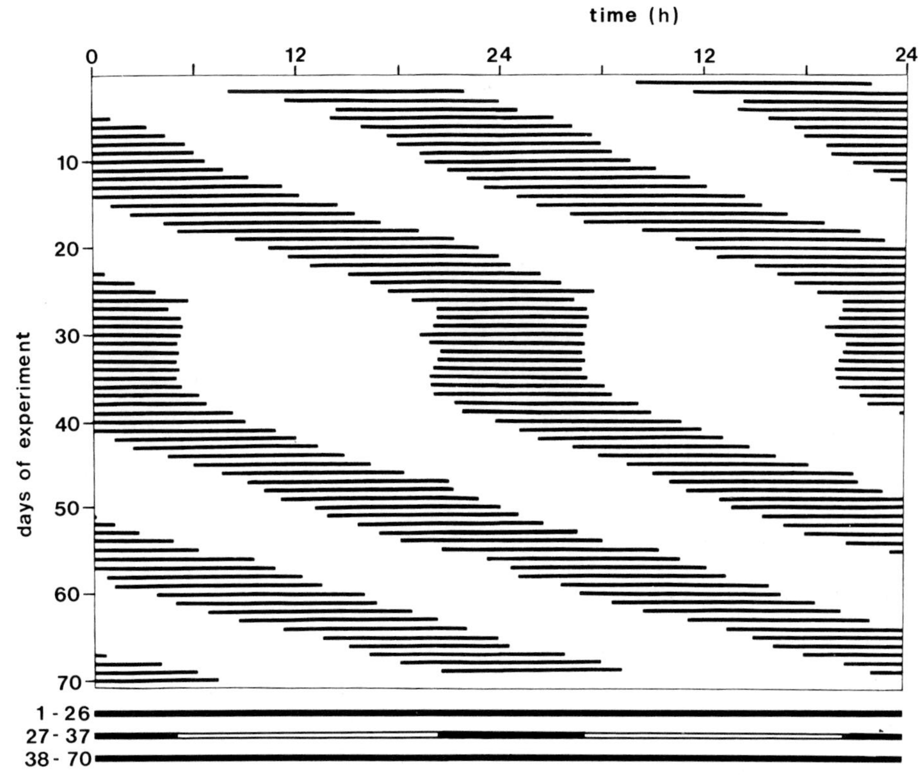

Fig. 2. Circadian rhythm of sensitivity in a median eye during continuous darkness (DD: days *1–26* and *38–70*) and light–dark cycles (LD: days *27–37*) (*bars below the graph*). The circadian night state of the eye is represented by *black stripes* in the graph. LD: natural light regime (L not exceeding 3 mlx) applied via a light guide to median and lateral eyes of one side. Note clear entrainment to 24 h during LD in spite of the very low light intensity. In DD the rhythm is free-running with a period length $\tau = 25.3$ h (before LD) and $\tau = 25.6$ h (after LD)

information (Schliwa and Fleissner 1980) in favor of functions similar to those of exposure meters; they are more sensitive and exhibit neglectably weak circadian fluctuation in their sensitivity. The combination of these two types of light receptor could enable the brain to calculate the time of dusk and dawn more precisely than only one of them would be able to do (Fleissner 1975, 1977d).

All five species of the scorpion families Buthidae and Scorpionidae we have studied so far have shown this ERG rhythm, which would therefore seem to be a general feature of scorpions. Surprisingly, all publications of other authors dealing with vision of the scorpions (Angermann 1957; Bedini 1967; Belmonte and Stensaas 1975; Carricaburu et al. 1982; Carricaburu and Cherrak 1968; Fouchard and Carricaburu 1971; Machan 1967; Yinon 1969) have neglected both the circadian sensitivity rhythm in the visual system and the fact that the scorpions, with only rare exceptions, are strictly night-active animals (Cloudsley-Thompson 1956; Wuttke 1966). It follows that all of this work has been done on eyes in their biologically less meaningful day state. The only exception is the paper of Machan (1968) in which, based on 16-h dark adaptation, the possibility of a diurnal sensitivity change was mentioned.

3 In Search of the Clock

Using the scorpion as a model system, our present aim is to try to understand how a circadian clock system is organized, both functionally and anatomically, at the level of its neuronal connections. The strategy we are following may appear rather simple, but applied consequently enough to this system it has proven to be rather effective: considering the ERG rhythm as a "hand" of the clock we just have to follow "cog by cog" until we find the "balance-wheel", i.e., the pace-making oscillator. The following questions have to be answered:

- What are the mechanisms of the endogenous sensitivity changes in the retina? (see Sect. 3.2)
- What is the circadian signal which controls these mechanisms? (see Sect. 3.3)
- Which pathways are taken by the circadian signal? (see Sect. 3.4)
- Where is the circadian oscillator located? (see Sect. 3.5)
- How is the scorpion's circadian clock organized as a whole? (see Sect. 3.6)

Before we proceed to answer these questions, however, we need some fundamental knowledge on the visual system of the scorpion.

3.1 Anatomy of the Visual System

The anatomy of the ocellar-like median and lateral eyes of the scorpion has been the subject of some earlier papers (Scheuring 1913; review Kaestner 1941). Several fundamental features of the eyes described in the older literature have to be corrected and extended with new cellular elements like the arhabdomeric cells (Fleissner and Siegler 1978; Schliwa and Fleissner 1979) and the efferent neurosecretory fibers (Fleissner and Schliwa 1977). Some of the more recent data are summarized by Root (1985).

3.1.1 Types of Eyes

Androctonus australis has 12 eyes. The two median eyes with a lens diameter of little more than 500 µm, are positioned centrally on the carapace. Two groups of lateral eyes, 5 on each side, are located at the anteriormost lateral edge of the prosoma (see Fig. 1c). Three lateral eyes in each group have a lens diameter of about 300 µm, the remaining two measure between 100 and 200 µm. Apart from ocular photoreception, an extraocular light sense has also been found in the metasoma of the scorpions *Urodacus novae-hollandiae* (Zwicky 1968) and *Heterometrus fulvipes* (Geethabali and Rao 1973). The visual systems of other chelicerates like *Limulus* contain several other types of photoreceptors such as median and lateral rudimentary eyes. A ventral eye and additional receptor cells scattered along the optic ganglia have also been described (for review Fahrenbach 1975). In the spiders *Argiope bruennichii* and *A. amoena* photosensitivity of the brain has been demonstrated (Yamashita and Tateda 1983, see also Yamashita, Chap. VI, this Vol.), so it would therefore not be surprising to find similar receptors in the scorpion as well. Sometimes clusters of pigmented cells can be observed at the surface of the first median eye ganglion, which might, for example, be such receptors. In this chapter, however, we will only consider the median and the lateral eyes.

1. The Median Eyes. Dioptric Apparatus. The prominent spherical cuticular lens, together with a well-designed vitreous body, form the dioptric apparatus of the median eyes. Carricaburu (1968) found these eyes to be very hyperopic in *Androctonus.* According to his own data, however, these eyes are well focused for long distances, as long as the vitreous body is also taken into account, the existence of which he has simply overlooked. This vitreous body is bordered by cells full of screening pigment granules (Fleissner 1971, 1974), resulting in a sharply defined iris-like aperture which improves image-processing in the median eye. The large aperture of this iris with a f-number of < 0.5 contributes to the physiologically demonstrated high sensitivity of the scorpion eyes (Fleissner 1977c). For further details on the optics of arthropod eyes refer to Land (1981) and to Land, Chapter IV, and Blest, Chapter V, this Volume.

Retinae. Both the median eye retinae have a common envelop formed by a pre- and postretinal membrane (see Fig. 7a), which separates them from the hemocoel in a fashion similar to the way a brain-hemolymph-barrier ensheaths the CNS (Lane and Harrison 1980). The retina mainly consists of the elongated visual cells which are surrounded over their whole length by very thin and richly fenestrated pigment cells. Arhabdomeric cells (Fleissner and Siegler 1978; Schliwa and Fleissner 1979) and neurosecretory fibers (Fleissner and Schliwa 1977) are further constituent parts of the retina. The postretina consists of a second type of pigment cells. All the axons of the visual and arhabdomeric cells and also the neurosecretory fibers form one common optic nerve.

Retinulae. Five visual cells make up a retinular unit (Fig. 3a, b). These retinulae, 500 to 600 in one median eye, are arranged parallel to each other in a hexagonal pattern similar to the ommatidia of compound eyes. In the distal plane of the retina the period length of this pattern is about 25 μm (Fleissner 1971). In cross-sections each visual cell contributes a V-shaped rhabdomere to the fused star-like rhabdome (Fig. 3a). The distal tip of the rhabdome is rather sharply pointed in the light-adapted eye and ends 20 to 30 μm away from the preretinal membrane (see Fig. 5b). Thus the screening pigment granules, which are densely packed in the distal part of the visual cells, effectively shield the rhabdomes against incident light. One arhabdomeric cell is always present in each retinula (Fig. 3b). Its soma, which is located below the perikarya of the visual cells in the center of the retinula, gives rise to an axon proximally and sends one or two dendrites up to the base of the rhabdome distally. Except for its uppermost part, the entire arhabdomeric cell is ensheathed by pigment cells. The naked distal dendrite sends arborizations into the visual cells, suggesting areas of electrical contacts. The visual cells only produce graduated receptor

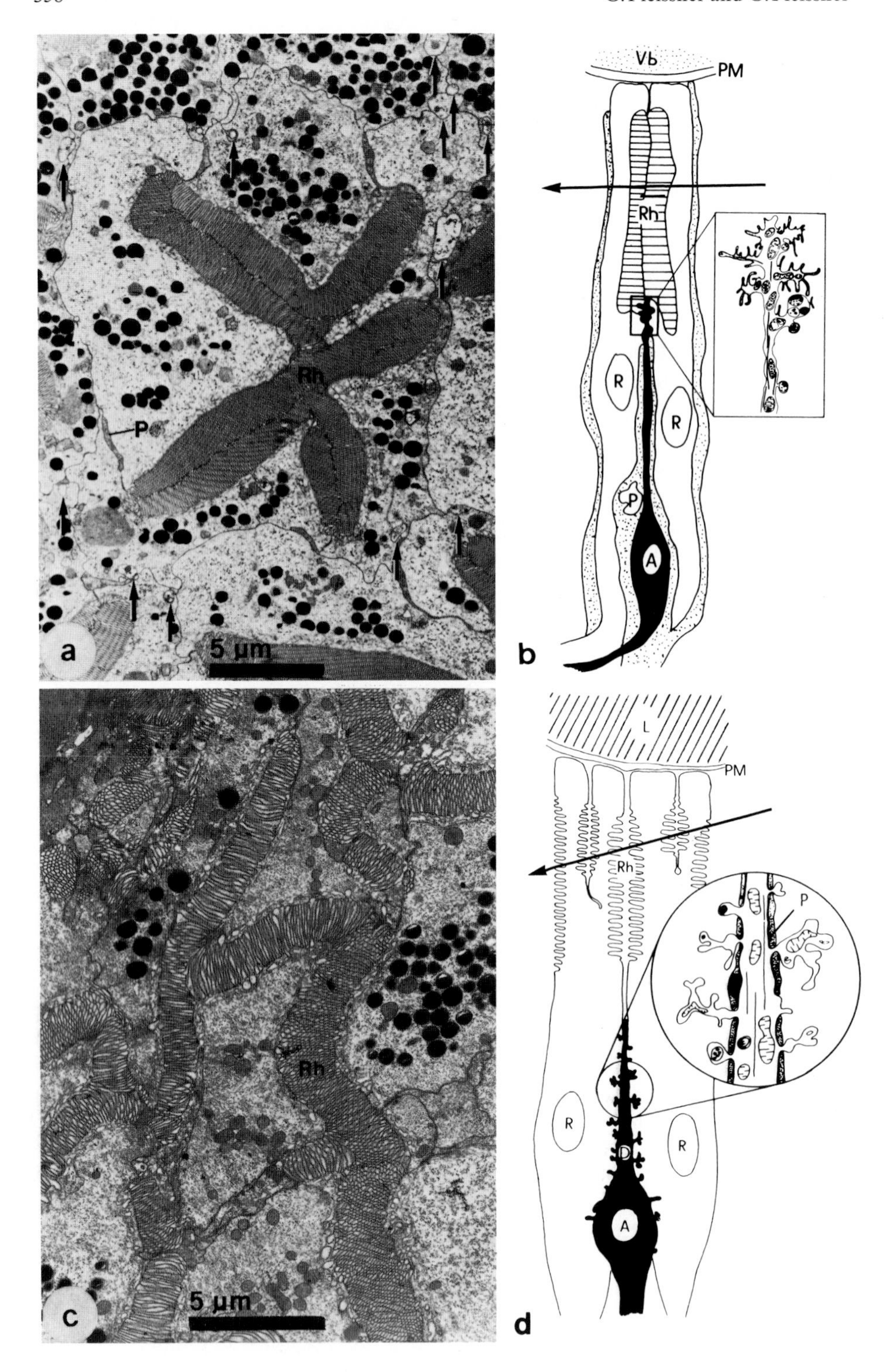
a
5 µm
Rh
P
b
Vb
PM
Rh
R
R
P
A
c
5 µm
Rh
d
L
PM
Rh
P
R
R
D
A

Fig. 3a–d. Fine structural details of the retina of the median (**a** and **b**) and of the lateral eyes (**c** and **d**) of *Androctonus australis.* **a** and **c** Cross-sections through the retinulae at locations as indicated by *arrows* in the respective scheme (**b** and **d**). **b** and **d** Schematized longitudinal sections through a retinular unit. *Insets* distal part of the dendrite of the arhabdomeric cell at higher magnification. *A* arhabdomeric cell; *D* dendrite; *L* lens; *P* pigment cell; *PM* preretinal membrane; *R* receptor cell; *Rh* rhabdome; *Vb* vitreous body; *arrows in* **a,** profiles of neurosecretory fibers. (**a** Fleissner and Schliwa 1977; **b** Schliwa and Fleissner 1979; **d** Schliwa and Fleissner 1980)

potentials, whereas the arhabdomeric cells give rise to propagated action potentials, which encode the time course of the receptor potentials (Fleissner 1971, 1985a). Regarding its retinal anatomy and function, the neuronal organization of the scorpion median eyes reveals striking similarities to the median (Jones et al. 1971; Nolte and Brown 1972) and lateral eyes of *Limulus* (for review Fahrenbach 1975).

2. The Lateral Eyes. The lateral eyes differ from the median eyes both in their absolute sensitivity (Fleissner 1977c) and in image processing. Although both types of eye contain the same retinal elements (Fig. 3c, d), their morphology shows characteristic differences: The cornea does not form a focusing lens (Carricaburu 1968), a vitreous body is lacking, and the rhabdomeres of all visual cells are fused to a net-like rhabdome over the whole retina. The retinular units consist of merely two or three visual cells, one arhabdomeric cell and several neurosecretory fibers (Schliwa and Fleissner 1980). The lack of image information in this type of eye obviously appears to enhance the measurement of brightness (Fleissner 1975).

3.1.2 The Optic Ganglia

Before entering the brain, the optic nerve shows a thickening, the first optic ganglion or lamina (Fig. 4). Our tracing experiments with cobalt and Lucifer Yellow CH in the periphery of the visual system established that the axons of the receptor cells, which constitute the bulk of the fibers in the optic nerve, terminate within the lamina. A smaller number of fibers (those arising from the arhabdomeric cells) terminate in the second ganglion or the medulla (Fleissner 1985a). This ganglion is shared by the median and lateral eyes of one side. Between the lamina and medulla there is a chiasma. Optic ganglia of higher order are also shared by median and lateral eyes (Babu 1954, also Chap. I, this Vol.). Further proximally, commissures to the contralateral side and connections to neuropils, like the central body, are formed (see Fig. 10).

3.2 Mechanisms of Endogenous Adaptation in the Retina

In principle, a photoreceptor has many possibilities to change its sensitivity (for review see Autrum 1981). In the median eyes of *Androctonus* we found periodic

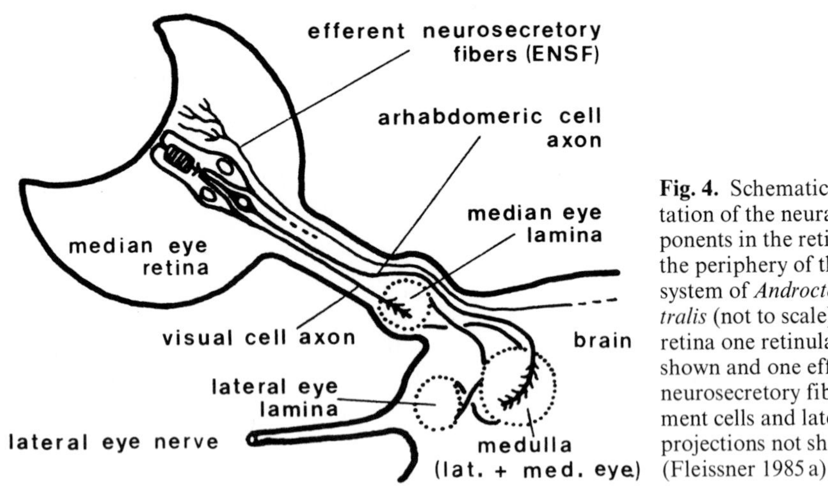

Fig. 4. Schematic representation of the neural components in the retina and the periphery of the visual system of *Androctonus australis* (not to scale). In the retina one retinular unit is shown and one efferent neurosecretory fiber. Pigment cells and lateral eye projections not shown. (Fleissner 1985 a)

migration of the screening pigment within the visual cells (Fleissner 1972, 1974): during the circadian night state, it is largely retracted into the depth of the retina, fully exposing the rhabdomes to the incident light (Fig. 5a). During the circadian day state, the screening pigment fills the distal part of the visual cells, shielding the rhabdomes against the light (Fig. 5b). In this position it functions as a filter with a transmission of 10^{-3} to 10^{-4} (Fleissner and Betz, unpublished). This corresponds exactly to the difference between the sensitivities in the circadian day and night state. The screening pigment migration may therefore be regarded as the main circadian adaptation factor in the median eyes.

As an additional mechanism of adaptation there is most probably a movement of pigment granules in the radial direction superimposed on the pigment migration along the visual cell axis. By opening small light channels, this mechanism could change the light flux to the rhabdomes within minutes. The speed with which the eye can reach an artificial night state triggered by electrical stimulation of the optic nerve (see Fig. 8a, Fleissner and Fleissner 1985) and the circadian variation of the acceptance angle of single visual cells (Marschall, unpublished) indicate the existence of such a mechanism.

The size and shape of the rhabdomes also appear to change. During the night the rhabdomes approach the preretinal membrane with their distal ends, and their tips are broadened, thus enlarging the aperture of the retinula (Fleissner 1974). Further study is required in order to estimate the contribution of this mechanism to the endogenous sensitivity control. In other arthropods a voluminous membrane turn-over as the structural basis of circadian sensitivity rhythms has been well established (for review Williams 1982; see also Blest, Chap. V, this Vol.).

Fig. 5a, b. Longitudinal sections through the median eye retina after 40 days in darkness during the circadian night state (**a**) and circadian day state (**b**) revealing the different positions of the screening pigment within the visual cells and the changed size and shape of the rhabdomes. (Abbreviations see Fig. 3.) (After Fleissner 1974)

3.3 The Circadian Signal

There are at least three possible sources of the circadian signal controlling these adaptation mechanisms in the eyes:

1. The circadian oscillation is generated within the retina as demonstrated for the eye of *Aplysia* (Jacklet 1969);
2. the oscillation is generated in the CNS and controls the effectors via a hormone (Rao and Gropalakrishnareddy 1967);
3. the oscillation is generated within the CNS but mediated along neuronal pathways.

For scorpion eyes the third possibility applies: whenever the optic nerve is cut, the sensitivity of the eye either remains at the low day-state level (Fig. 6) or, if the lesion has been performed during a circadian night, the high sensitivity immediately drops to the day-state level. The histology of the deefferented median eyes shows the corresponding day-state position of the screening pigment (Fig. 7). The circadian oscillation never comes back in the respective eye (Fleissner 1977b; Fleissner and Fleissner 1977, 1978). It can therefore be concluded: (1) No circadian oscillator is present in the retina, but must instead be located within the brain. (2) The circadian signal is delivered through the optic nerve. (3) The signal is tonic and has to remain "switched on" for as long as the night-state lasts. (4) The pigment distribution in the day state can be seen as a resting position.

The sensitivity decrease after cutting the optic nerve is neither due to manipulations during the operation, e.g., effect of light and CO_2 narcosis (Fleissner and Fleissner 1978) nor to wound reactions of the receptor or the arhabdomeric cell axons (Fleissner 1983; Fleissner and Fleissner 1985).

Spontaneous and rhythmic bursts of "efferent" action potentials can be recorded from the eye nerve, similar to those also found in a spider (Yamashita and Tateda 1981; see Yamashita, Chap. VI, this Vol.). In the case of *Androctonus*, however, this activity does not follow the circadian time pattern and is not actively conducted by the optic nerve. Its source might be the same as that of the noncircadian "prosomian nerve activity" described by Goyffon et al. (1975) for four species of scorpion. In our opinion *Limulus* therefore remains the first, and so far the only, arthropod where efferent action potentials which are mediating the circadian signal to the eyes could be convincingly demonstrated (Barlow et al. 1977; Barlow 1983).

The deefferented median eyes of the scorpion react, both in vivo and in vitro, immediately upon electrical stimulation of the optic nerve (Fig. 8). Within less than 10 min (much faster than we have observed in the undisturbed circadian rhythm) their sensitivity reaches its typical night state as indicated by the shape and amplitude of the ERG waves (Fig. 8a, Fleissner and Fleissner

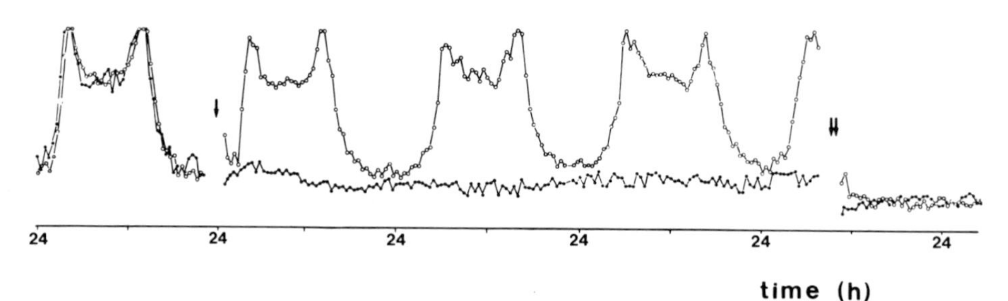

Fig. 6. Circadian sensitivity rhythm in both median eyes during a transection experiment (in DD). *Single arrow* optic nerve of the right median eye (●) cut; *double arrow* same operation with the left median eye (○). Cutting the nerve irreversibly abolishes the circadian oscillation in the respective median eye and keeps the sensitivity at the low day-state level. (After Fleissner and Fleissner 1978)

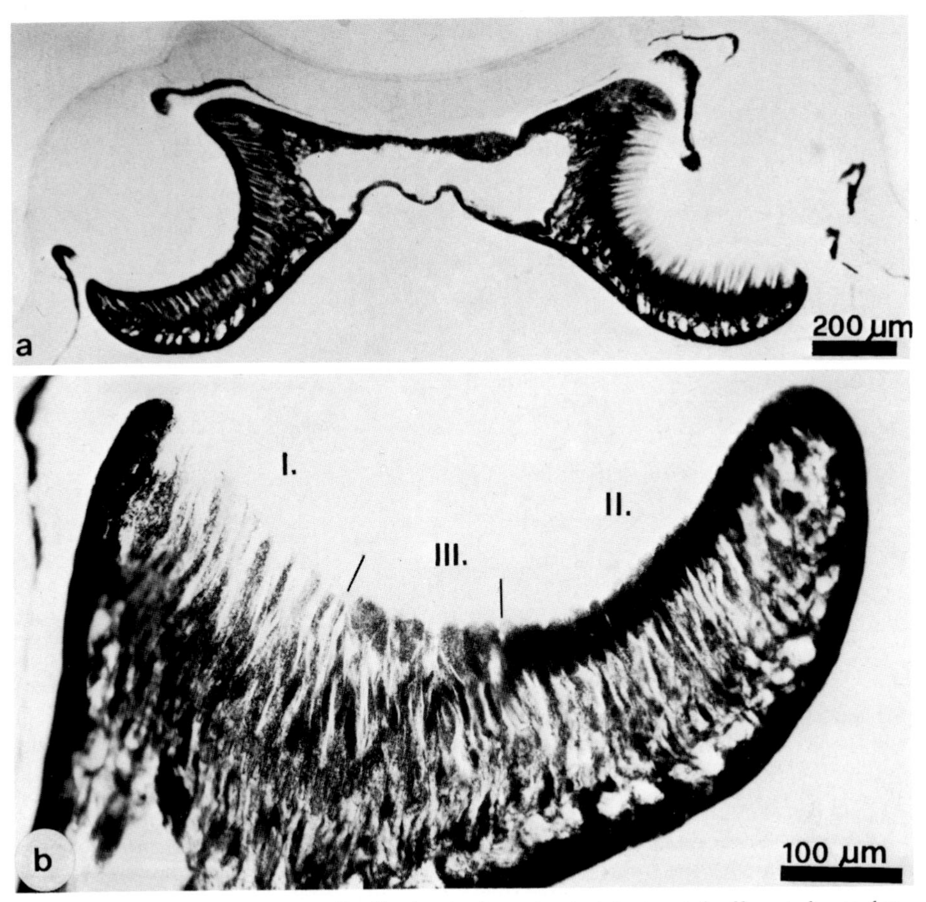

Fig. 7 a, b. Histology of pigment distribution in the retina in intact and deefferented eyes during circadian night state. **a** Transverse section through the dorsal part of the prosoma at the level of the median eyes, optic nerve of the left eye cut about 10 h previously. Note: In the deefferented eye on the left side the shielding pigment is positioned distally as in a day-state eye. **b** As **a**, but only one half of the eye nerve is cut. Note: Three zones of pigment distribution, *I.* night-state position in the intact dorsal part of the retina; *II.* day-state position in the deefferented ventral part; *III.* intermediate position as a consequence of partial deefferentiation of single visual cells

1985). To what extent this is also true for morphological aspects is still under investigation. Stopping the current pulses causes an immediate decrease of sensitivity to the day-state level, an effect resembling that of severing the nerve during the circadian night. Electrical activity in the optic nerve is obviously a necessary prerequisite for night-state conditions in the retina. The question remains as to whether it is really a completely sufficient one. Stimulation of the distal part of the cut optic nerve for more than 24 h can answer this question. The artificially increased sensitivity drops down to the day-state level after about 10 to 16 h, in spite of continued electrical stimulation (Fig. 8b). The time

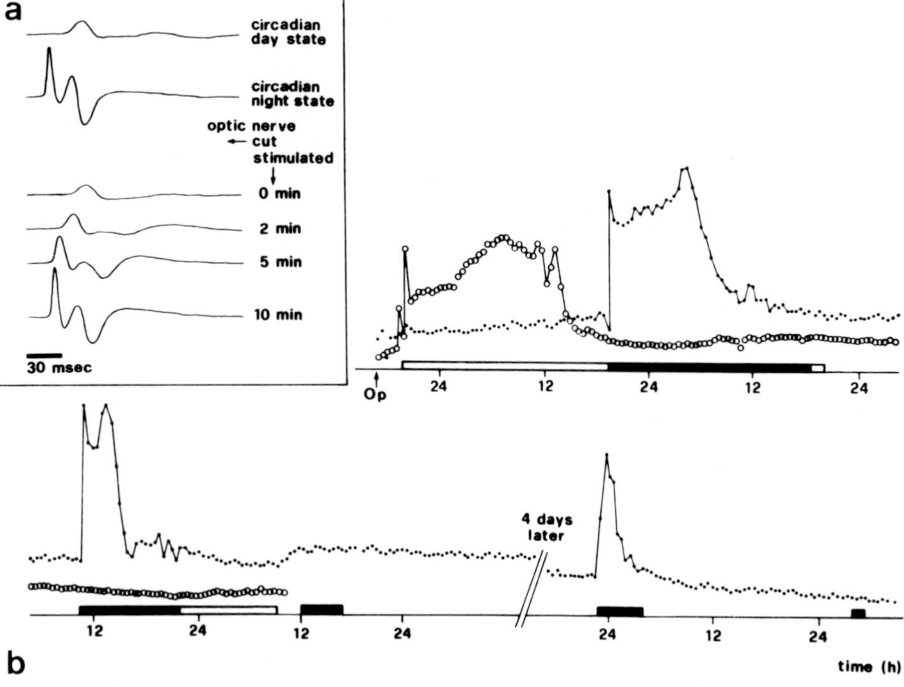

Fig. 8a, b. Artificial "night-state" in a median eye caused by electrical stimulation of the cut optic nerve. **a** ERG from a median eye in DD before and after cutting the optic nerve and during electrical stimulation of the nerve. Note: The ERG, which typically goes into the day state after the nerve is cut, changes to a typical night-state less than 10 min after nerve stimulation commenced. (AC recording.) **b** ERG-on-effect amplitudes of the two deefferented median eyes of the same scorpion in DD and their response to nerve stimulation. Note: Prolonged stimulation can exhaust a "night-state factor", 9 days after cutting the nerve the preparation is still able to react. *Bars* period of nerve stimulation, open and black bars refer to the corresponding curves (constant current pulses, 7 pps, 5 ms, 10 μA, pos. polarity). (Fleissner and Fleissner 1985)

course of this decrease resembles that of the end of the circadian night state. Degeneration processes do not explain the limited temporal effectiveness of electrical stimulation, because the same result is obtained when the stimulation is first started several days after cutting the nerve. Exhaustion of a factor, the synthesis of which depends on the soma, is a more plausible explanation. Since generally in neurosecretory cells the electrical nerve activity controls the release of the neurosecretory substance from the terminal structures (Gersch and Richter 1981, p 164), our findings seem to suggest that the circadian signal is of neurosecretory nature.

As almost nothing is known about the pharmacology of scorpion eyes, we have to refer to findings in other arthropods when searching for a likely transmitter candidate. In *Limulus* eyes octopamine has been demonstrated to partially simulate the effect of the circadian signal (Battelle et al. 1982; Kass and

Barlow 1984). Mancillas and Selverston (1984) could show by means of histology and electrophysiology that a Substance P-like peptide (10 out of 11 amino acids are identical with those of Substance P, Mancillas 1983) is another transmitter which seems to be involved in the efferent sensitivity control of the *Limulus* eyes. The effect of Substance P, as well as the normal circadian oscillation in the eye, could be prevented by a Substance P-blocker. Fibers with Substance P-like immunoreactivity could be identified (Mancillas 1983; Mancillas and Brown 1984). Because of the close phylogenetic relationship between *Limulus* and scorpions it was worthwhile testing the effect of Substance P and octopamine on the sensitivity of the median eyes of *Androctonus*. Preliminary results (Bhatti, unpublished) clearly demonstrate: octopamine applied to the inter-retinal space, generally produces changes of the ERG waves and increases the sensitivity as it is typical for the circadian night state. The possible role of Substance P-like proteins, so far, remains less clear. On the one hand, it seems to produce effects comparable to those of octopamine (Fleissner, unpublished), but on the other hand, the first immunocytological test in the retina and brain of the scorpion rules out Substance P itself as the transmitter of the circadian signal (Heinrichs and Schwab, unpublished).

3.4 Efferent Neurosecretory Fibers (ENSF): Link Between the Clock and Its Hand

The anatomical identification of the structures responsible for the transmission of the circadian signal from the brain to the retina would represent a further step toward the more central parts of the clock. The results reported in the previous sections seem to indicate that the search for neurosecretory structures in the retina and the optic nerve of the median eyes would prove promising.

Ultrastructural analysis of the retina and of the optic nerve of *Androctonus* has revealed the presence of numerous neurosecretory fibers (Fig. 3a). These terminate in synaptoid contacts on the visual cells in the median (Fleissner and Schliwa 1977) as well as the lateral eyes (Schliwa and Fleissner 1980). The synaptoid terminations contain both clear "synaptic" vesicles and neurosecretory granules, which points to the possibility of two neuroregulators within the same cell (Fig. 9). The fibers are characterized by their many varicosities and neurosecretory granules of about 100 nm in diameter, which have connections to all elements of the retinula unit. An individual fiber may terminate in several visual cells of the same retinula, as shown by dyade-like contacts to the receptor cells. It may also contact several receptor cells of neighboring retinula units, and the individual visual cell is innervated by several branches of different fibers.

The fibers are terminal arborizations of an efferent system. Their cell bodies lie somewhere in the brain proximal to the optic nerve. Serial sections through the retina did not reveal any perikarya containing neurosecretory substance; cutting the optic nerve results in the degeneration of these fibers distally and in an accumulation of neurosecretory granules proximally to the cut (Schliwa and Fleissner 1977). Forthwith we will refer to this system as the efferent neurosecretory fiber system (ENSF).

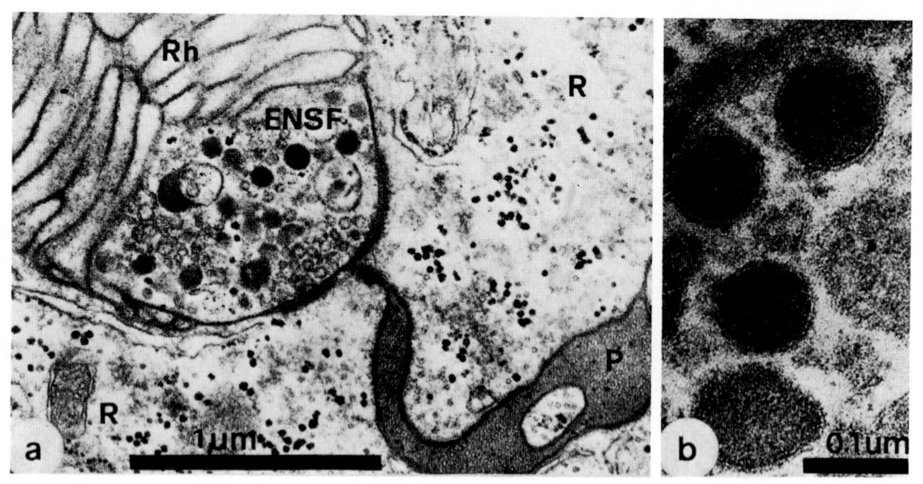

Fig. 9a, b. Cross-section of an efferent neurosecretory fiber (ENSF) in the median eye retina. **a** A presynaptic element of an ENSF (note the presence of clear vesicles beside the neurosecretory granules) forms a dyade-like connection with two visual cells of the same retinula. **b** The neurosecretory granules at higher magnification. (Abbreviations see Fig. 3.) (Fleissner and Schliwa 1977)

The central course of this fiber system, although outlined in previous papers, has not yet been fully described (Fleissner and Heinrichs 1982; Heinrichs and Fleissner 1982). Included here are the latest results obtained in our lab by Stefan Heinrichs (Heinrichs 1983, 1985; Heinrichs and Fleissner 1985a, b). The final anatomical analysis is still under investigation. The ENSF's, although randomly distributed throughout the cross-section of the optic nerve, then converge to form one bundle which passes next to the lamina and medulla and continues superficially along the dorsolateral edge of the supraesophageal ganglion (Fig. 10). It appears that they then make contact with the central body and send branches to the contralateral side, which extend to the contralateral median eye nerve. On either side of the central body the fibers descend to end in two groups of 10 to 20 cells near the circumoesophageal connectives. The cells, stained via the optic nerve, could be identified in ultrathin sections. They possess all the characteristics of neurosecretory cells and the neurosecretory granules are of the same size as those in the retinal ENSF. Gabe (1966) and Habibulla (1970) have described several types of neurosecretory cells in the brain of scorpions, but none of them are identical with the somata of the ENSF that we found.

The number of NS granules in the soma is remarkably small compared to the number to be observed, for instance, in the soma of a neuron which terminates in a neurohemal organ. However, since the terminal arborizations of the ENSF in the retina are in close contact with the effectors, this system is able to function with small neurosecretory quantities, and is highly effective.

On each side of the brain the ENSF sends fine branches toward the optic ganglia. These fine terminals resemble spines and synaptic buttons, suggesting areas of information-processing. Other similar contact zones are visible within the central body. Recent results with double staining methods (Heinrichs 1985) reveal that most, if not all, of the cells project into both median eyes at the same time. The complex interlacing of this fiber system between the left and right side may be the anatomical basis for the reported strong coupling between the circadian oscillations in the median eyes (see Fig. 1 d).

The *lateral eye* retinae also have efferent neurosecretory fibers with all the characteristics described above for the median eyes (Schliwa and Fleissner 1980). Interestingly, there seem to be only very few neurosecretory cells on each side positioned within the cluster of the somata of the median eye ENSF. The central course of the fibers differs only slightly from that of the median eyes, but the terminations can be observed in exactly the same areas where arborizations of the median eyes ENSF are found (Heinrichs and Fleissner 1985 b).

3.5 The Physiological Importance of the ENSF System

Knowing which course the ENSF's follow in the CNS provided the basis for investigating their physiological significance for the clock. Their fairly superficial position allowed the fibers to be cut in several places without damaging too many other tracts or neuropils (Fleissner 1983). In general, two conclusions can be drawn from these experiments (Fig. 11 and see Figs. 6, 7): (1) Whenever and wherever the ENSF's have been interrupted, the circadian rhythm in the median eyes was seriously damaged or even obliterated. (2) Details of the effect strongly depend on the position of the cut. The more distal the fibers were cut, the more direct and serious was the effect in the ipsilateral eye and the weaker

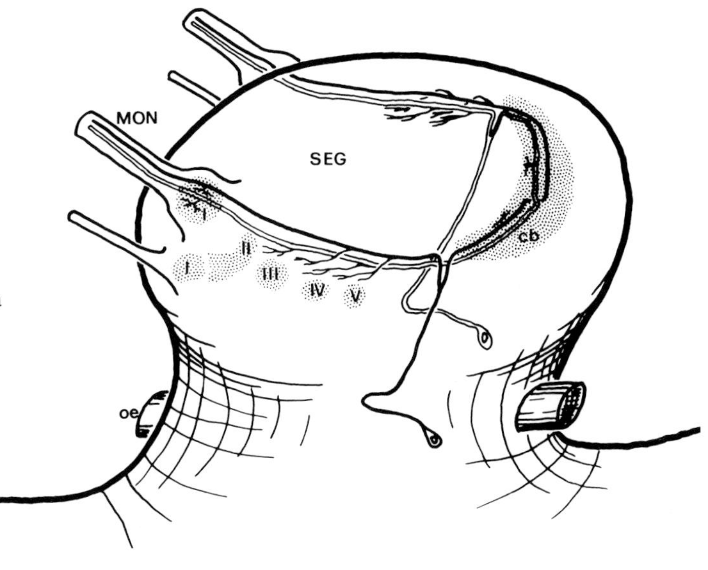

Fig. 10. Course taken by the efferent neurosecretory fiber system in the CNS. Schematic reconstruction of the supraesophageal ganglion (*SEG*) showing the fibers and cell bodies of two symmetrical ENSF units. All nerves except the lateral and median eye nerves (*MON*) omitted. The reconstruction is based on data obtained using several different staining techniques: Lucifer Yellow, cobalt, HRP, double staining with propidium iodide and HRP, and selective degeneration. (*oe* esophagus; *I, II, III, IV, V* optic ganglia; *cb* central body). (After Heinrichs and Fleissner 1985 a)

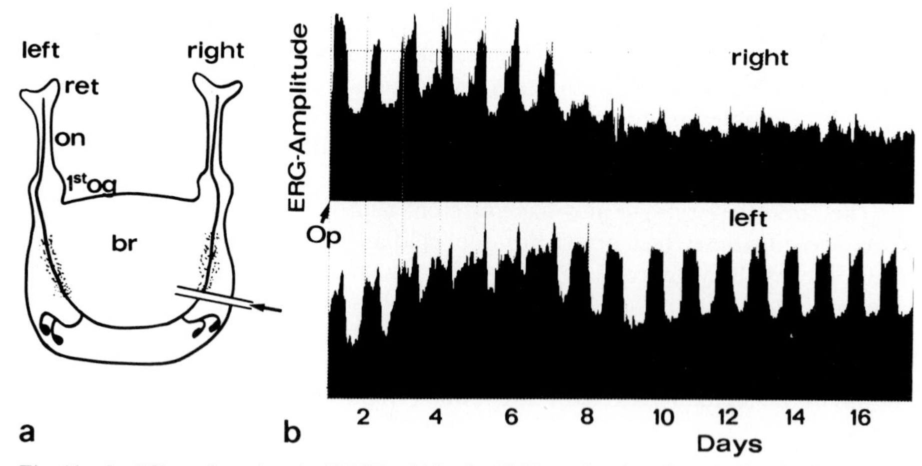

Fig. 11 a, b. Effect of cutting the ENSF within the CNS on the circadian rhythm in the median eyes of the scorpion. **a** Schematic drawing of the supraesophageal ganglion and the ENSF of the median eyes showing the position of the lesion (dorsal view, drawn in one plane). **b** Time course of the ERG amplitudes after the operation (*Op*). Note: The circadian oscillation, which is only slightly disturbed immediately after the nerve-cut, disappears on the 7th day in the ipsilateral (*right*) eye. In the contralateral eye, the rhythm which had become progressively obscured in the beginning also returned to normal on the 7th day. (Fleissner 1983)

it was in the contralateral eye. The more centrally the lesion was placed, the later the effect appeared ipsilaterally and the more pronounced was the disturbance of the rhythm in the contralateral eye. Any disconnection of the ENSF's from their cell bodies finally obliterated the circadian rhythm in the ipsilateral eye, whereas the rhythm in the contralateral eye was only temporarily affected.

These results, together with those of the electrical stimulation of the optic nerve, are strong evidence for the circadian mediator function of the ENSF's. In addition to this they tell us the following about the nature of the clock: (1) The circadian signal cannot come from the neurosecretory cell bodies, because the circadian rhythm in the eyes continued for a whole week after disconnection of the cell bodies from the ENSF's (Fig. 11 b). (2) It seems more likely that the ENSF's do not produce a circadian rhythm themselves, but receive it from another structure. (3) Several pieces of evidence point to the lateral part of the supraesophageal ganglion as the site where the circadian signal is probably coupled into the ENSF. (4) In the clock system there are contralateral inhibitory connections. (5) Rather complex interactions between left and right side, perhaps involving feedback signals from the retina, are possible.

3.6 The Organization of the Clock

How is the circadian clock, which drives the ERG rhythm, organized? Does it consist of one circadian oscillator or a combination of at least two or even more? In molluscs (Jacklet 1969), crayfish (Sanchez and Fuentes-Pardo 1977), and insects (Köhler and Fleissner 1978; Fleissner 1982; Tomioka and Chiba 1982) the circadian ERG clock has been proven to be a bilaterally symmetrical multioscillator system, and in general this seems to be a fundamental feature of any circadian clock (Pittendrigh 1974). We therefore might expect the circadian clock of the scorpion to be a multioscillator as well. Surprisingly, it always behaves like one functional unit: desynchronization of the rhythms in the various eyes of an individual scorpion never occurred spontaneously, nor was it possible to induce it by unilateral LD cycles (Fig. 1d). The rhythms of both median eyes remain tightly coupled to each other. However, there are also several clues which suggest multioscillator organization of the scorpions' clock: in synchronization experiments we observed "phase-jumps" and complex oscillations following the exposure of the lateral eyes to LD cycles (Fleissner 1977b). We interpret these reactions as indicative of a multioscillator system. Dissociation of the ERG rhythm into ultradian components could also be seen when the body temperature of the scorpion was lowered (e.g., to 20 °C, Fig. 12) (Michel and Fleissner 1983). In addition to this, a few of our lesion experiments (e.g., cutting the ENSF within the CNS) sometimes revealed slight phase-angle differences of some degrees between left and right median eyes occasionally even of 180°. However, internal desynchronization (which is based on different period lengths) such as we find in beetles (Köhler and Fleissner 1978) was never observed. Obviously further experimental evidence is necessary in this case to resolve the multioscillator question in the scorpion.

What role does this clock play in the organism as a whole? *Androctonus australis,* like nearly all the other scorpions, is a nocturnal animal (Cloudsley-Thompson 1956; Constantinou and Cloudsley-Thompson 1983). The pattern of its *locomotor activity* can vary between a clear uni- or bimodal circadian rhythm and near arrhythmicity (Dube, in preparation). Since so far we could not ob-

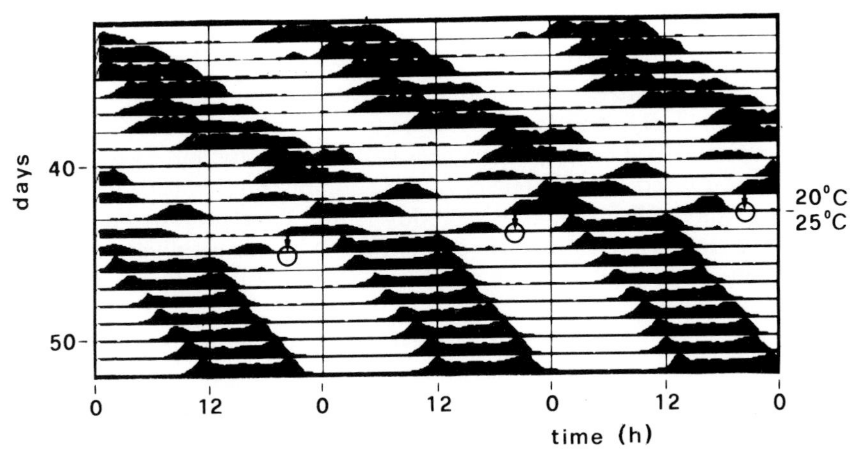

Fig. 12. Circadian ERG rhythm at different levels of constant temperature in continuous darkness. The night state dissociates into several peaks after a few days in 20 °C. (↓, time of temperature change). (Michel and Fleissner unpublished)

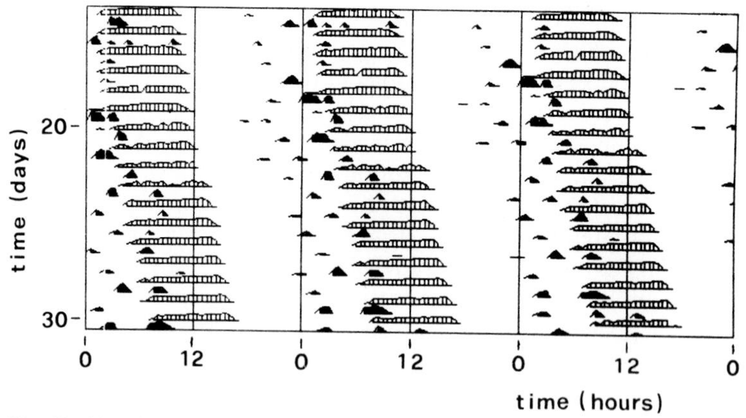

Fig. 13. Simultaneous recording of the circadian ERG rhythm (▥) of one median eye and the locomotor activity (■) of the same scorpion. The animal, which was tethered above a glass plate, was able to move its legs freely. The graph shows the amount of activity per half hour performed by one leg. The locomotor activity is obviously gated by a circadian rhythm with the same period length as that of the ERG rhythm. Note: The blocks of locomotor activity have an ultradian period length, which does not match that of the circadian rhythm.

serve this broad variability in the ERG rhythms, it seemed uncertain whether both functions were controlled by the same circadian oscillator. This situation has, however, now been changed due to recently recorded data from our laboratory: (1) Simultaneously recording both parameters in the same animal (Fig. 13, Fleissner 1985 b) demonstrates that the locomotor activity is basically driven by an ultradian oscillator (with about 5-h period length) but it is gated by a circadian oscillation with the same period length as the ERG rhythm. (2) C. Dube discovered that under LL conditions (about 1 lx), the pattern of locomotor activity can reach exactly the same clearness, shape, and precision as the ERG rhythm usually shows at temperatures above 25 °C (Michel and Fleissner 1983). (3) On the other hand, S. Michel demonstrated that the almost rectangular-shaped night state of the ERG rhythm sometimes dissociates into several blocks of high sensitivity with a time pattern which looks very similar to that of the ultradian blocks of activity.

It may therefore be that differences between the rhythms of ERG and locomotion are caused by differences in the coupling mechanisms of the two parameters. Rao and Gropalakrishnareddy (1967) have proposed a blood-borne factor as being responsible for coupling the locomotor activity to the oscillator and we have demonstrated that the ERG rhythm is driven via neurosecretory fibers.

As a result of these observations we propose the following hypothesis:

- In the scorpion the ERG rhythm and the rhythm of locomotor activity possess a common oscillator, in the form of a circadian master oscillator, which in turn gates several ultradian oscillators.
- The differences in the time patterns of the respective overt rhythms can be accounted for by the different nature of the link between the circadian oscillator and the effectors.

- These coupling mechanisms are sensitive to temperature changes and differ in their optimal temperature.
- Light, and possibly other external inputs, may additionally vary the time pattern of the overt rhythm.

4 Conclusions

Specialization in a particular field of research has led to the fact that neurobiology and circadian rhythmicity have often been viewed as completely separate topics. Even recent papers on these subjects sometimes fail to link the one aspect with the other. In spite of the fact that the subjective condition of an animal can alter its sensory behaviour, neurobiological work often does not take the circadian factor into consideration. On the other hand, much of the circadian research work still lacks a neurobiological basis.

When trying to analyse the circadian clock controlling the sensitivity in the eyes of arthropods, the interdependence of these two aspects becomes clear:

1. The sensitivity of the scorpion's eyes is controlled by a circadian clock located in the CNS. During the subjective night the median eyes are 1000 times more sensitive than during the subjective day. As scorpions are nocturnal animals, it would therefore seem more sensible to focus future experiments on the vision of these animals in their naturally active condition, i.e., the subjective night.

2. Efferent input to receptors seems to be a very widespread phenomenon and an important part of sensory physiology (see also Yamashita, Chap. VI, this Vol.). The visual system could provide a model which might enable further investigation of this principle.

3. Efferent neurosecretory fibers (ENSF's) mediating the circadian signal from the CNS to the periphery appear to be present in other arthropods as well as in scorpions. It is possible that similar structures are generally responsible for the transmission of circadian signals.

4. By following the course of the ENSF's, the neural organization of the circadian clock may be unraveled. These fibres are probably not the source of the oscillation, but merely mediate the circadian signal. The input regions to the ENSF system in scorpions seem to lie near the third or fourth optic ganglion.

5. The circadian clock controling the ERG rhythm may well serve a pacemaker function for the multioscillator system of the whole animal.

6. Several observations have revealed similarities in the neuronal organization of the circadian systems in scorpions and other arthropods, indicating perhaps a common principle in the neurobiology of the circadian systems in these animals.

Acknowledgments. We want to express our appreciation to Mr. Saadallah, le Directeur de Forêts, Ministère de l'agriculture in Tunis for giving us permission to collect specimens of *Androctonus* in Tunesia. We would also like to thank Mrs. Elke Nöring for her careful maintenance of the animals. We gratefully acknowledge the technical assistance provided by Karl-Heinz Schmitt, and thank Mrs. Susan Bhatti for the revision of the English text. Last, but not least, we want to thank all members of the Arbeitskreis Neurobiologie circadianer Rhythmen for their friendship and collaboration during the past years.

This investigation was supported by the Deutsche Forschungsgemeinschaft: Priority programs *Biologie der Zeitmessung, Mechanismen biologischer Uhren,* and *Vergleichende Neurobiologie des Verhaltens* (SFB 45).

References

Angermann H (1957) Über Verhalten, Spermatophorenbildung und Sinnesphysiologie von *Euscorpius italicus Hbst.* und verwandte Arten (Scorpiones, Chactidae). Z Tierpsychol 14:276–302

Arechiga H (1977) Circadian rhythmicity in the nervous system of crustaceans. Fed Proc 36(7):2036–2041

Arechiga H, Wiersma CAG (1969) Circadian rhythm of responsiveness in crayfish visual units. J Neurobiol 1:71–85

Arechiga H, Fuentes B, Barrera B (1973) Circadian rhythm of responsiveness in the visual system of crayfish. In: Salanky (ed) Neurobiology of invertebrates. Tihany 1971, pp 403–421

Aschoff J (ed) (1981) Handbook of behavioral neurobiology 4: Biological rhythms. Plenum Press, New York

Autrum HJ (1981) Light and dark adaptation in invertebrates. In: Autrum HJ (ed) Handbook of sensory physiology, vol VII/6 C. Springer, Berlin Heidelberg New York, pp 1–91

Babu KS (1965) Anatomy of the central nervous system of arachnids. Zool Jahrb Anat 82:1–154

Barlow RB (1983) Circadian rhythms in the *Limulus* visual system. J Neurosci 3:856–870

Barlow RB, Bolanowski SJ, Brachman ML (1977) Efferent optic nerve fibers mediate circadian rhythms in the *Limulus* eye. Science 197:86–89

Battelle BA, Evans JA, Chamberlain SC (1982) Efferent fibers to *Limulus* eyes synthesize and release octopamine. Science 216:1250–1252

Bedini C (1967) The fine structure of the eyes of *Euscorpius carpathicus L* (Arachnida, Scorpiones). Arch Ital Biol 105:361–378

Belmonte C, Stensaas JL (1975) Repetitive spikes in photoreceptor axons of the scorpion eye. Invertebrate eye structure and tetrodotoxin. J Gen Physiol 66:649–655

Bennett R (1983) Circadian rhythm of visual sensitivity in *Manduca sexta* and its development from an ultradian rhythm. J Comp Physiol 150:165–174

Block GD, Page TL (1978) Circadian pacemakers in the nervous system. Annu Rev Neurosci 1:19–34

Brady J (1974) The physiology of insect circadian rhythms. Adv Insect Physiol 10:1–115

Bünning E (1973) The physiological clock, rev 3rd edn. Springer, Berlin Heidelberg New York

Carricaburu P (1968) Dioptrique oculaire du scorpion *Androctonus australis.* Vision Res 8:1067–1072

Carricaburu P, Cherrak M (1968) Analysis of the elctroretinogram of the scorpion *Androctonus australis (L).* Z Vergl Physiol 61:386–393

Carricaburu P, Biwer G, Goyffon M (1982) Effects of Carbaryl on the spontaneous and evoked potentials of the scorpion prosomian nervous system. Comp Biochem Physiol 73 C:201–204

Cloudsley-Thompson JL (1956) Studies in diurnal rhythms. VI. Bioclimatic observations in Tunesia and their significance in relation to the physiology of the fauna especially woodlice, centipedes, scorpions and beetles. Ann Mag Nat Hist (12) 9:305–329

Cloudsley-Thompson JL (1978) Biological clocks in arachnida. Bull Br Arachnol Soc 4:184–191

Constantinou C, Cloudsley-Thompson JL (1980) Circadian rhythms in scorpions. Proc 8th Int Arachnol Soc 1980, pp 53–55

Enright JT (1980) The timing of sleep and wakefulness. In: Barlow HB et al. (eds) Studies of brain function, vol III. Springer, Berlin Heidelberg New York

Fahrenbach WH (1975) The visual system of the horseshoe crab *Limulus polyphemus.* Int Rev Cytol 41:285–349

Fleissner G (1971) Über die Sehphysiologie von Skorpionen. Inaug Dissertation, Univ Frankfurt/Main

Fleissner G (1972) Circadian sensitivity changes in the median eyes of the North African scorpion, *Androctonus australis.* In: Wehner R (ed) Information processing in the visual system of arthropods. Springer, Berlin Heidelberg New York, pp 133–139
Fleissner G (1974) Circadiane Adaptation und Schirmpigmentverlagerung in den Sehzellen der Medianaugen von *Androctonus australis L* (Buthidae, Scorpiones). J Comp Physiol 91:399–416
Fleissner G (1975) A new biological function of the scorpion's lateral eyes as receptors of Zeitgeber stimuli. Proc 6th Int Arachnol Congr 1974, pp 176–182
Fleissner G (1977a) Entrainment of the scorpion's circadian rhythm via the median eyes. J Comp Physiol 118:93–99
Fleissner G (1977b) Scorpion lateral eyes: Extremely sensitive receptors of Zeitgeber stimuli. J Comp Physiol 118:101–108
Fleissner G (1977c) The absolute sensitivity of the median and lateral eyes of the scorpion, *Androctonus australis L.* (Buthidae, Scorpiones). J Comp Physiol 118:109–120
Fleissner G (1977d) Differences in the physiological properties of the median and the lateral eyes and their possible meaning for the entrainment of the scorpion's circadian rhythm. J Interdiscipl Cycle Res 8:15–26
Fleissner G (1982) Isolation of an insect circadian clock. J Comp Physiol A 149:311–316
Fleissner G (1983) Efferent neurosecretory fibres as pathways for the circadian clock signals in the scorpion. Naturwissenschaften 70:366
Fleissner G (1985a) Intracellular recordings of light responses from spiking and nonspiking cells in the median and lateral eyes of the scorpion *Androctonus australis.* Naturwissenschaften 72:46–48
Fleissner G (1985b) The circadian clock of the eyes. A pacemaker for locomotor activity rhythms in invertebrates. Experientia (in press)
Fleissner G, Fleissner G (1977) Steuerung der circadianen Pigmentwanderung in den Medianaugen des Skorpions. Verh Dtsch Zool Ges 1977, 231
Fleissner G, Fleissner G (1978) The optic nerve mediates the circadian pigment migration in the median eyes of the scorpion. Comp Biochem Physiol 61A:69–71
Fleissner G, Fleissner G (1985) Efferent electrical activity – necessary, but not sufficient for circadian sensitivity control of the scorpion eyes. Naturwissenschaften (submitted)
Fleissner G, Heinrichs S (1982) Neurosecretory cells in the circadian clock system of the scorpion *Androctonus australis.* Cell Tissue Res 224:233–238
Fleissner G, Schliwa M (1977) Neurosecretory fibers in the median eyes of the scorpion *Androctonus australis L.* Cell Tissue Res 178:189–198
Fleissner G, Siegler W (1978) Arhabdomeric cells in the retina of the median eyes of the scorpion. Naturwissenschaften 65:210
Fouchard R, Carricaburu P (1970) Quelques aspects de la physiologie visuelle chez le scorpion *Buthus occitanus;* étude électrorétinographique. Bull Soc Hist Nat Afr Nord 61:57–67
Fouchard R, Carricaburu P (1971) L'électrorétinogramme de l'oeil médian du scorpion *Buthus occitanus* en function de l'adaptation à l'obscurité. C R Acad Sci 271 D:446–448
Gabe M (1966) Neurosecretion. Pergamon Press, Oxford New York
Geethabali, Rao KP (1973) A metasomic neural photoreceptor in the scorpion. J Exp Biol 58:189–196
Gersch M, Richter K (eds) (1981) Das peptiderge Neuron. VEB Fischer, Jena (refer to pp 164)
Goyffon M, Lluyckx J, Vachon M (1975) Sur l'existence d'une activité électrique rythmique spontanée du système nerveux céphalique de scorpion. C R Acad Sci 280D:873–876
Goyffon M, Vachon M, Broglio N (1982) Epidemiological and clinical characteristics of the scorpion evenomation in Tunesia. Toxicon 20:337–344
Habibulla M (1970) Neurosecretion in the scorpion, *Heterometrus swammerdami.* J Morphol 131:1–16
Heinrichs S (1983) Identification of neurons in the transmission electron microscope after retrograde labelling with the dye Lucifer Yellow. Mikroskopie 40:79–86
Heinrichs S (1985) Oxidases and fluorescent dyes for double and differential retrograde labelling. J Neurosci Meth (in press)
Heinrichs S, Fleissner G (1982) Anatomy of the efferent neurosecretory cells in the circadian oscillator system of the scorpion. Verh Dtsch Zool Ges 75:312

Heinrichs S, Fleissner G (1985 a) Studies on the efferent neurosecretory fiber (ENSF) system of the scorpion's circadian clock. I. Organization in the CNS. Cell Tissue Res (submitted)

Heinrichs S, Fleissner G (1985 b) Studies on the efferent neurosecretory fiber (ENSF) system of the scorpion's circadian clock. II. Different neurons of the same group supply median and lateral eyes. Cell Tissue Res (submitted)

Jacklet JW (1969) Circadian rhythm of optic nerve impulses recorded in darkness from isolated eye of *Aplysia*. Science 164:562–563

Jacklet JW, Rolerson C (1982) Electrical activity and structures of retinal cells of the *Aplysia* eye. II. Photoreceptors. J Exp Biol 99:381–395

Jacklet JW, Schuster L, Rolerson C (1982) Electrical activity and structure of retinal cells of the *Aplysia* eye. I. Secondary neurons. J Exp Biol 99:369–380

Jahn TJ, Crescitelli F (1940) Diurnal changes in the electrical response of the compound eye. Biol Bull 78:42–52

Jander R (1965) Die Phylogenie von Orientierungsmechanismen der Arthropoden. Verh Dtsch Zool Ges, 266–306

Jones C, Nolte J, Brown JE (1971) The anatomy of the median ocellus of *Limulus*. Z Zellforsch Mikrosk Anat 118:297–309

Kaestner A (1941) Arachnida, Scorpiones. In: Kükenthal W, Krumbach T (eds) Chelicerata, part 1. Handbook of zoology, vol III, De Gruyter, Berlin, pp 117–240

Kass L, Barlow RB jr (1984) Efferent neurotransmission of circadian rhythms in *Limulus* lateral eye. I. Octopamine-induced increases in retinal sensitivity. J Neuroscience 4:908–917

Kiesel A (1894) Untersuchungen zur Physiologie des facettierten Auges. S B Akad Wiss Wien Math-Nat Kl 103:97–139

Köhler WK, Fleissner G (1978) Internal desynchronisation of bilaterally organized circadian oscillators in the visual system of insects. Nature (London) 274:708–710

Land MF (1981) Optics and vision in invertebrates. In: Autrum HJ (ed) Comparative physiology and evolution of vision in invertebrates, Handbook of sensory physiology, vol VII/6B. Springer, Berlin Heidelberg New York, pp 471–592

Lane NJ, Harrison JB (1980) Unusual form of tight junction in the nervous system of the scorpion. Eur J Cell Biol 22:244

Larimer JL, Smith JFT (1980) Circadian rhythm of retinal sensitivity in crayfish: Modulation by the cerebral and optic ganglia. J Comp Physiol 136:313–326

Linsenmair KE (1968) Anemomenotaktische Orientierung bei Skorpionen (Chelicerata, Scorpiones). Z Vergl Physiol 60:445–449

Lohmann M (1964) Der Einfluß von Beleuchtungsstärke und Temperatur auf die tagesperiodische Laufaktivität des Mehlkäfers *Tenebrio molitor L.* Z Vergl Physiol 49:341–389

Machan L (1967) Studies of structure and electrophysiology of scorpion eyes. S Afr J Sci 63:512–520

Machan L (1968) The effect of prolonged dark-adaptation on sensitivity and the correlation of shielding pigment position in the median and lateral eyes of the scorpion. Comp Biochem Physiol 26:365–368

Mancillas JR (1983) Neuropeptide modulation of sensory and motor activity in simple nervous systems. Thesis, Univ Calif, San Diego

Mancillas JR, Brown MR (1984) Neuropeptide modulation of photosensitivity. I. Presence, distribution, and characterization of a substance P-like peptide in the lateral eye of *Limulus*. J Neurosci 4:832–846

Mancillas JR, Selverston AJ (1984) Neuropeptide modulation of photosensitivity. II. Physiological and anatomical effects of substance P on the lateral eye of *Limulus*. J Neurosci 4:847–859

Michel S, Fleissner G (1983) Einfluß der Temperatur auf eine circadiane Uhr des Skorpions. Verh Dtsch Zool Ges 1983, 308

Nolte J, Brown JE (1972) Electrophysiological properties of the cells in the median ocellus of *Limulus*. J Gen Physiol 59:167–185

Pittendrigh CS (1974) Circadian oscillations in cells and the circadian organization of multicellular systems. In: Schmitt TO, Warden FG (eds) The neurosciences. Third study program. MIT Press, Cambridge, Mass, pp 437–458

Rao KP, Gropalakrishnareddy T (1967) Blood borne factors in circadian rhythms of activity. Nature (London) 213:1047–1048
Root T (1985) Scorpion neurobiology. In: Polis G (ed) Biology of scorpions, in press
Rusak B, Zucker I (1979) Neural regulation of circadian rhythms. Physiol Rev 59:449–526
Sanchez JA, Fuentes-Pardo B (1977) Circadian rhythm in the amplitude of the electroretinogram in the isolated eye-stalk of the crayfish. Comp Biochem Physiol 56A:601–605
Saunders DS (1982) Insect clocks, 2nd edn. Pergamon Press, Oxford New York
Scheuring L (1913) Die Augen der Arachnoideen. I. Die Augen der Scorpioniden. Zool Jahrb Anat 33:533–588
Schliwa M, Fleissner G (1979) Arhabdomeric cells of the median eye retina of scorpions. I. Fine structural analysis. J Comp Physiol 130:265–270
Schliwa M, Fleissner G (1980) The lateral eyes of the scorpion, *Androctonus australis*. Cell Tissue Res 206:95–114
Tomioka K, Chiba Y (1982) Persistence of circadian ERG-rhythm in the cricket with optic tract severed. Naturwissenschaften 69:395–396
Vachon M (1952) Etudes sur les scorpions. Inst Pasteur d'Algerie, Alger
Vachon M (1953) The biology of scorpion. Endeavour 12:80–89
Wada S, Schneider G (1968) Circadianer Rhythmus der Pupillenweite im Ommatidium von *Tenebrio molitor*. Z Vergl Physiol 58:395–397
Welsh JH (1938) Diurnal rhythms. Q Rev Biol 13:123–139
Williams DS (1982) Ommatidial structure in relation to turnover of photoreceptor membrane in the locust. Cell Tissue Res 225:595–617
Winfree AT (1980) The geometry of biological time. Biomathematics, vol. VIII. Springer, Berlin Heidelberg New York
Wuttke W (1966) Untersuchungen zur Aktivitätsperiodik bei *Euscorpius carpathicus*. Z Vergl Physiol 53:405–448
Yamashita S, Tateda H (1981) Efferent neural control in the eyes of orb weaving spiders. J Comp Physiol 143:477–483
Yamashita S, Tateda H (1983) Cerebral photosensitive neurons in the orb weaving spiders, *Argiope bruennichii* and *A. amoena*. J Comp Physiol 150:467–472
Yinon U (1969) The electroretinogram of scorpion eyes. Comp Biochem Physiol 30:989–992
Zlotkin E, Miranda F, Rochat H (1978) Chemistry and pharmacology of Buthidae scorpion venoms. In: Bettini S (ed) Arthropod venoms. Springer, Berlin Heidelberg New York, pp 317–369
Zwicky KT (1968) A light response in the tail of *Urodacus*, a scorpion. Life Sci 7:257–262

Subject Index

Page numbers in ***bold italics*** refer to figures and tables

M. Heisenberg, R. Wolf

Vision in Drosophila

Genetics of Microbehavior

1984. 112 figures. IX, 250 pages
(Studies of Brain Function, Volume 12).
ISBN 3-540-13685-1

Contents: Introduction. – Eye, Brain, and Simple Behavior: The Compound Eye. Neuronal Architecture of the Visual System. Motion Sensitivity Under Open Loop Conditions. Toward Correlating Structure and Function. – The Behavioral Structure of the Visual System: Flying Straight. Endogenous Behavior in Yaw Torque Fluctuations. Orientation Toward Objects. Menotaxis. Foreground-Background Experiments. Visual Control in Free Flight. Selective Attention. Plasticity of Visuo-Motor Coordination. Valuation. – Synopsis. – Appendix 1: List of Neurological Mutants. – Appendix 2: Symbols, Dimensions, Abbreviations. – References. – Subject Index.

Springer-Verlag
Berlin
Heidelberg
New York
Tokyo

U. Bässler

Neural Basis of Elementary Behavior in Stick Insects

Translated from the German by C. M. Z. Strausfeld

1983. 124 figures. XI, 169 pages
(Studies in Brain Function, Volume 10)
ISBN 3-540-11918-3

Contents: Introduction. - Behavioral Components of Twig Mimesis - Experiments on the Femur-Tibia Joint. - Other Behaviors of the Stationary Animal. - Walking. - Orientation. - Anatomy of the Muscles, Nerves, and Sense Organs of the *Carausius* Thorax. - References. - Subject Index.

Research into the neural basis of behavior concentrates on invertebrates primarily because the number of neurons involved is smaller than in mammals.
An ideal subject for such research is the stick insect, which exhibits behavior patterns similar to those in mammals, in particular walking, catalepsy, rocking and joint positioning.
In this volume, the author summarizes the current status of neuroethological research on the diurnal behavior of the stick insect, *Carausius morsus.*
Drawing primarily upon studies conducted in West Germany, he begins with a quantitative description of stick insect behavior before describing the systems of which the behavior patterns are an expression, and the neural mechanisms underlying these systems. The investigatory methods used are drawn from behavioral physiology, systems theory and electrophysiology.

Springer-Verlag
Berlin
Heidelberg
New York
Tokyo